Applied Statistical Modeling and Data Mining

Applied Statistical Modeling and Data Mining

Editors

José Antonio Sáez Muñoz
José Luis Romero Béjar

Basel • Beijing • Wuhan • Barcelona • Belgrade • Novi Sad • Cluj • Manchester

Editors
José Antonio Sáez Muñoz
University of Granada
Granada
Spain

José Luis Romero Béjar
University of Granada
Granada
Spain

Editorial Office
MDPI
St. Alban-Anlage 66
4052 Basel, Switzerland

This is a reprint of articles from the Special Issue published online in the open access journal *Mathematics* (ISSN 2227-7390) (available at: https://www.mdpi.com/journal/mathematics/special_issues/Applied_Statistical_Modeling_Data_Mining).

For citation purposes, cite each article independently as indicated on the article page online and as indicated below:

Lastname, A.A.; Lastname, B.B. Article Title. *Journal Name* **Year**, *Volume Number*, Page Range.

ISBN 978-3-7258-1105-2 (Hbk)
ISBN 978-3-7258-1106-9 (PDF)
doi.org/10.3390/books978-3-7258-1106-9

© 2024 by the authors. Articles in this book are Open Access and distributed under the Creative Commons Attribution (CC BY) license. The book as a whole is distributed by MDPI under the terms and conditions of the Creative Commons Attribution-NonCommercial-NoDerivs (CC BY-NC-ND) license.

Contents

About the Editors vii

Preface ix

José A. Sáez and José L. Romero-Béjar
Impact of Regressand Stratification in Dataset Shift Caused by Cross-Validation
Reprinted from: *Mathematics* **2022**, *10*, 2538, doi:10.3390/math10142538 1

José A. Sáez
Noise Models in Classification: Unified Nomenclature, Extended Taxonomy and Pragmatic Categorization
Reprinted from: *Mathematics* **2022**, *10*, 3736, doi:10.3390/math10203736 15

Mohamed Ali Shabeeb Ali, Mohammed Abdullah Ammer and Ibrahim A. Elshaer
Determinants of Investment Awareness: A Moderating Structural Equation Modeling-Based Model in the Saudi Arabian Context
Reprinted from: *Mathematics* **2022**, *10*, 3829, doi:10.3390/math10203829 35

Francisco Javier Esquivel, José Antonio Esquivel, Antonio Morgado, José L. Romero-Béjar and Luis F. García del Moral
Preprocessing of Spectroscopic Data Using Affine Transformations to Improve Pattern-Recognition Analysis: An Application to Prehistoric Lithic Tools
Reprinted from: *Mathematics* **2022**, *10*, 4250, doi:10.3390/math10224250 53

Cesar Guevara and Matilde Santos
Smart Patrolling Based on Spatial-Temporal Information Using Machine Learning
Reprinted from: *Mathematics* **2022**, *10*, 4368, doi:10.3390/math10224368 67

Xiaofeng Xue, Haokun Mao, Qiong Li, Furong Huang and Ahmed A. Abd El-Latif
An Energy Efficient Specializing DAG Federated Learning Based on Event-Triggered Communication
Reprinted from: *Mathematics* **2022**, *10*, 4388, doi:10.3390/math10224388 94

Yeongmin Kim, Minsu Chae, Namjun Cho, Hyowook Gil and Hwamin Lee
Machine Learning-Based Prediction Models of Acute Respiratory Failure in Patients with Acute Pesticide Poisoning
Reprinted from: *Mathematics* **2022**, *10*, 4633, doi:10.3390/math10244633 111

Fahad Alsokhiry, Andres Annuk, Toivo Kabanen and Mohamed A. Mohamed
A Malware Attack Enabled an Online Energy Strategy for Dynamic Wireless EVs within Transportation Systems
Reprinted from: *Mathematics* **2022**, *10*, 4691, doi:10.3390/math10244691 135

David Payares-Garcia, Javier Platero and Jorge Mateu
A Dynamic Spatio-Temporal Stochastic Modeling Approach of Emergency Calls in an Urban Context
Reprinted from: *Mathematics* **2023**, *11*, 1052, doi:10.3390/math11041052 155

Raimundo Aguayo-Estremera, Gustavo R. Cañadas, Elena Ortega-Campos, Tania Ariza and Emilia Inmaculada De la Fuente-Solana
Validity Evidence for the Internal Structure of the Maslach Burnout Inventory-Student Survey: A Comparison between Classical CFA Model and the ESEM and the Bifactor Models
Reprinted from: *Mathematics* **2023**, *11*, 1515, doi:10.3390/math11061515 183

Erica Espinosa and Alvaro Figueira
On the Quality of Synthetic Generated Tabular Data
Reprinted from: *Mathematics* **2023**, *11*, 3278, doi:10.3390/math11153278 . . . . **204**

Elena Ortega-Campos, Gustavo R. Cañadas, Raimundo Aguayo-Estremera, Tania Ariza, Carolina S. Monsalve-Reyes, Nora Suleiman-Martos and Emilia I. De la Fuente-Solana
Evaluation of Convergent, Discriminant, and Criterion Validity of the Cuestionario Burnout Granada-University Students
Reprinted from: *Mathematics* **2023**, *11*, 3315, doi:10.3390/ math11153315 . . . . **222**

Fumin Zou, Weihai Wang, Qiqin Cai, Feng Guo and Rouyue Shi
Dynamic Generation Method of Highway ETC Gantry Topology Based on LightGBM
Reprinted from: *Mathematics* **2023**, *11*, 3413, doi:10.3390/math11153413 . . . . **238**

Yu-Chung Wei
BLogic: A Bayesian Model Combination Approach in Logic Regression
Reprinted from: *Mathematics* **2023**, *11*, 4353, doi:10.3390/math11204353 . . . . **268**

Gaohan Xiong, Wei Cai, Min Hu and Zhiyan Yu
Research on Emotional Infection of Passengers during the SRtP of a Cruise Ship by Combining an SIR Model and Machine Learning
Reprinted from: *Mathematics* **2023**, *11*, 4461, doi:10.3390/math11214461 . . . . **289**

About the Editors

José Antonio Sáez Muñoz

José Antonio Sáez Muñoz received his Ph.D. degree in Computer Science and Computing Technology from the University of Granada (Spain), where he is currently working as Associate Professor within the Department of Statistics and Operations Research. His research focuses on the field of Data Science, covering data modeling, preprocessing, and transformation across diverse domains. His work involves handling complex datasets in learning tasks, studying discretization methods, addressing imbalanced data, evaluating performance metrics, and exploring unsupervised techniques, among other areas. His main research line deals with noisy data in classification tasks, with more than 20 publications on this topic. In 2022, he was included in the ranking of the world's top 2% scientists in his research domain, published by Stanford University (California, United States).

José Luis Romero Béjar

José Luis Romero Béjar holds a Ph.D. degree in Applied and Mathematical Statistics from the University of Granada (Spain). He previously earned Bachelor's degrees in Mathematics and in Statistical Sciences and Techniques, and Master's degrees in Mathematics, Statistics, Applied Statistics, as well as Research Designs and Applications in Psychology and Health Sciences. He currently works as an Associate Professor in the Department of Statistics and Operations Research of the Faculty of Sciences in the University of Granada. His current research interests are related to risk assessment in spatial and spatiotemporal frameworks. In this regard, he has published a general methodology that combines classical risk measure theory, derived from applications in economics and finance, with integral and stochastic geometry, as the basis for analyzing extreme behavior in random fields. This methodology enables local risk mapping in spatio-temporal scenarios. Due to his background in formal and applied aspects of mathematics, statistics, and health sciences, he is also interested in interdisciplinary research in this field. Over the past two years, he has published nearly 30 articles through collaborations with various researchers in clinical and health areas, having a preferred authorship role in almost all these collaborations.

Preface

Welcome to this collection of 15 captivating research papers presented in the Special Issue "Applied Statistical Modeling and Data Mining" within the Mathematics journal. The fields of statistical modeling and data mining are fundamental disciplines that converge at the forefront of modern scientific research and its practical application. Statistical modeling forms the basis for understanding complex data relationships, allowing meaningful information to be extracted from vast and intricate datasets. Simultaneously, data mining allows researchers and practitioners to explore, analyze, and derive patterns or insights from these datasets, revealing invaluable information hidden in the data.

This compilation brings together various research articles that explore different facets of statistical modeling and data mining. They shed light on innovative approaches and diverse perspectives within each field. These 15 selected articles delve into critical and multifaceted topics of the moment, showing the importance of predictive modeling in healthcare, cybersecurity within transportation systems, data quality assessment, emergency call modeling, and innovative strategies for crime prevention, among other relevant topics. Covering a broad spectrum of topics, this compilation is intended to serve as a valuable resource for researchers, academics, and practitioners facing contemporary challenges across various domains.

Through the contributions presented in this book, readers will witness the practical implications and theoretical advances that arise from both statistical modeling and data mining. The main aim is to provide an understanding of their functions in addressing contemporary challenges in various sectors, thereby bridging the gap between theory and real-world applications. Therefore, this collection aims to inspire and assist researchers, academics, and professionals interested in exploring the nuances and potential of statistical modeling and data mining applications. Our aspiration is that this compendium will serve as a guide, encouraging further exploration and innovation in these fundamental domains.

As editors, we extend our sincere gratitude to the authors for their notable contributions, to the reviewers for their insightful evaluations, and to the dedicated MDPI team for their unwavering support in making this compilation a reality. We hope that this compilation generates new ideas, encourages the dissemination of knowledge, and contributes to future advances in statistical models, data mining, and their applications.

José Antonio Sáez Muñoz and José Luis Romero Béjar
Editors

Article

Impact of Regressand Stratification in Dataset Shift Caused by Cross-Validation

José A. Sáez [1] and José L. Romero-Béjar [1,2,3,*]

[1] Department of Statistics and Operations Research, University of Granada, Fuentenueva s/n, 18071 Granada, Spain; joseasaezm@ugr.es
[2] ibs.GRANADA —Instituto de Investigación Biosanitaria, 18012 Granada, Spain
[3] IMAG—Institute of Mathematics of the University of Granada, Ventanilla 11, 18001 Granada, Spain
* Correspondence: jlrbejar@ugr.es

Abstract: Data that have not been modeled cannot be correctly predicted. Under this assumption, this research studies how k-fold cross-validation can introduce dataset shift in regression problems. This fact implies data distributions in the training and test sets to be different and, therefore, a deterioration of the model performance estimation. Even though the stratification of the output variable is widely used in the field of classification to reduce the impacts of dataset shift induced by cross-validation, its use in regression is not widespread in the literature. This paper analyzes the consequences for dataset shift of including different regressand stratification schemes in cross-validation with regression data. The results obtained show that these allow for creating more similar training and test sets, reducing the presence of dataset shift related to cross-validation. The bias and deviation of the performance estimation results obtained by regression algorithms are improved using the highest amounts of strata, as are the number of cross-validation repetitions necessary to obtain these better results.

Keywords: cross-validation; dataset shift; target shift; stratification; regression

MSC: 62R07

Citation: Sáez, J.A.; Romero-Béjar, J.L. Impact of Regressand Stratification in Dataset Shift Caused by Cross-Validation. *Mathematics* **2022**, *10*, 2538. https://doi.org/10.3390/math10142538

Academic Editor: Catalin Stoean

Received: 29 June 2022
Accepted: 19 July 2022
Published: 21 July 2022

Publisher's Note: MDPI stays neutral with regard to jurisdictional claims in published maps and institutional affiliations.

Copyright: © 2022 by the authors. Licensee MDPI, Basel, Switzerland. This article is an open access article distributed under the terms and conditions of the Creative Commons Attribution (CC BY) license (https://creativecommons.org/licenses/by/4.0/).

1. Introduction

Knowing the performance of different models over a dataset or determining their best parameter setup are common tasks when facing a new problem in data science [1,2]. In order to address these aspects, k-fold cross-validation (*k*-fcv) [3,4] is one of the most simple and frequently used approaches (for a survey on cross-validation procedures, the reader may consult the work of Arlot and Celisse [5]). *k*-fcv creates *k* pairs of training and test sets from the original dataset, in such a way that the models are built from the training sets and validated on the test sets. After this, the mean of the test results is taken as the model performance estimate.

Even though *k*-fcv offers several advantages for performance estimation, such as a reduced computation time compared to leave-one-out [6], its application is not totally risk-free [7,8]. It may cause dataset shift [9,10], in which the data used to build and evaluate the model do not follow the same distribution. This fact usually involves wrong predictions when testing the system, implying an underestimation of the model performance [10].

There are different types of dataset shift that have been studied in the specialized literature, such as covariate shift [11] (which affects the distributions of the input variables), conditional shift [12] (which also affects the conditional distributions of the output variable given an input) and posterior shift [13] (which is produced when the conditional distributions of the output given an input vary but the input distributions do not). Among them, a common type of dataset shift, known as target shift or prior probability shift [11,14], occurs in the output variable. The problem of target shift has been widely studied in classification [15,16]. In this context, stratification [6] is employed to reduce target shift related to

k-fcv. It consists of having the same proportion of samples of each class in the training and test sets. This approach has provided successful results creating cross-validation folds for both model selection and evaluation in classification [6].

Nevertheless, in regression problems [17], the most common k-fcv scheme for performance estimation is the application of standard cross-validation (CV) [18–20], in which training and test sets are randomly built. Since the distribution of the output variable is not considered when partitioning a dataset, this approach has the inconvenience that it can potentially introduce target shift. Despite this, there are works that have applied stratification on the output variable to build more similar training and test sets [7,21,22]. These are mainly based on ordering the samples according to their regressand values, creating different strata of samples and evenly distributing the samples of each stratum among all the folds. Krstajic et al. [7] noted that there appears to be no clear consensus regarding the application of stratified cross-validation. Thus, Breiman and Spector [21] compared several partitioning approaches with regression datasets and concluded that there were no significant improvements using stratification. On the other hand, Baxter et al. [22] used stratification in the context of water treatment data as an effective alternative to make training and test sets illustrative of the problem domain. Their approach first determined the proportion of samples contained in each set and then iteratively assigned the previously ordered samples to each set based on such proportions. Other works are somewhere in the middle and concluded that stratification is not particularly useful when a large number of repeated k-fcv is used in model selection, whereas it is recommended for model assessment [7]. These facts highlight the importance of further studying the dataset shift induced by k-fcv and the usage of stratification in the field of regression to better understand their implications.

This paper deepens the understanding of the impacts of target shift induced by k-fcv in regression datasets. It analyzes the influence on target shift and its consequences of different stratification schemes in k-fcv with respect to CV. These schemes include a series of artificial strata of samples according to the values of the output variable, aiming to minimize the target shift between the distributions of the output variable in training and test sets. Thus, inspired by previous works [7,21,22], several stratification schemes have been designed and compared against CV, considering different amounts of strata: from the minimum amount of two strata to the maximum possible amount of strata equivalent to the quantity of samples in the dataset. The most common values of k in k-fcv in the literature (2, 5 and 10) have been considered along with these stratification schemes to study the effect of target shift in 28 real-world regression datasets using five algorithms belonging to several regression paradigms (such as decision trees and neural networks, among others [23,24]). The partitionings have been repeated thousands of times with each approach, leading to a total of more than 4 M results to analyze. The statistical tests recommended in the literature have been employed to contrast the conclusions derived from the analysis of test performance results and target shift between the training and test sets [25]. A webpage with the details of the experimentation, datasets, additional results and plots can be accessed at https://joseasaezm.github.io/scvreg/ (accessed on 29 June 2022). In summary, the main contributions of this paper are the following:

- Delving into the use of regressand stratification in k-fcv and analyzing whether, despite not being generalized, it should be recommended when dealing with regression data.
- Establishing a direct comparison between k-fcv with and without stratification at three levels (amount of dataset shift introduced, quality of performance estimation and convergence speed) to determine in which aspects stratification offers advantages and the degree of improvement in each of them.
- Studying different amounts of strata in the output variable in order to check if they significantly affect the results obtained and recommend the most appropriate values.
- Analyzing if the effects of stratification on the results depend on the number of folds k in k-fcv, through the study of the values of k commonly used in the literature (2, 5 and 10).

- Drawing conclusions through experimentation with different regression paradigms, both classic and more recent, including decision trees, extreme learning machines and ensembles, among others.

Note that, even though there are works in the literature dealing with regressand stratification, most of the research in this field has considered the distributions in the input space, thus addressing the presence of covariate shift in the data [26–28]. Some of these methods, such as *representative splitting cross-validation* (RSCV) [26], are based on the DUPLEX [29] algorithm to create partitions with k-fcv. Other works [27,30] are based on clustering to create training, validation and test sets for use with neural networks. For example, May et al. [27] created groups of samples using self-organizing maps, which were then distributed among the three sets. The proposal of Diamantidis et al. [28] is also based on clustering and one-center strategies, creating the folds deterministically using the distributions in the input space. A different process is followed by SPlit [31], which is based on the usage of support points and a sequential nearest neighbor method.

Unlike the above approaches, this paper focuses exclusively on target shift to delve into its impacts when partitioning using k-fcv. This fact will allow for knowing the degree of improvement in performance solely attributable to regressand stratification (considering different amounts of strata). Even though there are other works that have applied regressand stratification when using k-fcv (mainly to develop hydrological models [32–34]), to our best knowledge, this paper differs for simultaneously combining a comprehensive study on the impacts of regressand stratification while also considering different regression paradigms, dozens of datasets, stratification levels, numbers of folds k in k-fcv and a numerical study of the amount of target shift, performance and convergence speed for each of the k-fcv approaches studied dealing with regression problems.

The remainder of this manuscript is disposed as follows. Section 2 introduces how k-fcv can introduce dataset shift. Section 3 describes the partitioning methods employed in this paper and Section 4 details the experimental framework. Section 5 is devoted to the analysis of the results. Finally, Section 6 closes this work, summarizing the main findings.

2. On Dataset Shift Induced by Cross-Validation

Let x and y respectively be the input attributes and the output variable in a dataset, with $\mathcal{X}$ and $\mathcal{Y}$ their corresponding domains. In supervised learning, a function $f : \mathcal{X} \to \mathcal{Y}$ is usually estimated from a training set of m samples $D_{tra} = \{(x_i, y_i) \in \mathcal{X} \times \mathcal{Y}\}, i = 1, \ldots, m$, in order to predict the output variable in a different test set of m' samples $D_{tst} = \{(x'_j, y'_j) \in \mathcal{X} \times \mathcal{Y}\}, j = 1, \ldots, m'$. Commonly, it is assumed that the training and test sets have identical joint distributions, that is, $P_{tra}(x, y) = P_{tst}(x, y)$ [17]. However, if these sets are obtained by a k-fcv procedure without considering the distributions of the input and output variables, $P_{tra}(x, y) \neq P_{tst}(x, y)$ is likely to occur. This scenario is known as dataset shift [9,10] and occurs when the training and test sets follow different distributions [11]. In supervised data, either classification or regression, two main types of data shift are found:

1. *Target shift* [14,15], which affects the distributions of the output variable $P_{tra}(y) \neq P_{tst}(y)$, but it maintains the conditional distributions $P_{tra}(x|y) = P_{tst}(x|y)$;
2. *Covariate shift* [35,36], which affects the distributions of the input attributes $P_{tra}(x) \neq P_{tst}(x)$, but it maintains the conditional distributions $P_{tra}(y|x) = P_{tst}(y|x)$.

Among them, covariate shift has been widely studied in the specialized literature [26,27,37]. A common approach to reduce its negative impacts is to estimate a weight for each training sample relative to the test set [38], which is then used by learning algorithms. Examples of this strategy are the KLIEP [37] and uLSIF [39] methods. Other works are based on computing the weights by analyzing the means of the training and test sets in a kernel Hilbert space [40] or introducing a surrogate kernel matching [41]. There are also approaches to reduce covariate shift related to partitioning with k-fcv, such as the DB-SCV [42] and DOB-SCV [8] approaches in the field of classification or the aforementioned proposals for regression problems based on the DUPLEX algorithm [26], clustering [27] or support points [31]. These methods are generally based on creating training and test partitions by choosing close samples in the input space.

Even though it is recommended to address covariate and target shifts simultaneously in real-world applications, this research focuses on target shift since the output variable usually has a strong influence on the building and evaluation processes of the models. Note that, although other factors of the data apart from target shift (such as covariate shift) could affect the results obtained, these can potentially equally affect all the partition schemes in this paper since none of the k-fcv approaches studied deals with them specifically.

Figure 1 shows a regression dataset with varying degrees of target shift, x and y being the input and output variables, respectively. Figure 1b illustrates a high target shift, since those samples with $y > t$ used to validate the model are not considered to build it and, thus, they are probably wrongly predicted. This situation is partially corrected in Figure 1c, which shows that both sets have samples along the domain of y.

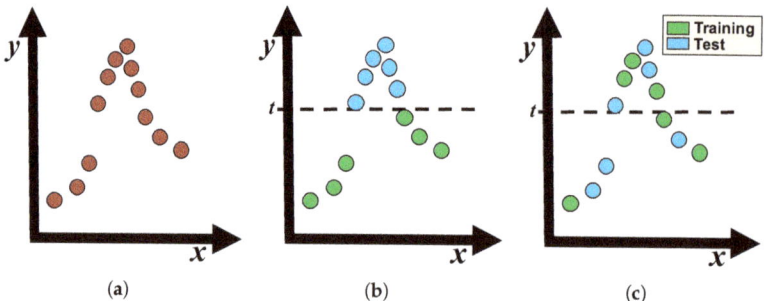

Figure 1. Examples of target shift in a regression dataset. (**a**) Original dataset. (**b**) High target shift. (**c**) Moderate target shift.

In real-world applications, target shift can occur because of the nature of the problem (e.g., when a model is built on past data and used to predict future data with different characteristics) or it can be unexpectedly introduced by cross-validation during the performance estimation of the models [8]. Using k-fcv, the data are divided into k separate folds. Then, each fold is used to test the model trained with the remaining $k-1$ folds and, finally, the k evaluation results are averaged to obtain an individual estimation. If the training and test sets are obtained without taking into account the distribution of the output variable, the data used to build the model may differ from those used to validate it.

In classification, target shift induced by k-fcv is commonly prevented by applying a stratification scheme [6]. However, the usage of k-fcv based on folds of random samples, usually employed in the field of regression, may imply that the impact of target shift is overlooked. This research focuses on the analysis of the presence and impact of target shift induced by k-fcv in regression datasets, studying how regressand stratification can help to reduce its negative consequences.

3. Cross-Validation in Regression Problems

This section describes the different k-fcv partitioning methods used in this research. They split a regression dataset D into k approximately equal-sized folds ($F_1, \ldots, F_k$). Each of these folds represents a test set and the remaining folds represent the corresponding training sets.

3.1. Standard Cross-Validation

Algorithm 1 shows the most widely used and simplest approach to partition a regression dataset with k-fcv: *standard cross-validation* (CV) [18,19]. First, it computes the number of samples per fold (line 2, being $|D|$ the number of samples). Then, the whole dataset D is split into k random folds of samples of equal size (lines 3–6). Note that, since this partitioning scheme does not consider the distribution of the output variable to create each fold, it can introduce target shift uncontrollably.

Algorithm 1 Standard cross-validation (CV).

Input: dataset D, number of folds k.
Output: folds $F_1, \ldots, F_k$

1: Set each fold $F_i = \emptyset$ $(i = 1, \ldots, k)$;
2: $n \longleftarrow |D|/k$;
3: **for** *each fold F_i* **do**
4: $\quad F_i \longleftarrow$ Randomly select n samples from D;
5: $\quad D \longleftarrow D \setminus F_i$;
6: **end for**

3.2. Totally Stratified Cross-Validation

Contrary to CV, which does not use the information of the regressand to create the folds, *totally stratified cross-validation* (TSCV) is one of the approaches used in this research to reduce target shift to the maximum degree. Algorithm 2 shows its pseudocode.

It introduces as many strata as samples in the dataset. The different strata are created according to the regressand distribution. First, it sorts the samples in D considering their output variable (line 2). Then, each sample is selected in order (line 3) and assigned to a fold with less samples (lines 4–5). If several folds are tied with less samples (line 4), one of them is arbitrarily chosen, which adds some randomness in the partitioning.

TSCV is based on the idea of assigning the closest samples (according to their output variables) to different folds. In this way, folds are intended to be as similar as possible to each other, while each of them contains the maximum possible diversity of values of the output variable. This fact finally implies that training and test sets have similar distributions of the output variables, reducing target shift.

Algorithm 2 Totally stratified cross-validation (TSCV).

Input: dataset D, number of folds k.
Output: folds $F_1, \ldots, F_k$

1: Set each fold $F_i = \emptyset$ $(i = 1, \ldots, k)$;
2: $L \longleftarrow$ Sort the samples in D by their output variables;
3: **for** *each sample $\delta \in L$* **do**
4: $\quad F_i \longleftarrow$ Select a fold with less samples;
5: $\quad F_i \longleftarrow F_i \cup \{\delta\}$;
6: **end for**

3.3. Stratified Cross-Validation

At an intermediate point between CV and TSCV, *t-stratified cross-validation* (SCV_t) is another approach used in this paper to introduce a variable stratification in the k-fcv process with regression problems. It is presented in Algorithm 3.

This procedure allows for creating the desired amount of strata t when building the k folds. First, SCV_t sorts all the samples according to the value of the output variable (line 2) and computes the number n of samples per stratum (line 3). Afterwards, it starts an iterative process to assign samples to each fold (lines 4–13): it selects blocks of n samples conforming each stratum (lines 5–6) and, then, each of the samples of that block (line 8) is assigned to a fold with less samples (line 9) until there are no more available samples to assign. Thus, SCV_t considers the same number of samples from each stratum in each of the folds.

SCV$_t$ is a generalization of the previous k-fcv schemes: CV and TSCV. If $t=1$, no stratification is considered and, thus, SCV$_t$ is equivalent to CV. If $t=|D|$, the maximum number of strata are considered and SCV$_t$ is equivalent to TSCV.

Algorithm 3 t-stratified cross-validation (SCV$_t$).

Input: dataset D, number of folds k, number of strata t.
Output: folds $F_1, \ldots, F_k$

1: Set each fold $F_i = \emptyset$ $(i = 1, \ldots, k)$;
2: $L \longleftarrow$ Sort the samples in D by their output variables;
3: $n \longleftarrow |D|/t$;
4: **for** each stratum S_j **do**
5:     $S_j \longleftarrow$ Select the first n samples from L;
6:     $L \longleftarrow L \setminus S_j$;
7:     **while** $S_j \neq \emptyset$ **do**
8:       $\delta \longleftarrow$ Randomly select a sample from S_j;
9:       $F_i \longleftarrow$ Select a fold with less samples;
10:      $F_i \longleftarrow F_i \cup \{\delta\}$;
11:      $S_j \longleftarrow S_j \setminus \{\delta\}$;
12:     **end while**
13: **end for**

---

4. Experimental Framework

Next, Sections 4.1 and 4.2 introduce the datasets and the parameter setup for the regression algorithms, respectively. Then, Section 4.3 describes the methodology of analysis.

4.1. Real-World Datasets

This research considers 28 real-world regression datasets taken from the *UCI machine learning* and *KEEL-dataset* repositories (https://archive.ics.uci.edu/, http://www.keel.es; both accessed on 29 June 2022). In order to study the impact of stratification in k-fcv with regression problems regardless of the characteristics of the data, datasets belonging to different applications and areas (including fields such as biology, geology, chemistry and so on) and with different numbers of attributes and samples are selected. Table 1 presents them, along with their number of attributes (*at*) and samples (*sa*). Those samples containing missing values in these datasets are removed before their usage. Furthermore, both the input attributes and the output variables are normalized to the interval $[0, 1]$.

Table 1. Regression datasets employed in the experimentation.

Dataset	at	sa	Dataset	at	sa
abalone	8	4177	friedman	5	1200
airfoil	5	1503	laser	4	993
anacalt	7	4052	machinecpu	6	209
autompg8	7	392	mortgage	15	1049
baseball	16	337	plastic	2	1650
concrete	8	1030	quake	3	2178
coolingeff	8	768	realestate	6	414
dailerons	5	7129	stock	9	950
dee	6	365	traffic	17	135
delevators	6	9517	treasury	15	1049
elength	2	495	wankara	9	321
emaintenance	4	1056	watertoxicity	8	546
fish	6	908	wizmir	9	1461
forest	12	517	yacht	6	308

4.2. Regression Algorithms

In order to build models on the above datasets, 5 algorithms belonging to different regression paradigms are chosen, including decision trees [23], distance-based models [43], neural networks [24], multiple linear regression [44] and ensemble-based models [45]. They are briefly described below along with their main parameters, which are shown in Table 2. In order to delve into the specific characteristics of each algorithm, the reader can consult the reference associated with each of them.

1. *Recursive partitioning and regression trees* (RPART) [23]. It builds a decision tree from the dataset, in which the nodes are successively split into subnodes using a homogeneity-based threshold attribute value. The process stops when the last subset of samples is included in the tree or the maximum number of leaves is reached (known as *tree pruning*).
2. *k-nearest neighbors* (NN) [43]. To estimate the output value for a sample, it computes the distances between such sample and all the training samples. Then, it selects the k closest samples to the query and averages their regressand values to obtain a single prediction.
3. *Extreme learning machine* (ELM) [24]. It is a feedforward neural network with a hidden layer of nodes whose parameters do not need to be tuned. Its main advantage is that it produces good generalization performance in less time compared to traditional neural networks trained with backpropagation.
4. *Multivariate adaptive regression spline* (MARS) [44]. It is a non-parametric algorithm based on two main stages. In the *forward* stage, it splits the data in several subsets and runs a linear regression model on each partition. In the *backward* stage, the model is pruned to avoid overfitting by removing the functions that contribute the least to performance.
5. *Generalized boosted regression modeling* (GBM) [45]. It iteratively builds decision trees based on random subsets of the training samples using boosting. For each new tree, those samples poorly modeled by previous trees have a higher probability of being selected.

Table 2. Regression methods.

Method	Parameters
RPART	min. split = 20; min. leaf = 6; complexity = 0.01; max. depth = 30
NN	$k = 3$; distance: *Euclidean*
ELM	neurons = 20; activation: *radial basis*; input weights: $\mathcal{N}(0,1)$
MARS	degree = 1; pruning = *backward*
GBM	distribution = *Gaussian*; trees = 100; learning rate = 0.1; bag = 0.8

4.3. Methodology of Analysis

To study the effects of target shift induced by k-fcv in regression datasets and how stratification can help to reduce its impacts, the following experimental study is performed. Each dataset in Table 1 is partitioned using three different values of k in k-fcv (2, 5 and 10). These folds are obtained with eight partitioning schemes (see Section 3), each one with a different stratification degree:

- CV, which does not consider any stratification;
- TSCV, which considers a total stratification of the samples;
- SCV$_t$ with six different values of t (2, 5, 10, 20, 50 and 100), which allows for controlling the stratification degree.

Once the datasets are split into k folds with the aforementioned schemes, the regression methods in Table 2 are evaluated over them, obtaining their test performance results with the RMSE metric:

$$\text{RMSE} = \sqrt{\frac{\sum_{i=1}^{n}(y_i' - y_i)^2}{n}}$$

with n the amount of samples and y_i and y'_i the real and predicted regressand values for the i-th sample, respectively.

Additionally, the *Kolmogorov–Smirnov* statistic (D_n) [46] is used to estimate the amount of target shift between the training and test sets, that is, the difference between the distributions of the regressand values in both sets. Given the samples of regressand values in the training and test sets, X and Y, and their empirical distribution functions F_X and F_Y, D_n is computed as:

$$D_n = \sup_x |F_X - F_Y|$$

The above procedure is repeated 1000 times with different seeds to generate random numbers, obtaining, thus, different partitions in each run. Table 3 shows a summary of the experiment performed, in which # indicates the amount of values of each variable. Thus, the experimentation of this research entails the analysis of more than 4 M results. The conclusions derived from them are contrasted using the statistical tests recommended in the specialized literature [25]. Specifically, *Wilcoxon*'s test [25] is used for rejecting the null hypothesis of the equality of means in pairwise comparisons, implying the superiority of one of the methods. A significance level $\alpha = 0.1$ is assumed in this paper.

Table 3. Details of the experimentation.

Parameter	Values	#
Datasets	See Table 1	28
Folds	$k = 2, 5,$ and 10	3
Partitioning	CV, TSCV and SCV$_t$ ($t = 2, 4, 10, 20, 50, 100$)	8
Seed	Random without replacement in [1, 1,000,000]	1000
Regression	RPART, NN, ELM, MARS, GBM	5
Metric	RMSE (performance) and D_n (target shift)	2

5. Analysis of Results

The analysis of results is divided into three main parts. First, Section 5.1 analyzes the amount of induced target shift by the different k-fcv schemes. Then, Section 5.2 focuses on the effect of stratification on the error estimation of the regression methods with k-fcv. Finally, Section 5.3 studies the convergence speed of the stratification schemes with respect to the performance estimated using CV, that is, the number of repetitions necessary by each stratification approach to reach a stable (better) behavior with respect to CV.

5.1. Analysis of Induced Target Shift by Cross-Validation Schemes

In order to measure the amount of target shift existing between the training and test sets created by each k-fcv approach, the *Kolmogorov–Smirnov* [46] non-parametric test is used. This test calculates a statistic $D_n \in [0, 1]$, which can be taken as an indicator of the difference between two samples. D_n is measured considering the distributions of the output variable in the different training and test sets created by each k-fcv scheme for each dataset. The lower the value of D_n is, the more similar the training and test distributions and the lower amount of target shift introduced by k-fcv are.

Table 4 shows the averaged D_n values when measuring target shift between training and test sets for all the datasets considering each one of the k-fcv schemes (with $k = 2, 5$ and 10). The results of each partitioning scheme are compared against two reference methods using *Wilcoxon*'s test, obtaining their associated p-values:

1. CV, which does not consider any stratification (row vs. CV);
2. TSCV, which considers a maximum stratification (row vs. TSCV).

The p-value of *Wilcoxon*'s test allows for rejecting the null hypothesis that the mean results of the two algorithms involved in the comparison are equal, that is, they have a similar behavior on average in all the datasets. Given that a significance level of 0.1 is considered and p-values are in scientific notation in Table 4, those with exponent -1 do not allow for rejecting the null hypothesis, whereas the rest do (indicating that there are differences between the behaviors of the two methods compared).

The best results in Table 4 are underlined, whereas p-values lower than $\alpha = 0.1$ are remarked in bold. A darker background in the results indicates that these are better. In addition to the p-value, the sum of ranks [47] associated with each algorithm within *Wilcoxon*'s test is calculated as a way of representing their effectiveness. In order to do this, the differences between both methods in the results of each dataset are computed and a ranking is assigned to the absolute value of each difference. The sum of ranks associated with the positive differences is assigned to the first algorithm, whereas the sum of ranks of negative differences is assigned to the second method. A higher sum of ranks represents a greater effectiveness of the corresponding algorithm. Finally, those cases in Table 4 in which the method of the row obtains a higher sum of ranks than that of the column in *Wilcoxon*'s test are indicated with an asterisk.

Table 4. Induced target shift by different k-fcv schemes. A darker background in the results indicates that these are better. Those cases in which the method of the row obtains a higher sum of ranks than that of the column in *Wilcoxon*'s test are indicated with an asterisk.

Folds	CV	SCV$_2$	SCV$_4$	SCV$_{10}$	SCV$_{20}$	SCV$_{50}$	SCV$_{100}$	TSCV
2-fcv	0.0604	0.0491	0.0390	0.0280	0.0217	0.0150	0.0109	**0.0055**
vs. CV	X	7.45E-9	7.45E-9	7.45E-9	7.45E-9	7.45E-9	7.45E-9	7.45E-9
vs. TSCV	7.45E-9 *	7.45E-9 *	7.45E-9 *	7.45E-9 *	7.45E-9 *	7.45E-9 *	1.49E-8 *	X
5-fcv	0.0754	0.0615	0.0491	0.0354	0.0273	0.0195	0.0155	**0.0109**
vs. CV	X	7.45E-9	7.45E-9	7.45E-9	7.45E-9	7.45E-9	7.45E-9	7.45E-9
vs. TSCV	7.45E-9 *	7.45E-9 *	7.45E-9 *	7.45E-9 *	7.45E-9 *	7.45E-9 *	7.45E-9 *	X
10-fcv	0.1002	0.0820	0.0656	0.0480	0.0378	0.0287	0.0246	**0.0213**
vs. CV	X	7.45E-9	7.45E-9	7.45E-9	7.45E-9	7.45E-9	7.45E-9	7.45E-9
vs. TSCV	7.45E-9 *	7.45E-9 *	7.45E-9 *	7.45E-9 *	7.45E-9 *	7.45E-9 *	1.49E-8 *	X

The above results show that, for each value of k in k-fcv, a total ordering according to the increasing number of strata is observed: from CV (with the highest target shift value) to TSCV (with the lowest target shift value). This ordering is also observed in the results of most of the individual datasets.

The comparisons among k-fcv schemes using *Wilcoxon*'s test support the above conclusions. They show that those partitionings considering stratification provide better results than not considering it (CV). Similarly, using the maximum number of strata (TSCV) implies an improvement compared to considering a fewer amount of strata.

When comparing the results of each partitioning method for the different values of k, it is observed that a higher k involves an increment of the injected target shift by k-fcv. This fact may be due to that the greater the number of folds, the more difficult it is for all of them to be similar.

The above results show the positive effects of the stratification schemes to reduce target shift in regression datasets compared to the traditional approach, which does not consider any stratification (CV). Specifically, TSCV is the stratification that achieves further reducing the differences between the distributions of values of the output variables in the training and test sets, with clear differences compared to the rest of the approaches. The next section analyzes whether this reduction in target shift in the data leads to a better error estimation of the models, implying a lower bias in performance estimation.

5.2. Effect of Stratification in Error Bias Related to Target Shift

Table 5 shows the error estimation using RMSE and standard deviation results of each partitioning scheme, considering all the runs with different numbers of folds k (2, 5 and 10). Additionally, the p-value associated with *Wilcoxon*'s test after comparing the results of each partitioning method against the method with no stratification (CV) and the method with maximum stratification (TSCV) is computed. Due to the large amount of results obtained, just those for the RPART and NN regression methods are shown in this paper. The results for the rest of the regression techniques are found on the webpage of this paper and show conclusions similar to those presented here.

Analogously to the amount of target shift in the previous section, the error estimation results for RPART and NN show that, for each value of k in k-fcv, an ordering according to the increasing number of strata is observed: from CV (with the highest error values) to TSCV (with lowest error values). Although there are some exceptions, this fact is also true in standard deviation results.

The statistical comparisons for the error results also support that all the methods using stratification are generally better than CV (although no differences are observed between CV and its closest stratification levels in some cases). The comparisons with TSCV show similar behavior: TSCV is usually better than the methods using a lower stratification, although with the approaches closest in number of strata (50 and 100), no differences are observed in some cases. The analysis of standard deviation provides similar results.

As a conclusion, it is observed that the reduction in target shift shown in Section 5.1 by the stratification schemes is related to that when estimating the error made by the models. This fact implies that the application of stratification allows for obtaining better estimations of the performance of the models, so its usage can be recommended against not considering it.

5.3. Convergence Speed of Stratification Schemes against CV

Section 5.2 focuses on error estimation when a large number of k-fcv repetitions are performed (1000). The analysis of the p-values of *Wilcoxon*'s test shows that stratification generally improves error estimation compared to not considering it. This section studies the speed (number of k-fcv repetitions) required by each partitioning method to reach this p-value lower than 0.1 when compared to CV. To this end, the performance (from 5 to 150 repetitions, by increments of 5) of each regression method using stratified k-fcv against CV is compared with *Wilcoxon*'s test and the associated p-values are computed. This process is repeated 50 times to obtain more robust results. Figure 2 shows the results for RPART and NN—the rest of the methods can be found on the webpage of this paper.

The analysis of Figure 2 shows a certain ordering in the results of the different stratification strategies. The highest p-values are usually related to those approaches that use a smaller number of strata, whereas an increase in the number of strata implies that the p-values are reduced. A higher variability in the results between consecutive k-fcv repetitions is also observed in those approaches that use a lower number of strata. Thus, these results show that the usage of higher amounts of strata in k-fcv usually involve lower numbers of k-fcv repetitions to obtain better and stable results compared to CV. Furthermore, Figure 2 shows that, if higher values of k in k-fcv are considered, the p-values obtained are generally higher with all the partitioning methods. A higher variability in the results between consecutive repetitions is also presented with higher values of k in k-fcv.

Table 5. Effect of stratification in error estimation (RMSE) and standard deviation results. A darker background in the results indicates that these are better. Those cases in which the method of the row obtains a higher sum of ranks than that of the column in *Wilcoxon*'s test are indicated with an asterisk.

Metric		Error									Standard Deviation								
RPART		CV	SCV$_2$	SCV$_4$	SCV$_{10}$	SCV$_{20}$	SCV$_{50}$	SCV$_{100}$	TSCV		CV	SCV$_2$	SCV$_4$	SCV$_{10}$	SCV$_{20}$	SCV$_{50}$	SCV$_{100}$	TSCV	
2-fcv		9.2947E-2	8.6454E-2	9.2609E-2	9.2355E-2	9.2156E-2	9.1978E-2	9.1991E-2	9.1875E-2		3.3380E-3	3.3012E-3	3.1851E-3	3.0756E-3	2.9521E-3	2.8680E-3	2.8328E-3	2.7992E-3	
vs. CV		✗	2.44E-4	1.51E-5	7.45E-9	5.22E-7	2.46E-7	1.54E-6	2.76E-6		✗	1.71E-1	1.87E-3	5.83E-4	2.75E-5	1.88E-6	7.45E-8	1.26E-6	
vs. TSCV		2.76E-6 *	4.77E-6 *	7.96E-6 *	1.35E-4 *	7.16E-4 *	3.27E-1	1.26E-1 *	✗		1.26E-6 *	4.10E-7 *	1.20E-4 *	7.92E-4 *	3.66E-5 *	2.02E-1 *	9.73E-1	✗	
5-fcv		9.0102E-2	9.0086E-2	9.0023E-2	8.9920E-2	8.9815E-2	8.9701E-2	8.9705E-2	8.9653E-2		1.9316E-3	1.9204E-3	1.9172E-3	1.8594E-3	1.8275E-3	1.7749E-3	1.7593E-3	1.7456E-3	
vs. CV		✗	3.62E-1	4.21E-5	1.29E-3	1.88E-6	1.10E-5	1.77E-5	4.83E-5		✗	2.64E-1	3.05E-1	4.41E-3	4.73E-4	1.06E-4	3.66E-5	3.18E-5	
vs. TSCV		4.83E-5 *	9.32E-5 *	2.74E-4 *	3.81E-4 *	8.86E-3 *	5.37E-1	7.35E-2 *	✗		3.18E-5 *	1.20E-4 *	5.53E-5 *	4.41E-3 *	1.70E-3 *	5.64E-2 *	6.62E-1 *	✗	
10-fcv		8.9141E-2	8.9143E-2	8.9137E-2	8.9071E-2	8.9021E-2	8.8974E-2	8.8957E-2	8.8945E-2		1.5020E-3	1.5006E-3	1.4981E-3	1.4573E-3	1.4092E-3	1.3977E-3	1.3674E-3	1.3662E-3	
vs. CV		✗	9.55E-1	2.64E-2	4.06E-3	1.36E-2	1.67E-2	1.03E-2	1.36E-2		✗	8.67E-1	5.67E-1	3.18E-5	6.32E-5	3.66E-5	1.35E-4	5.68E-6	
vs. TSCV		1.36E-2 *	8.86E-3 *	2.64E-2 *	1.79E-2 *	8.15E-2 *	1.09E-1 *	6.30E-1 *	✗		5.68E-6 *	6.32E-5 *	2.76E-6 *	8.75E-4 *	6.55E-3 *	3.74E-3 *	9.73E-1	✗	
NN		CV	SCV$_2$	SCV$_4$	SCV$_{10}$	SCV$_{20}$	SCV$_{50}$	SCV$_{100}$	TSCV		CV	SCV$_2$	SCV$_4$	SCV$_{10}$	SCV$_{20}$	SCV$_{50}$	SCV$_{100}$	TSCV	
2-fcv		8.6534E-2	8.6454E-2	8.6336E-2	8.6196E-2	8.6106E-2	8.5954E-2	8.5849E-2	8.5689E-2		3.0571E-3	3.0134E-3	3.0113E-3	3.0113E-3	2.9897E-3	2.9681E-3	2.9352E-3	2.9453E-3	
vs. CV		✗	6.98E-2	7.92E-4	4.77E-6	6.56E-7	4.77E-6	2.46E-7	4.10E-7		✗	5.95E-2	1.57E-1	6.62E-2	1.56E-2	3.06E-2	3.66E-5	2.81E-2	
vs. TSCV		4.10E-7 *	4.77E-6 *	1.94E-4 *	3.42E-4 *	1.29E-3 *	2.47E-2 *	2.18E-1 *	✗		2.81E-2 *	4.51E-2 *	2.18E-2 *	1.67E-2 *	5.34E-2 *	9.93E-2 *	6.14E-1 *	✗	
5-fcv		8.1336E-2	8.1350E-2	8.1270E-2	8.1197E-2	8.1136E-2	8.1016E-2	8.0928E-2	8.0753E-2		1.8602E-3	1.8684E-3	1.8340E-3	1.8393E-3	1.7967E-3	1.7092E-3	1.6800E-3	1.6733E-3	
vs. CV		✗	4.51E-1	1.79E-2	6.32E-5	5.53E-5	1.20E-4	8.20E-5	6.32E-5		✗	5.22E-1 *	3.74E-1	5.64E-2	4.25E-4	1.55E-3	6.32E-5	1.17E-3	
vs. TSCV		6.32E-5 *	5.53E-5 *	1.72E-4 *	5.26E-4 *	8.86E-3 *	2.90E-3 *	1.55E-3 *	✗		1.17E-3 *	3.81E-4 *	9.54E-3 *	7.92E-4 *	9.00E-2 *	2.64E-2 *	9.55E-1 *	✗	
10-fcv		7.9092E-2	7.9089E-2	7.9002E-2	7.9071E-2	7.9063E-2	7.9006E-2	7.8983E-2	7.8874E-2		1.362SE-3	1.3489E-3	1.3409E-3	1.3100E-3	1.2730E-3	1.2442E-3	1.2360E-3	1.2129E-3	
vs. CV		✗	9.55E-1	2.95E-1	*1.06E-3	4.78E-3	2.04E-3	3.16E-3	4.73E-4		✗	6.14E-1	1.57E-1	2.90E-3	2.44E-4	1.54E-6	1.26E-6	3.18E-5	
vs. TSCV		4.73E-4 *	5.83E-4 *	1.70E-3 *	4.06E-3 *	9.54E-3 *	1.03E-2 *	4.73E-4	✗		3.18E-5 *	1.88E-6 *	1.42E-3 *	2.04E-3 *	2.04E-2 *	4.77E-3 *	3.99E-1 *	✗	

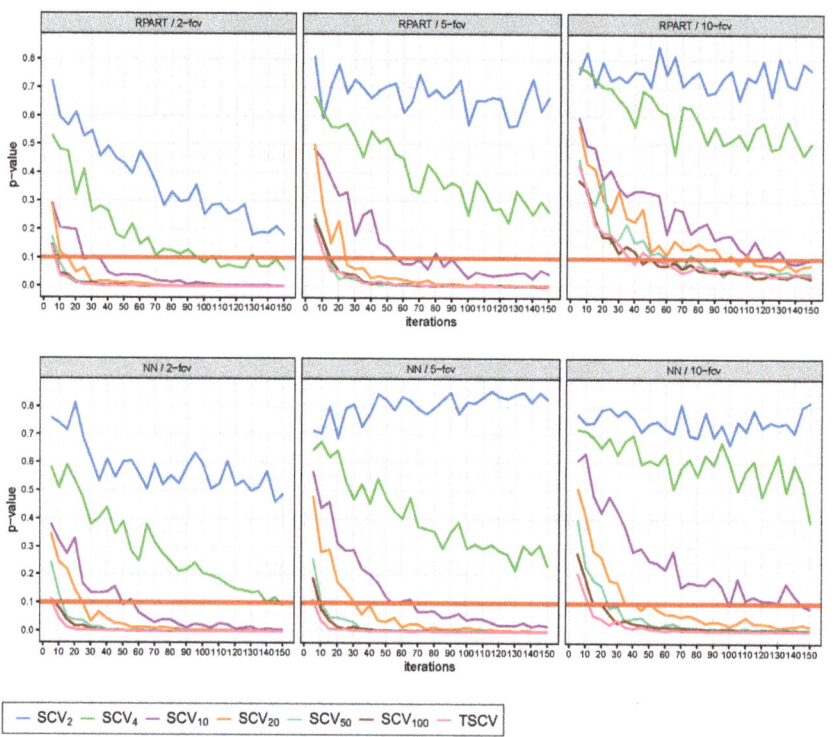

Figure 2. Evolution of *Wilcoxon*'s *p*-value across different iterations (k-fcv repetitions) after comparing each stratification scheme against not considering stratification with CV using RPART and NN.

6. Conclusions

This research has analyzed both how k-fcv can introduce target shift in regression datasets and its negative impact in performance estimation. Several stratification schemes have been analyzed to build more similar training and test distributions, decreasing the existence of target shift produced by cross-validation.

The experiments performed have shown that both dataset shift and bias are reduced by considering stratification. In general, the larger the number of strata is, the lower the target shift and error estimation results are. The convergence speed of the different stratification schemes when they are compared to CV shows that a larger number of strata usually implies that cross-validation provides a stable and better performance estimation faster. Among the stratification schemes studied, the usage of TSCV can be recommended, since it is the one that generally provides the best results in terms of the introduced target shift and estimation of the error made by the models built. Despite this, it should be noted that other regressand stratification schemes using a smaller number of strata also obtain good results compared to CV. Furthermore, the usage of lower stratification may imply some advantages in terms of computational cost when partitioning the dataset, particularly if the number of samples is high. Finally, even though this research has focused on the study of target shift in regression problems, it is also important to consider the presence of other types of dataset shift, such as that occurring in the input attributes. Their joint consideration may imply the need to further investigate the most appropriate synergy between the number of strata in the regressand with the different strategies to reduce other forms of dataset shift.

In future works, it is planned to study the behavior of other regression methods, such as *Support Vector Regression* [48] or XGBoost [49], with the proposed stratification schemes, as well as to use these algorithms along with other k-fcv approaches considering different types of dataset shift simultaneously, such as target and covariate shift.

Author Contributions: J.A.S. and J.L.R.-B. have contributed equally to this work. All authors have read and agreed to the published version of the manuscript.

Funding: J.L. Romero has been partially supported by grants MCIU/AEI/ERDF, UE PGC2018-098860-B-I00, grant A-FQM-345-UGR18 cofinanced by ERDF Operational Programme 2014–2020 and the Economy and Knowledge Council of the Regional Government of Andalusia, Spain, and grant CEX2020-001105-M MCIN/AEI/10.13039/501100011033.

Data Availability Statement: The datasets used in this research are taken from the *UCI machine learning* (https://archive.ics.uci.edu/) and *KEEL-dataset* (http://www.keel.es) repositories (both accessed on 29 June 2022).

Conflicts of Interest: The authors declare no conflict of interest. The funders had no role in the design of the study; in the collection, analyses, or interpretation of data; in the writing of the manuscript or in the decision to publish the results.

References

1. Liu, Y.; Liao, S.; Jiang, S.; Ding, L.; Lin, H.; Wang, W. Fast cross-validation for kernel-based algorithms. *IEEE Trans. Pattern Anal. Mach. Intell.* **2020**, *42*, 1083–1096. [CrossRef] [PubMed]
2. Rad, K.; Maleki, A. A scalable estimate of the out-of-sample prediction error via approximate leave-one-out cross-validation. *J. R. Stat. Soc. Ser. B Stat. Methodol.* **2020**, *82*, 965–996. [CrossRef]
3. Qi, C.; Diao, J.; Qiu, L. On estimating model in feature selection with cross-validation. *IEEE Access* **2019**, *7*, 33454–33463. [CrossRef]
4. Jiang, G.; Wang, W. Error estimation based on variance analysis of k-fold cross-validation. *Pattern Recognit.* **2017**, *69*, 94–106. [CrossRef]
5. Arlot, S.; Celisse, A. A survey of cross-validation procedures for model selection. *Stat. Surv.* **2010**, *4*, 40–79. [CrossRef]
6. Kohavi, R. A study of cross-validation and bootstrap for accuracy estimation and model selection. In Proceedings of the 14th International Joint Conference on Artificial Intelligence, Montreal, QC, Canada, 20–25 August 1995; Volume 2, pp. 1137–1143.
7. Krstajic, D.; Buturovic, L.; Leahy, D.; Thomas, S. Cross-validation pitfalls when selecting and assessing regression and classification models. *J. Cheminform.* **2014**, *6*, 10. [CrossRef]
8. Moreno-Torres, J.; Sáez, J.; Herrera, F. Study on the impact of partition-induced dataset shift on k-fold cross-validation. *IEEE Trans. Neural Netw. Learn. Syst.* **2012**, *23*, 1304–1312. [CrossRef]
9. Maldonado, S.; López, J.; Iturriaga, A. Out-of-time cross-validation strategies for classification in the presence of dataset shift. *Appl. Intell.* **2022**, *52*, 5770–5783. [CrossRef]
10. Wei, T.; Wang, J.; Chen, H.; Chen, L.; Liu, W. L2-norm prototypical networks for tackling the data shift problem in scene classification. *Int. J. Remote Sens.* **2021**, *42*, 3326–3352. [CrossRef]
11. Moreno-Torres, J.G.; Raeder, T.; Alaíz-Rodríguez, R.; Chawla, N.V.; Herrera, F. A unifying view on dataset shift in classification. *Pattern Recognit.* **2012**, *45*, 521–530. [CrossRef]
12. Nikzad-Langerodi, R.; Andries, E. A chemometrician's guide to transfer learning. *J. Chemom.* **2021**, *35*, e3373. [CrossRef]
13. Huyen, C. *Designing Machine Learning Systems: An Iterative Process for Production-Ready Applications*; O'Reilly Media: Sebastopol, CA, USA, 2022.
14. Li, Y.; Murias, M.; Major, S.; Dawson, G.; Carlson, D. On target shift in adversarial domain adaptation. In Proceedings of the 22nd International Conference on Artificial Intelligence and Statistics, Naha, Japan, 16–18 April 2019; Volume 89, pp. 616–625.
15. Redko, I.; Courty, N.; Flamary, R.; Tuia, D. Optimal transport for multi-source domain adaptation under target shift. In Proceedings of the 22nd International Conference on Artificial Intelligence and Statistics, Naha, Japan, 16–18 April 2019; Volume 89, pp. 849–858.
16. Podkopaev, A.; Ramdas, A. Distribution-free uncertainty quantification for classification under label shift. In Proceedings of the 37th Conference on Uncertainty in Artificial Intelligence, Online, 27–30 July 2021; pp. 844–853.
17. Hastie, T.; Tibshirani, R.; Friedman, J. *The Elements of Statistical Learning*; Springer Series in Statistics; Springer: New York, NY, USA, 2001.
18. Kang, S.; Kang, P. Locally linear ensemble for regression. *Inf. Sci.* **2018**, *432*, 199–209. [CrossRef]
19. Carrizosa, E.; Mortensen, L.; Romero Morales, D.; Sillero-Denamiel, M. The tree based linear regression model for hierarchical categorical variables. *Expert Syst. Appl.* **2022**, *203*, 117423. [CrossRef]
20. Dhanjal, C.; Baskiotis, N.; Clémençon, S.; Usunier, N. An empirical comparison of V-fold penalisation and cross-validation for model selection in distribution-free regression. *Pattern Anal. Appl.* **2016**, *19*, 41–53. [CrossRef]
21. Breiman, L.; Spector, P. Submodel selection and evaluation in regression. The x-random case. *Int. Stat. Rev.* **1992**, *60*, 291–319. [CrossRef]
22. Baxter, C.W.; Stanley, S.J.; Zhang, Q.; Smith, D.W. Developing artificial neural network models of water treatment processes: A guide for utilities. *J. Environ. Eng. Sci.* **2002**, *1*, 201–211. [CrossRef]

23. Breiman, L.; Friedman, J.; Olshen, R.; Stone, C. *Classification and Regression Trees*; Chapman and Hall/CRC: Boca Raton, FL, USA, 2017.
24. Ding, S.; Zhao, H.; Zhang, Y.; Xu, X.; Nie, R. Extreme learning machine: Algorithm, theory and applications. *Artif. Intell. Rev.* **2015**, *44*, 103–115. [CrossRef]
25. Baringhaus, L.; Gaigall, D. Efficiency comparison of the Wilcoxon tests in paired and independent survey samples. *Metrika* **2018**, *81*, 891–930. [CrossRef]
26. Xu, L.; Hu, O.; Guo, Y.; Zhang, M.; Lu, D.; Cai, C.; Xie, S.; Goodarzi, M.; Fu, H.; She, Y. Representative splitting cross validation. *Chemom. Intell. Lab. Syst.* **2018**, *183*, 29–35. [CrossRef]
27. May, R.; Maier, H.; Dandy, G. Data splitting for artificial neural networks using SOM-based stratified sampling. *Neural Netw.* **2010**, *23*, 283–294. [CrossRef]
28. Diamantidis, N.; Karlis, D.; Giakoumakis, E. Unsupervised stratification of cross-validation for accuracy estimation. *Artif. Intell.* **2000**, *116*, 1–16. [CrossRef]
29. Snee, R. Validation of regression models: Methods and examples. *Technometrics* **1977**, *19*, 415–428. [CrossRef]
30. Sahoo, A.K.; Zuo, M.J.; Tiwari, M.K. A data clustering algorithm for stratified data partitioning in artificial neural network. *Expert Syst. Appl.* **2012**, *39*, 7004–7014. [CrossRef]
31. Joseph, V.R.; Vakayil, A. SPlit: An optimal method for data splitting. *Technometrics* **2022**, *64*, 166–176. [CrossRef]
32. Wu, W.; May, R.; Dandy, G.C.; Maier, H.R. A method for comparing data splitting approaches for developing hydrological ANN models. In Proceedings of the International Congress on Environmental Modelling and Software, Leipzig, Germany, 1–5 June 2012; p. 394.
33. Wu, W.; May, R.; Maier, H.; Dandy, G. A benchmarking approach for comparing data splitting methods for modeling water resources parameters using artificial neural networks. *Water Resour. Res.* **2013**, *49*, 7598–7614. [CrossRef]
34. Zheng, F.; Maier, H.; Wu, W.; Dandy, G.; Gupta, H.; Zhang, T. On lack of robustness in hydrological model development due to absence of guidelines for selecting calibration and evaluation data: Demonstration for data-driven models. *Water Resour. Res.* **2018**, *54*, 1013–1030. [CrossRef]
35. Chapaneri, S.; Jayaswal, D. Covariate shift adaptation for structured regression with Frank-Wolfe algorithms. *IEEE Access* **2019**, *7*, 73804–73818. [CrossRef]
36. Chen, X.; Monfort, M.; Liu, A.; Ziebart, B. Robust covariate shift regression. In Proceedings of the 19th International Conference on Artificial Intelligence and Statistics, Cadiz, Spain, 9–11 May 2016; Volume 51, pp. 1270–1279.
37. Sugiyama, M.; Nakajima, S.; Kashima, H.; Buenau, P.; Kawanabe, M. Direct importance estimation with model selection and its application to covariate shift adaptation. *Adv. Neural Inf. Process. Syst.* **2007**, *20*, 1–8.
38. Shimodaira, H. Improving predictive inference under covariate shift by weighting the log-likelihood function. *J. Stat. Plan. Inference* **2000**, *90*, 227–244. [CrossRef]
39. Kanamori, T.; Hido, S.; Sugiyama, M. A least-squares approach to direct importance estimation. *J. Mach. Learn. Res.* **2009**, *10*, 1391–1445.
40. Huang, J.; Smola, A.J.; Gretton, A.; Borgwardt, K.M.; Schölkopf, B. Correcting sample selection bias by unlabeled data. *Adv. Neural Inf. Process. Syst.* **2006**, *19*, 601–608.
41. Zhang, K.; Zheng, V.W.; Wang, Q.; Kwok, J.T.; Yang, Q.; Marsic, I. Covariate shift in Hilbert space: A solution via sorrogate kernels. In Proceedings of the 30th International Conference on Machine Learning, Atlanta, GA, USA, 17–19 June 2013; Volume 28, pp. 388–395.
42. Zeng, X.; Martinez, T.R. Distribution-balanced stratified cross-validation for accuracy estimation. *J. Exp. Theor. Artif. Intell.* **2000**, *12*, 1–12. [CrossRef]
43. Curteanu, S.; Leon, F.; Mircea-Vicoveanu, A.M.; Logofatu, D. Regression methods based on nearest neighbors with adaptive distance metrics applied to a polymerization process. *Mathematics* **2021**, *9*, 547. [CrossRef]
44. Raj, N.; Gharineiat, Z. Evaluation of multivariate adaptive regression splines and artificial neural network for prediction of mean sea level trend around northern australian coastlines. *Mathematics* **2021**, *9*, 2696. [CrossRef]
45. Boehmke, B.; Greenwell, B. Gradient Boosting. In *Hands-On Machine Learning with R*; Chapman and Hall/CRC: Boca Raton, FL, USA, 2019; pp. 221–246.
46. Dimitrova, D.; Kaishev, V.; Tan, S. Computing the Kolmogorov-Smirnov distribution when the underlying CDF is purely discrete, mixed, or continuous. *J. Stat. Softw.* **2020**, *95*, 1–42. [CrossRef]
47. Derrac, J.; García, S.; Molina, D.; Herrera, F. A practical tutorial on the use of nonparametric statistical tests as a methodology for comparing evolutionary and swarm intelligence algorithms. *Swarm Evol. Comput.* **2011**, *1*, 3–18. [CrossRef]
48. Smola, A.J.; Schölkopf, B. A tutorial on support vector regression. *Stat. Comput.* **2004**, *14*, 199–222. [CrossRef]
49. Chen, T.; Guestrin, C. XGBoost: A scalable tree boosting system. In Proceedings of the 22nd International Conference on Knowledge Discovery and Data Mining, San Francisco, CA, USA, 13–17 August 2016; pp. 785–794.

Review

Noise Models in Classification: Unified Nomenclature, Extended Taxonomy and Pragmatic Categorization

José A. Sáez

Department of Statistics and Operations Research, University of Granada, Fuente Nueva s/n, 18071 Granada, Spain; joseasaezm@ugr.es

Abstract: This paper presents the first review of noise models in classification covering both label and attribute noise. Their study reveals the lack of a unified nomenclature in this field. In order to address this problem, a tripartite nomenclature based on the structural analysis of existing noise models is proposed. Additionally, a revision of their current taxonomies is carried out, which are combined and updated to better reflect the nature of any model. Finally, a categorization of noise models is proposed from a practical point of view depending on the characteristics of noise and the study purpose. These contributions provide a variety of models to introduce noise, their characteristics according to the proposed taxonomy and a unified way of naming them, which will facilitate their identification and study, as well as the reproducibility of future research.

Keywords: noise models; nomenclature; taxonomy; noisy data; classification

MSC: 62R07

Citation: Sáez, J.A. Noise Models in Classification: Unified Nomenclature, Extended Taxonomy and Pragmatic Categorization. *Mathematics* **2022**, *10*, 3736. https://doi.org/10.3390/math10203736

Academic Editor: Liangxiao Jiang

Received: 4 September 2022
Accepted: 6 October 2022
Published: 11 October 2022

Publisher's Note: MDPI stays neutral with regard to jurisdictional claims in published maps and institutional affiliations.

Copyright: © 2022 by the author. Licensee MDPI, Basel, Switzerland. This article is an open access article distributed under the terms and conditions of the Creative Commons Attribution (CC BY) license (https://creativecommons.org/licenses/by/4.0/).

1. Introduction

Human nature and limitations of measurement tools mean that the data that real-world applications rely on often contain, to a greater or lesser extent, errors or noise [1–3]. In classification [4,5], it can affect both output class labels [6,7] and input attributes [8,9] in the form of corruptions in the corresponding values. Learning from noisy data implies that classifiers are less accurate and more complex, since errors may be modeled [10,11]. These facts have caused the study of noisy data to have an important rise in current data science research [12–14]. In this context, since noise in real-world data is normally not quantifiable and its characteristics are unknown, noise models [15,16] have been proposed to introduce errors into them in a controlled way. They allow conclusions to be drawn from experimentation based on the type of noise, its frequency and characteristics [17].

In the current literature on noisy data in classification, there are no reviews on noise introduction models dealing with both label and attribute noise. On the other hand, there are two main proposals of taxonomies for noise models [18,19]. The first one, proposed by Frénay and Verleysen [18], can be used to categorize label noise models depending on whether they use class or attribute information to decide which samples are mislabeled. The second one, proposed by Nettleton et al. [19], divided noise models considering the affected variables, the distribution of the errors and their magnitude. Note that, despite the introduction of these taxonomies, these works mainly focused on different aspects of noisy data, such as noise preprocessing or robustness of algorithms.

This paper provides a review of relevant noise models for classification data, paying special attention to recent works found in the specialized literature [15,20,21]. A total of 72 noise models will be presented, considering schemes to introduce label noise [22,23] but also attribute noise [24,25] and both in combination [26,27], which tend to receive less attention in research works. The revision of the literature on noisy data shows some ambiguity in the terminology related to noise models and that there is no unified criteria for naming them [28–30]. In this regard, one might ask: is it possible to characterize the operation of noise models in such a way that it allows to define an approach to name and categorize them? For this, as a consequence of the study of existing noise models, their structure will be analyzed, reaching conclusions about the components that allow their characterization. Then, a tripartite nomenclature will be proposed to name both existing and future models in a descriptive way, referring to their main components. This nomenclature will ease their identification, avoiding terminological differences among works and the need to provide complete descriptions each time they are used; this will also help simplify the reproducibility of the experiments carried out. Additionally, the current taxonomies [18,19] will be revised and adapted according to relevant characteristics of existing noise models. This will result in a single expanded taxonomy to better reflect the nature of each noise model and its different components, allowing them to be categorized regardless of the type of noise they introduce. Finally, noise models will also be classified into different groups according to the characteristics of the noise to be studied and the available knowledge of the problem domain. All these aspects add practical value to this paper since it offers a wide range of alternatives as a basis for research, describing and suggesting noise models based on the needs and objective of the study. In summary, the main contributions of this work are the following:

1. Presentation of the first review of noise models for classification, including label noise, attribute noise and both in combination.
2. Analysis of the structure of noise models, which is usually overlooked in the literature, identifying the fundamental components that allow their characterization.
3. Detection of the absence and lack of uniformity in the nomenclature of noise models in the literature.
4. Proposal of nomenclature to name noise models in a descriptive way, referring to their main structural components.
5. Unification of existing taxonomies in the literature and updates to better reflect the types of noise models and their characteristics.
6. Categorization of noise models from a practical point of view, depending on the characteristics of noise and the available knowledge of the problem domain.

The rest of this paper is organized as follows. Section 2 provides the background on the current terminology and taxonomies of noise models. Sections 3 and 4 present the proposed nomenclature and taxonomy, respectively. Then, Sections 5 and 6 introduce the label and attribute noise models, organizing them from a practical point of view. Finally, Section 7 concludes this work and offers ideas about future research.

2. Background

This section focuses on the current state of terminology and taxonomies of noise models in classification. Section 2.1 highlights the difficulties and discrepancies in the nomenclature of noise models. Then, Section 2.2 describes the taxonomies to categorize them.

2.1. Need for a Unified Nomenclature

Nomenclature can be defined as the set of principles and rules that are applied for the unequivocal and distinctive naming of a series of related elements. In any discipline, it is of crucial importance for scientific advancement and to allow researchers to communicate unambiguously about them. Nevertheless, an in-depth study of the literature on noisy data reflects that there is no unified criterion for naming noise models due to two main causes [28,29,31]:

1. Many models are not assigned an identifying name,
2. There are discrepancies when naming known models.

The first and very frequent cause of a lack of nomenclature is that, in fact, most of the noise models used in the literature are only described, and a name is not usually associated with them [32,33]. This situation occurs both with models traditionally used [31,32,34] and recent proposals [33,35], among other examples [36,37].

The second cause involves discrepancies in the name of the noise models, as well as in the terminology related to them [28,29]. In order to delve into this aspect, the notation compiled in Table 1 is used, which is also employed in the remainder of this work. Let D be a classification dataset composed by n samples x_i ($i \in \{1, \ldots, n\}$), m input attributes v_j ($j \in \{1, \ldots, m\}$) and one output class v_0 taking one among c possible labels in $\mathcal{L} = \{l_1, \ldots, l_c\}$. The value of the variable v_j, either input or output, in the sample x_i is denoted as $x_{i,j}$.

Table 1. Notation used.

Notation	Description	Notation	Description
D	Original dataset to be corrupted with noise.	v_j	j-th attribute ($j \in \{1, \ldots, m\}$) in dataset D.
n	Number of samples contained in dataset D.	$max(v_j)$	Maximum value of the j-th attribute ($j \geq 1$).
m	Number of attributes contained in dataset D.	$min(v_j)$	Minimum value of the j-th attribute ($j \geq 1$).
c	Number of class labels contained in dataset D.	$mean(v_j)$	Mean value of the j-th attribute ($j \geq 1$).
x_i	i-th sample ($i \in \{1, \ldots, n\}$) in dataset D.	$median(v_j)$	Median value of the j-th attribute ($j \geq 1$).
$\mathcal{S}$	Set of indices of samples $\{1, \ldots, n\}$ in D.	$var(v_j)$	Variance value of the j-th attribute ($j \geq 1$).
v_0	Output class corresponding to dataset D.	$x_{i,j}$	Original value of j-th variable v_j in sample x_i.
$\mathcal{L}$	Set of class labels $\{l_1, \ldots, l_c\}$ in dataset D.	$\tilde{x}_{i,j}$	Noisy value of j-th variable v_j in sample x_i.
l_k	k-th output class label ($k \in \{1, \ldots, c\}$) in $\mathcal{L}$.	$\mathcal{Z}$	Set of indices of variables $\{0, \ldots, m\}$ in D.
π_k	Proportion of samples with class label l_k in D.	ρ	Noise level in $[0, 1]$ used by the noise model.

As an example of ambiguous nomenclature, consider one of the most widely used noise models [38], in which the label of each sample $x_{i,0}$ can be corrupted to a different label $\tilde{x}_{i,0} \in \mathcal{L} \setminus x_{i,0}$, with $\rho/(c-1)$ being the probability of choice of each class label and ρ being known as the noise level [6] in the dataset. This simple noise model is referred to in different ways in the literature, such as *symmetric* [28,39,40], *uniform* [29,41,42] or *random* [30,43]. Some of these terms, such as *symmetric* and *uniform*, are also used to designate other different models, such as that in which the label of each sample $x_{i,0}$ can be corrupted to any class label $\tilde{x}_{i,0} \in \mathcal{L}$, being the probability of each alternative ρ/c [44–46].

Other commonly used terms with discrepancies are *asymmetric* and *class-conditional*, which sometimes are used to mention a different noise level for each class [43,47,48]. Both terms, along with other ones such as *flip-random*, are also used to refer to different probabilities among classes when choosing the noisy label for a sample [45,49,50]. They are also used to indicate different noise models [51,52]. For example, *asymmetric* or *pair noise* [51,53,54], along with other names [29,45], are used to designate the model in which each class label l_i can be corrupted to any other prefixed label $l_j \in \mathcal{L} \setminus l_i$. Simultaneously, these terms are also used to refer to the noise model consisting of corrupting the label l_i to the next label l_j, $j = i + 1 \mod c$, when an order between class labels is assumed [52,55,56]. Finally, note that the term *asymmetric* has also been used in binary classification to indicate that one class can change to the other but not vice versa [22,57] and even to designate the fact that noise levels in the training and field data are different [58].

The concept *non-uniform* is also presented in several contexts [59,60]. It is used to identify models in which the probability of noise of each sample x_i depends on its attributes $x_{i,j}$ ($j \geq 1$) [59,61,62] but also to mention different probabilities of noise for each class [60,63]. This terminological ambiguity is a consequence, among other aspects, of the existence of a dichotomy between those studies that use terms such as *symmetric/uniform* and *asymmetric/class-conditional* referring to the probability that each class contains noise [43,47,48] and those referring to, when noise for a certain class occurs, the probability of choosing each label [52,64,65]. The aforementioned facts make the terminology related to noise models not unique throughout the literature: different models are known by the same name, whereas different names design the same model depending on the research work, which can lead to confusing scenarios.

2.2. Current Taxonomies

There exist two main proposals of taxonomies for noise models [18,19]. The most recent one [18], although originally proposed to categorize label noise from a statistical point of view, can be translated into models that introduce this type of noise. Thus, it considers a single dimension to classify label noise models, which are divided into three types depending on the information used to determine if a sample is mislabeled:

1. *Noisy completely at random* (NCAR). These are the simplest models to introduce noise. In NCAR models, the probability of mislabeling a sample does not depend on the information in classes or attributes [38].
2. *Noisy at random* (NAR). In NAR models, the mislabeling probability of a sample x_i depends on its class label $x_{i,0}$. This type of model allows considering a different noise level for each of the classes in the dataset. NAR is usually modeled by means of the transition matrix [56] (Equation (1)), where ρ_{ij} ($i, j \in \{1, \ldots, c\}$) is the probability of a sample with label l_i to be mislabeled as l_j:

$$T = \begin{pmatrix} \rho_{11} & \rho_{12} & \cdots & \rho_{1c} \\ \rho_{21} & \rho_{22} & \cdots & \rho_{2c} \\ \vdots & \vdots & \vdots & \vdots \\ \rho_{c1} & \rho_{c2} & \cdots & \rho_{cc} \end{pmatrix} \quad (1)$$

3. *Noisy not at random* (NNAR). These use both the information in class labels $x_{i,0}$ and attribute values $x_{i,j}$ ($j \geq 1$) to determine mislabeled samples. They constitute a more realistic scenario consisting of mislabeling samples in specific areas, such as decision boundaries, where classes share similar characteristics [22].

Note that these categories were also used to show the dependency of the sample's noisy label on the original label (in NCAR and NAR) or on the label and attribute values (in NNAR). In contrast to the previous taxonomy for label noise models, the work in [19] proposed another one for any type of noise (label and attributes) based on three dimensions:

1. *Affected variables*. This divides the noise models according to whether they introduce label noise, attribute noise or their combination.
2. *Error distribution*. This classifies the models by considering whether the introduced errors follow some known probability distribution, such as Gaussian.
3. *Magnitude of errors*. This divides the models according to whether the magnitude of the generated errors is relative to the values of each sample or to the minimum, maximum or standard deviation of each variable.

3. A New Unified Nomenclature for Noise Models

Due to the lack of a unified terminology for noise models, this section presents a procedure for naming them with a twofold objective: (i) assigning an unequivocal name to each noise model and (ii) being as descriptive as possible, providing information on the noise introduction process.

The analysis of existing noise models, as well as the dichotomy in research works when naming noise models discussed in Section 2.1, reveal that they have three main components that allow them to be described (see Figure 1):

1. *Noise type*. It is a characteristic that indicates the variables affected by the noise model. Thus, any noise model introduces noise into class labels [15,23,66], attribute values [25,67,68] or both in combination [26,27].
2. *Selection procedure*. Let $\mathcal{S} = \{1, \ldots, n\}$ be the set of indices of samples and $\mathcal{Z} = \{0, \ldots, m\}$ be the set of indices of variables (output class label and input attributes) in a dataset D to be corrupted. The selection procedure creates a set of pairs $\mathcal{P} \subseteq \mathcal{S} \times \mathcal{Z} = \{(s, z) \mid s \in \mathcal{S} \text{ and } z \in \mathcal{Z}\}$ with the values to be altered. An element $(s, z) \in \mathcal{P}$ indicates that the variable v_z in the sample x_s (that is, the value $x_{s,z}$) must

be corrupted. Note that $z = 0$ for label noise models, $z \geq 1$ for attribute noise models and $z \geq 0$ for combined noise models.

The set $\mathcal{P}$ can be seen as the indices $s \in \mathcal{S}$ of the samples to corrupt for each variable v_z ($z \in \mathcal{Z}$) in which noise is introduced. There are multiple ways to select such samples [20,69,70]. For example, all samples can have the same probability of being chosen [38,71], a different probability can be defined for the samples according to their class label [20,72], the probability of choosing a sample can depend on its proximity to the decision boundaries [69,70,73], the samples in a certain area of the domain can be altered [16,74] or even all the samples in the dataset can be selected to be corrupted [67].

3. *Disruption procedure*. Given the set $\mathcal{P}$ indicating the samples $s \in \mathcal{S}$ to be corrupted for each noisy variable v_z ($z \in \mathcal{Z}$), this procedure allows altering their original values $x_{s,z}$ by changing them to new noisy ones $\tilde{x}_{s,z}$. As in the case of the selection procedure, there are different alternatives to modify the original values by the disruption procedure [24,71,75]. For example, a new value within the domain can be chosen including the original value [26,76] or excluding it [38,77], a default value can be chosen dependently from the original value [75,78,79] or independently [71,72], additive noise following a Gaussian distribution can be considered [17,67], among others [80,81].

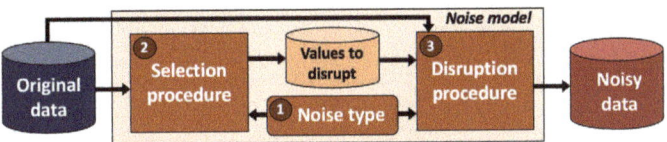

Figure 1. Structure and components of noise models.

Based on these three components, a tripartite nomenclature for noise models is proposed. Each part is formed by an identifier that evokes a characteristic of each of the previous components. Table 2 shows examples of identifiers defined for each component, together with their description and a research work using it. In those terms traditionally used in the literature where there are discrepancies (such as *symmetric* [28,39], *asymmetric* [51,53] or *uniform* [41,42]), one of their most used meanings has been adopted.

Table 2. Examples of identifiers defined for the nomenclature of noise models.

Component	Identifier	Description	Ref.
Type	label	Noise affects the class labels $x_{i,0}$ of some of the samples x_i ($i \in \{1, \ldots, n\}$) in the dataset.	[22]
	attribute	Noise affects the attribute values $x_{i,j}$ ($j \in \{1, \ldots, m\}$) of some samples x_i ($i \in \{1, \ldots, n\}$).	[78]
	combined	Noise affects the labels and attributes $x_{i,j}$ ($j \in \{0, \ldots, m\}$) of some samples x_i ($i \in \{1, \ldots, n\}$).	[26]
Selection	symmetric	Samples in all classes $\{l_1, \ldots, l_c\}$ or attributes $\{v_1, \ldots, v_m\}$ have equal probability of noise.	[38]
	asymmetric	Samples in each class $\{l_1, \ldots, l_c\}$ or attribute $\{v_1, \ldots, v_m\}$ have a different probability of noise.	[20]
	unconditional	Noise unconditionally affects all samples x_i ($i = 1, \ldots, n$) in the dataset to be corrupted.	[67]
	majority-class	Random choice of samples from the majority class within the dataset to be corrupted.	[21]
	Gaussian	A Gaussian distribution determines noise probabilities using distances to decision boundaries.	[22]
	Gamma	A Gamma distribution determines noise probabilities using distances to decision boundaries.	[70]
	one-dimensional	Given c intervals $[a_k, b_k]$ for v_j ($j \geq 1$), samples x_i with $x_{i,0} = l_k$ and $x_{i,j} \in [a_k, b_k]$ are corrupted.	[74]
Disruption	uniform	The noisy value is randomly chosen within the domain of the variable excluding the original value.	[38]
	completely-uniform	The noisy value is randomly chosen within the domain of the variable including the original value.	[44]
	default	Original clean values are replaced by a fixed noisy value within the domain of the variable to corrupt.	[71]
	Gaussian	A random value following a zero-mean Gaussian distribution is added to the original attribute value.	[17]
	natural-distribution	A random value with probability proportional to the original distribution replaces the original value.	[72]
	unit-simplex	The probability of choosing each value as noisy is determined by a k-dimensional unit-simplex.	[46]
	bidirectional	Given a pair of values (a, b) for a variable v_j, samples with $x_{i,j} = a$ change it to b and vice versa.	[33]

Once the identifier of each part is determined, the name of the model is formed from the type of noise it introduces, concatenating each part as follows:

$$\text{model name} = \{selection\} + \{disruption\} + \{type\} + \text{noise}$$

In order to facilitate recognition of the name of the noise model, it should be written in italics and with its first letter capitalized. Evoking the characteristics of each noise model makes the proposed nomenclature easier to remember. At the same time, the existence of noise models with the same *selection*, *disruption* or *type* names indicates that they are closely related to each other. The proposal for applying this nomenclature to the noise models considered in this work, together with their corresponding references, is found in Table 3. Note that, in some specific cases, there may be slight differences between the nomenclature and the exact noise model applied in the corresponding paper. This is mainly due to the fact that some works use noise models with specific configurations to achieve particular objectives and, in these cases, a name representing the model in a more general way is proposed. For example, in [78], some specific attributes of interest to the dataset used are chosen to corrupt. However, to provide a general name to this noise model, the identifier *symmetric* has been used for its selection procedure, referring to the fact that all attributes can be corrupted to the same degree. Other examples are the models used in [82], which introduce errors in binary attributes. In this case, even though the identifier *bidirectional* in Table 2 could be used for the disruption process, the identifier *uniform* has been chosen considering that these models can also be applied to attributes with more than two values. Thus, the above approach can be employed to increase consistency when naming new models so that the specific application can be separated from the noise model used.

Table 3. List of noise models and references.

Noise Model	Ref.	Noise Model	Ref.
Label noise models			
Asymmetric default label noise	[72]	PMD-based confidence label noise	[83]
Asymmetric sparse label noise	[84]	Quadrant-based uniform label noise	[62]
Asymmetric uniform label noise	[20]	Score-based confidence label noise	[85]
Attribute-mean uniform label noise	[86]	Sigmoid-bounded uniform label noise	[43]
Clustering-based voting label noise	[81]	Small-margin borderline label noise	[87]
Exponential borderline label noise	[69]	Smudge-based completely-uniform label noise	[88]
Exponential/smudge completely-uniform label noise	[76]	Symmetric adjacent label noise	[89]
Fraud bidirectional label noise	[35]	Symmetric center-based label noise	[90]
Gamma borderline label noise	[70]	Symmetric completely-uniform label noise	[44]
Gaussian borderline label noise	[22]	Symmetric confusion label noise	[63]
Gaussian-level uniform label noise	[61]	Symmetric default label noise	[71]
Gaussian-mixture borderline label noise	[22]	Symmetric diametrical label noise	[72]
Hubness-proportional uniform label noise	[16]	Symmetric double-default label noise	[91]
IR-stable bidirectional label noise	[33]	Symmetric double-random label noise	[80]
Laplace borderline label noise	[73]	Symmetric exchange label noise	[66]
Large-margin uniform label noise	[87]	Symmetric hierarchical label noise	[23]
Majority-class unidirectional label noise	[21]	Symmetric hierarchical/next-class label noise	[79]
Minority-driven bidirectional label noise	[92]	Symmetric natural-distribution label noise	[72]
Minority-proportional uniform label noise	[93]	Symmetric nearest-neighbor label noise	[41]
Misclassification prediction label noise	[81]	Symmetric next-class label noise	[75]
Multiple-class unidirectional label noise	[81]	Symmetric non-uniform label noise	[94]
Neighborwise borderline label noise	[15]	Symmetric optimistic label noise	[72]
Non-linearwise borderline label noise	[15]	Symmetric pessimistic label noise	[72]
One-dimensional uniform label noise	[74]	Symmetric uniform label noise	[38]
Open-set ID/nearest-neighbor label noise	[41]	Symmetric unit-simplex label noise	[46]
Open-set ID/uniform label noise	[41]	Uneven-Gaussian borderline label noise	[73]
Pairwise bidirectional label noise	[95]	Uneven-Laplace borderline label noise	[73]
Attribute noise models			
Asymmetric interval-based attribute noise	[58]	Symmetric scaled-Gaussian attribute noise	[25]
Asymmetric uniform attribute noise	[82]	Symmetric uniform attribute noise	[77]
Boundary/dependent Gaussian attribute noise	[68]	Symmetric/dependent Gaussian attribute noise	[67]
Importance interval-based attribute noise	[58]	Symmetric/dependent Gaussian-image attribute noise	[96]
Symmetric completely-uniform attribute noise	[26]	Symmetric/dependent random-pixel attribute noise	[96]
Symmetric end-directed attribute noise	[78]	Symmetric/dependent uniform attribute noise	[82]
Symmetric Gaussian attribute noise	[17]	Unconditional fixed-width attribute noise	[97]
Symmetric interval-based attribute noise	[58]	Unconditional vp-Gaussian attribute noise	[67]
Combined noise models			
Symmetric completely-uniform combined noise	[26]	Unconditional/symmetric Gaussian/uniform combined noise	[27]

As an example of the naming process, consider the noise model in which each sample has the same probability of noise and can be mislabeled with any class label other than the original [38]. Since the model introduces noise into class labels, its *type* identifier is *label*. The *selection* of samples to modify considers the same probability for all of them, so this identifier takes the value *symmetric*. Finally, since the *disruption* process chooses the noisy label following a uniform distribution excluding the original value, its identifier takes the value *uniform*. Therefore, the name of this noise introduction model is *Symmetric uniform label noise*.

As discussed above, it should be noted that although many of the noise models originally lacked a distinctive name, others were specifically named [46,71,82]. Some examples among label noise models are *Symmetric default label noise* (originally called *background flip noise* [71]), *Symmetric unit-simplex label noise* (called *random label flip noise* [46]), *Symmetric double-random label noise* (called *flip2 noise* [80]) or *Symmetric hierarchical label noise* (called *hierarchical corruption* [23]). Among attribute noise models, *Asymmetric uniform attribute noise* (originally called *asymmetric independent attribute noise* [82]) or *Symmetric Gaussian attribute noise* (called *Gaussian attribute noise* [17]) can be mentioned.

4. Proposal for an Extended Taxonomy of Noise Models

Even though the current taxonomies of noise models [18,19] fulfill the purpose for which they were designed, the analysis of existing models shows the need not only to consider them as a whole but also to modify them appropriately to better reflect the nature of any model. The taxonomy proposed by Frénay and Verleysen [18] allows classifying label noise models based on the information source (class and attributes) used to determine which samples are mislabeled. It is interesting to extend this idea to attribute noise models, as well as simultaneously consider other dimensions proposed by Nettleton et al. [19]. This joint taxonomy can be expanded to include other dimensions derived from the structure of noise models shown in Figure 1, which allow the development of a more detailed and descriptive categorization. Thus, the following dimensions are proposed to form part of the final taxonomy, each one described in a separate section:

- *Noise type* (Section 4.1). This classifies the models based on the introduction of label noise, attribute noise or both.
- *Selection source* (Section 4.2). This categorizes the models according to the information sources (class and/or attributes) used by the selection procedure.
- *Disruption source* (Section 4.3). This is similar to the selection source but aimed at the disruption procedure.
- *Selection distribution* (Section 4.4). The main probability distribution, if any, underlying the selection procedure.
- *Disruption distribution* (Section 4.5). The main probability distribution related to the disruption procedure.

Since the taxonomy defined by the dimensions above is based on previous approaches [18,19], these provide similar categorization results in those cases where they are comparable. For example, the taxonomy in [18], which is designed for label noise, could be compared to the selection source dimension if label noise models were considered. Similarly, this occurs with the noise type dimension, which was previously used in [19]. This fact, along with the appropriate extensions that are detailed throughout this section, make the proposed taxonomy adapt to previous versions and incorporate features to better reflect the variability of noise models and their characteristics.

4.1. Noise Type

This dimension is essential to divide the models between those introducing label noise [15,22], attribute noise [8,78] or their combination [26,27]. The most representative models within each noise type (label, attributes and combined) are *Symmetric uniform label noise* [38], *Symmetric uniform attribute noise* [77] and *Symmetric completely-uniform combined noise* [26], respectively. These models allow noise to be studied in a generic way since they introduce random errors and do not require knowing the domain of the problem addressed or making assumptions about the data. Both the selection of samples to be corrupted and their noisy values (in class labels or attributes) are carried out considering that the alternatives have the same probability of being chosen [31,98]. In addition, they allow the percentage of errors in the dataset to be controlled by means of a noise level, facilitating the analysis of the consequences of noise as the number of errors in the data increases [77]. These characteristics make them some of the most widely used models [15,24,44].

The analysis of the 72 models in Table 3 with respect to the noise type shows that a high percentage (75%) is focused on label noise. Attribute noise models represent 22.2%, whereas combined noise models are 2.8%. These frequencies are consistent with the limited number of works involving attribute noise [78,82]. This fact is due to the greater complexity that is usually associated with the identification and treatment of errors in attribute values compared to those in class labels [8].

4.2. Selection Source

This dimension will allow the models to be divided according to the information sources (classes and/or attributes) used by the selection procedure. It is inspired by the taxonomy proposed in [18], properly adapted to categorize all the models and types of noise. Thus, the analysis of existing noise models shows that the categorization in [18] should be expanded because there are label noise models [62,86] whose selection procedure uses only attribute information (and not class information), which had not been previously contemplated. For example, *Quadrant-based uniform label noise* [62] allows determining the probability of mislabeling samples in four different regions of the domain based on the values of two attributes. Following the naming style in [18], the selection source of this new type of noise models is denoted as *noisy partially at random* (NPAR). On the other hand, it is necessary to update this dimension to cover attribute noise [58,78], making it valid for any model regardless of the type of noise addressed. Thus, the selection source finally divides the noise models, either label or attribute noise, into the following categories:

1. *Noisy completely at random* (NCAR). These are models that do not consider the information of labels or attributes to select the noisy samples. For example, *Symmetric uniform label noise* [38] corrupts the class labels in the dataset assigning to each sample the same probability ρ of being altered, whereas *Unconditional vp-Gaussian attribute noise* [67] corrupts all the samples.
2. *Noisy at random* (NAR). These are models whose selection procedure uses the same information source as the noise type they introduce. Therefore, NAR label [33,35] and attribute [58,82] noise models, respectively, use class information ($x_{i,0}$) and attribute information ($x_{i,j}$ or v_j with $j \geq 1$) to determine the samples x_i to corrupt. For example, *Asymmetric uniform label noise* [20] and *Asymmetric uniform attribute noise* [82] consider a different noise probability for each class label $\{l_1, \ldots, l_c\}$ and attribute $\{v_1, \ldots, v_m\}$, respectively.
3. *Noisy partially at random* (NPAR). These are models whose selection procedure uses the opposite information source to the noise type they introduce. Examples of this type of selection source are found in *Attribute-mean uniform label noise* [86], which gives a higher probability of corrupting the samples whose attributes are closer to the mean values, or in the aforementioned *Quadrant-based uniform label noise* [62].
4. *Noisy not at random* (NNAR). The selection procedure of these models uses class and attribute information to determine the noisy samples. For example, *Neighborwise borderline label noise* [15] determines the samples to corrupt by computing a noise score for each sample as a function of its distances to its nearest neighbors (using the information from attributes) of the same and different class (using the information from class labels).

Note that, in contrast to [18], the categories in this dimension do not indicate, for example, the dependency of a specific noisy label on the original label or attributes. These relations have been reflected through a new dimension (see Section 4.3) since the selection procedure (which selects the values in the dataset to corrupt) and disruption procedure (which chooses the noisy values for the selected data) do not necessarily use the same information sources for their purpose.

4.3. Disruption Source

Since the disruption process is different from the selection process, models can also be classified according to the information sources used to choose the noisy values. Thus, analogously to the selection source, models can be divided into the following categories according to the information sources used by the disruption procedure:

1. *Disruption completely at random* (DCAR). These are models whose disruption procedure does not use information from class labels or attribute values. For example, *Symmetric default label noise* [71] always chooses the same class label as the noisy value regardless of the class and attribute values of each sample, whereas *Symmetric completely-uniform label noise* [44] chooses the value of the noisy label uniformly among all the possibilities in the domain.
2. *Disruption at random* (DAR). These are models whose disruption procedure uses only the same information source as the noise type they introduce. For example, *Symmetric adjacent label noise* [89] chooses one of the classes adjacent to that of the sample to corrupt, whereas *Symmetric Gaussian attribute noise* [17] adds another one to the value of the original attribute that follows a zero-mean Gaussian distribution.
3. *Disruption partially at random* (DPAR). These are models whose disruption procedure uses the opposite information source to the noise type they introduce. Even though this characteristic is not usually considered in the literature, it is interesting to define it for potential noise models.
4. *Disruption not at random* (DNAR). These are models whose disruption procedure uses class and attribute information. For example, in *Symmetric nearest-neighbor label noise* [41], the noisy label for each sample is taken from its closest sample of a different class. Therefore, it uses class and attribute information to determine the new noisy labels.

4.4. Selection Distribution

This dimension refers to the categorization of noise models based on the main probability distribution on which the selection procedure is based. In this context, a great variety of probability distributions can be distinguished, such as multivariate Bernoulli (in *Symmetric uniform label noise* [38]), Gaussian (in *Gaussian-level uniform label noise* [61]), exponential (in *Exponential/smudge completely-uniform label noise* [76]), Laplace (in *Laplace borderline label noise* [73]) or gamma (in *Gamma borderline label noise* [70]). Other models introduce noise in all samples of the dataset unconditionally [67]. Among them, the most used is the multivariate Bernoulli distribution, from which it is interesting to distinguish the following subcategories for the selection distribution:

- *Symmetric* [44,71]. Each sample follows a Bernoulli distribution of parameter ρ to be corrupted. Thus, a multivariate Bernoulli distribution of parameter ρ on $\{0,1\}^n$ is followed by the n samples of all the classes/attributes (according to the type of noise introduced), which therefore have the same probability of being selected.
- *Asymmetric* [20,82]. A total of k different multivariate Bernoulli distributions of parameters $\rho_1, \ldots, \rho_k$ are applied separately to the samples of each class ($k = c$) or attribute ($k = m$) depending on the type of noise introduced. Therefore, samples of each class/attribute have a different probability of being corrupted.

4.5. Disruption Distribution

This dimension refers to the main probability distribution associated with the disruption process. As in the case of the selection distribution, there are multiple alternatives for the disruption distribution [38,71,72]. For example, the new noisy value can be chosen following a uniform distribution, including the original value [44] or excluding it [38], following the natural distribution of values in the original data [72], using an additive zero-mean Gaussian distribution [17], choosing the new value uniformly within a given interval [58] or choosing a default value [71], among others [41,80]. Note that this dimension can also consider models with a deterministic choice of noisy values [33,95].

5. Label Noise Models

Label noise [99–101] occurs when the samples in the dataset are labeled with wrong classes. In real-world data, label noise can proceed from several sources [43,69,102]. Thus, human mistakes due to weariness, routine or quick examination of each case, as well as the imprecise information or subjectivity during the labeling process can produce this type of noise [102]. Additionally, automated approaches to collect labeled data, such as data mining on social media and search engines, inevitably involve label noise [43]. Label noise models [70,87,92] are designed to simulate these real-world labeling errors. Depending on the characteristics of the noise and the application to be studied, the following main groups of label noise models are considered:

(1) *Unrestricted label noise* [38,44]. The objective of these models is to introduce label noise that affects all classes equally so that they allow the noise problem to be studied in a generalized way. The most widely used models in this group are *Symmetric uniform label noise* [38] and *Symmetric completely-uniform label noise* [44], which belong to the DAR and DCAR disruption procedures, respectively. Another option to consider is *Symmetric unit-simplex label noise* [46], in which the probability of choosing each of the class labels as noisy by the disruption procedure is based on a k-dimensional unit-simplex ($k = c - 1$).

(2) *Borderline label noise* [15,69]. These models represent common scenarios in real-world applications, where mislabeled samples are more likely to occur near the decision boundaries [22,70]. Garcia et al. [15] proposed two borderline label noise models: *Non-linearwise borderline label noise*, in which a noise metric for each sample is based on its distance to the decision limit induced by SVM [103], and *Neighborwise borderline label noise*, which calculates a noise measure $N(x_i)$ for each sample x_i based on the distances to its closest samples:

$$N(x_i) = \frac{d(x_i, x_j = NN(x_i) \mid x_{j,0} = x_{i,0})}{d(x_i, x_k = NN(x_i) \mid x_{k,0} \neq x_{i,0})} \quad (2)$$

with $NN(x_i)$ being the nearest neighbor of x_i and $d(x_i, x_j)$ the Euclidean distance between the samples x_i and x_j. Similarly, *Small-margin borderline label noise* [87] trains a *logistic regression* classifier [104] and selects to corrupt a percentage of the correctly classified samples closest to the decision boundary.

Other models are also based on the distances of the samples to the decision boundaries, which are then used to estimate the probability that each sample is mislabeled, employing different probability density functions (PDF) [69,73]. For example, once the distance d_i of a sample x_i to the decision boundary is computed, *Gaussian borderline label noise* [22] uses the PDF of a Gaussian distribution $\mathcal{N}(\mu, \sigma^2)$ to assign its noise probability $P(x_i)$, which could also be adapted for samples of different classes:

$$P(x_i) = f(d_i; \mu, \sigma^2) = \frac{1}{\sigma\sqrt{2\pi}} e^{-\frac{1}{2}(\frac{d_i - \mu}{\sigma})^2} \quad (3)$$

Other PDFs that have been used are that of the exponential distribution (in *Exponential borderline label noise* [69]), the gamma distribution (in *Gamma borderline label noise* [70]) or the Laplace distribution (in *Laplace borderline label noise* [73]). In some of these models, such as *Uneven-Laplace borderline label noise* and *Uneven-Gaussian borderline label noise* [73], noise probability affects samples unequally depending on which side of the decision boundary they are on.

There are also noise models that are not based on distances to the decision boundaries but on predictions of some classifier to determine borderline samples [81,85]. For example, *Misclassification prediction label noise* [81] selects incorrectly classified samples by a multi-layer preceptron to be corrupted using their predicted labels. *Score-based confidence label noise* [85] calculates a noise score $N(x_i)$ for each sample x_i based on the outputs S^q ($q \in \{1, \ldots, Q\}$) of a *Deep Neural Network* (DNN) [105] at Q different epochs:

$$N(x_i) = \max_{k \neq x_{i,0}} S_{i,k} \quad S = \frac{\sum_{q=1}^{Q} S^q}{Q} \in \mathbb{R}^{n \times c} \quad (4)$$

Finally, $\rho\%$ of samples with the highest noise scores are chosen to be corrupted with the more reliable different labels offered by DNN. Note that some of the previous models [69,70,73] were proposed for binary datasets and, therefore, the identifier *bidirectional* in Table 2 could be used for their disruption process. However, they have been assigned the identifier *borderline* considering their potential use in multi-class problems, where the disruption procedure could follow a strategy coherent with this type of noise and choose any label as noisy, such as the majority label among the closest samples with a label different from that of the sample to be modified.

(3) *Label noise in other domain areas* [16,62]. In addition to noise in class boundaries, models have been proposed to introduce label noise in other regions of the dataset [16,74,86]. Thus, *One-dimensional uniform label noise* [74] and *Quadrant-based uniform label noise* [62] allow the practitioner to focus on specific areas of the domain based on the values of one and two variables of interest, respectively. Other examples in this group are *Attribute-mean uniform label noise* [86], which makes samples closer to attribute mean values more likely to be mislabeled, and *Large-margin uniform label noise* [87], which mislabels a percentage of the correctly classified samples that are farthest from the decision boundaries. *Hubness-proportional uniform label noise* [16] gives samples closer to hubs in the dataset a higher chance of being corrupted. Thus, $n \cdot \rho$ samples are chosen to be mislabeled, with each sample x_i having a probability $P(x_i)$ of being selected:

$$P(x_i) = \frac{N_k(x_i)}{n \cdot k} \qquad (5)$$

with $N_k(x_i)$ being the *hubness* of x_i, that is, the number of times that x_i is among the k nearest-neighbors of any other sample.

(4) *Label noise based on synthetic attributes* [76,88]. Other models may consider additional attributes in the dataset to make noise dependent on them. For example, *Exponential/smudge completely-uniform label noise* [76] uses a new attribute v_{m+1} with values uniformly distributed in $[0,1]$ to determine the error probability of each sample $P(x_i)$ employing an exponential function of parameter λ:

$$P(x_i) = \lambda e^{-\lambda s(1-x_{i,m+1})} \qquad (6)$$

with s being a user parameter to scale the attribute v_{m+1}. On the other hand, *Smudge-based completely-uniform label noise* [88] uses a specific feature in the data to indicate the presence of label noise. This can be achieved, for example, by assigning a certain value to existing or new attributes of mislabeled samples.

(5) *Label noise using default values* [71,91]. This type of label noise typically occurs when the true class cannot be determined for certain samples and a default value is assigned to them (for example, the value *other*). Thus, whenever noise in a sample occurs, its label is incorrectly assigned to a previously established class. In order to introduce it, *Symmetric default label noise* [71] and *Asymmetric default label noise* [72] can be used. Instead of a single default class, *Symmetric double-default label noise* [91] randomly chooses one of two possible default classes.

(6) *Label noise assuming order among classes* [75,89]. In some data science applications, classes have a natural order relation [89]. For example, a reader can use the labels {*bad, good, excellent*} for a book review. There are several label noise models considering this peculiarity [75,89]. *Symmetric next-class label noise* [75] mislabels a sample of class l_i to the next label l_j (with $j = i+1 \mod c$), whereas *Symmetric adjacent label noise* [89] randomly picks one of the classes adjacent to l_i as noisy. Prati et al. [72] proposed several label noise models in this scenario. For example, *Symmetric diametrical label noise* makes distant classes more likely to be chosen as noisy. An alternative to defining the probabilities $\rho_{i,j}$ of a sample with label l_i to be mislabeled as l_j is:

$$\rho_{i,j} = \frac{|i-j|}{\sum_{k=1}^{c} |i-k|} \qquad (7)$$

Other models are *Symmetric optimistic label noise* (in which labels higher than that of the noisy sample are more likely to be chosen by the disruption procedure) or *Symmetric pessimistic label noise* (which gives a higher probability of choice to labels lower than that of the sample).

(7) *Label noise in out-of-distribution classes* [41]. Sometimes the set of labels used in a classification problem is not complete and some of them are not considered. This fact implies that the samples of the classes that are ignored (out-of-distribution, OOD) are mislabeled with the labels that were finally preserved (in-distribution, ID). The simplest model in this group is *Open-set ID/uniform label noise* [41], which chooses a random label among ID classes for samples of OOD classes. Another example of this type of model is *Open-set ID/nearest-neighbor label noise* [41], which is similar to the previous one, but it chooses the label from the closest sample of the ID classes.

(8) *Label noise in binary classification* [33,92]. Binary classification, in which $c = 2$, is common in certain applications, such as medical datasets [106] (where healthy and unhealthy patients are commonly distinguished) or fraud detection [35]. Even though other label noise models can be applied to this type of data, there are some that are specifically designed for them. One of the most studied problems in binary classification is that of class imbalance [107], in which one of the classes has fewer samples than the other. In order to introduce label noise in these problems while keeping the number of samples in both classes, *IR-stable bidirectional label noise* [33] has been proposed. *Minority-driven bidirectional label noise* [92] allows controlling both the total number of corrupted samples n_c by means of the noise level ρ, as well as the number of those samples in the minority class (n_m) and the majority class (n_M) by means of the parameter $\eta \in (0,1)$:

$$n_c = 2 \cdot \rho \cdot |S_m|; \quad n_m = \eta \cdot n_c; \quad n_M = (1-\eta) \cdot n_c; \tag{8}$$

with $|S_m|$ being the number of samples in the minority class. Finally, in the context of fraud detection, *Fraud bidirectional label noise* [35] has been used, which introduces noise mainly in the minority class and, to a much lesser extent, in the majority class. Specifically, when the number of majority samples is high enough, it introduces n_c noisy samples in the dataset:

$$n_c = \rho \cdot |S_m| + \mu \cdot \rho \tag{9}$$

with $\mu \in \mathbb{R}$ being a parameter to control the number of majority of noisy samples, $\rho \cdot |S_m|$ the amount of minority samples corrupted and $\mu \cdot \rho$ the relatively small amount of majority samples corrupted.

(9) *Label noise between pairs of classes* [21,95]. It is common for each of the classes to be more prone to being confused with another [81]. These models simulate this scenario, but they usually require some knowledge or assumptions about the problem since the practitioner commonly decides which classes are confused with each other. They all tend to work in a similar way [21,84]. Let l_i and l_j be two classes. Then, $\rho\%$ of the samples of l_i are randomly selected and labeled as l_j.

In *Majority-class unidirectional label noise*, which can be applied to both binary [21] and multi-class [24] data, l_i is the majority class and l_j is usually the second majority class. *Multiple-class unidirectional label noise* [81] is similar but allows defining multiple pairs of classes (l_i, l_j) for the valid transitions in the dataset. Finally, *Pairwise bidirectional label noise* [95] considers that noise can occur in both directions (from l_i to l_j and from l_j to l_i), whereas *Asymmetric sparse label noise* [84] requires determining a different noise level for each of the classes.

(10) *Label noise with class hierarchy* [23,79]. In certain applications, the set of labels can be grouped into different superclasses [79]. For example, when classifying the following animals in a certain region {*wolf, goat, snake, lizard*}, the first two belong to the superclass *mammal*, whereas the last two belong to the superclass *reptile*. In this type of problem, mislabeling between labels of the same superclass is the most natural scenario, and noise between labels of different superclasses usually does not occur [91]. In this context, in *Symmetric hierarchical label noise* [23], the samples can be randomly mislabeled as belonging to any class within the corresponding superclass. Similarly, when classes are ordered,

Symmetric hierarchical/next-class label noise [79] chooses the next class within its superclass as the new noisy label.

(11) *Choosing noisy labels in the disruption procedure* [90,94]. In addition to class labels being chosen uniformly within the domain (as in those models with identifiers *uniform* [16,38] and *completely-uniform* [44,76] for the disruption procedure), there are other noise models that allow more alternatives to select the new noisy labels [72,94]. For example, *Symmetric non-uniform label noise* [94] considers different probabilities in the disruption procedure to choose each noisy label according to the original class. On the other hand, in *Symmetric natural-distribution label noise* [72], when noise for a certain class occurs, another class with a probability proportional to the original class distribution replaces it:

$$\rho_{i,j} = \frac{\pi_j}{1-\pi_i} \cdot \rho \qquad (10)$$

with π_i being the class proportion of l_i. In *Symmetric nearest-neighbor label noise* [41], the noisy label for each sample is taken from its closest sample of a different class. *Symmetric confusion label noise* [63] considers that the probability of choosing each noisy label given the original one is based on a normalized confusion matrix obtained from the dataset. Finally, in *Symmetric center-based label noise* [90], closer classes l_i/l_j, with $i \neq j$, are more likely to be confused:

$$\rho_{i,j} = \frac{\sqrt{1/d_{i,j}}}{\sum_{k \neq i} \sqrt{1/d_{i,k}}} \cdot \rho \qquad (11)$$

with $d_{i,j}$ being the distance between the centers of classes l_i and l_j.

(12) *Different noise levels in different classes* [20,93]. These models are suitable in those applications where certain classes are more prone to errors than others. Some of these models adopt the asymmetric version (with NAR selection procedure and different noise levels in each class) of the models mentioned above, such as *Asymmetric uniform label noise* [20] or *Asymmetric default label noise* [72]. Another example is *Minority-proportional uniform label noise* [93], which considers a noise level ρ_i in each class l_i based on the number of samples it has relative to the minority class:

$$\rho_i = \frac{\pi_m}{\pi_i} \cdot \rho \mid \pi_m, \pi_i \in (0,1) \qquad (12)$$

with π_m being the proportion of samples in the minority class, π_i the proportion of samples in the class l_i and $\sum_{k=1}^{c} \pi_k = 1$. Note that the expression π_m/π_i accompanying ρ in Equation (12) is in the interval $[0,1]$, so the final noise ρ_i is also in $[0,1]$. Similarly, this situation occurs in Equations (10)–(11), where the factor accompanying ρ is in $[0,1]$.

6. Attribute Noise Models

Attribute noise [68,78] involves errors in the attribute values of the samples in a dataset. Even though it may have a human cause, its origin is usually related to instrumental measurement errors (for example, in sensor devices), limitations in the transmission media and so on [8]. In order to represent the variety of scenarios in which errors in attributes can occur, different models for their introduction have been used [25,67,96]. Thus, depending on the type of noise and the application to be studied, attribute noise models in the literature can be grouped into the following main types:

(1) *Unrestricted attribute noise* [24,77]. These models are appropriate if restrictions in the noise introduction process are not relevant. They allow studying attribute noise in a generic way, introducing errors affecting any part of the domain [24]. Under these premises, the most recommended models are *Symmetric uniform attribute noise* [77] and *Symmetric completely-uniform attribute noise* (note that the latter comes from considering only the attribute noise introduction process of the model proposed in [26]). Considering a noise level ρ, both independently introduce the same amount of errors $n \cdot \rho$ in each attribute and randomly select noisy values following a uniform distribution within the attribute domain.

The main difference is that the former excludes the original value $x_{i,j}$ as noisy. Thus, for a numeric attribute v_j, a noisy value $\tilde{x}_{i,j}$ satisfies the following expression:

$$\tilde{x}_{i,j} \sim \mathcal{U}[min(v_j), max(v_j)] \mid \tilde{x}_{i,j} \neq x_{i,j} \tag{13}$$

whereas the *completely-uniform* alternative can choose the original value $x_{i,j}$ as noisy.

(2) *Attribute noise in specific domain areas* [68,78]. These models are suitable for applications where attribute noise is known to affect certain parts of the domain. For example, *Symmetric end-directed attribute noise* [78] can be used to introduce extreme noise affecting the limits of the domain. For each attribute value $x_{i,j}$ to be corrupted, the following procedure is applied to determine its noisy value $\tilde{x}_{i,j}$:

- if $x_{i,j} < median(v_j)$, $\tilde{x}_{i,j} = max(v_j) + k$;
- if $x_{i,j} > median(v_j)$, $\tilde{x}_{i,j} = min(v_j) - k$;
- if $x_{i,j} = median(v_j)$, one of the above options is chosen;

with $k = s \cdot max(v_j)$ and $s \in (0,1)$. Another example is *Boundary/dependent Gaussian attribute noise* [68], which was originally proposed to simulate an outlier effect in the data. It corrupts a percentage of the samples close to the decision boundary in a classification problem by introducing additive noise into the original values using a zero-mean Gaussian distribution.

(3) *Attribute noise simulating small errors* [17,58]. In these cases, errors are desired to be introduced in attribute values with a controlled variation, generally small, with respect to the original value. These models are useful in applications where measurement tools can generate minor inaccuracies in their operation. Thus, *Symmetric Gaussian attribute noise* [17] makes smaller errors more likely to be introduced. It corrupts each attribute v_j ($j \in \{1, \ldots, m\}$) by adding random errors ε that follow a Gaussian distribution:

$$\tilde{x}_{i,j} = x_{i,j} + \varepsilon \mid \varepsilon \sim \mathcal{N}(0, k^2(max(v_j) - min(v_j))^2) \tag{14}$$

with $k \in (0,1]$. Another way to introduce this type of error is by choosing the noisy values following a uniform distribution within a bounded area close to the original value [58,97]. For example, *Symmetric interval-based attribute noise* [58] selects the noisy value from one of the intervals adjacent to that of the original value after dividing the attribute using an equal-height histogram. Based on a model originally used in the context of active learning [97], *Unconditional fixed-width attribute noise* sets a margin E around the original value $x_{i,j}$ to uniformly select the erroneous value:

$$\tilde{x}_{i,j} = x_{i,j} + \varepsilon \mid \varepsilon \sim \mathcal{U}[-E, E] \tag{15}$$

In contrast to attribute noise using a zero-mean Gaussian distribution, which gives more probability to errors close to the original value, these models [58,97] give the same probability to all errors within the stated intervals.

(4) *Attribute noise affecting all samples* [67]. In certain applications, for example, when attribute values are obtained by a faulty automatic procedure, all samples may experience errors. Therefore, the probability of error for each sample x_i is $P(x_i) = 1$. In these cases, models such as *Unconditional vp-Gaussian attribute noise* [67] (which considers an additive zero-mean Gaussian noise with variance proportional to the variance of the attribute to corrupt) or the aforementioned *Unconditional fixed-width attribute noise*, among others, can be used. For a given attribute v_j, these models satisfy:

$$(s, j) \in \mathcal{P}, \forall s \in \{1, \ldots, n\} \tag{16}$$

with $\mathcal{P}$ being the set of pairs with the values to be altered created by the selection procedure.

(5) *Dependent attribute noise* [68,82]. The selection process in attribute noise models is commonly applied for each attribute independently [77] so the samples that are corrupted in each attribute are usually different. In certain situations, when noise in a sample occurs, it affects all its attribute values simultaneously [82]. This type of noise is introduced by

models such as *Symmetric/dependent uniform attribute noise* [82] (which randomly chooses noisy values within the domain) or *Boundary/dependent Gaussian attribute noise* [68] (which uses an additive Gaussian noise in samples close to decision boundaries).

(6) *Attribute noise in image classification* [96,108]. Even though other attribute noise models can be used in dealing with image classification [109], there are some schemes designed for this application considering that these problems have their attributes in the same domain (the value of the pixels). For example, in *Symmetric/dependent Gaussian-image attribute noise* [96], each image is replaced with random values following a Gaussian distribution with the same mean and variance as the original image distribution. Another example is *Symmetric/dependent random-pixel attribute noise* [96], in which the pixels of each image are shuffled using independent random permutations.

(7) *Different noise levels in different attributes* [58,82]. These models may require knowledge of the domain of the problem addressed since they usually need to specify a different noise level for each of the attributes. Once the noise levels have been determined, the disruption process can randomly choose values within the domain, as in *Asymmetric uniform attribute noise* [82], or restricted to an area close to the original attribute value, as in *Asymmetric interval-based attribute noise* [58]. A special case of this type of model is *Importance interval-based attribute noise* [58], in which different noise levels are assigned to attributes based on their information gain.

7. Conclusions and Future Directions

In this work, a review of noise models in classification has been presented. The analysis of the literature has shown the lack of a unified terminology in this field, for which a tripartite nomenclature has been proposed to name noise models. Additionally, the current taxonomies for noise models [18,19] have been revised, combined and expanded, resulting in a single taxonomy that encompasses both label and attribute noise models. Finally, the noise models have been grouped from a pragmatic point of view, according to the possible application and the type of noise to be studied.

The proposed taxonomy helps to deepen the knowledge of noise models and their characteristics. It is easy to interpret as it uses concepts from the field of noisy data in classification. The categorization is mainly based on the information sources of all classification problems (class labels and attributes), as well as on the fundamental components of noise models. From these main aspects, the different dimensions are coherently defined. Thus, noise models can be classified in each of such dimensions. The dimensions of noise type, selection source and disruption source can be easily identified according to the information sources involved. This fact also occurs with the selection distribution and the disruption distribution, even though for those models using several distributions simultaneously, it may be useful to define a category considering them together. Finally, note that the dimensions defined in the proposed taxonomy can also be applied to future models. Three of them (noise type, selection source and disruption source) are based on the aforementioned information sources in the classification. Since such sources are present in all classification datasets, these dimensions and their categories can be used in the future. New noise models can also use the rest of the dimensions (selection distribution and disruption distribution), although considering distributions not used until now will imply the appearance of new categories in them.

It is important to note that most of the existing models have been developed to introduce label noise [21,22,38]. The number of attribute noise models [77,78] is more limited, and the number of models combining label and attribute noise [26,27] is even smaller. This fact clearly shows that label noise is being widely studied in the literature, whereas attribute noise receives less recognition [8]. Therefore, it is necessary to pay more attention to this type of noise and develop new models for its introduction in a controlled way since it is frequent but often overlooked in classification problems [58,78].

Furthermore, in relation to attribute noise, other models closer to the noise produced in real-world datasets can be studied. Thus, even though there are different models introducing realistic label noise (for example, those focusing on borderline label noise [69,73]), these types of models are scarcer in attribute noise. On the other hand, most attribute noise models consider the same noise level in each attribute [17,77]. The exception could be asymmetric models [58,82], which allow defining a noise level in each attribute. Nevertheless, they may require knowledge of the problem to determine the noise level of each attribute and which of them are more or less altered. In order to overcome this problem, models based on the scheme employed in [110] (which was used for unsupervised problems) could be designed. It defines a single noise level, avoiding the difficulty of having multiple noise levels, and this is applied to all attributes using a salt-and-pepper procedure [110].

The separation of the selection and disruption procedures in each noise model also allows the creation of new models that combine the selection of a given model and the disruption of another. Finally, despite the large number of noise models proposed in the literature, there is a line for model design that has not yet been sufficiently explored, consisting of the imitation of real-world datasets with noisy values. This requires further study of datasets in which errors can be identified [111,112]. The design of new models could be based on the similarity they achieve by imitating the errors in these data. In this way, it will be possible to approximate the errors that usually occur in real-world datasets, which can be an important aspect in the design of noise models.

Funding: This research received no external funding.

Conflicts of Interest: The author declares no conflict of interest.

References

1. Yu, Z.; Wang, D.; Zhao, Z.; Chen, C.L.P.; You, J.; Wong, H.; Zhang, J. Hybrid incremental ensemble learning for noisy real-world data classification. *IEEE Trans. Cybern.* **2019**, *49*, 403–416. [CrossRef] [PubMed]
2. Gupta, S.; Gupta, A. Dealing with noise problem in machine learning data-sets: A systematic review. *Procedia Comput. Sci.* **2019**, *161*, 466–474. [CrossRef]
3. Martín, J.; Sáez, J.A.; Corchado, E. On the regressand noise problem: Model robustness and synergy with regression-adapted noise filters. *IEEE Access* **2021**, *9*, 145800–145816. [CrossRef]
4. Liu, T.; Tao, D. Classification with noisy labels by importance reweighting. *IEEE Trans. Pattern Anal. Mach. Intell.* **2016**, *38*, 447–461. [CrossRef] [PubMed]
5. Xia, S.; Liu, Y.; Ding, X.; Wang, G.; Yu, H.; Luo, Y. Granular ball computing classifiers for efficient, scalable and robust learning. *Inf. Sci.* **2019**, *483*, 136–152. [CrossRef]
6. Nematzadeh, Z.; Ibrahim, R.; Selamat, A. Improving class noise detection and classification performance: A new two-filter CNDC model. *Appl. Soft Comput.* **2020**, *94*, 106428. [CrossRef]
7. Zeng, S.; Duan, X.; Li, H.; Xiao, Z.; Wang, Z.; Feng, D. Regularized fuzzy discriminant analysis for hyperspectral image classification with noisy labels. *IEEE Access* **2019**, *7*, 108125–108136. [CrossRef]
8. Sáez, J.A.; Corchado, E. ANCES: A novel method to repair attribute noise in classification problems. *Pattern Recognit.* **2022**, *121*, 108198. [CrossRef]
9. Adeli, E.; Thung, K.; An, L.; Wu, G.; Shi, F.; Wang, T.; Shen, D. Semi-supervised discriminative classification robust to sample-outliers and feature-noises. *IEEE Trans. Pattern Anal. Mach. Intell.* **2019**, *41*, 515–522. [CrossRef]
10. Tian, Y.; Sun, M.; Deng, Z.; Luo, J.; Li, Y. A new fuzzy set and nonkernel SVM approach for mislabeled binary classification with applications. *IEEE Trans. Fuzzy Syst.* **2017**, *25*, 1536–1545. [CrossRef]
11. Yu, Z.; Lan, K.; Liu, Z.; Han, G. Progressive ensemble kernel-based broad learning system for noisy data classification. *IEEE Trans. Cybern.* **2022**, *52*, 9656–9669. [CrossRef] [PubMed]
12. Xia, S.; Zheng, S.; Wang, G.; Gao, X.; Wang, B. Granular ball sampling for noisy label classification or imbalanced classification. *IEEE Trans. Neural Netw. Learn. Syst.* **2021**, in press. [CrossRef] [PubMed]
13. Xia, S.; Zheng, Y.; Wang, G.; He, P.; Li, H.; Chen, Z. Random space division sampling for label-noisy classification or imbalanced classification. *IEEE Trans. Cybern.* **2021**, in press. [CrossRef] [PubMed]
14. Huang, L.; Shao, Y.; Zhang, J.; Zhao, Y.; Teng, J. Robust rescaled hinge loss twin support vector machine for imbalanced noisy classification. *IEEE Access* **2019**, *7*, 65390–65404. [CrossRef]
15. Garcia, L.P.F.; Lehmann, J.; de Carvalho, A.C.P.L.F.; Lorena, A.C. New label noise injection methods for the evaluation of noise filters. *Knowl.-Based Syst.* **2019**, *163*, 693–704. [CrossRef]
16. Tomasev, N.; Buza, K. Hubness-aware kNN classification of high-dimensional data in presence of label noise. *Neurocomputing* **2015**, *160*, 157–172. [CrossRef]
17. Sáez, J.A.; Galar, M.; Luengo, J.; Herrera, F. Analyzing the presence of noise in multi-class problems: Alleviating its influence with the One-vs-One decomposition. *Knowl. Inf. Syst.* **2014**, *38*, 179–206. [CrossRef]

18. Frénay, B.; Verleysen, M. Classification in the presence of label noise: A survey. *IEEE Trans. Neural Netw. Learn. Syst.* **2014**, *25*, 845–869. [CrossRef]
19. Nettleton, D.; Orriols-Puig, A.; Fornells, A. A study of the effect of different types of noise on the precision of supervised learning techniques. *Artif. Intell. Rev.* **2010**, *33*, 275–306. [CrossRef]
20. Zhao, Z.; Chu, L.; Tao, D.; Pei, J. Classification with label noise: A Markov chain sampling framework. *Data Min. Knowl. Discov.* **2019**, *33*, 1468–1504. [CrossRef]
21. Li, J.; Zhu, Q.; Wu, Q.; Zhang, Z.; Gong, Y.; He, Z.; Zhu, F. SMOTE-NaN-DE: Addressing the noisy and borderline examples problem in imbalanced classification by natural neighbors and differential evolution. *Knowl.-Based Syst.* **2021**, *223*, 107056. [CrossRef]
22. Bootkrajang, J.; Chaijaruwanich, J. Towards instance-dependent label noise-tolerant classification: A probabilistic approach. *Pattern Anal. Appl.* **2020**, *23*, 95–111. [CrossRef]
23. Hendrycks, D.; Mazeika, M.; Wilson, D.; Gimpel, K. Using trusted data to train deep networks on labels corrupted by severe noise. *Adv. Neural Inf. Process. Syst.* **2018**, *31*, 10477–10486.
24. Shanthini, A.; Vinodhini, G.; Chandrasekaran, R.M.; Supraja, P. A taxonomy on impact of label noise and feature noise using machine learning techniques. *Soft Comput.* **2019**, *23*, 8597–8607. [CrossRef]
25. Koziarski, M.; Krawczyk, B.; Wozniak, M. Radial-based oversampling for noisy imbalanced data classification. *Neurocomputing* **2019**, *343*, 19–33. [CrossRef]
26. Teng, C. Polishing blemishes: Issues in data correction. *IEEE Intell. Syst.* **2004**, *19*, 34–39. [CrossRef]
27. Kazmierczak, S.; Mandziuk, J. A committee of convolutional neural networks for image classification in the concurrent presence of feature and label noise. In Proceedings of the 16th International Conference on Parallel Problem Solving from Nature, Leiden, The Netherlands, 5–9 September 2020; Volume 12269, LNCS, pp. 498–511.
28. Mirzasoleiman, B.; Cao, K.; Leskovec, J. Coresets for robust training of deep neural networks against noisy labels. *Adv. Neural Inf. Process. Syst.* **2020**, *33*, 11465–11477.
29. Kang, J.; Fernandez-Beltran, R.; Kang, X.; Ni, J.; Plaza, A. Noise-tolerant deep neighborhood embedding for remotely sensed images with label noise. *IEEE J. Sel. Top. Appl. Earth Obs. Remote Sens.* **2021**, *14*, 2551–2562. [CrossRef]
30. Koziarski, M.; Wozniak, M.; Krawczyk, B. Combined cleaning and resampling algorithm for multi-class imbalanced data with label noise. *Knowl.-Based Syst.* **2020**, *204*, 106223. [CrossRef]
31. Xia, S.; Wang, G.; Chen, Z.; Duan, Y.; Liu, Q. Complete random forest based class noise filtering learning for improving the generalizability of classifiers. *IEEE Trans. Knowl. Data Eng.* **2019**, *31*, 2063–2078. [CrossRef]
32. Zhang, W.; Wang, D.; Tan, X. Robust class-specific autoencoder for data cleaning and classification in the presence of label noise. *Neural Process. Lett.* **2019**, *50*, 1845–1860. [CrossRef]
33. Chen, B.; Xia, S.; Chen, Z.; Wang, B.; Wang, G. RSMOTE: A self-adaptive robust SMOTE for imbalanced problems with label noise. *Inf. Sci.* **2021**, *553*, 397–428. [CrossRef]
34. Pakrashi, A.; Namee, B.M. KalmanTune: A Kalman filter based tuning method to make boosted ensembles robust to class-label noise. *IEEE Access* **2020**, *8*, 145887–145897. [CrossRef]
35. Salekshahrezaee, Z.; Leevy, J.L.; Khoshgoftaar, T.M. A reconstruction error-based framework for label noise detection. *J. Big Data* **2021**, *8*, 1–16. [CrossRef]
36. Abellán, J.; Mantas, C.J.; Castellano, J.G. AdaptativeCC4.5: Credal C4.5 with a rough class noise estimator. *Expert Syst. Appl.* **2018**, *92*, 363–379. [CrossRef]
37. Wang, C.; Shi, J.; Zhou, Y.; Li, L.; Yang, X.; Zhang, T.; Wei, S.; Zhang, X.; Tao, C. Label noise modeling and correction via loss curve fitting for SAR ATR. *IEEE Trans. Geosci. Remote Sens.* **2022**, *60*, 1–10. [CrossRef]
38. Wei, Y.; Gong, C.; Chen, S.; Liu, T.; Yang, J.; Tao, D. Harnessing side information for classification under label noise. *IEEE Trans. Neural Netw. Learn. Syst.* **2020**, *31*, 3178–3192. [CrossRef]
39. Chen, P.; Liao, B.; Chen, G.; Zhang, S. Understanding and utilizing deep neural networks trained with noisy labels. In Proceedings of the 36th International Conference on Machine Learning, Long Beach, CA, USA, 9–15 2019; Volume 97, *PMLR*, pp. 1062–1070.
40. Song, H.; Kim, M.; Lee, J.G. SELFIE: Refurbishing unclean samples for robust deep learning. In Proceedings of the 36th International Conference on Machine Learning, Long Beach, CA, USA, 9–15 June 2019; Volume 97, *PMLR*, pp. 5907–5915.
41. Seo, P.H.; Kim, G.; Han, B. Combinatorial inference against label noise. *Adv. Neural Inf. Process. Syst.* **2019**, *32*, 1171–1181.
42. Wu, P.; Zheng, S.; Goswami, M.; Metaxas, D.N.; Chen, C. A topological filter for learning with label noise. *Adv. Neural Inf. Process. Syst.* **2020**, *33*, pp. 21382–21393.
43. Cheng, J.; Liu, T.; Ramamohanarao, K.; Tao, D. Learning with bounded instance and label-dependent label noise. In Proceedings of the 37th International Conference on Machine Learning, virtual, 3–18 July 2020; Volume 119, *PMLR*, pp. 1789–1799.
44. Ghosh, A.; Lan, A.S. Contrastive learning improves model robustness under label noise. In Proceedings of the 2021 IEEE Conference on Computer Vision and Pattern Recognition Workshops, Virtual, 19–25 July 2021; pp. 2703–2708.
45. Wang, Z.; Hu, G.; Hu, Q. Training noise-robust deep neural networks via meta-learning. In Proceedings of the 2020 IEEE/CVF Conference on Computer Vision and Pattern Recognition, Virtual, 14–19 June 2020; pp. 4523–4532.
46. Jindal, I.; Pressel, D.; Lester, B.; Nokleby, M.S. An effective label noise model for DNN text classification. In Proceedings of the 2019 Conference of the North American Chapter of the Association for Computational Linguistics: Human Language Technologies, Minneapolis, MA, USA, 9–14 June 2019; pp. 3246–3256.

47. Scott, C.; Blanchard, G.; Handy, G. Classification with asymmetric label noise: Consistency and maximal denoising. In Proceedings of the 26th Annual Conference on Learning Theory, Princeton, NJ, USA, 12–14 June 2013; Volume 30, *JMLR*, pp. 489–511.
48. Yang, P.; Ormerod, J.; Liu, W.; Ma, C.; Zomaya, A.; Yang, J. AdaSampling for positive-unlabeled and label noise learning with bioinformatics applications. *IEEE Trans. Cybern.* **2019**, *49*, 1932–1943. [CrossRef]
49. Feng, L.; Shu, S.; Lin, Z.; Lv, F.; Li, L.; An, B. Can cross entropy loss be robust to label noise? In Proceedings of the 29th International Joint Conference on Artificial Intelligence, Yokohama, Japan, 11–17 July 2020; pp. 2206–2212.
50. Ghosh, A.; Kumar, H.; Sastry, P. Robust loss functions under label noise for deep neural networks. In Proceedings of the 31st AAAI Conference on Artificial Intelligence, San Francisco, CA, USA, 4–9 February 2017; pp. 1919–1925.
51. Tanaka, D.; Ikami, D.; Yamasaki, T.; Aizawa, K. Joint optimization framework for learning with noisy labels. In Proceedings of the 2018 IEEE/CVF Conference on Computer Vision and Pattern Recognition, Salt Lake City, UT, USA, 18–22 June 2018; pp. 5552–5560.
52. Sun, Z.; Liu, H.; Wang, Q.; Zhou, T.; Wu, Q.; Tang, Z. Co-LDL: A co-training-based label distribution learning method for tackling label noise. *IEEE Trans. Multimed.* **2022**, *24*, 1093–1104. [CrossRef]
53. Li, J.; Wong, Y.; Zhao, Q.; Kankanhalli, M.S. Learning to learn from noisy labeled data. In Proceedings of the 2019 IEEE/CVF Conference on Computer Vision and Pattern Recognition, Long Beach, CA, USA, 15–20 June 2019; pp. 5051–5059.
54. Harutyunyan, H.; Reing, K.; Steeg, G.V.; Galstyan, A. Improving generalization by controlling label-noise information in neural network weights. In Proceedings of the 37th International Conference on Machine Learning, Online, 13–18 July 2020; Volume 119, *PMLR*, pp. 4071–4081.
55. Han, B.; Yao, Q.; Yu, X.; Niu, G.; Xu, M.; Hu, W.; Tsang, I.W.; Sugiyama, M. Co-teaching: Robust training of deep neural networks with extremely noisy labels. *Adv. Neural Inf. Process. Syst.* **2018**, *31*, 8536–8546.
56. Nikolaidis, K.; Plagemann, T.; Kristiansen, S.; Goebel, V.; Kankanhalli, M. Using under-trained deep ensembles to learn under extreme label noise: A case study for sleep apnea detection. *IEEE Access* **2021**, *9*, 45919–45934. [CrossRef]
57. Bootkrajang, J.; Kabán, A. Learning kernel logistic regression in the presence of class label noise. *Pattern Recognit.* **2014**, *47*, 3641–3655. [CrossRef]
58. Mannino, M.V.; Yang, Y.; Ryu, Y. Classification algorithm sensitivity to training data with non representative attribute noise. *Decis. Support Syst.* **2009**, *46*, 743–751. [CrossRef]
59. Ghosh, A.; Manwani, N.; Sastry, P.S. On the robustness of decision tree learning under label noise. In Proceedings of the 21th Conference on Advances in Knowledge Discovery and Data Mining, Jeju, Korea, 23–26 May 2017; Volume 10234, *LNCS*, pp. 685–697.
60. Arazo, E.; Ortego, D.; Albert, P.; O'Connor, N.E.; McGuinness, K. Unsupervised label noise modeling and loss correction. In Proceedings of the 36th International Conference on Machine Learning, Beach, CA, USA, 9–15 June 2019; Volume 97, *PMLR*, pp. 312–321.
61. Liu, D.; Yang, G.; Wu, J.; Zhao, J.; Lv, F. Robust binary loss for multi-category classification with label noise. In Proceedings of the 2021 IEEE International Conference on Acoustics, Speech and Signal Processing, Toronto, ON, Canada, 6–11 June 2021; pp. 1700–1704.
62. Ghosh, A.; Manwani, N.; Sastry, P.S. Making risk minimization tolerant to label noise. *Neurocomputing* **2015**, *160*, 93–107. [CrossRef]
63. Ortego, D.; Arazo, E.; Albert, P.; O'Connor, N.E.; McGuinness, K. Towards robust learning with different label noise distributions. In Proceedings of the 25th International Conference on Pattern Recognition, Milan, Italy, 10–15 January 2020; pp. 7020–7027.
64. Fatras, K.; Damodaran, B.; Lobry, S.; Flamary, R.; Tuia, D.; Courty, N. Wasserstein adversarial regularization for learning with label noise. *IEEE Trans. Pattern Anal. Mach. Intell.* **2021**, in press. [CrossRef]
65. Qin, Z.; Zhang, Z.; Li, Y.; Guo, J. Making deep neural networks robust to label noise: Cross-training with a novel loss function. *IEEE Access* **2019**, *7*, 130893–130902. [CrossRef]
66. Schneider, J.; Handali, J.P.; vom Brocke, J. Increasing trust in (big) data analytics. In Proceedings of the 2018 Advanced Information Systems Engineering Workshops, Tallinn, Estonia, 11–15 June 2018; Volume 316, *LNBIP*, pp. 70–84.
67. Huang, X.; Shi, L.; Suykens, J.A.K. Support vector machine classifier with pinball loss. *IEEE Trans. Pattern Anal. Mach. Intell.* **2014**, *36*, 984–997. [CrossRef]
68. Bi, J.; Zhang, T. Support vector classification with input data uncertainty. *Adv. Neural Inf. Process. Syst.* **2004**, *17*, 161–168.
69. Bootkrajang, J. A generalised label noise model for classification in the presence of annotation errors. *Neurocomputing* **2016**, *192*, 61–71. [CrossRef]
70. Bootkrajang, J. A generalised label noise model for classification. In Proceedings of the 23rd European Symposium on Artificial Neural Networks, Bruges, Belgium, 22–23 April 2015; pp. 349–354.
71. Ren, M.; Zeng, W.; Yang, B.; Urtasun, R. Learning to reweight examples for robust deep learning. In Proceedings of the 35th International Conference on Machine Learning, Stockholm Sweden, 10–15 July 2018; Volume 80, *PMLR*, pp. 4331–4340.
72. Prati, R.C.; Luengo, J.; Herrera, F. Emerging topics and challenges of learning from noisy data in nonstandard classification: A survey beyond binary class noise. *Knowl. Inf. Syst.* **2019**, *60*, 63–97. [CrossRef]
73. Du, J.; Cai, Z. Modelling class noise with symmetric and asymmetric distributions. In Proceedings of the 29th AAAI Conference on Artificial Intelligence, Austin, TX, USA, 25–30 January 2015; pp. 2589–2595.

74. Görnitz, N.; Porbadnigk, A.; Binder, A.; Sannelli, C.; Braun, M.L.; Müller, K.; Kloft, M. Learning and evaluation in presence of non-i.i.d. label noise. In Proceedings of the 17th International Conference on Artificial Intelligence and Statistics, Reykjavik, Iceland, 22–25 April 2014; Volume 33, *PMLR*, pp. 293–302.
75. Gehlot, S.; Gupta, A.; Gupta, R. A CNN-based unified framework utilizing projection loss in unison with label noise handling for multiple Myeloma cancer diagnosis. *Med Image Anal.* **2021**, *72*, 102099. [CrossRef] [PubMed]
76. Denham, B.; Pears, R.; Naeem, M.A. Null-labelling: A generic approach for learning in the presence of class noise. In Proceedings of the 20th IEEE International Conference on Data Mining, Sorrento, Italy, 17–20 November 2020; pp. 990–995.
77. Sáez, J.A.; Galar, M.; Luengo, J.; Herrera, F. Tackling the problem of classification with noisy data using multiple classifier systems: Analysis of the performance and robustness. *Inf. Sci.* **2013**, *247*, 1–20. [CrossRef]
78. Khoshgoftaar, T.M.; Hulse, J.V. Empirical case studies in attribute noise detection. *IEEE Trans. Syst. Man Cybern. Part C Appl. Rev.* **2009**, *39*, 379–388. [CrossRef]
79. Kaneko, T.; Ushiku, Y.; Harada, T. Label-noise robust generative adversarial networks. In Proceedings of the 2019 IEEE/CVF Conference on Computer Vision and Pattern Recognition, Long Beach, CA, USA, 16–20 June 2019; pp. 2462–2471.
80. Ghosh, A.; Lan, A.S. Do we really need gold samples for sample weighting under label noise? In Proceedings of the 2021 IEEE Winter Conference on Applications of Computer Vision, Waikoloa, HI, USA, 3–8 January 2021; pp. 3921–3930.
81. Wang, Q.; Han, B.; Liu, T.; Niu, G.; Yang, J.; Gong, C. Tackling instance-dependent label noise via a universal probabilistic model. In Proceedings of the 35th AAAI Conference on Artificial Intelligence, Online, 2–9 February 2021; pp. 10183–10191.
82. Petety, A.; Tripathi, S.; Hemachandra, N. Attribute noise robust binary classification. In Proceedings of the 34th AAAI Conference on Artificial Intelligence, New York, NY, USA, 7–12 February 2020; pp. 13897–13898.
83. Zhang, Y.; Zheng, S.; Wu, P.; Goswami, M.; Chen, C. Learning with feature-dependent label noise: A progressive approach. In Proceedings of the 9th International Conference on Learning Representations, Online, 3–7 May 2021; pp. 1–13.
84. Wei, J.; Liu, Y. When optimizing f-divergence is robust with label noise. In Proceedings of the 9th International Conference on Learning Representations, Online, 3–7 May 2021; pp. 1–11.
85. Chen, P.; Ye, J.; Chen, G.; Zhao, J.; Heng, P. Beyond class-conditional assumption: A primary attempt to combat instance-dependent label noise. In Proceedings of the 35th AAAI Conference on Artificial Intelligence, Online, 2–9 February 2021; pp. 11442–11450.
86. Nicholson, B.; Sheng, V.S.; Zhang, J. Label noise correction and application in crowdsourcing. *Expert Syst. Appl.* **2016**, *66*, 149–162. [CrossRef]
87. Amid, E.; Warmuth, M.K.; Srinivasan, S. Two-temperature logistic regression based on the Tsallis divergence. In Proceedings of the 22nd International Conference on Artificial Intelligence and Statistics, Naha, Japan, 16 April 2019; Volume 89, *PMLR*, pp. 2388–2396.
88. Thulasidasan, S.; Bhattacharya, T.; Bilmes, J.A.; Chennupati, G.; Mohd-Yusof, J. Combating label noise in deep learning using abstention. In Proceedings of the 36th International Conference on Machine Learning, Long Beach, CA, USA, 9–15 June 2019; Volume 97, *PMLR*, pp. 6234–6243.
89. Cano, J.R.; Luengo, J.; García, S. Label noise filtering techniques to improve monotonic classification. *Neurocomputing* **2019**, *353*, 83–95. [CrossRef]
90. Pu, X.; Li, C. Probabilistic information-theoretic discriminant analysis for industrial label-noise fault diagnosis. *IEEE Trans. Ind. Inform.* **2021**, *17*, 2664–2674. [CrossRef]
91. Han, B.; Yao, J.; Niu, G.; Zhou, M.; Tsang, I.W.; Zhang, Y.; Sugiyama, M. Masking: A new perspective of noisy supervision. *Adv. Neural Inf. Process. Syst.*, **2018**, *31*, 5841–5851.
92. Folleco, A.; Khoshgoftaar, T.M.; Hulse, J.V.; Bullard, L.A. Software quality modeling: The impact of class noise on the random forest classifier. In Proceedings of the 2008 IEEE Congress on Evolutionary Computation, Hong Kong, 1–6 June 2008; pp. 3853–3859.
93. Zhu, X.; Wu, X. Cost-guided class noise handling for effective cost-sensitive learning. In Proceedings of the 4th IEEE International Conference on Data Mining, Brighton, UK, 1–4 November 2004; pp. 297–304.
94. Kang, J.; Fernández-Beltran, R.; Duan, P.; Kang, X.; Plaza, A.J. Robust normalized softmax loss for deep metric learning-based characterization of remote sensing images with label noise. *IEEE Trans. Geosci. Remote Sens.* **2021**, *59*, 8798–8811. [CrossRef]
95. Fefilatyev, S.; Shreve, M.; Kramer, K.; Hall, L.O.; Goldgof, D.B.; Kasturi, R.; Daly, K.; Remsen, A.; Bunke, H. Label-noise reduction with support vector machines. In Proceedings of the 21st International Conference on Pattern Recognition, Munich, Germany, 30 July–2 August 2012; pp. 3504–3508.
96. Huang, L.; Zhang, C.; Zhang, H. Self-adaptive training: Beyond empirical risk minimization. *Adv. Neural Inf. Process. Syst.* **2020**, *33*, 19365–19376.
97. Ramdas, A.; Póczos, B.; Singh, A.; Wasserman, L.A. An analysis of active learning with uniform feature noise. In Proceedings of the 17th International Conference on Artificial Intelligence and Statistics, Reykjavik, Iceland, 22–25 April 2014; Volume 33, *JMLR*, pp. 805–813.
98. Yuan, W.; Guan, D.; Ma, T.; Khattak, A. Classification with class noises through probabilistic sampling. *Inf. Fusion* **2018**, *41*, 57–67. [CrossRef]
99. Ekambaram, R.; Fefilatyev, S.; Shreve, M.; Kramer, K.; Hall, L.; Goldgof, D.; Kasturi, R. Active cleaning of label noise. *Pattern Recognit.* **2016**, *51*, 463–480. [CrossRef]

100. Zhang, T.; Deng, Z.; Ishibuchi, H.; Pang, L. Robust TSK fuzzy system based on semisupervised learning for label noise data. *IEEE Trans. Fuzzy Syst.* **2021**, *29*, 2145–2157. [CrossRef]
101. Berthon, A.; Han, B.; Niu, G.; Liu, T.; Sugiyama, M. Confidence scores make instance-dependent label-noise learning possible. In Proceedings of the 38th International Conference on Machine Learning, Online, 18–24 July 2021; Volume 139, *PMLR*, pp. 825–836.
102. Sáez, J.A.; Krawczyk, B.; Woźniak, M. On the influence of class noise in medical data classification: Treatment using noise filtering methods. *Appl. Artif. Intell.* **2016**, *30*, 590–609. [CrossRef]
103. Baldomero-Naranjo, M.; Martínez-Merino, L.; Rodríguez-Chía, A. A robust SVM-based approach with feature selection and outliers detection for classification problems. *Expert Syst. Appl.* **2021**, *178*, 115017. [CrossRef]
104. Hosmer, D.W.; Lemeshow, S. *Applied Logistic Regression*; John Wiley and Sons: Hoboken, NJ, USA, 2000.
105. Dombrowski, A.K.; Anders, C.J.; Müller, K.R.; Kessel, P. Towards robust explanations for deep neural networks. *Pattern Recognit.* **2022**, *121*, 108194. [CrossRef]
106. Belarouci, S.; Chikh, M. Medical imbalanced data classification. *Adv. Sci. Technol. Eng. Syst. J.* **2017**, *2*, 116–124. [CrossRef]
107. Bao, F.; Deng, Y.; Kong, Y.; Ren, Z.; Suo, J.; Dai, Q. Learning deep landmarks for imbalanced classification. *IEEE Trans. Neural Netw. Learn. Syst.* **2020**, *31*, 2691–2704. [CrossRef]
108. Zhang, C.; Bengio, S.; Hardt, M.; Recht, B.; Vinyals, O. Understanding deep learning (still) requires rethinking generalization. *Commun. ACM* **2021**, *64*, 107–115. [CrossRef]
109. Pradhan, A.; Mishra, D.; Das, K.; Panda, G.; Kumar, S.; Zymbler, M. On the classification of MR images using "ELM-SSA" coated hybrid model. *Mathematics* **2021**, *9*, 2095. [CrossRef]
110. Iam-On, N. Clustering data with the presence of attribute noise: A study of noise completely at random and ensemble of multiple k-means clusterings. *Int. J. Mach. Learn. Cybern.* **2020**, *11*, 491–509. [CrossRef]
111. Zhang, Z.; Zhang, H.; Arik, S.Ö.; Lee, H.; Pfister, T. Distilling effective supervision from severe label noise. In Proceedings of the 2020 IEEE/CVF Conference on Computer Vision and Pattern Recognition, Seattle, WA, USA, 14–19 June 2020; pp. 9291–9300.
112. Lee, K.; He, X.; Zhang, L.; Yang, L. CleanNet: Transfer learning for scalable image classifier training with label noise. In Proceedings of the 2018 IEEE/CVF Conference on Computer Vision and Pattern Recognition, Salt Lake City, UT, USA, 18–22 June 2018; pp. 5447–5456.

Article

Determinants of Investment Awareness: A Moderating Structural Equation Modeling-Based Model in the Saudi Arabian Context

Mohamed Ali Shabeeb Ali [1,2,3], Mohammed Abdullah Ammer [1,4] and Ibrahim A. Elshaer [1,5,6,*]

1. The Saudi Investment Bank Scholarly Chair for Investment Awareness Studies, The Deanship of Scientific Research, The Vice Presidency for Graduate Studies and Scientific Research, King Faisal University, Al-Ahsa 31982, Saudi Arabia
2. Department of Accounting, School of Business, King Faisal University, Al-Ahsa 31982, Saudi Arabia
3. Accounting Department, Faculty of Commerce, South Valley University, Qena 1464040, Egypt
4. Department of Finance, School of Business, King Faisal University, Al-Ahsa 31982, Saudi Arabia
5. Department of Management, School of Business, King Faisal University, Al-Ahsa 31982, Saudi Arabia
6. Hotel Studies Department, Faculty of Tourism and Hotels, Suez Canal University, Ismailia 41522, Egypt
* Correspondence: ielshaer@kfu.edu.sa

Citation: Ali, M.A.S.; Ammer, M.A.; Elshaer, I.A. Determinants of Investment Awareness: A Moderating Structural Equation Modeling-Based Model in the Saudi Arabian Context. *Mathematics* **2022**, *10*, 3829. https://doi.org/10.3390/math10203829

Academic Editors: José Luis Romero Béjar and Jose Antonio Sáez Muñoz

Received: 26 September 2022
Accepted: 14 October 2022
Published: 17 October 2022

Publisher's Note: MDPI stays neutral with regard to jurisdictional claims in published maps and institutional affiliations.

Copyright: © 2022 by the authors. Licensee MDPI, Basel, Switzerland. This article is an open access article distributed under the terms and conditions of the Creative Commons Attribution (CC BY) license (https://creativecommons.org/licenses/by/4.0/).

Abstract: In line with today's economy, investment and financial awareness are necessary for success and an individual's well-being, specifically for the younger generations. Therefore, this study aims to examine the relationships between financial literacy, saving behavior, a lack of self-control, family financial socialization, and investment awareness. Further, it investigates the moderating role of both family financial socialization and the lack of self-control in these relationships. Employing a quantitative study technique and partial least squares structural equation modeling (PLS-SEM), we analyzed a sample of 409 students representing young adults at King Faisal University, specifically in the School of Business. Our results indicate that financial literacy, saving behavior, and family financial socialization are significantly and positively related to investment awareness. Interestingly and as expected, a lack of self-control negatively and significantly affects investment awareness. For the moderating impact, it was found that the connection between financial literacy, saving behavior, and investment awareness is positively and strongly moderated by family financial socialization. Likewise, a lack of self-control significantly and negatively moderated the association between financial literacy, saving behavior, and investment awareness. The results of this study provide substantial implications for regulators, educational organizations, individuals, and their families.

Keywords: family financial socialization; financial literacy; investment awareness; lack of self-control; saving behavior; Saudi Arabia

MSC: 91C99

1. Introduction

The financial position and well-being of individuals depend on their actions. Even though these actions can be affected by outside factors, such as economic policies implemented by private sectors or governments, individuals eventually make financial decisions. Further, there is a growing diversification in financial services and products, which increases the complications for individuals in making investment decisions [1]. In effect, these diversified services and products involve many choices and tools for individual investors to decide where and how to invest. In such financial circumstances, a lack of investment awareness could substantially affect an individual's financial outcomes [2].

Nowadays, the concept of investment awareness has received great importance as a financial concern [3]. Hastings and Mitchell [4] revealed that constant growth in the level of investment awareness had been witnessed globally for different age groups. Financial and investment awareness are two vital factors for the sound improvement of the financial market. Communities categorized with a high degree of financial awareness are more inclined to have individuals with strong skills in making sound investment decisions [5]. Investors with deficient appropriate financial awareness may make unreasonable financial decisions [6]. Further, the decisions of investors are considerably impacted by their behavior and commonly, individual investors display illogical and ineffective behavior in the market [7].

The existence of poor financial awareness in Saudi Arabia as reported by Alshebami and Aldhyani [8,9], emphasizes the necessity for recognizing the crucial elements influencing investment awareness. Saving signifies an essential foundation for an individual's investments that will lead to the growth and development of the country's economy [10]. Agarwalla et al. [2] advocated that the participation of families in businesses leads to their children being more aware of the fundamentals of personal finances. Indeed, children and adults obtain financial abilities inside the family via different socialization practices such as noticing the financial behavior of their parents. A fundamental emotion, a lack of self-control, is the individual thinking concerning the accomplishment of one's investment decisions [11]. To this end, understanding the link between the factors, i.e., financial literacy, saving behavior, family financial socialization, a lack of self-control, and an individual's investment awareness, is increasingly recognized as a critical financial issue.

Social influence, i.e., family, can form the principles and behaviors of an individual and thus could influence the individual's decisions [12]. Saving behavior can be one of the behaviors shaped by family socialization. Parents are important in guiding and instructing their children toward being financially literate [13], that is, families represent a vital source of inspiration and education for their children about financial knowledge and behaviors such as spending and saving. For children to live well without bad financial problems, their families must educate them about financial matters and investments [14]. Thus, examining the moderating role of family financial socialization in the association between financial literacy, saving behavior, and investment awareness is desired. Further, when investigating the link between financial literacy, saving behavior, and investment awareness, it is imperative to consider the impact of a lack of self-control on these links. It has been reported that an individual's self-control can moderate the connection between financial literacy and saving behavior [15]. A lack of self-control can affect the individual's capability to monitor his/her requirements, thoughts, and actions to accomplish precise goals, i.e., saving, spending, investing, and a suitable retirement plan [16]. Similarly, preceding work by Romal and Kaplan [17] has also linked self-control to savings, financial management, and credit problems. It was reported by [8] that exercising a substantial degree of self-control is needed to increase the numerous benefits of financial literacy. To this end, examining the moderating role of a lack of self-control on the relationships between financial literacy, saving behavior, and investment awareness is important.

Financial knowledge and investment awareness are receiving great interest in Saudi Arabia. For example, the Ministry of Education in Saudi Arabia is about to integrate a financial literacy course into the curriculum of the secondary stage for the current academic year 2022–2023. The General Authority for Statistics in Saudi Arabia [18] revealed that young Saudi people aged 15 to 34 years represent 36.7% of the whole population. Young individuals lack financial awareness regarding financial planning, services, and products [19]. Regardless of the critical modifications and reforms that happened to the Saudi economy, a low level of financial awareness still exists [20]. It is evident from Sedais and Al Shahab [21] that the degree of financial awareness among Saudi youth (below 37 years) is low. According to the Organization for Economic Cooperation and Development (OECD), Saudi young individuals' financial literacy represents only 9.6 out of 21, signifying a weak degree of financial literacy compared to other countries [22]. The OECD also mentioned that this degree is the lowest among the G20 countries. Alyahya [23] pointed out that most university students in Saudi Arabia are financially illiterate. Furthermore, the Saudi Vision 2030 is considering the importance of the financial sector as it works to increase the awareness and culture of investment in the Saudi market via the Capital Market Authority (CMA) and other bodies. The Saudi Vision 2030 also aims to increase the level of saving among Saudi families to 10%. To conclude, certain awareness and capabilities are required for young individuals to make sound investment decisions. Thus, prior discussions draw attention to the existence of a gap in the investment awareness level in Saudi Arabia, which our study aims to investigate.

This study proposes to contribute to and progress the existing literature in some ways. First, it enlarges and deepens the current literature on the link between investors' awareness and financial literacy, behaviors (saving, self-control), and family financial socialization in an emergent country, Saudi Arabia. This link has been mostly overlooked, especially in Saudi Arabia and other developing countries [24]. Second, this study is unique as it examines the impact of both cognitive factors, i.e., financial literacy, and non-cognitive variables, such as a lack of self-control and saving behavior, on investment awareness. The prior works focused mainly on exploring cognitive factors [25,26]. Third, so far, there are no documented works relating to the moderating impact of both family financial socialization and a lack of self-control on the relationship between financial literacy, saving behavior, and investment awareness. Consequently, this study is pioneering in providing a distinctive perception of such moderating impacts and filling the gaps in the existing literature on a developing country, Saudi Arabia. Finally, this study has important implications for policymakers, regulators, universities, families, and individuals by considering financial and investment awareness at an early age.

The remaining parts of our study are structured as follows. The Section 2 discusses the related literature and hypotheses building. The Section 3 discusses the study's methodology involving the procedures and measurements. The Section 4 shows the data analysis and results. The discussion is presented in Section 5. Finally, in conclusion, the implications in addition to the limitations and guidelines for future works are presented in Section 6.

2. Literature Review and Hypotheses Development

2.1. Financial Literacy and Investment Awareness

Financial literacy is becoming even more crucial. As a result, it is now essential to have a deeper understanding of financial concepts and goods to acquire the ability to make wise financial decisions and increase financial well-being [27]. Financial literacy is defined by The Organization for Economic Co-operation and Development (OECD) as "A combination of awareness, knowledge, skill, attitude, and behavior necessary to make sound financial decisions and ultimately achieve individual financial well-being" [28]. Mitchell and Lusardi [29] defined financial literacy as the capacity of an individual to obtain, comprehend, and use financial data to make effective and complete financial judgments regarding financial matters (i.e., planning, investment, and liabilities management), that is, if people or families want to be financially successful, they must have the necessary level of financial literacy.

The growth in bankruptcy and social issues among the younger generation are said to be caused by financial instability and a lack of financial literacy [30]. An individual's attitudes about various topics such as investment can be changed by gaining financial management knowledge. Azhar et al. [26] discovered that respondents' financial literacy in information gathering, investment types, and investment methods substantially impacted their investment awareness. Likewise, Heniawan and Dewi [25] indicated that respondents' financial literacy considerably influenced investing awareness. These results are consistent with Azizah et al. [31], who emphasized the value of financial knowledge when it comes to investing money. According to Lusardi and Mitchell [32], financial literacy enables investors to succeed with their investments and earn greater returns. In addition, Abreu, and Mendes [33] confirmed that people with good financial understanding might diversify their assets to reduce risk. Financially literate people are also seen to borrow and save money sensibly and invest and spend money prudently [34,35]. Chen and Volpe [36] supposed that students with more advanced financial education were more driven to make the right financial decisions than students with a less developed financial education. Aren and Zengin [37] revealed a strong correlation between financial literacy and investing choices. They added that less financially confident investors are more likely to invest in deposits. Conversely, investors with higher financial literacy prefer to invest in equities. These results align with Mazumdar [38], who stated that financially literate investors often put more money into hazardous ventures.

The fundamental argument for financial literacy is that better financial management stems from greater financial knowledge, which improves investment awareness, that is, individuals with higher financial literacy will have better investment awareness. Based on the above discussion, we suggest the following hypothesis:

Hypothesis 1 (H1). *Financial literacy has a positive impact on the investment awareness of the young Saudi generation.*

2.2. Saving Behavior and Investment Awareness

Saving behaviors are crucial to both economic development and growth. Every person must understand how to manage their finances in terms of investing and saving [39]. In line with Denton [40], saving behavior reflects potential requirements, saving decisions, and wealth-creating behaviors. Mpaata et al. [15] claimed that saving money is a basic human need that enables people to handle difficult financial choices in their life. Controlling one's consumption and being knowledgeable about how to spend money wisely are two benefits of saving money [41]. In their study, Mohamad et al. [42] indicated that young Malaysian persons, particularly students, could not use educational loans for saving because they lacked financial knowledge. The ability to do calculations and develop a saving strategy are two crucial skills required for saving behavior [43].

According to Bandura's theory [44] of social cognition, knowledge acquisition and learning are processes that occur in a social environment. It explains how environmental factors have an impact on people throughout their lives. According to the theory, the settings might influence how people behave. This affirms the necessity for understanding financial management, which will improve saving behaviors, investing knowledge, and abilities. Therefore, saving is a behavior that can significantly affect the choices made about investments. We, therefore, propose the following hypothesis:

Hypothesis 2 (H2). *Saving behavior has a positive impact on the investment awareness of the young Saudi generation.*

2.3. Family Financial Socialization and Investment Awareness

Parental financial practices can be transmitted to their offspring through the use of financial socialization. Financial socialization is the acquiring and developing of knowledge, skills, norms, standards, and attitudes related to money and money management, including understanding the fundamental terms and concepts of investing, saving, banking, insurance, using credit cards, and other financial matters [45]. Children acquire financial knowledge within their families at an early age via various socialization methods such as seeing their parents' financial activities or conversing with them about money and investments [46]. In this way, young adults enhance their financial capacities and competencies, increasing their financial independence and awareness. Despite this, only a handful of research has examined family financial socialization techniques [47]. In order to fill this gap, we evaluate the function of family financial socialization in increasing the investment knowledge of youngsters in the present research.

Teens need to obtain financial knowledge through the family environment during their youth, where they can acquire and retain good money attitudes throughout their lives [48,49]. Moreover, family socialization is essential in informing and guiding students to move from illiteracy to financial literacy [41,50]. In their research on the association between financial socialization experiences and favorable financial practices in early life,

Kim and Chatterjee [51] demonstrated that holding a savings account during infancy is favorably associated with possessing financial assets as an adult. The investment awareness of a person and the investment decisions he/she makes can be influenced by the positive financial habits they exhibited during their childhood. Mpaata et al. [12] argued that social factors influence financial literacy.

Financial Socialization Theory [52] proposes that child–parent interactions and observations regarding money primarily predict financial success as a young adult during the formative years. In addition, the theory of social learning demonstrates how young individuals' financial decisions are influenced by their social surroundings including their families. This theory has shown that young adults obtain expertise in financial matters via their parents' deliberate guidance and observation [53]. Thus, financial education gained from families throughout infancy could favorably affect future financial choices such as investing. Specifically, we hypothesize that family financial socialization can positively contribute to improving investment awareness. Therefore, we propose the following hypothesis:

Hypothesis 3 (H3). *Family financial socialization has a positive impact on the investment awareness of the young Saudi generation.*

2.4. Lack of Self-Control and Investment Awareness

Bernheim et al. [54] defined self-control as "Controlling one's behavior in situations when there is a simple trade-off between long-term goals and immediate enjoyment". Mpaata et al. [12] argued that people who lack self-control need plenty of financial literacy to have a favorable impact on their savings behavior. To increase our understanding of how people make financial choices such as investment decisions, it is necessary to investigate the psychological factors that influence their financial behavior and welfare [55]. Thus, it is necessary to investigate the relationship between self-control and investment awareness.

In their research, Stromback et al. [55] found that those with more self-control are more likely to have better financial behavior, save more money, and feel financially secure in the present and future. This conclusion was confirmed by the findings of Younas et al. [56], who found that those with stronger self-control exhibit improved financial conduct and capacity to manage their finances. Thus, families with difficulties in self-control accumulate less wealth [57]. Evidence indicates that an individual's self-control may predict their future intellectual and self-regulatory skills and well-being [58], as well as their investment awareness. Miotto and Parente [59] discovered that those with established self-control and future-planning tendencies might handle their assets more effectively.

Under the behavioral life-cycle theory articulated by Shefrin and Thaler [60], the financial conduct of people throughout their lives is determined by their capacity to regulate urges and the costs involved in exercising such self-discipline. It is proposed that young people with more self-control should have established financial behavior and a greater capacity to manage their assets and money, hence enhancing their investment knowledge. Following the above discussion, the following hypothesis is formulated:

Hypothesis 4 (H4). *A lack of Self-control is positively related to the investment awareness of the Saudi young generation.*

2.5. The Moderating Role of Family Financial Socialization in the Young Saudi Generation's Investment Awareness

Financial literacy refers to the skills of managing money and financial affairs, i.e., investments, with the intention of making operative financial decisions to achieve the goals of individuals and families [61]. It is a critical problem when the young generations are characterized by low financial knowledge, which may lead to poor investment decisions. Parents are inclined to positively impact their children's financial attitudes and knowledge [14]. Indeed, parents do this to guarantee the successful financial lives of their children.

Social exchange theory hypothesizes that social relations, i.e., family, could impact individuals, particularly during financial decision making [3]. Gallery et al. [62] indicated that individuals count on others and strive for guidance from relatives and friends before making financial and investment decisions due to their lack of financial literacy in analyzing financial plans. Concerning financial decisions, the central aim of exchange behavior is to increase profits and reduce losses [3]. The social exchange theory adds that actions based on social interactions are greatly determined by the strength of the association between the consultant and the customer. Thus, family is considered one of the key variables that affect an individual's financial decisions [63].

On the empirical front, Alshebami and Aldhyani [8] showed that parents influence the financial literacy of Saudi youth. Hellström et al. [64] reported that family members' investment portfolios encouraged other family members to participate in the investments. Likewise, other evidence recognized parents' impacts on their children's financial behavior [65]. In the same vein, students coming from families that do not practice so-called family budgets are likely to have less financial awareness [2].

Based on the social exchange theory and previous studies, we thus theorize that family financial socialization can enhance the relationship between an individual's financial literacy and their investment awareness. Thus, we articulate the following hypothesis:

Hypothesis 5 (H5). *The financial literacy–investment awareness relationship is positively moderated by family financial socialization.*

2.6. The Moderating Role of Family Financial Socialization in the Saving Behavior–Investment Awareness Relationship

The perception of financial socialization has been developed based on the socialization theory [66]. Financial socialization displays the learning of various financial abilities, such as savings and investments, to increase an individual's financial well-being. Parents can provide opportunities for financial socialization by educating their children on how to save and spend money [20]. The role of parents is dominant among all other financial socialization agents, i.e., teachers and friends. This is in line with the findings of Pinto et al. [67], which showed that parents' financial interactions are negatively related to the financial difficulties of their children.

Ariffin et al. [41] reported that parental socialization is considered an important indicator of saving behavior proficiency. They added that encouraging children to save and invest is the role of their guardians. The findings of Manfrè [68] showed that once children receive financial socialization early in life, their saving habits will be better. Indeed, by observing the financial behavior of their parents and talking with them about financial issues, children will have a better chance of obtaining financial skills [46]. Children can be obligated by their parents to save [69]. Cude et al. [70] reported a positive association between parents' and students' saving behavior. Further, parental socialization positively influences students' saving behavior in Bangladesh [71]. Similarly, the social variable of family members has been documented to influence investors' investment behavior [72]. Families can bring about substantial change in the attitudes and behaviors of their young members in terms of investment, saving, and borrowing [2]. To this end, in line with socialization theory and the outcomes of prior works, it is expected that family financial socialization can boost the link between saving behavior and investment awareness. Hence, we can state the following hypothesis:

Hypothesis 6 (H6). *Saving behavior–investment awareness relationship is positively moderated by family financial socialization.*

2.7. The Moderating Role of Lack of Self-Control in the Financial Literacy–Investment Awareness Relationship

Psychological factors such as self-control are expected to be better depending on the financial power of individuals [2]. When selecting investment opportunities, self-control is considered an important emotion in an individual's selections [11]. Younas et al. [56] revealed that the financial behavior of economic agents is related to this emotion. Iram et al. [73] investigated the financial awareness of women and reported that women could make better investment decisions when they are related to good self-control and high financial literacy.

How an individual manages and controls themselves in their current financial cases will influence their financial future [56]. Self-control is supposed to help an individual deal with personal money management issues [73]. However, if the individual lacks self-control, their investments and other financial affairs will be adversely impacted. This suggestion is supported by the findings reported by [8] that showed that the association between financial literacy and saving behavior was negatively moderated by self-control. They added that Saudi young adults (university students) have a lack of self-control. Building on these claims, it can be specified that a lack of self-control can adversely impact an individual's financial literacy and hence the ability to make good investment decisions. Thus, we hypothesize that:

Hypothesis 7 (H7). *The financial literacy–investment awareness relationship is negatively moderated by a lack of self-control.*

2.8. The Moderating Role of Lack of Self-Control in the Saving Behavior–Investment Awareness Relationship

Based on behavioral life cycle theory (BLCT), an individual's saving behavior can be enhanced by their self-control [15]. Self-control can improve an individual's determination, judgments, and practices concerning realizing their objectives such as constructing appropriate retirement plans, managing spending attitudes, and saving [15,16]. Agarwalla et al. [2] indicated that mainstream Indian students lack self-control. Alshebami and Seraj [10] reported that Saudi students with a high level of self-control can simply monitor their spending and expenses.

Individuals need to implement a specific degree of self-control to make investments; it is essential for individuals to implement a specific degree of self-control. However, when individuals suffer from a lack of self-control, they may spend more and save less and will ultimately encounter financial troubles [74]. These arguments are supported by prior evidence such as Alshebami and Seraj [10], who found that self-control is negatively and significantly related to saving behavior. In the same way, the difficulties of self-control, i.e., a lack of self-control, can delay saving through more spending [75]. Thus, individuals tend to save less since they lack the self-discipline and determination to do so. Accordingly, self-control difficulties lead to more expenditure and finally, less wealth [76].

Further, Gathergood [77] showed that clients who lack self-control are among those who highly utilize quick-access financial products. Ameriks et al. [76] showed that self-control difficulties are experienced less often by mature individuals than younger ones. Gathergood and Weber [77] advocated that individuals who lack self-control typically select investment tools that offer greater instant value due to their spending behaviors, which are determined by short-term and impulsive motivations. Based on the above discussions and prior evidence, we formulate the following hypothesis:

Hypothesis 8 (H8). *The saving behavior–investment awareness relationship is negatively moderated by a lack of self-control.*

3. Methodology

3.1. Participants and Procedures

To achieve the objectives of this study, we targeted a sample of students at King Faisal University. Specifically, an online questionnaire created via Google Forms was distributed to bachelor's, master's, and diploma students at the School of Business. For the diploma students, we targeted those who studied courses related to finance and accounting and could transfer to the School of Business to obtain their bachelor's degrees. The master's and bachelor's respondents have all studied accounting and finance courses, which is in line with [78]. Those respondents can better read and comprehend the questions regarding investment awareness and other variables. As the respondents were Arabic speakers, we translated the questionnaire into Arabic since all our measurements were adopted from English sources. Following some prior studies such as [8], we employed convenience sampling owing to its advantages concerning reaching respondents and saving time and costs. We sent the questionnaire link to all respondents via the Blackboard learning management system and standard electronic mail of the university in addition to WhatsApp groups. After leaving the link available to the respondents (students) during June 2022, 409 responses were received and collected. Since privacy is an important issue, we presented an introductory segment at the beginning of the survey to clarify the purpose of the survey and the confidentiality of the data collected.

3.2. Measurement

This study adopted the measurements of its variables from the common and key prior studies related to this study's issues. Table 1 represents the study's variables and the sources of their measurements.

Table 1. Study variables' measurement sources.

Construct	Source of Measurement
Investment Awareness (IA)	Azhar et al. [26]
Financial literacy (FL)	Azhar et al. [26]
Saving Behavior (SB)	Agarwalla et al. [2]
Family Financial Socialization (FFS)	Ariffin et al. [41]
Lack of Self-Control (LSC)	Ariffin et al. [41]

3.3. Data Analysis Techniques

The technique that was selected to test the proposed relationships and the impacts of their inter-relationships (direct and moderating) was the PLS-SEM. PLS-SEM is considered the most appropriate data analysis technique for our study as it requires a small sample size than the CB-SEM and the data do not need to have a normal distribution [79]. Furthermore, PLS-SEM enables the retention of a greater number of variables per factor. In the model, the connections between the latent and observed variables were categorized as reflective. This is due to the fact that changes in the latent variables affect the measurement of the observed variables [80]. The present study employed "Structural Equation Modeling" (SEM) via "Partial least squares" (PLS) to test the study hypotheses with the SmartPLS-4.0 program. Following Leguina [79]'s two-step approach, the proposed theoretical model was evaluated after evaluating the outer measurement model. Several criteria were employed to evaluate the outer measurement model and the inner structure model recommended by Hair et al. [80], as shown in Table 2.

Table 2. Recommended threshold values for the outer and inner models.

Inner Measurement Model			Outer Model					
Criteria	SFL	a	CR	AVE	R^2	Q^2	SRMR	NFI
threshold value	>0.70	>0.60	>0.70	>0.50	≥0.10	>0.0	<0.08	>0.90

SFL: Standardized Factor Loading; a: Cronbach's alpha value; CR: Composite reliability; AVE: Average Variance Extracted; R^2: Coefficient of determination; Q^2: Stone–Geisser Q2; SRMR: Standardized root mean square error; NFI: Normed Fit Index.

4. Data Analysis Results

4.1. Evaluation of the Outer Measurement Model

To assess the reliability and validity of the outer model, composite reliability (CR), internal consistency reliability (Cronbach's alpha), convergent validity, and discriminant validity were all evaluated. Cronbach's alpha (α) values ranged from 0.944 to 0.974, and the CR values from 0.949 to 0.975, as shown in Table 3, indicating that the scale had satisfactory internal reliability.

Table 3. Evaluation of the Outer Measurement Model.

	Abbreviation	SFL	α	CR	AVE
	Family Financial Socialization		0.974	0.975	0.904
FFS1:	"My family is good example for me when it comes to financial management".	0.952			
FFS2:	"I always talk about financial management with my family".	0.985			
FFS3:	"I appreciate it when my family gives me advice about what to do with my money".	0.911			
FFS4:	"Saving is something I do regularly because my family wanted me to save when I was little".	0.953			
	Financial Literacy		0.952	0.954	0.869
FL1:	"I know the different type of investments".	0.939			
FL2:	"I know investment has good and bad effects".	0.969			
FL3:	"I know where to get the information regarding investment".	0.887			
	Investment Awareness		0.950	0.954	0.793
IA1:	"I am aware of different investment avenues".	0.826			
IA2:	"I am aware that investment is important for the future".	0.862			
IA3:	"I am aware that investments are good for financial planning".	0.838			
IA4:	"I am aware that investment can give more income".	0.972			
IA5:	"I am aware that investment has high risk".	0.944			
	Lack of Self Control		0.944	0.949	0.768
LSC1:	"When I get money, I always spend it immediately (within 1 or 2 days)".	0.927			
LSC1:	"I don't save, because I think it's too hard".	0.819			
LSC1:	"('I see it, I like it, I buy it') describe me".	0.755			
LSC1:	"I always failed to control myself from spending money".	0.884			
LSC1:	"When I set saving goals for myself, I rarely achieve them".	0.981			
	Saving Behavior		0.951	0.958	0.752
SB1:	"I save for living expenses".	0.742			
SB2:	"I save for my dependents".	0.871			

Table 3. Cont.

Abbreviation	SFL	α	CR	AVE
SB3: "I save for future sources of income".	0.912			
SB4: "I save for future obligations".	0.920			
SB5: "I save in order to meet the expected high rates of inflation (high prices)".	0.920			
SB6: "In order to save, I always follow a careful monthly budget".	0.910			
SB7: "I save for living expenses".	0.777			

Second, all of the utilized dimensions had "Standardized Factor Loading" (SFL) scores greater than 0.70, indicating that the factors had adequate reliability. Third, convergent validity was supported as the AVE values for all the employed dimensions were greater than 0.50 [80]. The discriminant validity of the constructs was then evaluated using three criteria, as suggested by Leguina [79]. These methods included "cross-loading", the "Fornell-Larcker criterion", and the "heterotrait-monotrait" ratio (HTMT). First, for each latent variable, the outer-loading (in bold) was greater than the cross-loading (with other measurements), as shown in Table 4.

Table 4. Factors' Cross-loading.

	FFS	FL	IA	LSC	SB
FFS1	0.952	0.401	0.301	0.031	0.200
FFS2	0.985	0.408	0.312	0.038	0.230
FFS3	0.911	0.361	0.288	0.010	0.226
FFS4	0.953	0.339	0.302	0.047	0.224
FL1	0.363	0.939	0.440	−0.002	0.234
FL2	0.389	0.969	0.454	−0.006	0.231
FL3	0.358	0.887	0.415	−0.004	0.177
IA1	0.228	0.387	0.826	−0.135	0.267
IA2	0.284	0.390	0.862	−0.082	0.292
IA3	0.289	0.371	0.838	−0.103	0.284
IA4	0.289	0.466	0.972	−0.079	0.304
IA5	0.317	0.464	0.944	−0.065	0.251
LSC1	−0.016	−0.030	−0.095	0.927	−0.266
LSC2	0.003	−0.037	−0.084	0.819	−0.235
LSC3	0.053	0.035	−0.077	0.755	−0.195
LSC4	0.070	0.006	−0.090	0.884	−0.177
LSC5	0.038	0.011	−0.100	0.981	−0.198
SB1	0.225	0.225	0.233	−0.314	0.742
SB2	0.189	0.216	0.273	−0.217	0.871
SB3	0.198	0.219	0.286	−0.216	0.912
SB4	0.196	0.177	0.288	−0.217	0.920
SB5	0.210	0.210	0.288	−0.209	0.920
SB6	0.211	0.206	0.285	−0.206	0.910
SB7	0.179	0.146	0.244	−0.110	0.777

Second, Table 5 shows that the diagonal AVE values in bold were higher than the inter-variable correlation coefficient, indicating high discriminant validity [80]. Third, the HTMT values should be less than 0.90, as stated by Leguina [79]. In the study, the HTMT levels were significantly lower than 0.90 (see Table 5). The findings support and confirm the measurement model's reliability and discriminant and convergent validity. Consequently, the outer measurement model's results were considered adequate to move forward with the structural model's assessment for hypotheses testing.

Table 5. Inter-construct correlations, the square root of AVE, and the HTMT results.

	AVEs Values					HTMT Results				
	FFS	Fl	IA	LSC	SB	FFS	Fl	IA	LSC	SB
1−FFS	0.951									
2−FL	0.397	0.932				0.397				
3−IA	0.317	0.468	0.891			0.316	0.467			
4−LSC	−0.233	−0.204	−0.102	0.876		0.243	0.229	0.103		
5−SB	0.231	0.230	0.314	−0.244	0.867	0.235	0.232	0.317	0.248	

4.2. Assessment of the Structural Model

Then, a structural equation analysis was used to test the hypotheses. In particular, we looked at the model's ability to predict and explain the variation in the exogenous variables and the endogenous variables [80]. The VIF values of the observed variables ranged from 2.554 to −4.751, all of which were less than 5, confirming that the structural model lacked multicollinearity. In addition, [81] recommended an R^2 value of at least 0.10 to ensure a satisfactory model fit. As a result, the endogenous variable investment awareness had an R^2 value of 0.51, indicating that the study model adequately represented the data (Table 6). Moreover, The Stone–Geisser Q^2 assessment showed a value (0.537) of more than zero (Table 6), demonstrating the adequate predictive power of the model [82].

Table 6. Coefficient of determination (R^2) and (Q^2) and model fit (SRMR-NFI).

Endogenous Latent Construct	(R^2)	(Q^2)
Investment awareness	0.51	0.537
Model Fit	SRMR	NFI
	0.061	0.912

Finally, a bootstrapping method was used in smart PLS4 to examine the path coefficient and t-value of the direct and moderating relationships. The current study proposes four direct hypotheses and four moderating hypotheses. The smart PLS output showed that financial literacy had a positive and significant impact on investment awareness ($\beta = 0.31$, t-value = 0.569, $p < 0.001$), supporting hypothesis H1, as shown in Figure 1 and Table 7. Similarly, saving behavior was found to have a positive and significant impact ($\beta = 0.29$, t-value = 0.387, $p < 0.01$) on investment awareness, and hypothesis H2 was supported. The Smart PLS findings showed a positive and significant impact of family financial socialization on investment awareness ($\beta = 0.39$, t-value = 0.246, $p < 0.001$), which supports hypothesis H3. On the other hand, a lack of self-control was found to have a significant negative impact on investment awareness ($\beta = -0.41$, t-value = −0.855, $p < 0.001$), supporting hypothesis H4. The findings also provided data about the moderating effects, where family financial socialization strengthened the impact of financial literacy and investment behavior ($\beta = 0.24$, t-value = 0.264, $p < 0.01$), as well as the impact of saving behavior, on investment awareness ($\beta = 0.29$, t-value= 0.265, $p < 0.01$); hence, H6 and H6 were supported. Conversely, a lack of self-control as a moderating variable was found to dampen the impact of financial literacy on investment awareness ($\beta = -0.38$, t-value$= -6.56$, $p < 0.01$) and the impact of saving behavior on investment awareness ($\beta = -0.32$, t-value $= -4.76$, $p < 0.01$), therefore supporting hypotheses H7 and H8.

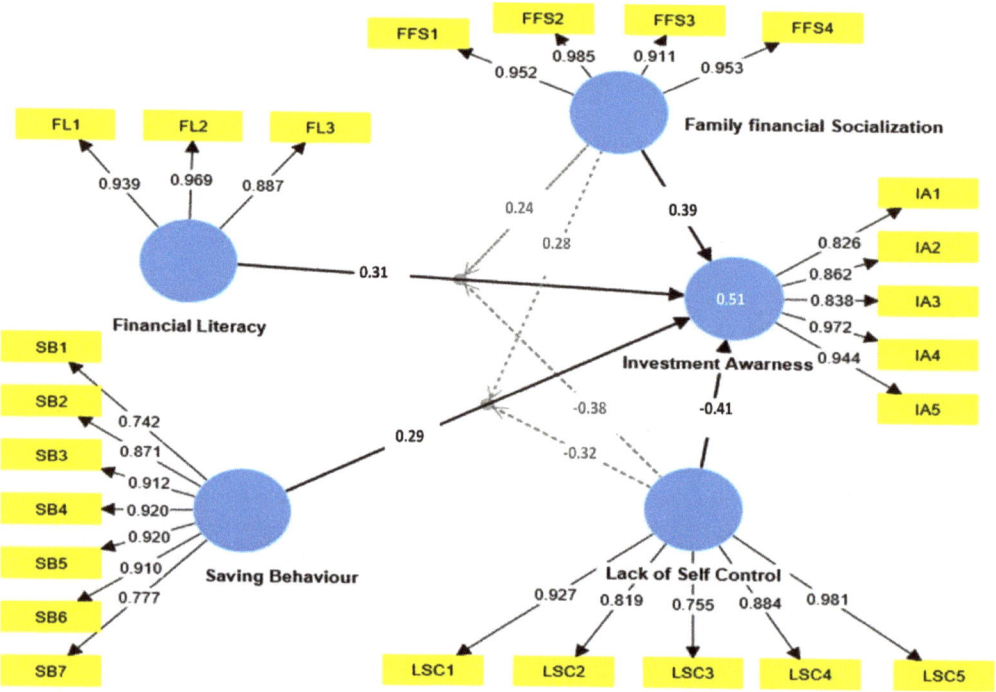

Figure 1. The study's structural and measurement model.

Table 7. The structural model's results.

	Hypothesis	Beta (β)	(t-Value)	p Value	Result
H1	Financial Literacy -> Investment Awareness	0.31	5.69	0.000	Accepted
H2	Saving Behaviour -> Investment Awareness	0.29	3.87	0.003	Accepted
H3	Family financial Socialization -> Investment Awareness	0.39	2.46	0.000	Accepted
H4	Lack of Self-Control -> Investment Awareness	−0.41	−8.55	0.000	Accepted
H5	Family financial Socialization × Financial Literacy -> Investment Awareness	0.24	2.64	0.001	Accepted
H6	Family financial Socialization × Saving Behaviour -> Investment Awareness	0.29	2.65	0.009	Accepted
H7	Lack of Self Control × Financial Literacy -> Investment Awareness	−0.38	−6.56	0.000	Accepted
H8	Lack of Self Control × Saving Behaviour -> Investment Awareness	−0.32	−4.76	0.000	Accepted

5. Discussion

We conducted this study in Saudi Arabia, a state experiencing interest in developing its economy. In addition, young individuals represent the mainstream of the population. In line with Vision 2030, the Saudi government is enhancing its financial sector, financial awareness, investment culture, and savings rates. Therefore, Saudi students are expected to experience a considerable enhancement relating to financial and investment awareness and saving to support their well-being. Thus, the study provides noteworthy insights into the body of knowledge and practical implications by examining the direct impact of financial literacy, saving behavior, family financial socialization, and a lack of self-control on investment awareness. Further, it examines the moderating impact of family financial socialization and a lack of self-control on the links between financial literacy, saving behavior, and investment awareness.

The results of this study indicate that financial literacy is positively and significantly related to investment awareness, thus confirming our first hypothesis (H1). The findings are in line with the argument that financial literacy is a key factor in managing and dealing with money, resulting in higher financial well-being and more efficient investment decisions. A certain degree of financial knowledge is needed for better investment awareness. The results are in line with those of Azhar et al. [26] and Azizah et al. [31], who showed that investment awareness is positively and significantly influenced by financial literacy; Aren and Zengin [37], who showed that financial literacy is related to preferences in investment; and Lusardi and Mitchell [43], who showed that financial literacy enables investors to attain greater returns. In the Saudi context, our results support Saber [3], who reported that Saudi individuals' financial awareness has an impact on their investment choices. Concerning saving behavior, the results provide support for the hypothesis (H2) that saving behavior is positively related to students' investment awareness. The results show that Saudi students tend to save more money. This good saving behavior will allow students to make investments in the future and solve financial problems. Based on this, the theory of social cognition is supported. Our students deal with savings as a financial source that can be reinvested in the future [72]. Likewise, Mugo [83] exposed that to conduct investment decisions, saving practices are needed.

Regarding family financial socialization, our theoretical hypothesis (H4) was found to be significantly supported in that the family positively impacts a student's investment awareness. These results support the theory of social learning, which posits that the financial actions of young adults are influenced by their social surroundings including their families. Our results are also in accordance with prior studies, for example, Mpaata et al. [15], who found that social surroundings can impact financial awareness, and Ariffin et al. [41], who revealed that family socialization changed children's situations from financial illiteracy into great literacy, that is, individuals are better at investing their money when their financial awareness is enhanced by their families [15]. For a lack of self-control, our results indicate that when students are characterized by problems with self-control, this negatively impacts their investment awareness and decisions. Hence, our hypothesis (H4) is strongly supported. Liu et al. [84] mentioned that mental accounting is a self-control tool that avoids unnecessary consumption. Hence, when individuals lack this tool, they spend more and then no more money can be saved for investments. Our results support those of Chia et al. [85] and Esenvalde [86], which showed that students with a lack of self-control are found to be related to more spending and less control concerning their desires. In contrast, our results are inconsistent with Mpaata et al. [15] and Griesdorn et al. [87] in that individuals with robust self-control have the needed skills to succeed in their financial matters.

For the moderating impact of family financial socialization on the association between financial literacy and investment awareness, our results show that family financial socialization can positively and significantly moderate this association. Thus, our hypothesis (H5) is confirmed. The reported moderating impact is in agreement with social exchange theory. This finding is in line with [14], in that families have a better tendency to favorably influence their children's financial awareness and attitudes. Further, Hellström et al. [64] advocated that financial literacy and the decisions of individuals are affected by the trusted information they obtain from social relationships. In the Saudi context, Saudi students' financial awareness is influenced by their parents [8], which supports our results.

Regarding the other moderating impact of family financial socialization on the connection between saving behavior and investment awareness, we found that family financial socialization can strongly and positively enhance this connection. Thus, our hypothesis (H6) is accepted. Based on these results, we claim that the socialization theory is supported. The results are in accordance with prior works, for instance, Kaur and Vohra [72], who showed that a social family variable influences investors' investment behavior. Similar findings were found in a study on students in Bangladesh by Khatun [71], which showed that parental socialization and saving behavior are positively related.

Finally, in terms of the moderating influence of a lack of self-control on the financial literacy–investment awareness relationship, the results signify that a lack of self-control negatively and significantly impacts this relationship. Hence, hypothesis (H7) is supported, that is, a student's investments and financial activities are negatively affected if they are associated with a lack of self-control. In the Saudi context, our results are in line with those of [8], which showed that the association between financial literacy and saving behavior was negatively moderated by self-control. In the same way, a lack of self-control was reported to negatively and significantly moderate the association between saving behavior and investment awareness. These results are in line with those of prior works, for example, Ameriks et al. [76], who advocated that more spending and less wealth are related to a lack of self-control; Gathergood [77], who stated that quick access to financial products is related to customers with a lack of self-control, and finally, Ameriks et al. [76], who mentioned that a mature individual is associated with fewer problems with self-control than a young adult.

6. Implications and Conclusions

This study was motivated by witnessing a dearth of studies on the factors affecting financial and investment awareness, in addition to the occurrence of deprived financial awareness, specifically in Saudi Arabia. The absence or lack of financial awareness can lead young individuals to select inappropriate financial products and services and save less than they should. An awareness of finance and investment issues forms accountable attitudes and behaviors concerning managing financial matters, leading to a successful young adult life. Thus, our study proposed to examine this important issue and aimed to investigate, in the Saudi context, financial literacy, saving behavior, family financial socialization, and a lack of self-control as the determinants of investment awareness among university students (young generation). The reported results indicate that all the determinants were found to be strongly related to investment awareness with a positive relationship in the case of financial literacy, saving behavior, and family financial socialization and a negative relationship in relation to a lack of self-control.

Further, our study provided unique insights by examining the moderating role of both family financial socialization and a lack of self-control in the relationships between financial literacy and investment awareness and saving behavior and investment awareness. The results were as expected: family financial socialization has a significant and positive impact on these relationships. In contrast, a lack of self-control negatively and significantly influences these relationships.

This study builds on the prevailing studies by providing distinctive evidence of the determinants of investment awareness. It is one of the limited works available for the Saudi Arabian context. It fills existing gaps in the literature since there is insufficient work on investment awareness in emerging countries, principally Saudi Arabia. The results provide robust evidence of the significant impact of financial literacy, saving behavior, family financial socialization, and a lack of self-control on investment awareness levels. The further theoretical contribution is that the results support the social learning theory. Moreover, this study adds to the body of knowledge by considering the moderating role of both family financial socialization and a lack of self-control in the study's theoretical framework.

In terms of the practical implications, first, the results are important for policymakers as they discuss the importance of the factors that can enhance the country's economic growth such as increasing financial and investment awareness. Further, the low level of investment awareness among the young generation will result in the appropriate involvement of policymakers through rules, initiatives, and recommendations that can enhance this level among the adults who represent the highest percentage of the population in Saudi Arabia. Additionally, the findings will inform regulators of the effectiveness of the Saudi Vision 2030 programs that aim to enhance investment culture and saving. Concerning educational institutions (schools and universities), training programs, workshops, and courses in financial and investment awareness are needed to improve the financial knowledge of students. Finally, parents and families are advised to discuss and educate their children on financial concerns and good behaviors regarding saving and investing.

7. Limitations and Directions for Future Research

There are some limitations related to this study. Since this study examined investment awareness among Saudi university students, it does not represent all fields in the country. Hence, future works could focus on sectors other than the academic sector for a better generalization of the study findings. In the same vein, the generalization of the results to other countries should be made with caution due to differences in culture, systems, and economic situations. Further, to increase the generalization of the results, future works could conduct a longer study rather than a cross-sectional one. We conducted our study on one group (students). Thus, future studies could consider the comparison of various groups in the community in terms of levels of financial awareness and the fundamental determinants. It is also important for future works to examine the role of financial policies, i.e., introducing financial knowledge courses concerning financial awareness. A valuable opportunity for future studies could be to examine and compare financial awareness and behaviors among males and females to explore the impact of gender.

Author Contributions: Conceptualization, I.A.E., M.A.S.A. and M.A.A.; methodology, I.A.E.; software, I.A.E.; validation, I.A.E., M.A.S.A. and M.A.A.; formal analysis, I.A.E.; investigation, I.A.E., M.A.S.A. and M.A.A.; data curation, I.A.E. and M.A.A.; writing—original draft preparation, I.A.E., M.A.S.A. and M.A.A.; writing—review and editing, I.A.E., M.A.S.A. and M.A.A.; visualization, I.A.E.; supervision, I.A.E.; project administration, I.A.E., M.A.S.A. and M.A.A.; funding acquisition, I.A.E., M.A.S.A. and M.A.A. All authors have read and agreed to the published version of the manuscript.

Funding: This work was supported by The Saudi Investment Bank Scholarly Chair for Investment Awareness Studies, the Deanship of Scientific Research, Vice Presidency for Graduate Studies and Scientific Research, King Faisal University, Saudi Arabia (Grant No. CHAIR151).

Informed Consent Statement: Informed consent was obtained from all subjects involved in the study.

Data Availability Statement: Data are available upon request from researchers who meet the eligibility criteria. Kindly contact the first author privately through e-mail.

Conflicts of Interest: The authors declare no conflict of interest.

References

1. Arora, S.; Marwaha, K. Financial literacy level and awareness regarding stock market: An empirical study of individual stock investors of Punjab. *Manag. Labour Stud.* **2013**, *38*, 241–253. [CrossRef]
2. Agarwalla, S.K.; Barua, S.; Jacob, J.; Varma, J.R. A Survey of Financial Literacy Among Students, Young Employees and the Retired in India. 2012. Available online: https://faculty.iima.ac.in/~{}iffm/literacy/youngemployessandretired2012.pdf (accessed on 9 September 2022).
3. Saber, A. The Impact of Financial Literacy on Household Wealth in the Kingdom of Saudi Arabia. Ph.D. Thesis, Victoria University, Melbourne, Australia, 2020.
4. Hastings, J.; Mitchell, O.S. How financial literacy and impatience shape retirement wealth and investment behaviors. *J. Pension Econ. Financ.* **2020**, *19*, 1–20. [CrossRef] [PubMed]
5. Alekam, E.; Mohammed, J.; Salleh, B.; Salniza, M. The Effect of Family, Peer, Behavior, Saving and Spending Behavior on Financial Literacy among Young Generations. *Int. J. Organ. Leadersh.* **2018**, *7*, 309–323. [CrossRef]
6. Tuffour, J.K.; Amoako, A.A.; Amartey, E.O. Assessing the effect of financial literacy among managers on the performance of small-scale enterprises. *Glob. Bus. Rev.* **2020**, *23*, 1200–1217. [CrossRef]
7. Raut, R.K.; Das, N.; Mishra, R. Behaviour of individual investors in stock market trading: Evidence from India. *Glob. Bus. Rev.* **2020**, *21*, 818–833. [CrossRef]

8. Alshebami, A.S.; Aldhyani, T.H.H. The Interplay of Social Influence, Financial Literacy, and Saving Behaviour among Saudi Youth and the Moderating Effect of Self-Control. *Sustainability* **2022**, *14*, 8780. [CrossRef]
9. Madinga, N.W.; Maziriri, E.T.; Chuchu, T.; Magoda, Z. An Investigation of the Impact of Financial Literacy and Financial Socialization on Financial Satisfaction: Mediating Role of Financial Risk Attitude. *Glob. J. Emerg. Mark. Econ.* **2022**, *14*, 60–75. [CrossRef]
10. Alshebami, A.S.; Seraj, A.H.A. The antecedents of saving behavior and entrepreneurial intention of Saudi Arabia University students. *Educ. Sci. Theory Pract.* **2021**, *21*, 67–84.
11. Tangney, J.P.; Baumeister, R.F.; Boone, A.L. High self-control predicts good adjustment, less pathology, better grades, and interpersonal success. *J. Personal.* **2004**, *72*, 271–324. [CrossRef]
12. Mpaata, E.; Koskei, N.; Saina, E. Financial literacy and saving behavior among micro and small enterprise owners in Kampala, Uganda: The moderating role of social influence. *J. Econ. Financ. Account. Stud.* **2020**, *2*, 22–34.
13. Amari, M.; Salhi, B.; Jarboui, A. Evaluating the effects of sociodemographic characteristics and financial education on saving behavior. *Int. J. Sociol. Soc. Policy.* **2020**, *40*, 1423–1438. [CrossRef]
14. Jorgensen, B.L. Financial Literacy of College Students: Parental and Peer Influences. Ph.D. Thesis, Virginia Polytechnic Institute and State University, Blacksburg, VA, USA, 2007.
15. Mpaata, E.; Koske, N.; Saina, E. Does self-control moderate financial literacy and savings behavior relationship? A case of micro and small enterprise owners. *Curr. Psychol.* **2021**. [CrossRef]
16. Fujita, K.; Han, H.A. Moving Beyond Deliberative Control of Impulses. *Psychol. Sci.* **2015**, *20*, 799–804. [CrossRef] [PubMed]
17. Romal, J.B.; Kaplan, B.J. Difference in self-control among spenders and savers. *Psychol. A J. Hum. Behav.* **1995**, *32*, 8–17.
18. The General Authority for Statistics. Saudi Youth in Numbers "Report for the World Youth Day 2020". 2020. Available online: https://www.stats.gov.sa/ar/news/397 (accessed on 4 May 2022).
19. Nga, J.K.H.; Yong, L.H.L.; Sellappan, R.D. A study of financial awareness among youths. *Young Consum. Insights Ideas Responsible Mark.* **2010**, *11*, 277–290. [CrossRef]
20. Ali, M.; Ali, L.; Badghish, S.; Soomro, Y. Determinants of Financial Empowerment Among Women in Saudi Arabia. *Front. Psychol.* **2021**, *12*, 4049. [CrossRef]
21. Sedais, K.I.; Al Shahab, O. The Impact of COVID-19 on the Banking Sector of Saudi Arabia. 2020, pp. 1–19. Available online: https://home.kpmg/sa/en/home/insights/2020/04/the-impact-of-covid-19-on-the-banking-sector.html (accessed on 6 September 2022).
22. OECD. G20/OECD INFE Report on Adult Financial Literacy in G20 Countries. 2017. Available online: https://www.oecd.org/daf/fin/financial-education/G20-OECD-INFE-report-adult-financial-literacy-in-G20-countries.pdf (accessed on 8 September 2022).
23. Alyahya, R.Y. Financial literacy among College Students in Saudi Arabia. Master's Thesis, School of Economics, The University of Queensland, Brisbane, Australia, 2017.
24. Alshebami, A.S.; Al Marri, S.H. The Impact of Financial Literacy on Entrepreneurial Intention: The Mediating Role of Saving Behavior. *Front. Psychol.* **2022**, *13*, 911605. [CrossRef]
25. Heniawan, D.A.; Dewi, A.S. Factors Affecting Investment Awareness: Case Study on Productive Age in Surabaya City. *Asian J. Res. Bus. Manag.* **2021**, *3*, 147–152.
26. Azhar, Z.; Azilah, N.; Syafiq, A. Investment awareness among young generation. In *International Conference on Business and Management Research*; Atlantis Press: Paris, France, 2017; pp. 126–135.
27. Philippas, N.D.; Avdoulas, C. Financial literacy and financial well-being among generation-Z university students: Evidence from Greece. *Eur. J. Finance.* **2020**, *26*, 360–381. [CrossRef]
28. OECD/INFE. High-Level Principles on National Strategies for Financial Education. 2012. Available online: https://www.oecd.org/finance/financial-education/OECD-INFE-Principles-National-Strategies-Financial-Education.pdf (accessed on 6 September 2022).
29. Mitchell, O.; Lusardi, A. Financial Literacy and Economic Outcomes: Evidence and Policy Implications. *J. Retire* **2015**, *3*, 107–114. [CrossRef]
30. Rahim, N.M.; Ali, N.; Adnan, M.F. Students' Financial Literacy: Digital Financial Literacy Perspective. *J. Financ. Bank. Rev.* **2022**, *6*, 18–25. [CrossRef]
31. Azizah, N.; Nurfadhilah, R.; Mior, A. Financial Literacy: A study among the university students. *Interdiscip. J. Contemp. Res. Bus.* **2013**, *5*, 279–299.
32. Lusardi, A.; Mitchell, O.S. Planning and financial literacy: How do women fare? *Am. Econ. Rev.* **2008**, *98*, 413–417. [CrossRef]
33. Abreu, M.; Mendes, V. Financial literacy and portfolio diversification. *Quant. Financ.* **2010**, *10*, 515–528. [CrossRef]
34. Atkinson, A.F.; Messy, F. Measuring financial literacy: Results of the OECD/International Network on Financial Education (INFE) pilot study. *Financ. Insur. Priv. Pensions* **2012**, *15*, 1–73.
35. Grohmann, A.; Kouwenberg, R.; Menkhoff, L. *Financial Literacy and Its Consequences in the Emerging Middle Class*; Working Paper No 1943; Kiel Institute for the World Economy: Kiel, Germany, 2014. Available online: https://www.ifw-kiel.de/fileadmin/Dateiverwaltung/IfW-Publications/Lukas_Menkhoff/financial-literacy-and-its-consequences-in-the-emerging-middle-class/KWP_1943.pdf (accessed on 1 May 2022).
36. Chen, H.; Volpe, R. An analysis of personal financial literacy among college students. *Financ. Serv. Rev.* **1998**, *7*, 107–128. [CrossRef]

37. Aren, S.; Zengin, A.N. Influence of financial literacy and risk perception on choice of investment. *Procedia—Soc. Behav. Sci.* **2016**, *235*, 656–663. [CrossRef]
38. Mazumdar, S. Individual Investment Behavior with Respect to financial Knowledge and Investment Risk Preference. *Int. J. Manag. Res. Bus. Strategy* **2014**, *3*, 47–55.
39. Elshaer, I.A.; Saad, S.K. Entrepreneurial resilience and business continuity in the tourism and hospitality industry: The role of adaptive performance and institutional orientation. *Tour. Rev.* **2021**, *77*, 1365–1384. [CrossRef]
40. Denton, F. *Independence and Economic Security in Old Age*; UBC Press: Vancouver, BC, Canada, 2011.
41. Ariffin, M.R.; Sulong, Z.; Abdullah, A. Students' perception towards financial literacy and saving behavior. *World Appl. Sci. J.* **2017**, *35*, 2194–2201.
42. Mohamad, S.; MacDonald, M.; Jariah, M.; Laily, P.; Tahira, K. Financial behavior and problems among college students in Malaysia: Research and education implication. *Consum. Interests Annu.* **2020**, *54*, 166–170.
43. Lusardi, A.; Mitchell, O.S. The economic importance of financial literacy. *Theory Evid. J. Econ. Lit.* **2014**, *52*, 5–44. [CrossRef]
44. Bandura, A. Self-efficacy: Toward a unifying theory of behavioral change. *Psychol. Rev.* **1977**, *84*, 191–215. [CrossRef]
45. Ismail, S.; Koe, W.-L.; Halim Mahphoth, M.; Abu Karim, R.; Yusof, N.; Ismail, S. Saving Behavior Determinants in Malaysia: An Empirical Investigation. *KnE Soc. Sci.* **2020**, *4*, 731–743. [CrossRef]
46. Solheim, C.A.; Zuiker, V.S.; Levchenko, P. Financial socialization family pathways: Reflections from college students' narratives. *Fam. Sci. Rev.* **2011**, *16*, 97–112.
47. Vosylis, R.; Erentaitė, R. Linking family financial socialization with its proximal and distal outcomes: Which socialization dimensions matter most for emerging adults' financial identity, financial behaviors, and financial anxiety? *Emerg. Adulthood* **2020**, *8*, 464–475. [CrossRef]
48. Bucciol, A.; Manfrè, M.; Veronesi, M. Family Financial Socialization and Wealth Decisions. *BE J. Econ. Anal. Policy* **2022**, *22*, 281–309. [CrossRef]
49. Allsop, D.B.; Boyack, M.N.; Hill, E.J.; Loderup, C.L.; Timmons, J.E. When Parenting Pays Off: Influences of Parental Financial Socialization on Children's Outcomes in Emerging Adulthood. *J. Fam. Econ. Issues* **2021**, *42*, 545–560. [CrossRef]
50. Elshaer, I.A.; Sobaih, A.E.E. FLOWER: An Approach for Enhancing E-Learning Experience Amid COVID-19. *Int. J. Environ. Res. Public Health* **2022**, *19*, 3823. [CrossRef]
51. Kim, J.; Chatterjee, S. Childhood financial socialization and young adults' financial management. *J. Financ. Couns. Plan.* **2013**, *24*, 61–79.
52. Gudmunson, C.G.; Danes, S.M. Family financial socialization: Theory and critical review. *J. Fam. Econ. Issues* **2011**, *32*, 644–667. [CrossRef]
53. Lachance, M.J.; Choquette-Bernier, N. College students' consumer competence: A qualitative exploration. *Int. J. Consum. Stud.* **2004**, *28*, 433–442. [CrossRef]
54. Bernheim, B.D.; Ray, D.; Yeltekin, Ş. Poverty and Self-Control. *Econometrica* **2015**, *83*, 1877–1911. [CrossRef]
55. Strömbäck, C.; Lind, T.; Skagerlund, K.; Västfjäll, D.; Tinghög, G. Does self-control predict financial behavior and financial well-being? *J. Behav. Exp. Financ.* **2017**, *14*, 30–38. [CrossRef]
56. Younas, W.; Javed, T.; Kalimuthu, K.R.; Farooq, M.; Khalil-ur-Rehman, F.; Raju, V. Impact of self-control, financial literacy and financial behavior on financial well-being. *J. Soc. Sci. Res.* **2019**, *5*, 211–218.
57. Biljanovska, N.; Palligkinis, S. Control thyself: Self-Control Failure and Household Wealth. SAFE Working Paper No. 69. 2015. Available online: https://ssrn.com/abstract=2509080 (accessed on 5 May 2022).
58. Moffitt, T.E.; Arseneault, L.; Belsky, D.; Dickson, N.; Hancox, R.J.; Harrington, H.; Sears, M.R. A gradient of childhood self-control predicts health, wealth, and public safety. *Proc. Natl. Acad. Sci. USA* **2011**, *108*, 2693–2698. [CrossRef]
59. Miotto, A.P.S.; Parente, J. Antecedents and consequences of household financial management in Brazilian lower-middle-class. *Rev. De Adm. De Empresas* **2015**, *55*, 50–64. [CrossRef]
60. Shefrin, H.M.; Thaler, R.H. The behavioral life-cycle hypothesis. *Econ. Inq.* **1988**, *26*, 609–643. [CrossRef]
61. Mavlutova, I.; Fomins, A.; Spilbergs, A.; Atstaja, D.; Brizga, J. Opportunities to increase financial well-being by investing in environmental, social and governance with respect to improving financial literacy under COVID-19: The case of Latvia. *Sustainability* **2021**, *14*, 339. [CrossRef]
62. Gallery, N.; Newton, C.; Palm, C. Framework for Assessing Financial Literacy and Superannuation Investment Choice Decisions. *Australas. Account. Bus. Financ. J.* **2011**, *5*, 3–22.
63. Hilgert, M.A.; Hogarth, J.M.; Beverly, S.G. Household Financial Management: The Connection between Knowledge and Behavior. *Fed. Reserve Bull.* **2003**, *89*, 309–323.
64. Hellström, J.; Zetterdahl, E.; Hanes, N. Loved Ones Matter: Family Effects and Stock Market Participation. 2013. Available online: https://www.diva-portal.org/smash/record.jsf?pid=diva2%3A658287&dswid=-4666 (accessed on 16 September 2022).
65. Putri, D.N.; Wijaya, C. Analysis of Parental Influence, Peer Influence, and Media Influence Towards Financial Literacy at University of Indonesia Students. *SSRG Int. J. Humanit. Soc. Sci.* **2020**, *7*, 66–73.
66. Moschis, G.P. *Consumer Socialization: A Life-Cycle Perspective*; Lexington Books: Lexington, MA, USA, 1987.
67. Pinto, M.B.; Mansfield, P.M.; Parente, D.H. Relationship of credit attitude and debt to self-esteem and locus of control in college-age consumers. *Psychol. Rep.* **2004**, *94*, 1405–1418. [CrossRef]

68. Manfrè, M. Saving Behavior: Financial Socialization and Self-Control. 2017. Available online: https://siecon3-607788.c.cdn77.org/sites/siecon.org/files/media_wysiwyg/288-manfre.pdf (accessed on 14 September 2022).
69. Furnham, A. The saving and spending habits of young people. *J. Econ. Psychol.* **1999**, *20*, 677–697. [CrossRef]
70. Cude, B.; Lawrence, F.; Lyons, A.; Metzger, K.; LeJeune, E.; Marks, L.; Machtmes, K. College students and financial literacy: What they know and what we need to learn. Proceedings of the Eastern Family. *Econ. Resour. Manag. Assoc.* **2006**, *102*, 106–109.
71. Khatun, M. Effect of financial literacy and parental socialization on students savings behavior of Bangladesh. *Int. J. Sci. Res. Publ.* **2018**, *8*, 296–305. [CrossRef]
72. Kaur, M.; Vohra, T. Understanding Individual Investors Behaviour—A Review Empirical Evidences. *Pac. Bus. Rev. Int.* **2012**, *5*, 10–18.
73. Iram, T.; Bilal, A.R.; Latif, S. Is awareness that powerful? Women's financial literacy support to prospects behaviour in prudent decision-making. *Glob Bus Rev.* **2021**. [CrossRef]
74. Thaler, R.H.; Benartzi, S. Save More Tomorrow: Using Behavioral Economics to Increase Employee Saving. *J. Political Econ.* **2004**, *112*, 64–87. [CrossRef]
75. Thaler, R.H.; Shefrin, H.M. An economic theory of self-control. *J. Political Econ.* **1981**, *89*, 392–406. [CrossRef]
76. Ameriks, J.; Caplin, A.; Leahy, J.; Tyler, T. Measuring self-control problems. *Am. Econ. Rev.* **2007**, *97*, 966–972. [CrossRef]
77. Gathergood, J. Self-control, financial literacy and consumer over-indebtedness. *J. Econ. Psychol.* **2012**, *33*, 590–602. [CrossRef]
78. Sadiq, M.N.; Khan, R.A.A. Impact of personality traits on investment intention: The mediating role of risk behavior and the moderating role of financial literacy. *J. Financ. Econ. Res.* **2019**, *4*, 1–18. [CrossRef]
79. Leguina, A. A Primer on Partial Least Squares Structural Equation Modeling (PLS-SEM). *Int. J. Res. Method Educ.* **2015**, *38*, 220–221. [CrossRef]
80. Hair, J.F., Jr.; Hult, G.T.M.; Ringle, C.; Sarstedt, M. *A Primer on Partial Least Squares Structural Equation Modeling (PLS-SEM)*; SAGE Publications: Thousand Oaks, CA, USA, 2016.
81. Chin, W.W. The Partial Least Squares Approach for Structural Equation Modeling. *Mod. Methods Bus. Res.* **1998**, *295*, 295–336.
82. Henseler, J.; Ringle, C.M.; Sinkovics, R.R. The Use of Partial Least Squares Path Modeling in International Marketing. In *Advances in International Marketing*; Sinkovics, R.R., Ghauri, P.N., Eds.; Emerald Group Publishing Limited: Bingley, UK, 2009; Volume 20, pp. 277–319. [CrossRef]
83. Mugo, E. Effect of Financial Literacy on Investment Decision Cooperative Societies Member in Nairobi. Master's Dissertation, KCA University, Nairobi, Kenya, 2016.
84. Liu, F.; Yilmazer, T.; Loibl, C.; Montalto, C. Professional financial advice, self–control and saving behavior. *Int. J. Consum. Stud.* **2019**, *43*, 23–34. [CrossRef]
85. Chia, Y.K.; Chai, M.T.; Fong, S.N.; Lew, W.C.; Tan, C.T. Determinants of Saving Behaviour among the University Students in Malaysia. 2011. Available online: http://eprints.utar.edu.my/607/1/AC-2011-0907445.pdf (accessed on 12 September 2022).
86. Esenvalde, I. *Psychological Predictors of Savings Behavior: Contrasting the Impact of Optimism and Burnout on Self-Control, Achievement Motivation and Savings Behavior*; Alliant International University: Los Angeles, CA, USA, 2011.
87. Griesdorn, T.S.; Lown, J.M.; DeVaney, S.A.; Cho, S.H.; Evans, D.A. Association between Behavioral Life-cycle Constructs and Financial Risk Tolerance of Low-to-moderate-income Households. *J. Financ. Couns. Planning.* **2014**, *25*, 27–40.

Article

Preprocessing of Spectroscopic Data Using Affine Transformations to Improve Pattern-Recognition Analysis: An Application to Prehistoric Lithic Tools

Francisco Javier Esquivel [1,2], José Antonio Esquivel [2,3], Antonio Morgado [3], José L. Romero-Béjar [1,4,5,*] and Luis F. García del Moral [6,7]

1 Department of Statistics and Operations Research, University of Granada, 18011 Granada, Spain
2 Laboratory of 3D Archaeological Modelling, University of Granada, 18011 Granada, Spain
3 Department of Prehistory and Archaeology, University of Granada, 18011 Granada, Spain
4 Instituto de Investigación Biosanitaria (ibs.GRANADA), 18014 Granada, Spain
5 Institute of Mathematics, University of Granada (IMAG), Ventanilla 11, 18001 Granada, Spain
6 Department of Plant Physiology, University of Granada, 18011 Granada, Spain
7 Institute of Biotechnology, University of Granada, 18011 Granada, Spain
* Correspondence: jlrbejar@ugr.es

Abstract: The analysis of spectral reflectance data is an important tool for obtaining relevant information about the mineral composition of objects and has been used for research in chemistry, geology, biology, archaeology, pharmacy, medicine, anthropology, and other disciplines. In archaeology, the use of spectroscopic data allows us to characterize and classify artifacts and ecofacts, to analyze patterns, and to study the exchange of materials, etc., as well as to explain some properties, such as color or post-depositional processes. The spectroscopic data are of the so-called "big data" type and must be analyzed using multivariate statistical techniques, usually principal component analysis and cluster analysis. Although there are different transformations of the raw data, in this paper, we propose preprocessing by means of an affine transformation. From a mathematical point of view, this process modifies the values of reflectance for each spectral signature scaling them into a [0, 1] interval using minimum and maximum values of reflectance, thus highlighting the features of spectral curves. This method optimizes the characteristics of amplitude and shape, reduces the influence of noise, and improves results by highlighting relevant features as peaks and valleys that may remain hidden using the raw data. This methodology has been applied to a case study of prehistoric chert (flint) artifacts retrieved in archaeological excavations in the Andévalo area located in the Archaeological Museum of Huelva (Huelva, Andalusia). The use of transformed data considerably improves the results obtained with raw data, highlighting the peaks, valleys, and the shape of spectral signatures.

Keywords: affine transformation; archaeology; flint (chert); multivariate statistics; pattern recognition; spectroscopy

MSC: 62H99

1. Introduction

Dolmens are the oldest stone architecture erected by humans to monumentalize their funerary spaces. They are collective or individual tombs with a trousseau associated with the megalithic phenomenon from the 5th millennium to the 3rd millennium BC [1], with permanence and/or reuse reaching up to the 2nd and 1st millennium BC [2–4]. The trousseaus are made up mainly of chert (flint) and, in some places, obsidian, and other minerals.

Chert is a sedimentary rock made up of 70–99.9% microcrystalline quartz (SiO_2), containing small percentages of water and various associated oxides (Ca, C, Fe, K, Al, and Mg) [5,6]. These features allow us to compare different rocks to determine similarities and differences. Chert from different origins can often be distinguished visually but using only visual identification per se is inappropriate because samples from the same provenance can exhibit a high degree of visual diversity [7]. Petrological characterization of chert has been based mainly on its mineralogical composition, texture, fossil content, and the environment in which it was formed [8]. At present, reflectance spectroscopy is being used for the identification and characterization of minerals and rocks since they are non-destructive techniques at medium cost and allow for quite rapid acquisition of raw data. Nevertheless, the great amount of data obtained presents great variability and requires the application of complex methods of mathematical big data analysis [9–11].

In Andalusia, the megalithic phenomenon has great relevance, with a large number of dolmens used as individual or collective funerary tombs [12]. In this work, we obtained the spectral signatures of lithic material found inside the excavated dolmens in the Andévalo region (Figure 1):

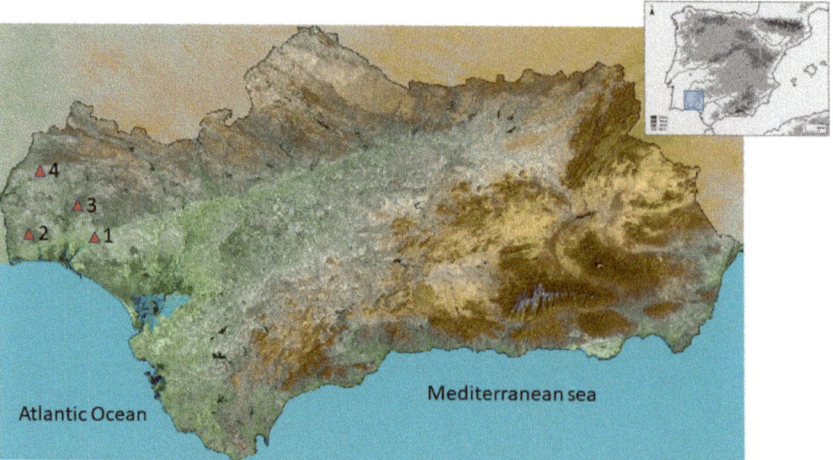

Figure 1. Sites analyzed in the Andévalo region (Andalusia, Spain). (1) El Moro tholos, (2) San Bartolomé tholos, (3) El Tejar dolmen, and (4) La Zarcita tholos.

A spectral library was created containing the spectral signatures of these artifacts. The material was analyzed using preprocessing transformation and big data methods to characterize, classify, establish similarities, and discriminate between chert artifacts. In this paper, we propose preprocessing by means of an affine transformation.

The interaction between light and matter is a complex process, and, in addition, data collection is limited by the accuracy of the instrument used, the wavelength range used, and the distortion caused by noise in the data acquisition. It is usual to use previous data preprocessing, mainly applying mathematical techniques of smoothing, baseline removal, and data normalization [13–15]. This transformation uses the specific parameters Maximum and Minimum of the spectral signature and enhances the values of each sample by sharpening the shapes and highlighting the peaks and valleys, obtaining important details that remain masked using raw data. In this work, preprocessing data using the affine transformation to highlight the amplitudes and shapes of the spectral signatures is proposed.

2. Materials and Methods

Spectroscopy records the interaction between light and matter, and the study of the reflected radiation after its interaction with matter is known as reflectance spectroscopy. Light reflected by a material gives rise to specular reflectance and diffuse reflectance. In specular reflectance, the incident rays are reflected without interacting with the sample, with the incident angle equal to the reflected angle. In diffuse reflectance, part of the incident rays interacts with the sample components and are absorbed and diffused according to the intensity and spectral composition of the radiation, reflecting in different directions. Only diffuse reflectance is relevant to reflectance spectroscopy. Each sample, depending on its chemical composition and optical properties, will return a characteristic reflectance curve called spectral signature as a function of absorption and reflection due to the sample composition.

The spectra in the visible and near-infrared ranges contain considerable information about the physical and molecular composition of materials (e.g., flint), as they detect molecules from the absorbed wavelengths such as water, hydroxyls, phosphates, nitrates, carbonates, sulfates, and metal oxides and hydroxides [3]. Therefore, infrared spectroscopy has been proposed as a new technique to detect heat-induced effects within lithic artifacts and to quantify OH and water ions in archaeological flints.

The siliceous rock quarries linked to the production of specialized flint flakes in Andalusia are grouped into four large areas [11]: (1) the Pyritic Belt of Huelva with rhyolites and rhyodacites mainly from the Paleozoic; (2) the Middle Sub-Baetic of External Areas, centered in Granada province, with flints characteristic of the Upper Jurassic Milanos Formation; (3) flint quarries in Málaga province and the Campo de Gibraltar Complex (Cádiz province); and (4) the flint quarries of the Malaver Formation (Ronda, Málaga) from the Tertiary.

In this study, the samples analyzed are prehistoric lithic artifacts retrieved in archaeological excavations in the Andévalo area (Huelva, Andalusia). The chert artifacts come from four locations corresponding to San Bartolomé de la Torre dolmen (SBn° with 5 samples), El Tejar dolmen from Higuera de la Sierra (Tn° having 5 samples), La Zarcita dolmen in the town of Santa Bárbara de Casa (Zn° with 15 samples), and the tholos of El Moro in Niebla (TMn° with 1 sample) (Figure 2).

Figure 2. Some of the lithic tools from the tholos of La Zarcita: (1) Z-132, (2) Z-134, and (3) Z-215.

Reflectance spectra were collected using an Analytical Spectrometer Device (ASD) Portable Spectroradiometer FieldSpec4STD (Malvern Panalytical, Malvern, UK), which records the amplitude value of electromagnetic waves between 350 nm and 2500 nm (spectral range). This spectroradiometer is designed with three channels, and in each channel, it can distinguish a very small difference in wavelengths (spectral resolution), with a spectral resolution between 3 nm at 700 nm and 10 nm at 1400 and 2100 nm, with a total number of 2151 spectral bands. The measuring interval was 1.4 nm in the spectral range of 350–1000 nm and 2 nm in the 1001–2500 nm range, using a high-intensity contact probe A122307 (Analytical Spectral Devices, Inc., Boulder, CO, USA) with a halogen light source, a measurement surface area equivalent to a circle 2 cm in diameter, and a maximum specular reflectance of 5%. The white level was calibrated on a Spectralon of 3.62″ diameter (Analytical Spectral Devices, Inc.), providing a nearly 100% reflective Lambertian surface across the entire spectrum. Dark correction (DCC) was applied to remove the electrical current generated by thermal electrons and was added to that generated by incoming photons. Raw data returned are 16-bit numbers corresponding to the output of each element in the VIS/NIRS detector array and each 2 nm sample of the spectrum to generate a relative reflectance. This system reduces noise using a spectrum averaging technique (average of 20 spectra per quantification). Three spectra were acquired for each chert sample, and their mean was used. According to the ASD manual, the white level was calibrated after every 25 spectra.

The results obtained in the quarries in the case of Andévalo have been analyzed by using preprocessing mathematical transformations to analyze spectroscopy data, including logarithms, standardizations of different types (the most usual one is Z-score), etc. The affine transformation is based on the use of specific parameters obtained from the spectral signature of the data used (Maximum and Minimum). This transformation allows enhancement of the values of each sample by sharpening the shapes and highlighting the peaks and valleys. The scale is [0,1] because spectrometers usually provide the recorded reflectance values in this interval. Figure 3 shows a graphic diagram of the proposed methodology.

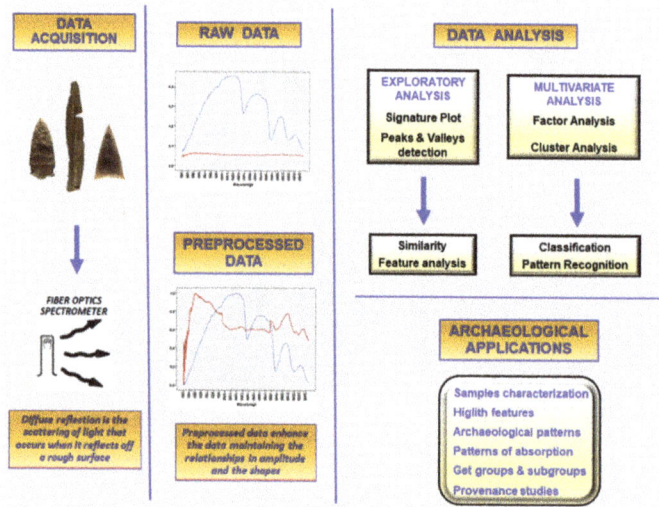

Figure 3. Graphic diagram of the proposed methodology.

3. Preprocessing Data

The analysis of spectral data from raw data presents some problems, mainly due to the characteristics of the interaction process of light and matter, the limitations of the instruments used, and the distortion produced by noise. However, the use of preprocessing methods allows us to improve the results and attenuate the influence of noise by applying mathematical techniques of smoothing, elimination of baseline, scaling, and normalization [13–15]. The suitable preprocess transforms the reflectance values, maintaining the relationships in amplitude and between the shapes, features, and other details and allowing them to stand out when they would otherwise remain hidden or masked [16].

Preprocess transformations belong to three basic groups: functional, statistical, and geometric. Functional transformations are based on applying a function, generally defined in implicit coordinates, to the spectral signatures. Among the most common functional transformations, those belonging to the family of logarithmic functions $X'_i = \log_a(X_i)$ and potentials $X'_i = a^{X_i}$ (usually $a = e$) stand out, although there are some quite complex functions, such as the sigmoid function $X' = \frac{1-e^{(-aX)}}{1+e^{(-aX)}}$ that equalizes the reflectance data with distortion proportional to the a parameter [17].

Statistical transformations modify the scale of the distribution of a variable to make comparisons between elements, sets of elements, or different parameters, homogenizing the units of measurement between variables. Among the most used are the typified Z or standardized scores from a normal distribution ($Z_i = \frac{X_i - \mu}{\sigma}$), related to range ($X'_i = \frac{X_i}{X_{max} - X_{min}}$), to range 0–1 ($X'_i = \frac{X_i - \mu}{X_{max} - X_{min}}$), to the maximum magnitude ($X'_i = \frac{X_i}{X_{max}}$), to the mean ($X'_i = X_i - \mu$), or to the standard deviation $\left(X'_i = \frac{X_i}{\sigma}\right)$ [18].

A very important class is composed of geometric transformations based on the concepts of Euclidean geometry (adjusted to the Euclid postulates). Among them, affine geometry stands out as a generalization of Euclidean geometry with properties applicable in a Minkowski space. These concepts were formalized in the language of Felix Klein's Erlangen Program [19,20]. Among the important geometric transformations are the affine transformations in the $\mathbb{R} \times \mathbb{R}$ vector space $\vec{q} = \alpha \left(A \vec{s} + b\right)$, where A is a real and non-singular 2×2 matrix, b a 2×1 real vector, and α a real number. In the case of spectral curves, these values are simplified since A and B are real numbers, and the standard affine transformation is expressed by $f : [r_{min}, r_{max}] \rightarrow [r'_{min}, r'_{max}] = [0, 1]$

$$f(x) = \frac{x - r_{min}}{r_{max} - r_{min}}$$

providing a min–max normalization (MMN) [17,21]. The classification performance can hardly be improved by this method [22].

In this work, the affine transformation is proposed in order to simultaneously maintain the coherence of the original reflectance values (*raw data*) and the shape of the graphical representation, highlighting as well the peaks, valleys, and trends of each signature. Raw data and transformed plots are shown in Figure 4.

Signatures of samples from San Bartolomé are homogeneous in shape with a decreasing trend at the end of the spectrum. The standard affine transformation highlights the aggrupation of these samples and points out materials with the same diagenetic processes and, possibly, similar post-depositional alterations of environmental and/or anthropic origin. However, SB1357 shows a very flat curve with an almost constant reflectance and a gently increasing trend, making it very difficult to distinguish its shape (Figure 4a). The preprocessed values highlight a peak at ~530 nm, the maximum of the signature, as well as a doublet at ~2200 nm. The petrochemical analysis assigns this sample to the group of rhyolitic rocks in the Odiel river area [23], having differences from the other samples at 550–2500 nm (Figure 4b).

Four samples from El Tejar agree with the general patterns of chert with high reflectance. Sample T3874 is similar to SB1357, with a peak at ~530 nm, low reflectance, and a characteristic double valley shape at ~2200 nm. Both samples can be associated with crystalline and luminous minerals.

Fifteen lithic artifacts found in the Zarcita dolmen show great variability but maintain the general shape of signatures of chert. The Z215 sample is very different from the rest, having a small reflectance range with an absolute maximum at ~550 nm and doublet absorption at ~2200 nm.

The visual analysis of T3874 and Z215 show similar features even in the visible spectrum, but the adscription of SB1357 is not clear using the raw data (Figure 5).

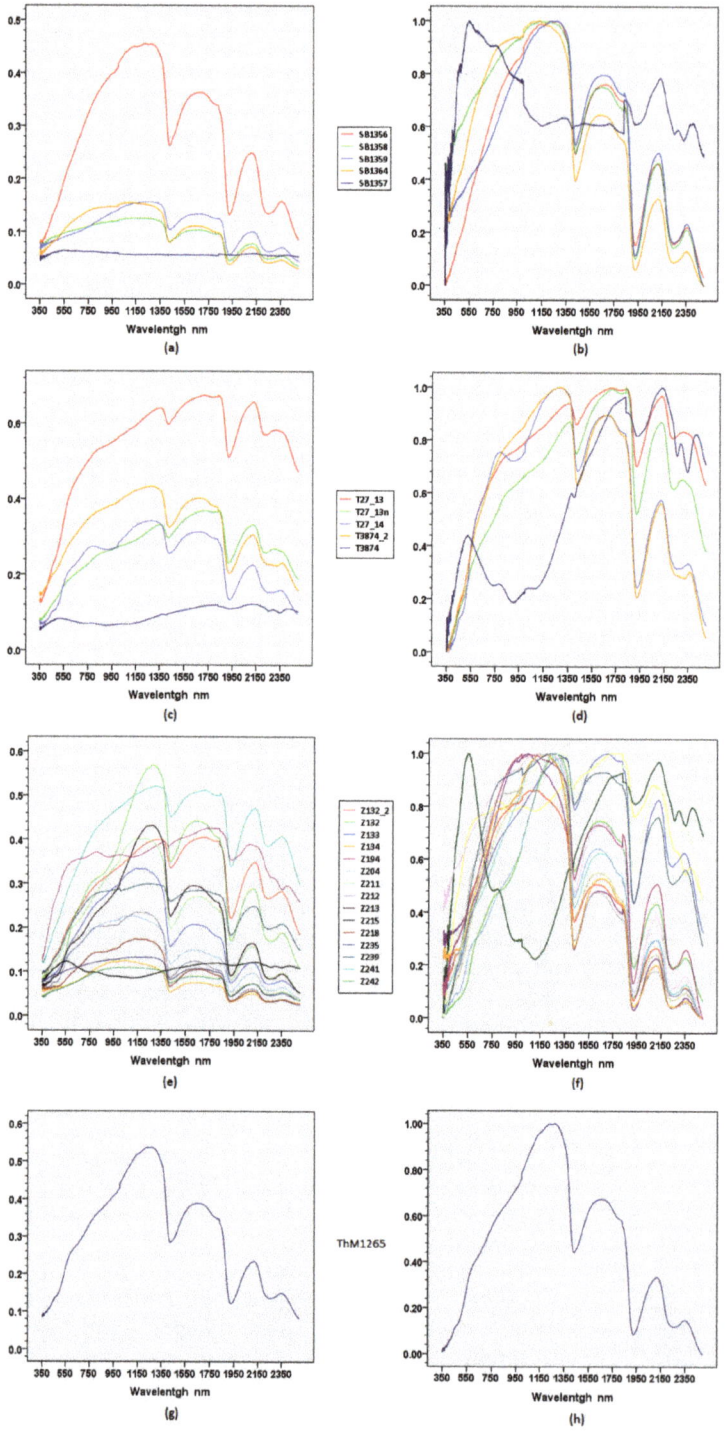

Figure 4. Signature plots of Andévalo region. Raw data (**left**) and data modified by affine transformation (**right**): Dolmen of San Bartolomé (**a**,**b**); Dolmen of El Telar (**c**,**d**); Dolmen of La Zarcita (**e**,**f**); Tholos of El Moro (**g**,**h**).

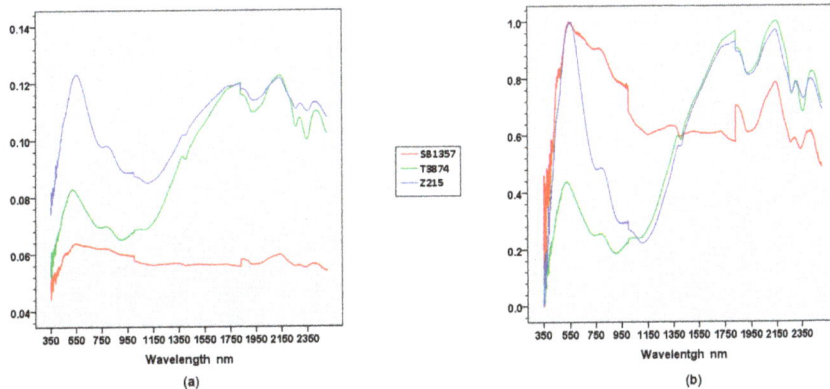

Figure 5. Plot of T3874, Z215, and SB1357 using (**a**) raw data, and (**b**) modified data by the affine transformation.

Again, the affine transformation highlights important underlying features. Samples T3894 and Z215 have similar shapes in the VNIR spectrum, while SB1357 has the same curve only in the 350–550 nm range. The three signatures have a decreasing trend in 1000–1700 nm, but SB1357 is different. These features point out diagenetic processes that are similar but with some differences at 800–1800 nm, possibly due to large areas of Europe having been subjected to similar processes in the formation of geological materials during a specific period, although with local variations.

4. Singular Statistical Parameters: Peaks and Valleys

Rock identification is dependent upon several variables, such as mineral associations, diagenetic processes and transformations through geologic time, mineral alterations, current large- and small-scale morphology, and recent climatic conditions [24]. Some authors propose a description of spectral curves based on: (1) the position of wavelengths minimizing the values of reflectance of absorption bands; (2) the depth, width, area, and asymmetry of the bands; and (3) the position of the slope changes and inflection points on the curve [25].

The spectral absorption patterns in the SWIR spectrum provide great homogeneity with very small coefficients of variation at important flint wavelengths (~1400, ~1900, ~2200, and ~2350) (Table 1).

Table 1. Statistical parameters of the most prominent absorption patterns in the SWIR range.

	~1400		~1900		~2200		~2350
	Peak	Valley	Peak	Valley	Peak	Valley	Peak
$\bar{x}$	1227.82	1424.66	1653.71	1923.56	2111.30	2245.20	2349.92
sd	87.99	6.08	29.15	1.90	41.21	9.42	13.82
cv %	7.2	0.4	1.5	0.1	1.9	0.4	0.6

Note: $\bar{x}$ = sample mean; sd = standard deviation; cv % = coefficient of variation.

The last peak (~2350 nm) is a specific diagnostic with a great homogeneity, identified by the peaks of various cations and anions in the SWIR region, mainly due to carbonates [26]. Moreover, some authors propose the characteristic bands of major absorption in the SWIR region for the common mineral groups corresponding to carbonates and Mg–OH, as well as Mg–OH vibration in amphiboles [24,27].

On the other hand, the shape in the visible spectrum of SB1357, T3874, and Z215 stand out from the rest, with a peak at ~540 nm, having the maximum reflectance in each signature. This peak is related to ferric and ferrous ions (mainly oxides), silicates, sulfates, sulfites, manganese, and chromium ions [26,28]. Other authors propose Fe_2 + chlorites, Ni Chrysoprase, and Cr-diopside [27], as unbound electrons whose atomic structure and mineral impurities the diagnostic attributes detected [29]. These three samples possibly correspond to rocks whose main components are serpentine, olivine, or malachite, along with other components. These common minerals in Western Andalusia have great reflectance at ~540 nm, with green or greenish colors.

5. Pattern Recognition

The large amount of data (*big data*) registered by spectroscopy requires the use of complex statistical methods of multivariate statistical analysis to obtain the possible patterns (*pattern recognition*) in the data. These methods include primarily factor analysis (FA), principal component analysis (PCA), discriminant analysis (DA), principal component regression (PCR), multiple linear regression (MLR), and partial least squares regression (PLS). DA and PCA are widely used tools for quantitative analysis. Additionally, CA (cluster analysis) is another important method that allows us to obtain a data classification using geometric and statistical methods [3,30].

In this paper, we have applied Principal Component Analysis and Cluster Analysis, as they are the most reliable methods for our analysis.

5.1. Principal Component Analysis

The essential concept is to consider each signature as a point in multidimensional vector space with the wavelengths as the dimensions, usually with 1 nm accuracy. Factorial analysis (FA) provides a new coordinate system constituted by a linear combination of the original variables, which are called components. The most used algorithm is the principal components algorithm (PCA), whose components are chosen so that PC1 includes the greatest variability of the data. Then, PC2 is chosen orthogonal to PC1, including the maximum remaining variability, and so on.

The PCA method is used with a great number of variables to drastically reduce their dimensionality in an interpretable way, preserving the greatest part of the information in the data. Normally, the first principal components involving a large amount of total variance (usually more than 80%) are taken into account. The remainder of the components then contains minor characteristics and noise. The importance of each original variable is measured by the matrix of loadings, and these values indicate the importance of each original variable in the data set. Usually, the first principal components are considered in order to study factor structure. Then, the trends underlying the data are drawn in different bivariate plots using PCAs [31,32]. The main uses of PCAs are exploratory rather than inferential to detect the underlying trends in the data and, in some cases, to obtain a visual prior data classification.

The application of exploratory PCAs to the samples from Andévalo shows some notable trends. Using the first three components, the results are similar in terms of total variation carried with the raw data (99.2%) or the transformed data (95.2%) (Table 2).

Table 2. Eigenvalues and accumulated variance: a) raw data, and b) transformed data.

	Eigenvalues (Raw Data)			Eigenvalues (Affine Transformed)	
	% Var	% Cum Var		% Var	% Cum Var
PC1	89.726	89.726	PC1	55.438	55.438
PC2	7.040	96.766	PC2	21.218	76.657
PC3	2.375	99.141	PC3	18.638	95.294

Raw variables are highly correlated, providing a very large first eigenvalue (size effect) with explained variance close to 100%. The transformed data show a factorial structure with three factors that have more defined matrix loadings. Matrix loading determines which variables are most important when analyzing the variability and is fundamental in deciding which minimum values are chosen, with the loadings interpreted using robust cut-offs in the presence of non-normal distributions [22]. Some authors propose absolute values ≥ 0.45 as relevant, ≥ 0.45 to 0.55 as good, and ≥ 0.63 as very good [33]. Other methods use resampling techniques, such as jack-knifing or bootstrapping [34,35], with similar results.

The instrument used is based on the FORS (Fiber Optics Reflectance Spectroscopy) technique with 2151 variables corresponding to wavelengths of 350–2500 nm, so choosing only a few factors (components) is quite difficult. We propose choosing continuous wavelength intervals with no isolated values, corresponding to most of the load factors when their absolute value is ≥ 0.55 (Figure 6).

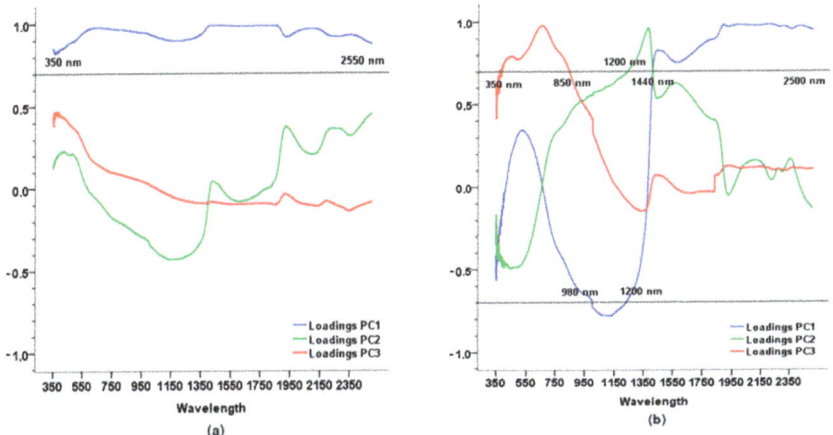

Figure 6. Plot of loadings to first three components using (**a**) raw data, and (**b**) transformed data.

The factor structure with raw data is dominated by PC1, a "size factor" with loads greater than 0.8. The other factors have very small loads. Thus, obtaining detailed features is very difficult. The use of transformed data provides the emergence of a more detailed factorial structure. The PC1 factor highlights the NIR interval (1400–2500 nm) as one of the most important features when studying absorption bands of minerals, those being mainly -OH oxidrile (~1400 nm), H_2O (~1900 nm), and Metal-OH ions (~2200 nm) [6]. By separating the samples, the loads in PC2 show a bipolar factor due to the reflectance in the VIS + NIR and SWR2 intervals (e.g., SB1357) against high values in the full spectrum (e.g., Z239).

Plot PC1–PC2 shows that important details of diagnostic features are lost with raw data but transformed data highlight the differences much better, with the visual groups being more homogeneous and consistent. Thus, with raw data, only factor 1 is relevant, it being a "size factor"; the other factors do not practically discriminate (Figure 7a). With the exception of SB1357 and Z215, the transformed data highlight an important relationship between the samples from San Bartolomé and those at the Zarcita dolmen and also establish possible graphic associations (Figure 7b):

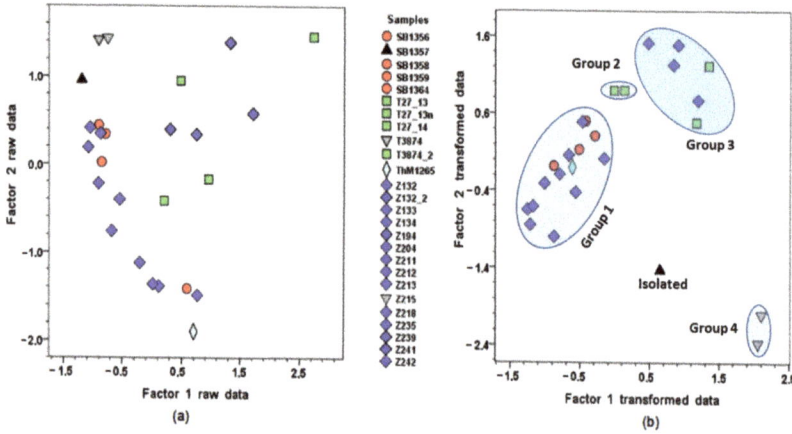

Figure 7. Plots of PC1–PC2: (**a**) raw data, and (**b**) transformed data.

5.2. Cluster Analysis

Using multivariate geometric and statistical algorithms, cluster analysis is focused on classifying objects into significant groups or clusters where there was no prior information about cluster membership in predefined groups. The most used clustering techniques are of type SHAN (Sequential Hierarchical Agglomerative Not overlapping), obtaining a classification dendrogram [36,37].

The algorithms are based on a previous similarity measure or distance in addition to an agglomeration criterion maximizing intragroup similarity and minimizing intergroup similarity. The single linkage and the average linkage algorithms are among the most used and highlight the Ward, the nearest neighbor. The Euclidean or squared distance is the most commonly used similarity measure for quantitative variables. A SHAN clustering is computed on the Andévalo samples with squared Euclidean distance as a similarity measure and Ward's clustering algorithm (minimum variance) as the clustering algorithm [38,39].

The Ward method is based on the loss of information between each object and the average of each cluster in which it is integrated, which is produced in each pass when calculating the total sum of the squares of the deviations between each object and the average of the cluster in which it is integrated. Some authors have found that this provides a more optimal classification than other methods, such as minimum distance, maximum distance, average distance, or centroid methods [40].

The determination of the number of clusters is based on the Elbow Method, which looks at the total WSS (Within Sum of Squares) to measure the compactness of the clustering and the goodness of the clustering structure. A curve plots the number of clusters versus the WSS measures. The location of an elbow ("knee") in the plot is considered an indicator of the appropriate number of clusters [41,42]. Application of that method to our data set provides that the best number of clusters is $n = 5$ (Figure 8).

CLUSTER 1 presents great variability with samples from all the areas studied, pointing to the fact that the objects are exchanged in the zone. CLUSTER 2 consists almost exclusively of samples from the tholos La Zarcita in Northern Andévalo, except one sample from the El Moro, and CLUSTER 3 is formed from samples from El Tejar and La Zarcita, possibly because they come from quarries with similar characteristics. CLUSTER 4 is composed of a rhyolite (extrusive igneous rock) very different from flint, which is a rock formed by a mixture of siliceous minerals from the closest volcanic area to the Andévalo, and points to exchanges throughout the south of the Iberian Peninsula. Finally, CLUSTER 5 is composed of two samples with great similarities between them. This group is similar to the rhyolite in CLUSTER 4, indicating that they come from the same quarry (Figure 5), although with small variations. Further, SB1357 appears isolated from the others, although it shares many similar characteristics with T3874 and Z215, maintaining the absorption in the visible spectrum but with very high reflectance values. Its singularity comes from the higher values in the visible spectrum, with a peak at ~535 nm and higher values than in previous cases. SB1357 is a volcanic siliceous rock mainly composed of rhyolite, very different from chert. Identification of these three samples corresponds to rocks whose main components are chlorite, serpentine, olivine, or malachite (peaks in ~535 nm), very abundant in this so-called "pyritic belt" area of Andalusia, in addition to other components.

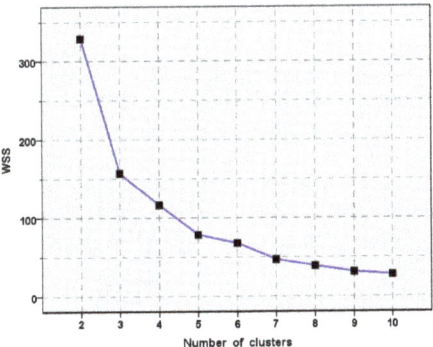

Figure 8. Elbow method to determine the number of clusters.

From the Andévalo samples: cluster analysis provides a classification into five clusters (Figure 9).

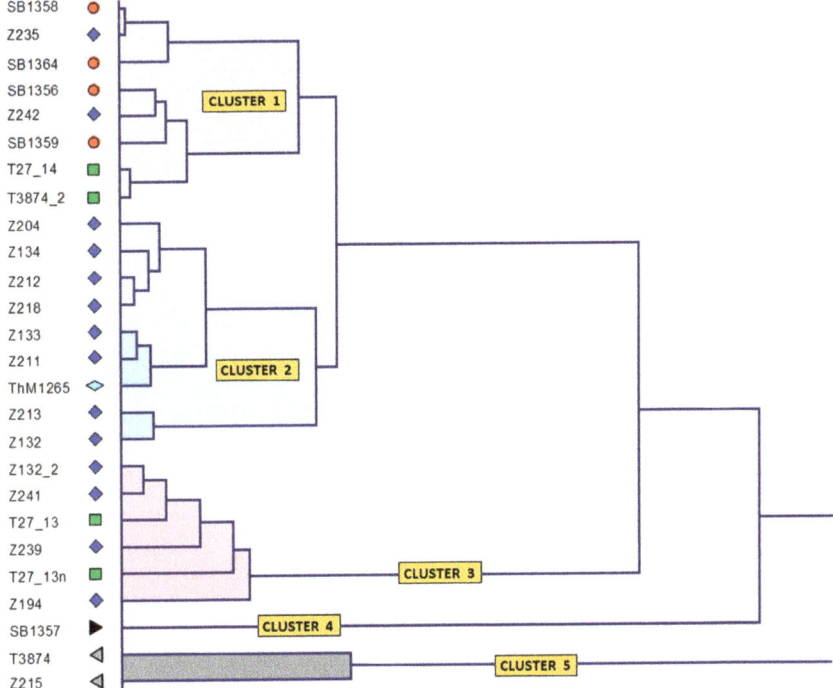

Figure 9. Dendrogram showing the clusters (Ward method) from the Andévalo samples (transformed data).

6. Conclusions

Spectroscopic methods allow us to obtain important information about the structure and composition of minerals in the VNIR range of the electromagnetic spectrum at the level of atoms, electrons, molecules, and crystal lattices. This information can be used to better understand the physical and chemical conditions of the mineral formations and explain various properties, such as color, and analyze the provenance of lithic tools manufactured by prehistoric people.

The samples analyzed correspond to prehistoric lithic artifacts obtained from archaeological excavations in the Andévalo region (Huelva, Andalusia) belonging to the Later Neolithic and Chalcolithic periods. Spectral signatures were obtained and analyzed, but their low reflectance and the small range of variation between minimum and maximum in some samples required previous preprocessing data to highlight their characteristics and obtain more reliable measurements of some features. Among the large number of existing transformations, the standard affine transformation has been used to typify data values and highlight their differences while keeping the shape of the curve. Further, the affine-transformed data greatly increased the effectiveness of statistical analysis with respect to the raw data.

The results confirm the circulation of lithic artifacts in the Andévalo region and allow us to draw conclusions about the places of extraction, the origin of lithic material, patterns of cultural transmission, circulation, and paths of exchange. Given this, we can deduce that during the Chalcolithic period, the southwest of Andalusia seems to have formed an area in which the exchange of information, objects, and, possibly, people took place.

It is important to note that two statistical methodologies, Principal Component Analysis and Cluster Analysis, have been applied to study the classification of the samples considering the transformed data, obtaining identical classifications that also coincide with the origin of the samples.

Author Contributions: Conceptualization, J.A.E. and L.F.G.d.M.; methodology, J.A.E., F.J.E., and J.L.R.-B.; software, F.J.E. and J.L.R.-B.; validation, A.M. and L.F.G.d.M.; formal analysis, J.A.E.; investigation, J.A.E., L.F.G.d.M., and A.M.; resources, L.F.G.d.M. and A.M.; data curation, F.J.E. and J.L.R.-B.; writing—original draft preparation, J.A.E. and F.J.E.; writing—review and editing, J.L.R.-B.; visualization, J.A.E., A.M., and L.F.G.d.M.; funding acquisition, J.A.E. and L.F.G.d.M. All authors have read and agreed to the published version of the manuscript.

Funding: This work has been funded by the Ministry of Economy and Competitiveness, Project UNGR15-CE-3531, and by the research groups AGR123 and ARCHAEOSCIENCE HUM1037 of the Junta de Andalucía.

Data Availability Statement: Data available upon request to the authors.

Acknowledgments: The authors thank the staff of the Huelva Museum for the facilities given to access the archaeological flint blades studied in this article. F.J. Esquivel and J.L. Romero have been partially supported by grant PID2021-128077NB-I00, funded by MCIN/ AEI/10.13039/501100011033/ ERDF A way of making Europe, EU.

Conflicts of Interest: The authors declare no conflict of interest.

References

1. Paulsson, B.S. Radiocarbon dates and Bayesian modelling support maritime diffusion model for megaliths in Europe. *Proc. Natl. Acad. Sci. USA* **2019**, *116*, 3460–3465. [CrossRef] [PubMed]
2. Aguayo, P.; García, L. The megalithic phenomenon in Andalusia. An overview. In *Origin and Development of the Megalithic Phenomenon of Western Europe, Proceedings of the International Symposium, Bougon, France, 26–30 October 2002*; Joussaume, R., Laporte, L., Scarre, C., Eds.; Conseil Général de Deux Sèvres: Bougon, France, 2006; pp. 452–472.
3. García del Moral, L.F.; Morgado, A.; Esquivel, J.A. Espectroscopia de Reflectancia de Fibra Óptica (FORS) de las principales canteras de rocas silíceas de Andalucía y su aplicación a la identificación de la procedencia de artefactos líticos tallados durante la Prehistoria. *Complutum* **2022**, *33*, 35–67. [CrossRef]

4. Esquivel, F.J.; Morgado, A.; Esquivel, J.A. 2017 La Arqueología de la Muerte y el Megalitismo en Andalucía. Una Aproximación a los Rituales de Enterramiento V Milenio BC-II Milenio B.C. In *La Muerte desde la Prehistoria a la Edad Moderna*; Espinar, M., Ed.; EPCCM: Granada, Spain, 2017; Volume 23, pp. 93–124.
5. Luedtke, B.E. *An Archaeologist's Guide to Flint and Flint Archaeological Research Tools*; Institute of Archaeology, University of California: Los Angeles, CA, USA, 1992.
6. Clark, R.N. 1999 Spectroscopy of Rocks and Minerals, and Principles of Spectroscopy. In *Manual of Remote Sensing, Remote Sensing for the Earth Sciences*; Chapter 1; Rencz, A.N., Ed.; John Wiley and Sons: New York, NY, USA, 1999; Volume 3, pp. 3–58.
7. Luedtke, B.E. The identification of sources of chert artifacts. *Am. Antiq.* **1979**, *44*, 744–756. [CrossRef]
8. Affolter, J. *Provenance des Silex Préhistoriques du Jura et des Regions Limitrophes. Archéologie Neuchâteloise, 28*; Service et Musée cantonal D'archéologie: Neuchâtel, France, 2002.
9. Parish, R.M. A Chert Sourcing Study Using Visible/Near-infrared Reflectance Spectroscopy at the Dover Quarry Sites, Tennessee. Unpublished. Master's Thesis, Department of Geosciences, Murray State University, Murray, Kentucky, 2009.
10. Beyer, M.A.; Laney, D. *The Importance of Big Data: A Definition*; Gartner Inc.: Stanford, CA, USA, 2012.
11. García del Moral, L.F.; Morgado, A.; Esquivel, J.A. Reflectance spectroscopy in combination with cluster analysis as tools for identifying the provenance of Neolithic flint artefacts. *J. Archaeol. Sci. Rep.* **2021**, *37*, 103041. [CrossRef]
12. Linares, J.A. El Megalitismo en el sur de la Península Ibérica. Arquitectura, Construcción y Usos de Los Monumentos del Área de Huelva, Andalucía Occidental. Ph.D. Thesis, University of Huelva: Huelva, Spain, 2017.
13. Gholizadeh, A.; Borůvka, L.; Saberioon, M.M.; Kozák, J.; Vašát, R.; Němeček, K. Comparing Different Data Preprocessing Methods for Monitoring Soil Heavy Metals Based on Soil Spectral Features. *Soil Water Res.* **2015**, *10*, 218–227. [CrossRef]
14. Fang, Q.; Houng, H.; Zhao, L.; Kukolich, S.; Yin, K.; Wang, C. Visible and Near-Infrared Reflectance Spectroscopy for Investigating Soil Mineralogy: A Review. *J. Spectrosc.* **2018**, *2018*, 3168974. [CrossRef]
15. Angelopoulou, T.; Balafoutis, A.; Zalidis, G.; Bochtis, S. From Laboratory to Proximal Sensing Spectroscopy for Soil Organic Carbon Estimation. *Sustainability* **2020**, *12*, 443. [CrossRef]
16. Sgavetti, M.; Pompilio, L.; Meli, S. Reflectance spectroscopy (0.3–2.5 μm) at various scales for bulk-rock identification. *Geosphere* **2006**, *2*, 142–160. [CrossRef]
17. Da Fontoura, L.; Marcontes, R. *Shape Classification and Analysis*, 2nd ed.; CRC Press, Taylor and Francis Group: Boca Ratón, FL, USA, 2009.
18. Dodge, Y. *The Oxford Dictionary of Statistical Terms*, 6th ed.; Oxford University Press: Oxford, UK, 2003.
19. Nomizu, K.; Sasaki, T. *Affine Differential Geometry: Geometry of Affine Immersions. (Cambridge Tracts in Mathematics, Series Number 111)*; Cambridge University Press: Cambridge, UK, 2008.
20. Vargas, J.G. *Differential Geometry for Physicists and Mathematicians. Moving Frames and Differential Forms: From Euclid Past Riemann*; World Scientific: New York, NY, USA, 2014.
21. Solomon, C.; Breckon, T. *Fundamentals of Digital Image Processing. A Practical Approach with Examples in Matlab*; John Wiley & Sons: Oxford, UK, 2011.
22. Wang, B.; Yan, X.; Jiang, Q. Loading-Based Principal Component Selection for PCA Integrated with Support Vector Data Description. *Ind. Eng. Chem. Res.* **2015**, *54*, 1615–1627. [CrossRef]
23. Donaire, T.; Toscano, M.; Valenzuela, A.; González, M.J.; Pascual, E. Alteración diferencial de las rocas volcánicas ácidas en el sector de Riotinto, Faja Pirítica Ibérica. *Geogaceta* **2010**, *48*, 147–150.
24. Longhi, I.; Sgavetti, M.; Chiari, R.; Mazzoli, C. Spectral analysis and classification of metamorphic rocks from laboratory reflectance spectra in the 0.4–2.5 mm interval: A tool for hyperspectral data interpretation. *Int. J. Remote Sens.* **2001**, *22*, 3763–3782. [CrossRef]
25. Grove, C.I.; Hook, S.J.; Paylor, E.D. *Laboratory Reflectance Spectra of 160 Minerals, 0.4 to 2.0 Micrometers*; JPL Publication 92-2; Jet Propulsion Laboratory: Pasadena, CA, USA, 1992.
26. Gupta, R.P. *Remote Sensing Geology*, 3rd ed.; Springer: Berlin, Germany, 2018.
27. Hauff, P. *An Overview of VIS-NIR-SWIR Field Spectroscopy as Applied to Precious Metals Exploration*; Spectral International Inc.: Arvada, CO, USA, 2008.
28. Hunt, G.R. Spectral signatures of particulate minerals in the visible and near infrared. *Geophysics* **1977**, *42*, 468–671. [CrossRef]
29. Parish, R.M. Reflectance Spectroscopy as a Chert Sourcing Method. *Archaeol. Pol.* **2016**, *54*, 115–128.
30. Esbensen, K.H.; Swarbrick, B. *Multivariate Data Analysis: An introduction to Multivariate Analysis, Process Analytical Technology and Quality by Design*, 6th ed.; CAMO Software AS: Oslo, Norway, 2018.
31. Ritz, M.; Vaculíková, L.; Plevová, E. Application of Infrared Spectroscopy and Chemometric Methods to Identification of Selected Minerals. *Acta Geodyn. Geomater.* **2011**, *8*, 47–58.
32. Izenman, A.J. *Modern Multivariate Statistical Techniques: Regression, Classification, and Manifold Learning. Springer Texts in Statistics*, 2nd ed.; Springer: New York, USA, 2013.
33. Finch, A.P.; Brazier, J.E.; Mukuria, C.; Bjorner, J.B. An Exploratory Study on Using Principal-Component Analysis and Confirmatory Factor Analysis to Identify Bolt-On Dimensions: The EQ-5D Case Study. *Value Health* **2017**, *20*, 1362–1375. [CrossRef] [PubMed]
34. Peres-Neto, P.R.; Jackson, D.A.; Somers, K.M. Giving Meaningful Interpretation to Ordination Axes: Assessing Loading Significance in principal Component Analysis. *Ecology* **2018**, *84*, 2347–2363. [CrossRef]

35. Timmerman, M.E.; Kiers, H.A.L.; Smilde, A.K. Estimating confidence intervals for principal component loadings: A comparison between the bootstrap and asymptotic results. *Br. J. Math. Stat. Psychol.* **2007**, *60*, 295–314. [CrossRef] [PubMed]
36. Sneath, P.H.A.; Sokal, R.R. *Numerical Taxonomy: The Principles and Practice of Numerical Classification, 2nd revised ed.*; W.H. Freeman: San Francisco, CA, USA, 1973.
37. Everitt, B.S.; Landau, S.; Leese, M.; Stahl, D. *Cluster Analysis*, 5th ed.; Wiley in Probability and Statistics: New York, NY, USA, 2011.
38. King, R.S. *Cluster Analysis and Data Mining: An Introduction*; Mercury Learning & Information: Vancouver, BC, Canada, 2014.
39. Wierzchon, S.T.; Klopotek, M.A. *Modern Algorithms of Cluster Analysis. Studies in Big Data 34*; Springer International Publishing AG: Cham, Switzerland, 2018.
40. Kuiper, F.K.; Fisher, L. A Monte Carlo comparison of six clustering procedures. *Biometrics* **1975**, *31*, 777–783. [CrossRef]
41. Raykov, Y.P.; Boukouvalas, A.; Baig, F.; Little, M.A. What to do when K-Means Clustering Fails: A Simple yet Principled Alternative Algorithm. *PLoS ONE* **2016**, *26*, e0162259. [CrossRef] [PubMed]
42. Patil, C.; Baidari, I. Estimating the Optimal Number of Clusters k in a Dataset Using Data Depth. *Data Sci. Eng.* **2019**, *4*, 132–140. [CrossRef]

Article

Smart Patrolling Based on Spatial-Temporal Information Using Machine Learning

Cesar Guevara [1,2,*] and Matilde Santos [3]

1. The Institute of Mathematical Sciences (ICMAT-CSIC), DataLab, 28049 Madrid, Spain
2. Centro de Investigación en Mecatrónica y Sistemas Interactivos—MIST, Universidad Indoamérica, Machala y Sabanilla, Quito 170103, Ecuador
3. Institute of Knowledge Technology, Complutense University of Madrid, 28040 Madrid, Spain
* Correspondence: cesar.guevara@icmat.es or cesarguevara@uti.edu.ec

Abstract: With the aim of improving security in cities and reducing the number of crimes, this research proposes an algorithm that combines artificial intelligence (AI) and machine learning (ML) techniques to generate police patrol routes. Real data on crimes reported in Quito City, Ecuador, during 2017 are used. The algorithm, which consists of four stages, combines spatial and temporal information. First, crimes are grouped around the points with the highest concentration of felonies, and future hotspots are predicted. Then, the probability of crimes committed in any of those areas at a time slot is studied. This information is combined with the spatial way-points to obtain real surveillance routes through a fuzzy decision system, that considers distance and time (computed with the OpenStreetMap API), and probability. Computing time has been analized and routes have been compared with those proposed by an expert. The results prove that using spatial–temporal information allows the design of patrolling routes in an effective way and thus, improves citizen security and decreases spending on police resources.

Keywords: security; crime prediction; police patrol routes; machine learning; artificial intelligence

MSC: 68Q25

Citation: Guevara, C.; Santos, M. Smart Patrolling Based on Spatial-Temporal Information Using Machine Learning. *Mathematics* **2022**, *10*, 4368. https://doi.org/10.3390/math10224368

Academic Editors: Jose Antonio Sáez Muñoz and José Luis Romero Béjar

Received: 3 October 2022
Accepted: 15 November 2022
Published: 20 November 2022

Publisher's Note: MDPI stays neutral with regard to jurisdictional claims in published maps and institutional affiliations.

Copyright: © 2022 by the authors. Licensee MDPI, Basel, Switzerland. This article is an open access article distributed under the terms and conditions of the Creative Commons Attribution (CC BY) license (https://creativecommons.org/licenses/by/4.0/).

1. Introduction

Citizen insecurity is a serious problem in all countries, especially in Latin America and the Caribbean, which have higher rates of criminal events, great insecurity, and a profound deterioration of trust in citizen surveillance institutions, such as the police and the army. An example of this situation is the prevalence of endemic levels of violence, where homicide rates have been growing steadily over time. The rate of homicides has remarkably increased from 25.9% in 2017 to 30% in 2020 in Central America and from 24.2% in 2017 to 29% in 2020 in South America [1].

Criminal events increase due to unemployment, social and economic problems, among other reasons. Ecuador is one of the countries with a rising crime rate, even though the National Police carries out a continuous surveillance of sectors with high crime incidences to reduce this problem. The homicide rate in Ecuador was 5.78% in 2017 and 7.78% in 2020, which represents a relevant increase [2].

With the motivation of improving citizen security, this study proposes an algorithm for forecasting locations with a high probability of crime. This prediction is then used for the generation of patrol routes in a given area. A database with spatial–temporal information of the crimes that occurred in Quito City, Ecuador, from January to December 2017 is used.

The patrol route planning is a complex problem as it involves location, hotspots, manpower, and scheduling of vehicles and human resources. To deal with it, the application of several computational methods to develop efficient algorithms is required. The combination of various artificial intelligence (AI) and machine learning (ML) techniques has proven effective in decision-making applications in diverse settings [3,4]. That is the strategy proposed in this research, that has provided succesful results in this specific application.

The hybrid system here designed combines some intelligent data processing techniques for decision making. First, geographic data from the most conflictive areas regarding crimes are grouped using the k-means clustering algorithm. Statistical methods are then used to identify critical points in each region defined by a cluster, and linear regression functions are used for the prediction of future crime points. Subsequently, an analysis of the probability of the occurrence of crimes is carried out according to geographical location and temporal information available. With this information, fuzzy logic is applied so to consider several criteria in the drawing of the patrol route. The algorithm has been tested on various real-world scenarios, with useful and satisfactory results. This strategy provides relevant information for the management of police resources, which allows better use of these means and the reduction of urban crimes.

The novelty of the proposal consists of showing how some well-known intelligent techniques are configured and combined to develop an efficient tool for patrol route generation. It has been proved in several fields that hybrid systems are necessary to address the complexity of real problems. In this work, the sequence of activities is key to design the route generation algorithm, together with the consideration of spacial-temporal information. The main contributions of this article can be then summarized as follows: (1) Geographic segmentation of crimes to obtain points with high crime rates (hotspots) in an automatic way. (2) Forecasting of future spatial crime locations based on historic information. (3) Design of optimal patrol routes based on crime probability considering spatio-temporal data. Fuzzy logic uses distance and travel time between surveillance way-points to minimize resources cost and response time emergencies.

The structure of the paper is as follows. In Section 2 some relevant related works on crime data analysis are discussed. The materials and methods used are described in Section 3. Sections 4 and 5 detail the flow and operation of the algorithm, as well as the application of ML and statistical techniques. In Section 6, simulation results and analysis of the data obtained are presented. The paper ends with the conclusions and future lines of research.

2. Related Works

The analysis and prediction of crimes have raised great interest because of their usefulness in improving citizen security. This topic has been approached from different points of view. Regression and AI techniques have been frequently applied although papers suing space–time information are not so common. To mention some examples, in the research developed by [5], the crimes forecasting is done by applying a deep neural network architecture, identifying evolutionary patterns of crimes and the relationship between space–time information. Crime points (geospatial) are then transformed into crime heat maps for the same sector of a city and time. Subsequently, convolutional hierarchical structures are applied to train a crime prediction model with heat maps. Esquivel [6] applied convolutional neural networks and a short- and a long-term memory network for crime prediction in the city of Baltimore, USA. It uses spatial and temporal correlations of historical crime data for future predictions. An inspirational paper was the one by Farjami [7], who proposed a genetic-fuzzy algorithm for spatial-temporal crime prediction. First, available information is processed, which consists of the geographical location (latitude, longitude) and the time of crimes. Then, a forecasting model is built to determine the time and place where future crimes will occur. Finally, the model is evaluated with simulated data from the city of Tehran, Iran. Hu [8] applied Bayesian spatial–temporal modeling for urban crime and analyzed its trend in the city of Wuhan, China. It uses socio-economic and population variables such as people agglomeration places, unemployment, tourist and residential places, and so on. Vural [9] used the Naïve Bayes theory to identify an offender with the highest probability of executing a criminal act based on the deliquent's history. The information conveyed is the date, geographical location, type, and some crime data. The proposed model works with a georeferenced information system that visualizes various characteristics of crimes and identifies patterns on a defined territory. Win [10] proposed a fuzzy grouping algorithm of criminal activities to detect patterns of criminal behavior. This algorithm predicts hotspots by using geospatial information in various cities in Iraq, Pakistan, Afghanistan, and India.

Regression is also commonly applied in this area. An interesting related paper is the one by Catlett [11], who applied auto regressive models on spatial–temporal information to automatically identify high-risk crime regions in highly populated urban areas. A spatial grouping of the dataset is carried out to detect these regions of high crime density. Finally, an integrated moving average auto-regressive prediction model is generated to forecast the reliably forecast crime trends in each region. Kadar [12] used space–time, socio-economic, meteorological, and temporal information from a real environment in the city of Aargau, Switzerland. For the predictive model, logistic regression with regularization, bagging (random forests), and boosting (AdaBoost) were applied. Cowen [13] analyzed the relationship between rates of theft and assault crimes. They combined ordinary least square regression models and statistical analysis of geospatial data patterns in different time periods in Miami-Dade, Florida, between 2007 and 2015. Piza [14] carried out a spatio-temporal analysis of residential thefts, automobiles, and other motor vehicles in Indianapolis City. Multimodal linear regression models were applied to predict crimes and their relationship with future events (search for the initial event in a chain of criminal events).

The Cokriging algorithm, a generalized form of multivariate linear regression model, has also been used in some recent works, such as that by [15]. This article analyzes historical crime data in urban areas of Cincinnati City, Ohio. The information is structured in time series, with time information being the main variable and urban areas as a secondary variable. The results show an increase in the correlation between urban areas and reported crimes. This space–time Cokriging prediction model was also used by [16], with historical crime movement data in Zigong, China. The temporary models are generated weekly, biweekly, and quarterly, with geospatial information of the offender, obtaining improved results in short periods. In [17] authors combine a logistic regression and a neural network to predict three crime categories in a certain spatial region in the city of Amsterdam, Netherlands. Monthly predictions are made in two periods of the day: day and night. The results show that the monthly predictions give better results than the biweekly ones.

Regarding some of the techniques applied in our article, Hu [18] used the kernel density algorithm with space–time data for the prediction of hotspots of residential robberies in Baton City, Louisiana, USA. A cross-validation threshold and statistical tests were used to identify false positives and negatives. Fuentes-Santos [19] also applied kernel density for the analysis of spatial–temporal patterns of shots in Rio de Janeiro City, Brazil. They applied first- and second-order non-parametric inference tools to the reported events and compared them with crime prediction hotspot models, identifying chronic critical points. Ristea [20] detailed the distribution and spatial correlation between the historical records of reported crimes and their geographical location, socio-economic and environmental variables, and messages published on social networks from Chicago City, Illinois, USA. The most suitable variables for the study are selected, and the kernel density method is applied for crime prediction with a linear regression.

Other ML algorithms have been used to obtain predictive models. Umair [21] analyzed social networks for crime prediction. Specifically, language recognition has been used to predict hotspots. Random forest and k-nearest neighbor were applied in this paper with good results.

Hajela [22] analyzed criminal events recorded in New York to generate a spatial–temporal predictive model. The prediction aimed to identify hotspots in delimited geographic sectors by applying k-means clustering. Another contribution in the prediction of hotspots is the research of Lee [23], wherein the proposed model uses criminal information from the cities of Portland and Cincinnati. The algorithm applies population heterogeneity to identify hotspot locations. Subsequently, a dependency model is applied in historical periods divided by months to efficiently determine points of high crime rates. In [24], it is proposed the use of statistical metrics and a geospatial grid of crime data for crime prediction in Portland City, USA.

On the other hand, route generation and path planning are strategies developed and applied in very different fields. Although it is possible to find many scientific papers that deals with this topic, mainly on any type of autonomous vehicles and mobile robots, it is not so common to find them regarding people routes. Concerning the first ones, in [25], a survey of the existing approaches for trajectory planning for Autonomous Vehicles (AVs) can be found. In [26], an offline route planning method and online navigation of AVs with reinforcement learning are analyzed. This proposal obtained encouraging results by building different routes based on some initial criteria. A survey on vehicle routing problems with time windows using meta-heuristic algorithms can be found in [27]. According to the authors, the most common methods applied to autonomous vehicles are Artificial Bee Colony algorithm (ABC), Ant Colony Optimization (ACO), Particle Swarm Optimization (PSO), among others. The paper by [28] presents a survey on some bio-inspired algorithms applied to robot route planning. The most widely used techniques are described, such as ant colony, evolutionary strategies and genetic algorithms, swarm algorithms, etc. Interestingly, the paper concludes that most of those bio-inspired algorithms do not give optimal results in real-time route planning problems, as it takes them a long time to generate an optimal route. Similarly, route generation algorithms are applied for machining processes and transport in [29]. The proposal by [30] details a method of planning tourist urban routes applying multi-objective genetic algorithms. This work improves the accuracy by combining internal and external tourist hot spots to optimize the route. The data used for this study have been obtained from a geographic information system (GIS) to generate a road network for the city of Chengdu, China.

But papers on patrol routes are very few as they deal with a complex process, as claimed by [31]. Despite its importance, the literature has not thoroughly studied patrol routing although patrolling is essential to handle insufficient police resources and reduce crime time response. Some recent papers have focused on well-known standard routing but they do not consider the crime data distribution in the spatio-temporal frame [32]. They only address car route patrolling following a defined pattern. Nevertheless, the exciting review paper by [31] analyzes different methods to define an efficient police patrol route. This survey describes many studies about the dynamic vehicle routing problem of the police to alleviate the detected knowledge gap on articles referring to policing.

In this survey, some hybrid methods to generate patrol routes, such as Genetic Algorithm (GA) and linear programming, are detailed. These hybrid models are more efficient in the local search and in the police patrol route problem.

In [33], a research on route optimization for community patrol is presented. This study develops a simulation multi-agent model with genetic algorithms, directed graph model, and GIS map. The GIS allows visualizing the environment of the patrol inspection geographic area but it does not allow representing the route information. Another related paper is [32], which develops a mathematical model to improve the planning of route patrolling and speeds up the time response to possible accidents of police vehicles. This model also minimizes the cost of vehicle resources. A hybrid solution approach that integrates genetic algorithms and continuous approximation (CA) is applied. In this case, the hotspots and patrol routes are represented in a graph with information about maximum response time. In a previous conference paper, [34] used the same crime database as in this paper to propose a clustering algorithm to identify the hotpots of high crime rate concentration and that way, to predict the future crime points.

Finally, the model proposed by [35] describes a visual based classification of crime activities in a street-level environment with the goal of identifying high and low crime areas. The model uses semantic categories such as roads, buildings, and others elements extracted from images of a GIS system. They use deep learning to image segmentation. The study was applied to two cities in the USA with high accuracy results, between 95% and 98%.

As it is possible to see, the patrol routing is a complex problem that usually requires merging different techniques to cover all the steps of this important task. Table 1 presents a summary of works (last five years) on crime prediction and patrolling routes generation. It shows the criminal event datasets, models and methodologies used. The nomenclature of the columns is as follows: C (city), S (spatial information), T (temporal information), LR (linear regression), Cl (clustering), NN (neural networks), FL (fuzzy logic), RF (random forest), St (statistical methods), KDE (kernel density estimation), RL (Reinforcement Learning), GA (genetic algorithms). The common objective of these works is to identify geospatial crime concentration points and thus, to develop strategies to improve security. We want to highlight that only few papers use spatial–temporal information. They commonly apply several AI and ML techniques for the analysis, grouping, and prediction of criminal events. In the summary shown in Table 1, the mark "✓" means the methodolgy applied; otherwise the symbol "-" is used to indicate that this specific technique was not used.

Table 1. Articles on crime prediction classified according to data information and applied techniques.

Articles	C	S	T	LR	Cl	NN	FL	RF	St	KDE	RL	GA	AV
[5]	New York USA	✓	✓	-	-	✓	-	-	-	-	-	-	-
[7]	Teheran Iran	✓	✓	-	-	-	✓	-	-	-	-	-	-
[11]	Chicago, New York USA	✓	✓	✓	-	-	-	-	✓	✓	-	-	-
[12]	Aargau Swiss	✓	✓	✓	-	-	-	✓	-	-	-	-	-
[17]	Amsterdam Netherland	✓	✓	✓	-	✓	-	-	-	-	-	-	-
[8]	Wuhan China	✓	✓	-	-	-	-	-	-	✓	-	-	-
[18]	Baton, Luisiana USA	✓	✓	-	-	-	-	-	✓	✓	-	-	-
[6]	Baltimore USA	✓	✓	-	-	-	-	-	-	✓	-	-	-
[9]	Austin, Atlanta USA	✓	✓	-	✓	-	-	-	-	✓	-	-	-
[13]	Miami USA	✓	✓	✓	-	-	-	-	-	✓	-	-	-
[19]	Rio de Janeiro Brazil	✓	✓	-	-	-	-	-	-	✓	-	-	-
[15]	Cincinnati, Ohio USA	✓	✓	-	-	-	-	-	-	✓	-	-	-
[21]	Pakistan	✓	✓	-	✓	-	-	✓	-	-	-	-	-
[14]	Indianapolis USA	✓	✓	✓	-	-	-	-	-	-	-	-	-
[16]	China	✓	✓	-	-	-	-	-	✓	-	-	-	-

Table 1. Cont.

Articles	C	S	T	LR	Cl	NN	FL	RF	St	KDE	RL	GA	AV
[23]	Portland, Cincinnati USA	✓	✓	-	-	-	-	-	✓	-	-	-	-
[20]	Chicago, Illinois USA	✓	✓	✓	-	-	-	-	✓	✓	-	-	-
[24]	Portland USA	✓	✓	-	-	-	-	-	✓	-	-	-	-
[10]	Irak, Pakistan, Afganistan, India	✓	✓	-	✓	-	✓	-	✓	-	-	-	-
[22]	New York USA	✓	✓	-	✓	-	-	-	-	-	-	-	-
[25]	-	-	-	-	-	-	-	-	-	-	-	-	✓
[28]	-	-	-	-	-	-	-	-	-	-	-	✓	✓
[26]	-	-	-	-	-	-	-	-	-	-	✓	-	✓
[27]	-	-	-	-	-	-	-	-	-	-	-	✓	✓
[30]	Chengdu China	-	-	-	-	-	-	-	-	-	-	✓	-
[33]	-	-	-	-	-	-	-	-	-	-	-	✓	-
[32]	-	-	-	-	-	-	-	-	-	-	-	✓	✓
[3]	-	-	-	✓	✓	-	-	-	-	-	-	-	-
[31]	-	-	-	✓	-	-	-	-	-	-	-	✓	✓
[35]	USA	-	-	-	-	✓	-	-	-	-	-	-	-

The main differences with the research here mentioned and our work can be summarized as follows. First, the objective of this article is to design optimal surveillance routes, not only the spatio-temporal prediction of crimes. Another significant contribution is that it determines the temporal order of the patrol way points, based on the temporal probability of the crimes, distance, and time with a real API. Finally, several AI and ML techniques are combined, specifically regression, kernel density, clustering, and fuzzy logic, to cover all the steps of the route generation process.

3. Materials and Methods

3.1. Materials: Dataset Description

The dataset used is a compilation of criminal information from the National Police of Ecuador, from January to December 2017. This information contains spatial–temporal data and other characteristics of the criminal events in each territory. This study focuses on Pichincha Province because of its high crime rate, with 17,365 crimes in 2017. Each record has nine attributes: four with temporal information (year, month, day, hour), three with spatial information (code subcircuit, latitude, longitude), and two with the mode (M) and type of crime (R).

The types of crime range from R1 (household robbery), R2 (motorcycle robbery), R3 (people robbery), to R6. They have been carried out with different modes, from M1 to M12 (M1, assault; M2, pickpockets; M3, false officials, and so on.).

In the database, the territorial division in each of the provinces is described by the variable *code − subcir*. The National Planning and Development Secretariat (Senplades) established the following territorial planning levels: provinces or zones, Z_n; districts, D_t; circuits, C_r; and subcircuits, S_c. Code Z_{17}-D_{07}-C_{03}-S_{01} is a territorial coding example.

3.2. Methods

3.2.1. Feature Selection

The selection of variables is one of the most important tasks in data processing and is crucial to obtain good models [36]. Two well-known feature selection (FS) techniques have been applied in this work, namely, Relief and Information gain ratio.

The FS Relief method is based on the estimation of the attributes [37]. It assigns a relevance grade to each data set feature. The features valued over a user-given threshold are selected. Relief-F generalizes the behavior of Relief to classification. It finds one nearest neighbor of every class. With these neighbors, Relief-F estimates the relevance of every feature $f \in F$, stored in vector $W[f]$. In this process, a hit H is the nearest neighbor from the same class C. The $diff$ function in (1) calculates the difference of the value of features between two instances. On the contrary, $M(C)$ is a different class from the same class C. Finally, $W[f]/m$ is calculated as the average in the interval $[-1, 1]$.

$$W[f] = W[f] - diff(f, E_1, H) + \sum_{(C \neq class(E_1))} P(C) \times diff(f, E_1, M(C)) \quad (1)$$

The information gain ratio method calculates the ratio between intrinsic information and information gain. Then, it decreases the information gain of a feature when the number of branching features is high. That is, the gain ratio takes into consideration the size and the number of branches to select a feature. This is a way to diminishes bias with multi-valued attributes when selecting a specific feature [38].

The information gain calculates the entropy of any attribute. When an only classification can be obtained for the obtained attribute, the relative entropies subtracted from the total entropy are then 0 [39].

3.2.2. Techniques Used in the Algorithm

Kernel Density Estimation (KDE)

The Kernel density estimation (KDE) is a handy statistical tool and a non-parametric form to estimate the probability density function of a dataset. The current way to implement it is an adaptive two-stage approach. This method works on the building of a local bandwidth factor, defined as λ_i, at each sample data. The local bandwidth factors have a unit mean that multiplies a global fixed bandwidth, defined as h. For this reason, h regulates the overall degree of smoothing, whereas λ_i extends or reduces sample data bandwidths to adapt to the density of the data [8]. The adaptive KDE is presented in Equation (2).

$$\hat{f}_h(x) = \frac{1}{\sum_{i=1}^n w_i} \sum_{i=1}^n \frac{w_i}{h_i} K(x - x_i/h_i) \quad (2)$$

where x_i are the data points associated to weights w_i, and K is the kernel function, and $h_i = h\lambda_i$. The local bandwidth factors are defined as $\lambda_i = \lambda(x_i) = \{G/\hat{f}(x_i)\}^{0.5}$, which are proportional to the square root of the underlying density functions at the sample data, where G is the geometric mean over all i of the density estimate $\hat{f}(x)$. The density estimate is a typical fixed bandwidth kernel density estimate acquired with h as bandwidth.

K-means clustering

K-means is an algorithm that generates groups in a datasetbased based on similar characteristics. Suppose a dataset of n data points $X_1, X_2, X_3, \ldots, X_n$, where X_i is in R^d, to group the data into k clusters is necessary to find k points $m_j (j = 1, 2, 3, \ldots, k)$ in R^d, such that

$$\frac{1}{n} \sum_{i=1}^n [min_j d^2(x_i, m_j)] \quad (3)$$

is reduced, where $d(x_i, m_j)$ is the Euclidean distance between x_i and m_j. The points m_j ($j = 1, 2, \ldots, k$) are the centroids of the clusters [40]. The key in (3) is to identify k cluster centroids so the distance between a data point and its nearest cluster centroid is reduced.

The K-means algorithm is considered a gradient descent iterative procedure that updates the centroids using an objective function (3). This algorithm converges to a local minimum [41].

Elbow method (Number of Clusters Optimization)

The elbow method is a heuristic procedure that determines an optimal number of clusters in a dataset. This algorithm uses the percentage of variance defined as a function of the number of subgroups (clusters). The first clusters will add much information to the analysis, but the marginal gain will be reduced while more clusters are added. The correct number of groups k is chosen at this point, that will give an angle in the graph of variance vs. number of clusters, i.e., an "elbow" [3,42].

Fuzzy logic

Fuzzy set theory is an artificial intelligence technique based on multivalued logic. It allows an element to have a (partial) membership to one or more sets, whereas classical set theory specifies that set membership is unique and crisp. Formaly, being S a universe of data with $x(S = x)$, the membership function of any element of S is $0 \leq f_i(x) \leq 1$, that is, a partial membership to the ith set [43,44].

Linear Regression

Linear regression is a statistical model that approximate the relationship between a dependent variable Y with m independent variables, X_i, where $m \in Z^+$ and ϵ is a independent term. This model is defined as shown in Equation (4).

$$Y = \beta_0 + \beta_1 X_1 + \ldots + \beta_m X_m + \epsilon \qquad (4)$$

where terms $\beta_0, \beta_1, \beta_2, \ldots, \beta_m$ are parameters of the model that measure the influence of the explanatory variables and are regression coefficients [45].

The previously mentioned techniques have been used for the following purposes:

- K-means clustering, using the Euclidean distance, has been used to spatial grouping the crimes of a circuit. To determine the optimal spatial distribution of crime data and obtain the optimal number of clusters we used the Elbow method.
- Kernel density estimation (KDE) has been used to estimate the probability density function of a random variable. Specifically, KDE has been applied to obtain the main point of concentrate crimes (hot spots) in each cluster.
- Linear Regression is used to fit a mathematical model to a crime dataset to predict a future crime point in each cluster.
- Fuzzy logic has been used to implement the making-decision system to determine the optimal route of patrolling based on the route crime probability, time, and distance.

4. Data Analysis and Feature Selection

A spatial–temporal data analysis of the criminal database is carried out to identify the most appropriate period (monthly, quarterly, semi-annual, or annual) and the geographic sector (district, circuit, or sub-circuit) where to focus the study. Furthermore, the most relevant characteristics of the database are identified, which will serve as a basis for the development of the crime prediction algorithm and subsequent patrolling (Figure 1).

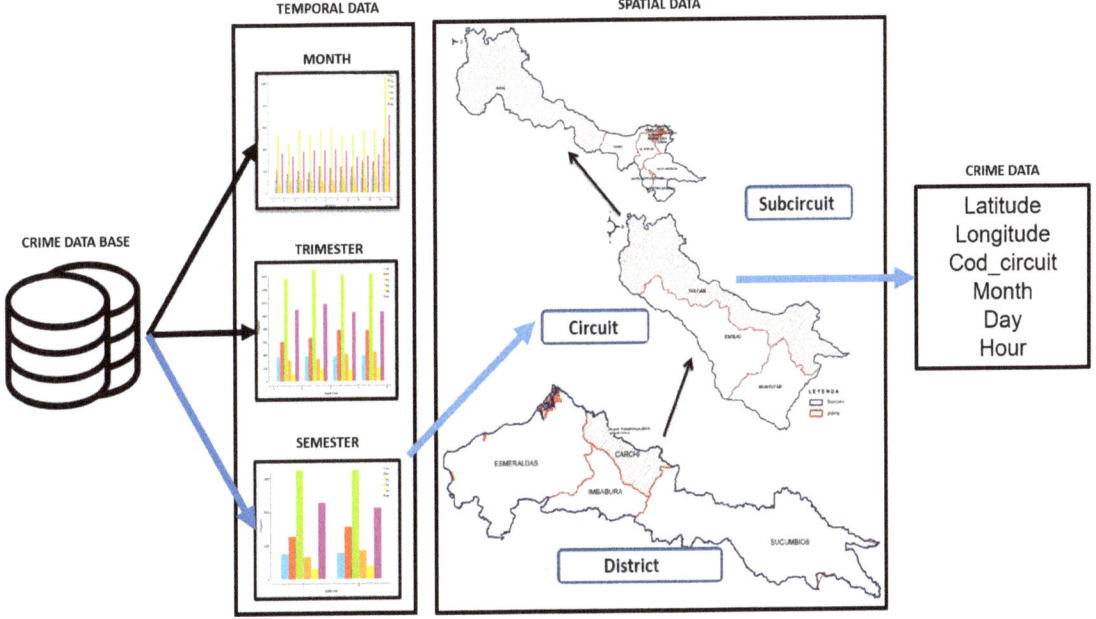

Figure 1. The analysis process of a Crime Database.

4.1. Temporal Analysis

Figure 2 shows the monthly, quarterly, and semi-annual frequency of crimes in 2017 for the different types of crimes (R1–R6). In the monthly distribution, the average per month is 1447 records, with a minimum of 32 records of each crime type. In Figure 2, the quarterly distribution is shown, with an average of 4471 crimes per month and a minimum of 107 records of each type of crime. Finally, in Figure 2, the mean by semester is 8682 records, with a minimum of 292 records by crime type. Based on these data, we have decided that the most appropriate period to identify criminal activity is 6 months. Thus, the necessary number of records is available for each type of crime. The distribution of data between 1, 3, and 6 months maintains the same proportion regarding the number of crimes. Therefore, 6 months was chosen to obtain enough information.

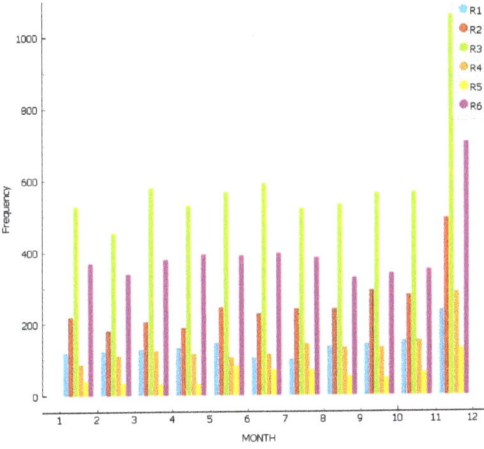

Figure 2. *Cont.*

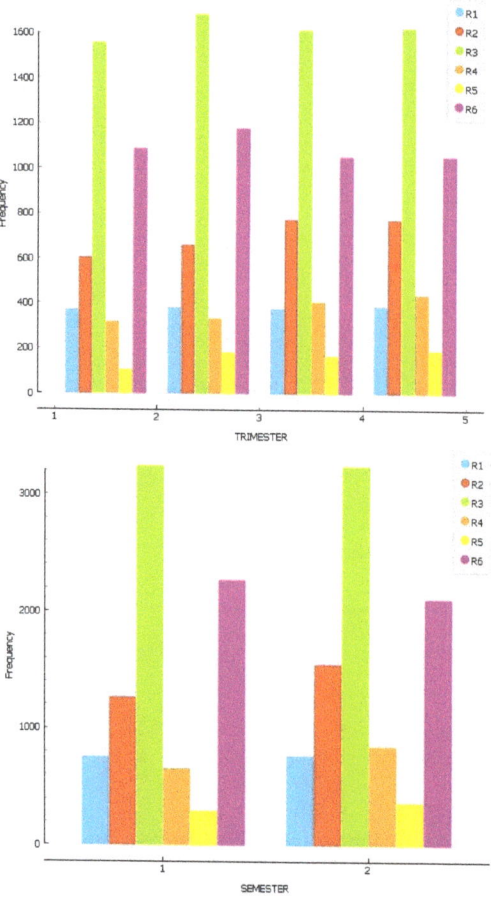

Figure 2. Data distribution by crime type in the periods: monthly, quarterly, and semi-annual.

4.2. Spatial Analysis

The geographical distribution of crime records in districts, circuits, and sub-circuits in Pichincha Province, Ecuador (Z_{17}) is also analyzed. In Figure 3, the frequency of crimes is shown according to crime type by district. Clearly, the D_{05} district has the highest number of crimes (with a maximum of 6,058). Among the crimes, those of type R3 (40.48%) and R6 (32.30%) stand out. Figure 3 represents the number of crimes according to type per circuit. In this case, circuits C_2 and C_{10} have the highest number of crimes. The most common are again R3 (45.74%) and R6 (32.94%).

From this study, district D_{05} is selected, given that it is the one with the highest number of reported crimes. Working with subcircuits is ruled out because the information is not enough to train the models. The selected circuit within the D_{05} district is C_{10}.

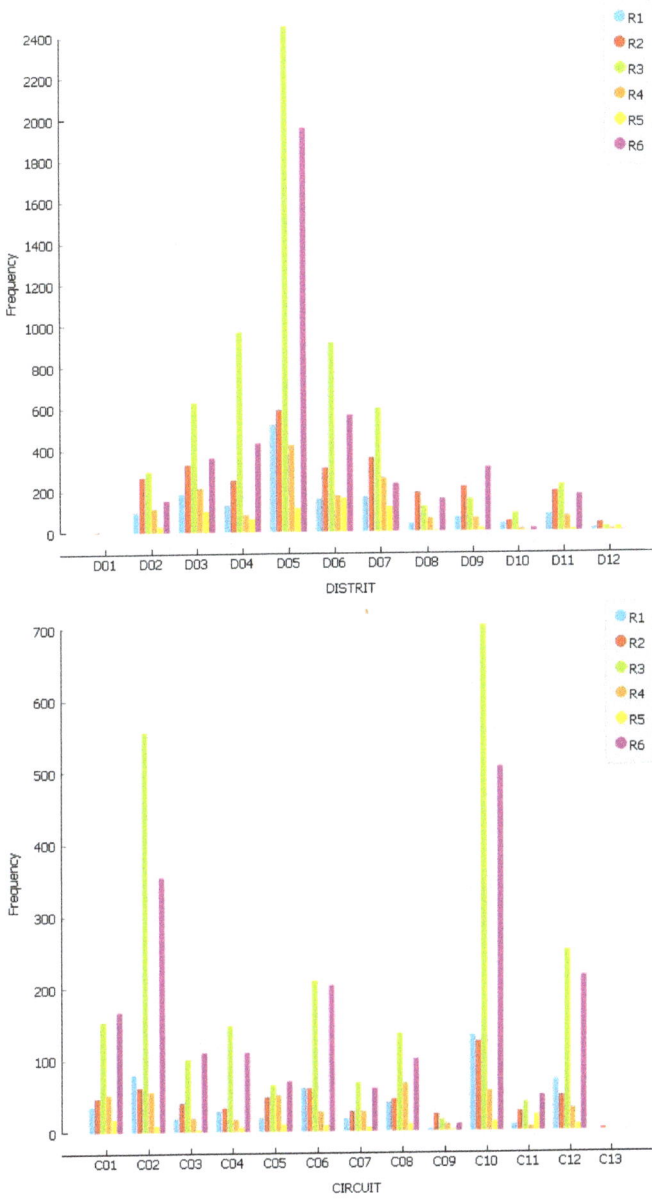

Figure 3. Frequency of Crime Types by District and Circuit.

4.3. Feature Selection

For the selection of the most relevant variables, the two previously mentioned techniques were applied, Relief (Equation (1)) and Information gain ratio.

According to the results shown in Figure 4, Year and $code - subcir$ characteristics are not enough relevant to be taken into account. Given that only the data from 2017 will be used and the circuit will be the smallest spatial unit for the study this result is reasonable. On the contray, the most important characteristics seem to be the temporal ones: *month*, *day*, and *hour*. The geospatial features, *latitude*, and *longitude*, also scored high in the selection. By contrast, crime and mode, which represent the type and manner of carrying out the crime respectively, obtained a medium–low relevance value, so they will not be considered; that is, any type of crime is important for police surveillance.

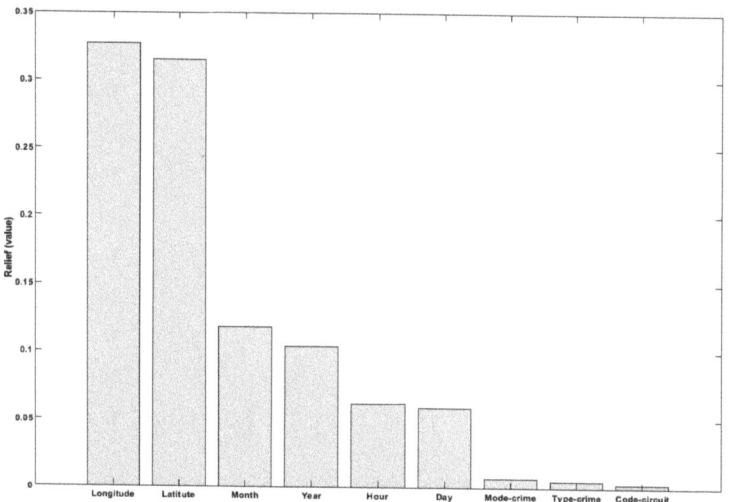

Figure 4. Feature selection using the Relief method.

Therefore, for this analysis, the variables month, mt; day, dy; hour, hr; latitude, lat; and longitude, $long$, are selected.

5. Crime Prediction Smart Patrol Algorithm (CPSPA)

The crime prediction smart patrol algorithm (CPSPA) uses the characteristics selected in the previous section. It works with circuits, C_n, where n is the number of the circuit. Each C_n circuit has a set of crimes, $de_{C_n} = \{de_1, de_2, de_3, \ldots, de_m\}$, where m is the number of registered crimes. Each offense is defined as $de_i = mt, dy, hr, lat, long$, where mt, dy, hr are the temporal information, and $lat, long$ are the spatial information.

The CPSPA algorithm is structured in five phases (Figure 5). Along them, different ML techniques and statistical analysis are applied depending on the objective, to finally determine the prediction of possible crimes and propose the optimal route that a police officer should take during a surveillance turn.

To clarify the algorithm, although the different steps are going to be described in detaile in the next sections, Figure 6 shows the activity UML diagrams. It shows the interaction among the Police Officer Device, the CPSPA algorithm and the database. The sequence of actions carried out to patrol the most conflictive points for each surveillance shift is shown. This diagram has as inputs the circuit that is being patrolled, C_n, and day and time of the patrol. The output is the police patrol route obtained, that consists of a starting point, some way-points along the route, and the final point. This diagram includes all the phases of the CPSPA algorithm.

In addition, Figure 7 presents the sequence diagram of the operation of the CPSPA algorithm. It is possible to see how some of the activities can be carried out in parallel.

The selected initial conditions of the experiments, after data analysis (Section 3), are the circuits $C_{n=2}$, $C_{n=10}$ y $C_{n=10}$, all of them belonging to D_{05} district, due to the high number of reported crime records. The Matlab 2021a software and the Nvidia Geforce GTX 1050 GPU with frame buffer: 4 GB GDDR5, 7 Gbps memory speed have been used for the simulation. The phases of the algorithm are described below.

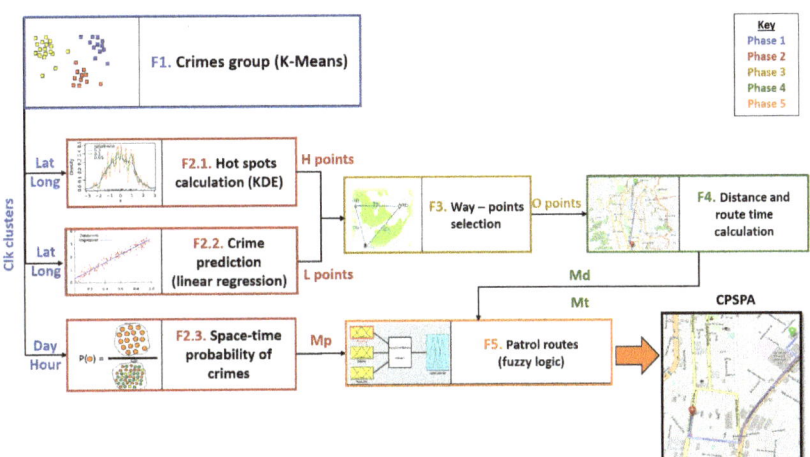

Figure 5. Diagram of Smart Patrol and Crime Prediction Algorithm.

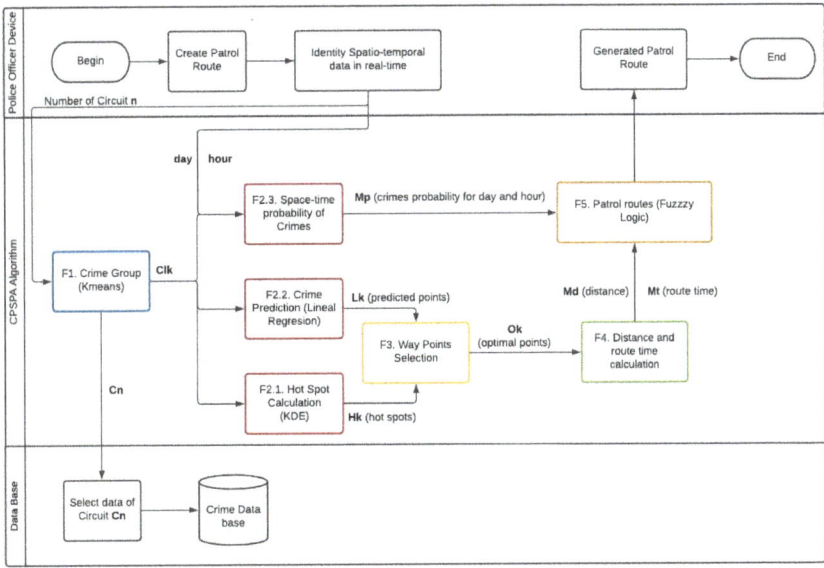

Figure 6. Activity diagram UML of Smart Patrol and Crime Prediction Algorithm.

5.1. Phase 1. Crime Grouping from Criminal Database

In this phase (see Figure 5), the crimes of a C_n circuit are grouped using the k-means algorithm. As it is an unsupervised clustering, the optimal number of clusters k is obtained with the Elbow method. The objective is to determine the most conflicting areas, Cl_k (clusters), in circuit C_n, to analyze crimes in more detail. A number of crimes r has been reported in the area that represents each cluster, that is, $de_{Cl_k} = \{de_1, de_2, de_3, \ldots, de_r\}$. As an example: within district D_{05} the circuit C_{10} is analyzed, which has 1539 crimes, de, reported in 2017. The k-means algorithm is applied with the Elbow method, and the optimal number of clusters is determined, $k = 14$ (Figure 8). In Figure 9, the k groups are represented for the C_{10} circuit in OpenStreetMap.

The k-means method has been applied because it is a well-know algorithtm that suits the objective of this phase, which is to determine crime concentration zones, including several geographic areas, as sub-circuits, although any other grouping algorithm may be used. Besides, the outcomes of its application gave similar results to those of the expert. Finally, k-means is simple, easy and quick to use and implement in an online real system.

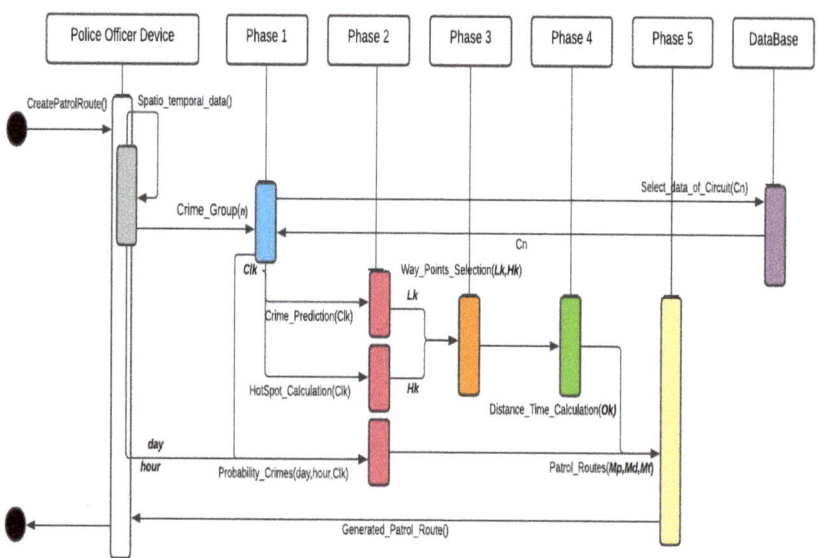

Figure 7. Sequence Diagram UML of Smart Patrol and Crime Prediction Algorithm.

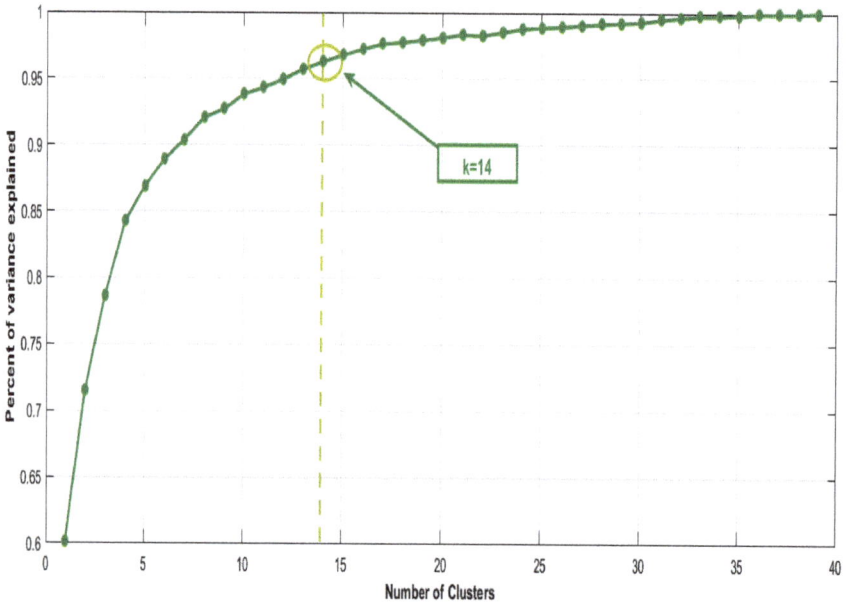

Figure 8. Application of k-means to C_{10} circuit with Elbow Method.

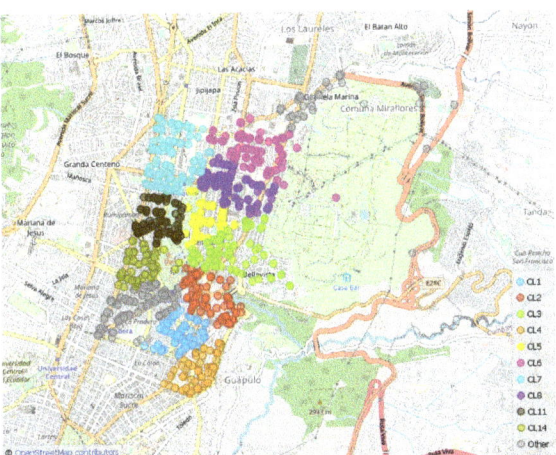

Figure 9. Application of k-means to C_{10} circuit with k Clusters.

5.2. Phase 2. Calculation of Hotspots, Prediction, and Probability of Crimes

At this stage, the spatial information, latitude, and longitude, is used to calculate hotspots (Phase 2.1, Section 5.2.1) and perform crime prediction in each cluster, (Phase 2.2, Section 5.2.2). With the temporal information, the spatial-temporal distribution of crimes is calculated (Phase 2.3, Section 5.2.3) (see Figure 5).

5.2.1. Phase 2.1. Calculation of Hotspots by Applying KDE

Hotspots are crime concentration points, defined as $H = \{H_1, H_2, H_3, \ldots, H_k\}$. That is, a hotspot is calculated for each cluster determined in Phase 1, Section 5.1. KDE Equation (3) is applied to the spatial data of each crime, $lat, long$, to obtain the hotspot vector H.

In the example, each of the 14 clusters in C_{05} circuit has several records (offenses), with a maximum of $r = 180$ in clusters Cl_7 and Cl_8, and a minimum of $r = 8$ in Cl_{12} (Figure 10). H points with the highest concentration (hotspots) of crimes that result from applying KDE are represented by the blue circles in Figure 11, one for each cluster, for the C_{05} circuit.

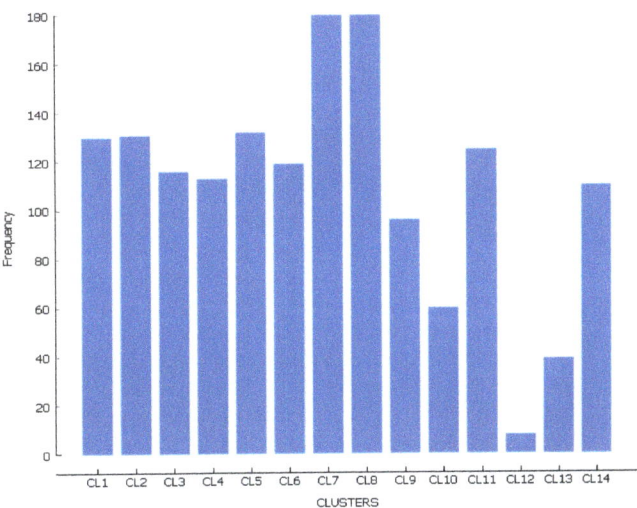

Figure 10. Crime Distribution by cluster.

The light blue circles in Figure 11 represent the position of the point with the highest crime concentration within each cluster.

Figure 11. Hotspot of each cluster Cl_k (blue circles).

5.2.2. Phase 2.2. Crime Point Prediction via Linear Regression

In Phase 2.2 (see Figure 5), linear regression is used to obtain models that can be used for crime prediction. To do so, Equation (4) is applied to the crime records of each cluster Cl_k. Thus, the L points where the concentration of crimes may be higher, $L = \{L_1, L_2, L_3, \ldots, L_k\}$, $k = 14$, are obtained.

The crimes were ordered chronologically, $de = \{mt, dy, hr\}$. Then, with the spatial information, two linear functions are generated for each cluster, that is, $f_{lat}(Cl_k = \{lat\})$ and $f_{long}(Cl_k = \{long\})$. That is, $L_k = \{lat, long\}$ predicts the future crime concentration point of the k cluster. Those two linear functions have been obtained for each of the 14 clusters Cl_k of the C_{05} circuit.

For instance, for cluster Cl_2 the equations obtained when applying the linear regression are: $f_{lat} = -3.187 \times 10^{-6}x - 0.1912$ for latitude and $f_{long} = -5.199 \times 10^{-6}x - 78.48$ for longitude. The R^2 obtained was 0.48 and the resulted regresion point was $L_k = (-0.194375911761479, -78.4791191630853)$. The R-squared values were between 0.40 and 0.53 for the 14 clusters.

5.2.3. Phase 2.3. Space-Time Probability of Crimes

In this last step of Phase 2 (see Figure 5), a crime probability matrix Mp is calculated for each Cl_k cluster for the days and hours of a patrol turn. The Mp matrix has dimensions (row, column), where $row = nd \times nh \times k$, $column = nc$, $nd =$ number of days, $nh =$ number of patrol hours, and $k =$ number of clusters. The four columns of Mp matrix represent the day, time, cluster, and probability of crimes; hence, $nc = 4$.

The Mp matrix is obtained by applying Algorithm 1. The variables and symbols used are as follows. The datasets are divided into three surveillance turns, as carried out by the Ecuadorian National Police for their patrols. Turn 1 covers the schedule from 00:00 to 7:59, turn 2 runs from 8:00 a.m. to 3:59 p.m., and turn 3 runs from 4:00 p.m. to 11:59 p.m.

The total amount of crime on a circuit is tde. Each offense is assigned to a day (dy), hour (hr), and cluster Cl_k. The crimes on a given day and time are $totCrCl$. The number of crimes per day, hour, and cluster is $repCr$. Equation (5) is applied to calculate the probability that a crime will take place that day, time, and in the spatial location of a given cluster.

$$P(de_j = (dy, hr, Cl_k)) = \frac{repCr}{totCrCl} \qquad (5)$$

That is, the quotient between all the crimes in a cluster and the total number of crimes in the circuit. For example, the probability of a crime on a Monday ($dy = 1$) at 6:00 p.m. ($hr = 18$) in cluster 3 ($Cl_{k=13}$) would be: $P(de = (dy = 1, hr = 18, Cl_{k=13})) = 5/24 = 0.2083$.

The Mp matrix has a dimension of 588 × 4. An example of a row is: $[dy_i, hr_j, Cl_k, P(de_j = (dy, hr, Cl_k))] = [2, 19, 3, 015]$.

5.3. Phase 3. Selection of Surveillance Points

In this Phase 3 (see Figure 5), the information obtained in previous subsections will be integrated to determine the way points that the patrol route must follow. First, hotspots H_k are obtained in each cluster, which may or may not coincide with the prediction points of maximum crime L_k. The distance between them is calculated so that if they are very distant from each other, a midpoint will be obtained. If they are very close to each other, the hotspot H_k will be selected.

Algorithm 1 Obtaining the Crime Probability Matrix Mp in a Cluster.

Inputs:
hr_{start} %turn start time%
hr_{end} %turn end time%
Output:
$Mp[]$ %Crime Probability Matrix%

    for $dy = 1$ to nd **do**
      for $hr = hr_{start}$ to hr_{end} **do**
        for $Cl_i, i \leftarrow 1$ to k **do**
          for $j \leftarrow 1$ to tde **do**
            if $(de_j(dy) ==$ day AND $de_j(hr) ==$ hour) **then**
              $totCrCl + 1$
            end if
            if $(de_j(dy) ==$ day AND $de_j(hr) ==$ hour AND $de_j(Cl) == i$) **then**
              $repCr + 1$
            end if
          end for
          $row + 1$;
          $Mp(row, column = 1) \leftarrow dy$;
          $Mp(row, column = 2) \leftarrow hr$;
          $Mp(row, column = 3) \leftarrow Cl$;
          $Mp(row, column = 4) \leftarrow repCr/(totCrCl)$;
        end for
      end for
    end for

The Euclidean distance between $H_k = \{lat, long\}$ and $L_k = \{lat, long\}$, is applied.

$$d_E(H_k, L_k) = \sqrt{(lat_H - lat_L)^2 + (long_H - long_L)^2} \qquad (6)$$

The distance must meet the following conditions to determine the waypoints:

$$O_k = \begin{cases} d_E, & d_E = 0 \longrightarrow O_k = L_k \\ d_E, & d_E > \epsilon \longrightarrow O_k = H_k \\ d_E, & \epsilon > d_E > 0 \longrightarrow O_k = P_{mean} = (\frac{lat_H - lat_L}{2}, \frac{long_H - long_L}{2}) \end{cases}$$

where ϵ is a constant (threshold) that limits the maximum distance between the hot spot H_k and the predicted point L_k, defined as $\epsilon = 0.01$ (1 km). With this expression, a vector is obtained with the way-points O_k, which the patrol route should go through.

In Figure 12, the points H_k(light blue circles) and L_k (x red markers) are shown. In Figure 13, the points O_k (gray circles) resulting from the selection of points for the routes of the C_{10} circuit are presented. Each cluster has one point.

As an example of this phase, in cluster Cl_1, the hotspot is H_1 ($lat = -0.196, long = -78.483$), and the prediction shows L_1 ($lat = -0.198, long = -78.484$), where $d_E = 0.0017 < \epsilon$. Hence, P_{mean} is calculated as follows:

$$P_{mean} = (\frac{(lat_{H_1} = -0.196) - (lat_{L_1} = -0.198)}{2}, (\frac{(long_{H_1} = -78.483) - (long_{L_1} = -78.484)}{2}).$$

The resulting way-point O_1 has the coordinates $(-0.196, -78.483)$.

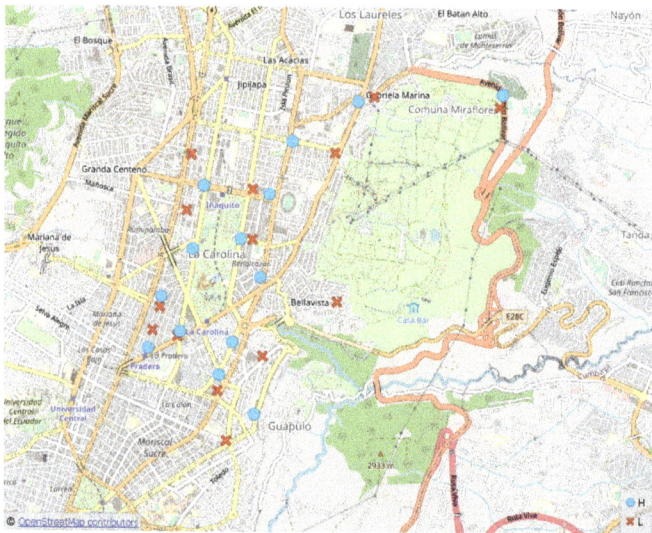

Figure 12. Points H_k (blue circles) and L_k (x red markers).

%vspace-6pt

5.4. Phase 4. Distances and Route Time Calculation with OpenStreetMap API

The distance and driving time between the way points must be determined to find the best patrol route. In a city, a distance may be small but it may take a long time to travel due to the conditions of the road or due to traffic at certain times. To obtain realistic values, the distance and travel time between the O_k way points are obtained with the API (Application Programming Interface) of the OpenStreetMap application. The result is represented in the square matrices $Md_{k,k}$ (distance in km) and $Mt_{k,k}$ (time in minutes) from the initial point O_i to the end point O_j.

The distanceAPI (distance of travel) function is applied to obtain the elements, $dApi_{i,j} = distanceAPI(O_i, O_j)$, of the matrix $Md_{k,k}$.

$$Md_{k,k} = \begin{bmatrix} dApi_{1,1} & dApi_{1,2} & ... & dApi_{1,j} \\ dApi_{1,2} & dApi_{2,2} & ... & dApi_{2,j} \\ ... & ... & ... & ... \\ dApi_{i,1} & dApi_{i,2} & ... & dApi_{i,j} \end{bmatrix}$$

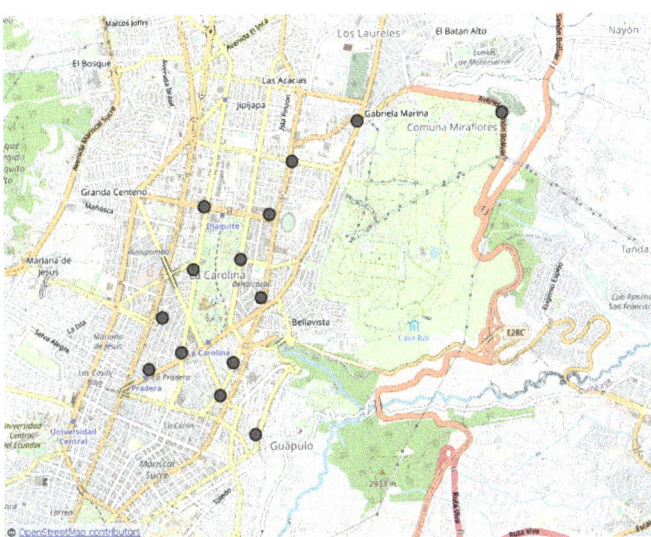

Figure 13. Way-points O_k (gray circles).

In our case, the matrix $Md_{14,14}$, has the results of the distance API(O_i, O_j) functions, where, for example, the distance between $O_1(-0.196, -78.483)$ and $O_2(-0.192, -78.482)$ is $dApi_{1,2} = 0.85$.

$$Md_{k,k} = \begin{bmatrix} dApi_{1,1} = 0 & dApi_{1,2} = 0.85 & dApi_{1,3} = 2.8 & \ldots & dApi_{1,14} = 1.8 \\ dApi_{2,1} = 1.4 & dApi_{2,2} = 0 & dApi_{2,3} = 1.9 & \ldots & dApi_{2,14} = 1.5 \\ dApi_{3,1} = 2.3 & dApi_{3,2} = 1.2 & dApi_{3,3} = 0 & \ldots & dApi_{3,14} = 1.7 \\ \ldots & \ldots & \ldots & \ldots & \ldots \\ dApi_{14,1} = 1.5 & dApi_{14,2} = 1.7 & dApi_{14,3} = 2.4 & \ldots & dApi_{14,14} = 0 \end{bmatrix}$$

In the same way, the OpenStreetMap timeAPI (travel time) function is used to calculate the route time, $tApi_{i,j} = timeAPI(O_i, O_j)$, which are the elements of matrix $Mt_{k,k}$.

$$Mt_{k,k} = \begin{bmatrix} tApi_{1,1} & tApi_{1,2} & \ldots & tApi_{1,j} \\ tApi_{1,2} & tApi_{2,2} & \ldots & tApi_{2,j} \\ \ldots & \ldots & \ldots & \ldots \\ tApi_{i,1} & tApi_{i,2} & \ldots & tApi_{i,j} \end{bmatrix}$$

For instance, in the matrix $Mt_{14,14}$, timeAPI(O_i,O_j) between the points $O_1(-0.196, -78.483)$ and $O_2(-0.192, -78.482)$ is $tApi_{1,2} = 4$.

$$Md_{k,k} = \begin{bmatrix} tApi_{1,1} = 0 & tApi_{1,2} = 4 & tApi_{1,3} = 9 & \ldots & tApi_{1,14} = 8 \\ tApi_{2,1} = 5 & tApi_{2,2} = 0 & tApi_{2,3} = 6 & \ldots & tApi_{2,14} = 6 \\ tApi_{3,1} = 11 & tApi_{3,2} = 8 & tApi_{3,3} = 0 & \ldots & tApi_{3,14} = 9 \\ \ldots & \ldots & \ldots & \ldots & \ldots \\ tApi_{14,1} = 5 & tApi_{14,2} = 7 & tApi_{14,3} = 10 & \ldots & tApi_{14,14} = 0 \end{bmatrix}$$

5.5. Phase 5. Application of Fuzzy Logic to Determine Patrol Routes

Finally, once the way-points O_k of route and the information on the probability of crimes and proximity, both in length (distance) and route time, have been determined, the route to be taken must be decided. A fuzzy decision-making system (FDSS) that uses probability matrices, Mp, distances, Md, and route times, Mt, has been designed.

Given that it is possible to obtain several solutions that can be equally valid, fuzzy logic is used because it allows the representation of variables including the uncertainty associated with route times, probability, and so on. The fuzzy decision-making system has three input variables, normalized to $[0, 1]$:

1. Route distance—Rd_k: distance from the initial point Pi to the final point Pf. This variable has been assigned three fuzzy sets with triangular membership functions: near (0 to 0.4), middle (0.1 to 0.9), and far (0.6 to 1.0). They correspond to the information in matrix Md, with $Rd_k = Md(Pi, Pf)$.
2. Route time—Rt_k: route time from the initial point Pi to the final point Pf. The same fuzzy sets have been assigned as for the previous variable. Travel times are found in matrix Mt, with $Rt_k = Mt(Pi, Pf)$.
3. Crime spot probability—$P(CS)_k$: the probability of crimes at the destination point Pf, on a given day dy and hour hr. Three fuzzy sets with triangular membership functions have been assigned to this variable: low (0 to 0.4), medium (0.1 to 0.9), and high (0.6 to 1.0). These probabilities are identified in matrix Mp.

The output variable is called point selection (PS), which determines if a way-point O_k is selected as the next way point of the route. It outputs two fuzzy singletons: selected and unselected.

The fuzzy decision-making system rules are, for instance: If $P(CS)$ is High and Rd is Middle and Rt is Near, then Point Selection is NOT SELECTED. The route implementation algorithm after applying the fuzzy decision system is shown in Algorithm 2.

Algorithm 2 Patrol route Implementation.

Inputs:
O_k % Way points
Mp % probability matrix
Md % distance matrix
Mt % route times matrix
   Output:
$rPoint$ % order of points of the patrol route

    for $i = 1$ to k **do**
      $Pi = O_i$;
      for $f = 1 \leftarrow k$ **do**
      $Pf = O_j$
      $Rd = Md(Pi, Pf)$;
      $Rt = Mt(Pi, Pf)$;
      $P(Cs) = Mp(P(de = (dy = dy_{start}, hr = hr_{start}, Cl_{k=f})))$;
      $rPoint = FDSS(P(Cs), Rd, Rt)$;
      for $rPoint ==$ SELECTED **then**
        Carry out patrol of O_i a O_j;
        $dy_{start} + Rt$;
      end if
    end if
end if

6. Results and Discussion

Numerous simulation experiments have been carried out with criminal records of the first (1729 reported crimes) and second (1498 reported crimes) semesters of 2017. The route algorithm is applied to circuits with a high incidence of crimes, specifically C_{02}, C_{06} y C_{10}. These simulations will be compared with those carried out by a citizen security expert from the national police of Ecuador.

The results of the application of the decision made for the patrol route for a specific case are shown in Figures 14 and 15 and Table 2, where the probabilities (Equation (5)), distances, route time, and time of the patrol between each of the points (blue points Figure 15) are listed. In this specific example, the patrol day is $dy = 6$ (Saturday), the start time is $hr_{start} = 18$, and the patrol starting point is $O_{k=1}$. The obtained patrol route is 30.6 km.

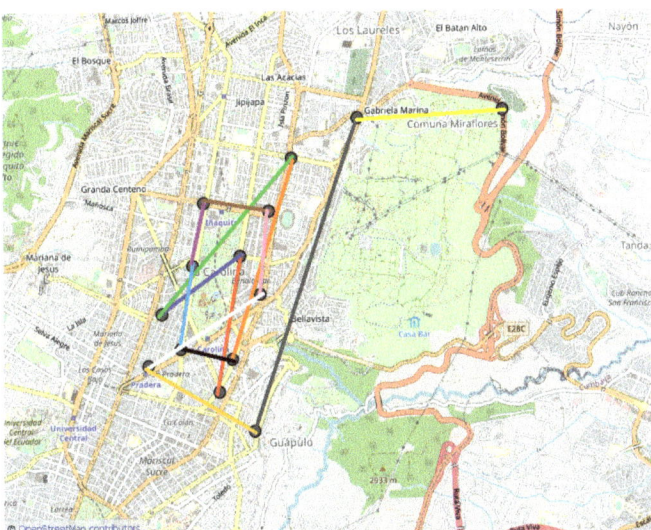

Figure 14. Way-points sequence obtain by the Fuzzy Decision-Making System (gray circles are way-points and color lines are the routes, Table 2).

The results obtained by the security expert (Table 3, Figures 16 and 17) showed lower probability rate for each generated route. The expert has estimated the probability of the routes using date, hour and a specific geographical area. The distance obtained by the expert was 21.65 km, smaller in comparison to the one obtained by the proposed algorithm, 30.6 km. The total time patrolling route was 175 min, that is, slower than the CPSPA algorithm, with 111 min. In general, the results show that the CPSPA algorithm works with the most relevant variables and finds valid routes that may help find new routes with some advantages.

In addition, as previously mentioned, the period of a semester (6 months) has been proved to have enough information of crimes to deal with, but still it may be not large enough for a more complete analysis of the data. Hence, to further test the algorithm, it has been applied to the second half of 2017 set of data. Table 4 shows data and results working with this new data set: the number of crimes broken down into patrol turns in the first column; the H_k hotspots of the circuits that are then calculated, and the L_i predictions obtained of where crimes will take place for each circuit (Phase 2.2). If the distance between them is smaller than 1 km, L_i will be set as way point of the route O_k.

$$Accuracy = \frac{deP}{(deP + deN)} 100\% \quad (7)$$

The results obtained regarding accuracy are quite similar to [7,35], proving that finding patterns and predicting future crimes with clustered crimes in space and time is possible. When the rest of the phases of the algorithm are applied, the way points of the patrol routes, O_k, are selected. The route distance, route time, and processing time of the proposed algorithm are then calculated (Table 4). For each circuit, between 8 and 15 crime concentration points have been identified, depending on the schedule in which the surveillance is carried out. This value depends on the result obtained with k-means and the Elbow method.

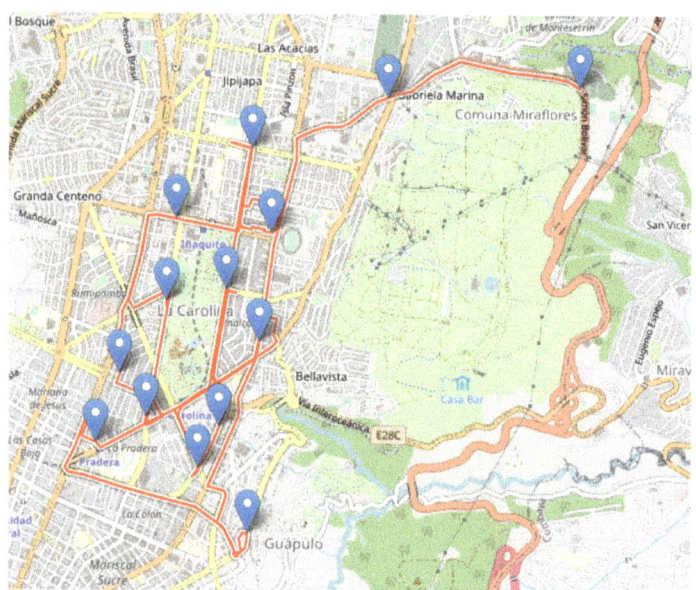

Figure 15. Resulting route of the algorithm. (blue icons are the way-points)

Table 2. Patrol Route Selection Data with CPSPA Algorithm.

Route	Color	Probability	Distance (km)	Time (min)	Hour
Route 1	Red	0.05	2.5	9	18:09
Route 2	Blue	0.14	2.3	12	18:21
Route 3	Green	0.06	2.0	12	18:33
Route 4	Orange	0.02	1.4	13	18:46
Route 5	Black	0.09	1.4	5	18:51
Route 6	Light blue	0.11	0.9	3	18:54
Route 7	Purple	0.07	1.8	4	18:58
Route 8	Brown	0.08	2.4	6	19:04
Route 9	Pink	0.03	2.7	9	19:13
Route 10	White	0.10	1.7	8	19:21
Route 11	Dark Yellow	0.04	2.8	9	19:32
Route 12	Gray	0.13	6.1	10	19:42
Route 13	Yellow	0.12	2.6	11	19:53
	Total		30.6	111	

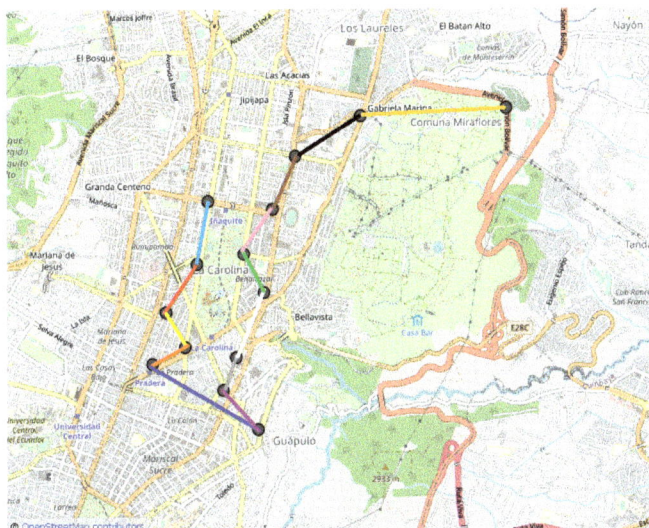

Figure 16. Way-points sequence obtain by Expert (gray circles are way-points and color lines are the routes, Table 3).

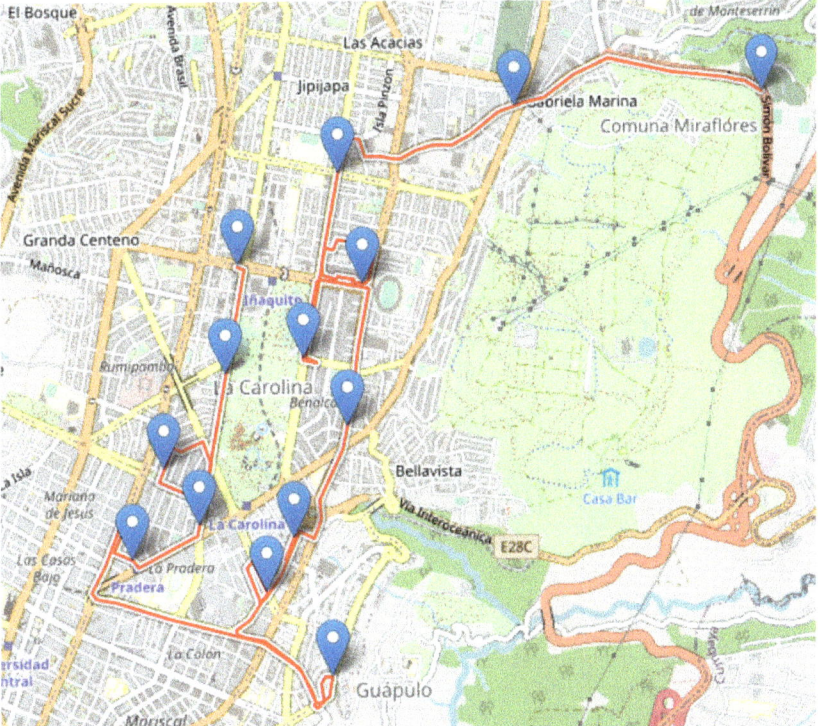

Figure 17. Resulting route by the Expert (blue icons are the way-points).

Table 3. Patrol route selection data from expert decision.

Route	Color	Probability	Distance (km)	Time (min)	Hour
Route 1	Light blue	0.0004	0.75	8	18h08'
Route 2	Red	0.0061	1.3	14	18h22'
Route 3	Yellow	0.0014	1.0	9	18h31'
Route 4	Orange	0.0026	0.6	3	18h34'
Route 5	Blue	0.0039	3.6	22	18h56'
Route 6	Purple	0.0018	2.3	12	19h08'
Route 7	Gray	0.0060	1.8	14	19h22'
Route 8	White	0.0009	1.5	16	19h38'
Route 9	Green	0.0079	2.2	20	19h58'
Route 10	Pink	0.0081	1.8	16	20h14'
Route 11	Brown	0.0086	1.0	9	20h25'
Route 12	Black	0.0093	1.7	12	20h37'
Route 13	Dark Yellow	0.0097	3.8	20	20h57'
	Total		21.65	175	

A direct relationship was found between the number of crimes reported and the number of surveillance points. The number of crimes in turns 1 and 2 is smaller than in turn 3.

As shown in Table 4, the time of the route in each circuit has an average of 140.77 min (2 h 34 min). As each turn is 8 h, the route takes 29.32% of the total shift time. Turns 2 and 3 require more time due to traffic. Furthermore, turn 3 has a higher number of crimes.

Table 4. Resultsin circuits C_{02}, C_{06}, y C_{10} with data from the second semester of 2017.

Datasets	Records	O_k	Distance (km)	Accuracy	Time (min)	Processing Time (s)
C_{02}- Turn 1	105	8	34.70	85.01	164	226.919
C_{02}- Turn 2	157	10	35.90	81.03	172	273.192
C_{02}- Turn 3	295	13	39.60	79.78	197	292.792
C_{06}- Turn 1	37	6	28.50	77.52	101	187.984
C_{06}- Turn 2	92	7	30.10	77.48	114	194.682
C_{06}- Turn 3	103	9	33.00	82.39	139	226.281
C_{10}- Turn 1	100	9	31.20	76.88	124	242.634
C_{10}- Turn 2	238	14	33.40	80.11	145	269.965
C_{10}- Turn 3	371	15	30.60	76.99	111	311.484
Total	1498	Average	33.00	79.69	140.77	247.32

The distance of the route has an average of 33.00 km, which is efficient for covering a geographical area of 36 km^2, with approximately 11,000 inhabitants. It depends on the number of way points of the route. The patrol distance increases for turn 2 and is even greater for turn 3.

For this reason, we initially analyzed the temporal data window to work with, as shown in Section 4.1, with the primary objective of determining an optimal period and the necessary historical crime data so that the efficiency of the proposed algorithm is high while time processing is not so demanding. It is worth it to note that adding more data is not relevant to the goal of the algorithm, it could improve the precision a bit but increasing computional time.

Indeed, the average processing time of the algorithm has been calculated, it is 247.32 s. Phases 4 and 5 require more computational time due to the execution of the OpenStreetMap API and the making decision system. The API depends on the speed of the internet connection and the processor of the local computer. The decision-making system uses many resources to generate optimal routes, both for points with high probability and for proximity. As expected, the larger the amount of data, the longer the algorithm processing time.

Therefore, this proposed algorithm provides appropriate routes for all selected waypoints in the order the fuzzy logic system determines. When compared with other routes designed by a security expert, results are similar in time and distance, though the proposed method tends to obtain quicker routes. The solutions have been projected on a map of the affected zones where the results have been verified, and the obtained routes are realistic. The analysis of results allows for testing the successful performance of this strategy.

7. Conclusions and Future Works

In this research, an algorithm for the detection of spatio-temporal sources of crime is proposed. The final goal is to design patrolling routes to optimize police resources, reduce crime time response and improve citizen security.

The algorithm consists of several phases and combines different artificial intelligent and ML techniques to deal with the available information. First, Relief and Information gain feature selection procedures are applied to obtain the relevant attributes. Hotspots with a high concentration of crime are identified by applying k-means clustering and KDE. A prediction of future crime points and the probability is obtained based on temporal information of the crimes. Spatial way points of routes are then calculated, and the real distance and travel time are computed with the OpenStreetMap API. Finally, a fuzzy inference system determines the order of the way-points in the route based on their probability, distance, and time.

The experimental evaluation was performed on a real crime dataset collected in Quito, Ecuador, in 2017. It allowed to conclude that this analysis makes it possible the use of information both in space and time of crimes committed in a region to determine policing more efficiently. Furthermore, the use of the OpenStreetMap API allows working with real measures and including traffic, giving more realistic solutions but at the cost of more computational time and resources.

The sequence of strategies applied along the procedure to determine the routes has been proved key to the succes of the algorithm. Besides, the hibridization of techniques has also been shown a must in order to address this type of complex problems. The complete development of the algorithm is presented, from the analysis and processing of the spatial and temporal information to the patrol routes generation. It allows to obtain automatically a patrolling route in a similar way to an expert.

To summarize, the main advantages of our proposal are that the route is automatically obtained, it is optimized in terms of time and distance, and the computational time is low. It is similar but more complete than the one proposed by the expert as it identifies the hot spots and future crime points, while the expert only works with hotspots in a specific area, so the available information is more limited. In addition, the routes obtained by the algorithm tend to be faster as they consider real time information of traffic.

For future research, we propose to study the generation of maps that incorporate temporal information of the surveillance zones to reduce the execution time of the algorithm and allows its use in real-time. It is also intended to continue analyzing the problem to improve patterns and prediction, incorporating information of the relationship between crimes and areas of population concentration, such as parks, shops, liquor stores, restaurants, and so on. From the algorithm point of view, other clustering techniques, such as Gaussian Mixture Models, Mean-Shift Clustering, and DBSCAN, may be tried to see how the tecnique chosed affects the results of the CPSPA algorithm.

Author Contributions: Conceptualization, C.G. and M.S.; methodology, C.G. and M.S.; software, C.G.; validation, C.G. and M.S.; formal analysis, C.G. and M.S.; investigation, C.G. and M.S.; resources, C.G.; writing—original draft preparation, C.G.; writing—review and editing, M.S.; project administration, C.G.; funding acquisition, C.G. All authors have read and agreed to the published version of the manuscript.

Funding: This work was supported by Universidad Tecnológica Indoamérica [project number: INV-0012-028; 2021–2025; Artificial Intelligence Applied to Engineering -IAAI]

Institutional Review Board Statement: Not applicable.

Informed Consent Statement: Not applicable.

Data Availability Statement: Our research data are the private information of the National Police of Ecuador.

Acknowledgments: With consent and support from the AXA-ICMAT Chair in Adversarial Risk Analysis. This work was supported by Universidad Tecnológica Indoamérica [project number: INV-0012-028; Project: Artificial Intelligence Applied to Engineering-IAAI].

Conflicts of Interest: The authors declare no conflict of interest.

References

1. UNODC. *GLOBAL STUDY ON HOMICIDE Understanding homicide: Typologies, Demographic Factors, Mechanisms and Contributors*; UNODC: Vienna, Austria, 2019.
2. UNODC. *GLOBAL STUDY ON HOMICIDE Gender-Related Killing of Women and Girls*; UNODC: Vienna, Austria, 2019.
3. Guevara, C.; Santos, M. Surveillance Routing of COVID-19 Infection Spread Using an Intelligent Infectious Diseases Algorithm. *IEEE Access* **2020**, *8*, 201925–201936. [CrossRef] [PubMed]
4. San Juan, V.; Santos, M.; Andújar, J.M. Intelligent UAV Map Generation and Discrete Path Planning for Search and Rescue Operations. *Complexity* **2018**, *2018*, 6879419. [CrossRef]
5. Huang, C.; Zhang, J.; Zheng, Y.; Chawla, N.V. DeepCrime: Attentive hierarchical recurrent networks for crime prediction. In Proceedings of the International Conference on Information and Knowledge Management, Turin, Italy, 22–26 October 2018; Association for Computing Machinery: New York, NY, USA, 2018; pp. 1423–1432. [CrossRef]
6. Esquivel, N.; Nicolis, O.; Peralta, B.; Mateu, J. Spatio-Temporal Prediction of Baltimore Crime Events Using CLSTM Neural Networks. *IEEE Access* **2020**, *8*, 209101–209112. [CrossRef]
7. Farjami, Y.; Abdi, K. A genetic-fuzzy algorithm for spatio-temporal crime prediction. *J. Ambient Intell. Humaniz. Comput.* **2021**, *1*, 3. [CrossRef]
8. Hu, T.; Zhu, X.; Duan, L.; Guo, W. Urban crime prediction based on spatiotemporal Bayesian model. *PLoS ONE* **2018**, *13*, e0206215. [CrossRef]
9. Vural, M.S.; Gök, M. Criminal prediction using Naive Bayes theory. *Neural Comput. Appl.* **2017**, *28*, 2581–2592. [CrossRef]
10. Win, K.N.; Chen, J.; Chen, Y.; Fournier-Viger, P. PCPD: A Parallel Crime Pattern Discovery System for Large-Scale Spatiotemporal Data Based on Fuzzy Clustering. *Int. J. Fuzzy Syst.* **2019**, *21*, 1961–1974. [CrossRef]
11. Catlett, C.; Cesario, E.; Talia, D.; Vinci, A. Spatio-temporal crime predictions in smart cities: A data-driven approach and experiments. *Pervasive Mob. Comput.* **2019**, *53*, 62–74. [CrossRef]
12. Kadar, C.; Maculan, R.; Feuerriegel, S. Public decision support for low population density areas: An imbalance-aware hyper-ensemble for spatio-temporal crime prediction. *Decis. Support Syst.* **2019**, *119*, 107–117. [CrossRef]
13. Cowen, C.; Louderback, E.R.; Roy, S.S. The role of land use and walkability in predicting crime patterns: A spatiotemporal analysis of Miami-Dade County neighborhoods, 2007–2015. *Secur. J.* **2019**, *32*, 264–286. [CrossRef]
14. Piza, E.L.; Carter, J.G. Predicting Initiator and Near Repeat Events in Spatiotemporal Crime Patterns: An Analysis of Residential Burglary and Motor Vehicle Theft. *Justice Q.* **2018**, *35*, 842–870. [CrossRef]
15. Yang, B.; Liu, L.; Lan, M.; Wang, Z.; Zhou, H.; Yu, H. A spatio-temporal method for crime prediction using historical crime data and transitional zones identified from nightlight imagery. *Int. J. Geogr. Inf. Sci.* **2020**, *34*, 1740–1764. [CrossRef]
16. Yu, H.; Liu, L.; Yang, B.; Lan, M. Crime prediction with historical crime and movement data of potential offenders using a spatio-temporal cokriging method. *ISPRS Int. J. Geo-Inf.* **2020**, *9*, 732. [CrossRef]
17. Rummens, A.; Hardyns, W.; Pauwels, L. The use of predictive analysis in spatiotemporal crime forecasting: Building and testing a model in an urban context. *Appl. Geogr.* **2017**, *86*, 255–261. [CrossRef]
18. Hu, Y.; Wang, F.; Guin, C.; Zhu, H. A spatio-temporal kernel density estimation framework for predictive crime hotspot mapping and evaluation. *Appl. Geogr.* **2018**, *99*, 89–97. [CrossRef]
19. Fuentes-Santos, I.; González-Manteiga, W.; Zubelli, J.P. Nonparametric spatiotemporal analysis of violent crime. A case study in the Rio de Janeiro metropolitan area. *Spat. Stat.* **2021**, *42*, 100431. [CrossRef]
20. Ristea, A.; Al Boni, M.; Resch, B.; Gerber, M.S.; Leitner, M. Spatial crime distribution and prediction for sporting events using social media. *Int. J. Geogr. Inf. Sci.* **2020**, *34*, 1708–1739. [CrossRef] [PubMed]

21. Umair, A.; Sarfraz, M.S.; Ahmad, M.; Habib, U.; Ullah, M.H.; Mazzara, M. Spatiotemporal analysis of web news archives for crime prediction. *Appl. Sci.* **2020**, *10*, 8220. [CrossRef]
22. Hajela, G.; Chawla, M.; Rasool, A. A Clustering Based Hotspot Identification Approach for Crime Prediction. *Procedia Comput. Sci.* **2020**, *167*, 1462–1470. [CrossRef]
23. Lee, Y.J.; SooHyun, O.; Eck, J.E. A Theory-Driven Algorithm for Real-Time Crime Hot Spot Forecasting. *Police Q.* **2020**, *23*, 174–201. [CrossRef]
24. Mohler, G.; Porter, M.D. Rotational grid, PAI-maximizing crime forecasts. *Stat. Anal. Data Min.* **2018**, *11*, 227–236. [CrossRef]
25. Sharma, O.; Sahoo, N.C.; Puhan, N.B. A Survey on Smooth Path Generation Techniques for Nonholonomic Autonomous Vehicle Systems. In Proceedings of the IECON 2019 - 45th Annual Conference of the IEEE Industrial Electronics Society, Lisbon, Portugal, 14–17 October 2019; pp. 5167–5172. [CrossRef]
26. Kozjek, D.; Malus, A.; Vrabič, R. Reinforcement-Learning-Based Route Generation for Heavy-Traffic Autonomous Mobile Robot Systems. *Sensors* **2021**, *21*, 4809. [CrossRef] [PubMed]
27. Dixit, A.; Mishra, A.; Shukla, A. Vehicle Routing Problem with Time Windows Using Meta-Heuristic Algorithms: A Survey. *Adv. Intell. Syst. Comput.* **2019**, *741*, 539–546. [CrossRef]
28. Li, J.; Yang, S.X.; Xu, Z. A survey on robot path planning using bio-inspired algorithms. In Proceedings of the IEEE International Conference on Robotics and Biomimetics, ROBIO, Dali, China, 6–8 December 2019; pp. 2111–2116. [CrossRef]
29. Zhang, Y.; Zhang, S.; Huang, R.; Huang, B.; Yang, L.; Liang, J. A deep learning-based approach for machining process route generation. *Int. J. Adv. Manuf. Technol.* **2021**, *115*, 3493–3511. [CrossRef]
30. Damos, M.A.; Zhu, J.; Li, W.; Hassan, A.; Khalifa, E. A Novel Urban Tourism Path Planning Approach Based on a Multiobjective Genetic Algorithm. *ISPRS Int. J. Geo-Inf.* **2021**, *10*, 530. [CrossRef]
31. Dewinter, M.; Vandeviver, C.; Vander Beken, T.; Witlox, F. Analysing the Police Patrol Routing Problem: A Review. *ISPRS Int. J. Geo-Inf.* **2020**, *9*, 157. [CrossRef]
32. Hajibabai, L.; Saha, D. Patrol Route Planning for Incident Response Vehicles under Dispatching Station Scenarios. *Comput.-Aided Civ. Infrastruct. Eng.* **2019**, *34*, 58–70. [CrossRef]
33. Fu, Y.; Zeng, Y.; Wang, D.; Zhang, H.; Gao, Y.; Liu, Y. Research on route optimization based on multiagent and genetic algorithm for community patrol. In Proceedings of the 2020 International Conference on Urban Engineering and Management Science, ICUEMS, Zhuhai, China, 24–26 April 2020; pp. 112–116. [CrossRef]
34. Guevara, C.; Santos, M. Crime Prediction for Patrol Routes Generation Using Machine Learning. *Adv. Intell. Syst. Comput.* **2019**, *1267*, 97–107. [CrossRef]
35. Amiruzzaman, M.; Curtis, A.; Zhao, Y.; Jamonnak, S.; Ye, X. Classifying crime places by neighborhood visual appearance and police geonarratives: A machine learning approach. *J. Comput. Soc. Sci.* **2021**, *4*, 813–837. [CrossRef]
36. Guevara, C.; Santos, M.; López, V. Data leakage detection algorithm based on task sequences and probabilities. *Knowl.-Based Syst.* **2017**, *120*, 236–246. [CrossRef]
37. Liu, W.J.; Gao, P.P.; Yu, W.B.; Qu, Z.G.; Yang, C.N. Quantum Relief algorithm. *Quantum Inf. Process.* **2018**, *17*, 280, [CrossRef]
38. Gong, F.; Jiang, L.; Zhang, H.; Wang, D.; Guo, X. Gain ratio weighted inverted specific-class distance measure for nominal attributes. *Int. J. Mach. Learn. Cybern.* **2020**, *11*, 2237–2246. [CrossRef]
39. Liu, Y.; Bi, J.W.; Fan, Z.P. Multi-class sentiment classification: The experimental comparisons of feature selection and machine learning algorithms. *Expert Syst. Appl.* **2017**, *80*, 323–339. [CrossRef]
40. Yuan, C.; Yang, H. Research on K-Value Selection Method of K-Means Clustering Algorithm. *J* **2019**, *2*, 226–235. [CrossRef]
41. Vaitkevicius, P.; Marcinkevicius, V. Comparison of Classification Algorithms for Detection of Phishing Websites. *Informatica* **2020**, *31*, 143–160. [CrossRef]
42. Liu, F.; Deng, Y. Determine the Number of Unknown Targets in Open World Based on Elbow Method. *IEEE Trans. Fuzzy Syst.* **2021**, *29*, 986–995. [CrossRef]
43. Prabakaran, G.; Vaithiyanathan, D.; Ganesan, M. Fuzzy decision support system for improving the crop productivity and efficient use of fertilizers. *Comput. Electron. Agric.* **2018**, *150*, 88–97. [CrossRef]
44. López, V.; Santos, M.; Montero, J. Fuzzy specification in real estate market decision making. *Int. J. Comput. Intell. Syst.* **2010**, *3*, 8–20.
45. Miranda-Vega, J.E.; Rivas-López, M.; Flores-Fuentes, W.; Sergiyenko, O.; Lindner, L.; Rodríguez-Quiñonez, J.C. Pattern recognition applying LDA and LR to optoelectronic signals of optical scanning systems. *RIAI—Rev. Iberoam. De Autom. E Inform. Ind.* **2020**, *17*, 401–411. [CrossRef]

Article

An Energy Efficient Specializing DAG Federated Learning Based on Event-Triggered Communication

Xiaofeng Xue [1], Haokun Mao [1], Qiong Li [1,*], Furong Huang [2] and Ahmed A. Abd El-Latif [3,4]

[1] Information Countermeauser Technique Institute, School of Cyberspace Science, Faculty of Computing, Harbin Institute of Technology, Harbin 150001, China
[2] School of International Studies, Harbin Institute of Technology, Harbin 150001, China
[3] EIAS Data Science Lab, College of Computer and Information Sciences, Prince Sultan University, Riyadh 11586, Saudi Arabia
[4] Department of Mathematics and Computer Science, Faculty of Science, Menoufia University, Menofia 32511, Egypt
* Correspondence: qiongli@hit.edu.cn

Abstract: Specializing Directed Acyclic Graph Federated Learning (SDAGFL) is a new federated learning framework with the advantages of decentralization, personalization, resisting a single point of failure, and poisoning attack. Instead of training a single global model, the clients in SDAGFL update their models asynchronously from the devices with similar data distribution through Directed Acyclic Graph Distributed Ledger Technology (DAG-DLT), which is designed for IoT scenarios. Because of many the features inherited from DAG-DLT, SDAGFL is suitable for IoT scenarios in many aspects. However, the training process of SDAGFL is quite energy consuming, in which each client needs to compute the confidence and rating of the nodes selected by multiple random walks by traveling the ledger with 15–25 depth to obtain the "reference model" to judge whether or not to broadcast the newly trained model. As we know, the energy consumption is an important issue for IoT scenarios, as most devices are battery-powered with strict energy restrictions. To optimize SDAGFL for IoT, an energy-efficient SDAGFL based on an event-triggered communication mechanism, i.e., ESDAGFL, is proposed in this paper. In ESDAGFL, the new model is broadcasted only in the event that the new model is significantly different from the previous one, instead of traveling the ledger to search for the "reference model". We evaluate the ESDAGFL on the FMNIST-clustered and Poets dataset. The simulation is performed on a platform with Intel®Core™ i7-10700 CPU (CA,USA). The simulation results demonstrate that ESDAGFL can reach a balance between training accuracy and specialization as good as SDAGFL. What is more, ESDAGFL can reduce the energy consumption by 42.5% and 51.7% for FMNIST-clustered and Poets datasets, respectively.

Keywords: energy efficient; federated learning; event-triggered communication; DAG-DLT

MSC: 68T99

Citation: Xue, X.; Mao, H.; Li, Q.; Huang, F.; Abd El-Latif, A.A. An Energy Efficient Specializing DAG Federated Learning Based on Event-Triggered Communication. *Mathematics* **2022**, *10*, 4388. https://doi.org/10.3390/math10224388

Academic Editor: Daniel-Ioan Curiac

Received: 9 October 2022
Accepted: 15 November 2022
Published: 21 November 2022

Publisher's Note: MDPI stays neutral with regard to jurisdictional claims in published maps and institutional affiliations.

Copyright: © 2022 by the authors. Licensee MDPI, Basel, Switzerland. This article is an open access article distributed under the terms and conditions of the Creative Commons Attribution (CC BY) license (https://creativecommons.org/licenses/by/4.0/).

1. Introduction

With the Internet of Things (IoT) development, data have become more diversified and distributed, stimulating the demand for privacy and security [1,2]. As a result, traditional centralized cloud-based machine learning is being challenged. A few machine learning techniques have been proposed to meet these challenges. A technique called Federated Learning (FL) [3] provides a promising solution that allows the clients to work together to build a global machine learning model without sharing the local data on their own devices. A typical federated learning framework consists of a server and some local clients. The clients in FL can access the same global model and train it using local data. Then, they update the trained local model to the server. Once the locally trained models are received, the server will aggregate them as a new global model and feed it back to the clients for the next local update. FL will repeat this process several rounds until the global model converges or the number of repetitions meets the predetermined target.

This way, the client can protect data privacy as FL implements model training without collecting user data. Nevertheless, there are still some challenges in Federated Learning, as follows:

- **Device Heterogeneity:** In FL, the server must wait for all the selected clients to complete the local training before aggregating the local models. However, the computing power and network bandwidth vary from client to client [4,5]. The clients with a higher computing power and network bandwidth can complete the local training faster than the others. Thus, the server will wait a long time until the slowest client completes the local training before starting the next round. This leads to the performance bottleneck of FL [6];
- **Data Heterogeneity:** A significant difference between FL and other machine learning is that model's training in FL is completed on the clients with the data of Non-IID. The Non-IID data distribution will cause a significant decrease in the global model's accuracy. The existing reach shows that the accuracy can be reduced by 55% of the neural network trained with the highly skewed Non-IID dataset [7];
- **Single Point of Failure:** In the traditional FL, there is a single central server to aggregate the local model and publish the new global model. Therefore, the hacker can attack the central server to cause a single point of failure, leading to a performance decrease or even a failure of FL training;
- **Poisoning Attack:** The clients in FL are not all honest, and some clients may use the poisoned local data to train the model and then send the poisoned model to the server. However, the server cannot detect the poisoned model and will aggregate it into the global model. The poisoning attack will cause a decrease in the global model's accuracy [8].

Researchers have conducted much research to solve the above problems and proposed various solutions. For example, blockchain technology has been introduced into FL [9–11] to address the problem of a single point of failure and poisoning attack. Some distributed federated learning frameworks [12–14] have been proposed to adapt to the devices' heterogeneity. The problem of data heterogeneity has been widely studied, and different algorithms of framework have been proposed [15,16]. However, the existing studies focus on solving one or more problems, but not all of them. The framework proposed by Beilharz et al., called Specializing Directed Acyclic Graph Federated Learning (SDAGFL), realizes adaptive data and device heterogeneity, and is robust for poisoning in a fully decentralized federated learning environment.

SDAGFL introduced the Tangle [17] into the FL. The Tangle is a typical DAG Distributed Ledger Technology (DLT) designed by the IOTA foundation for the devices of the Internet of Things (IoT) to participate in a low-energy network. The Tangle uses a DAG data structure to store the transactions, allowing multiple transactions to be added to the ledger simultaneously, and it can achieve consensus similar to the Nakamoto DLT [18], which is also called blockchain, in a distributed system. Furthermore, the Tangle has a higher Transaction Per Second (TPS) than the blockchain because the transactions in the Tangle can be confirmed within minutes. In summary, the Tangle has the advantages of high TPS, low energy usage, and decentralization. Therefore, the Tangle is considered to be suitable for the scenario that includes many distributed devices.

SDAGFL inherits the features of the Tangle. In the SDAGFL, the participating clients use the DAG-DLT for the communication of models and an accuracy-biased random walk to obtain the models from other devices with similar data distribution to update their local model. It does not only overcome the challenges of device heterogeneity, failure of a single point, and poisoning attack, but also creates a balance between reaching a consensus on a generalized model and personalizing the model to the clients, which is different from the traditional FL framework, where all the participating clients train and reach a consensus for a global model together.

With the above advantages of the SDAGFL, it is suitable for the FL in an IoT scenario. However, some devices in the IoT are usually powered by a battery with strict energy restrictions [19]. It is necessary to reduce the energy consumption of federated learning. The benefit of reducing energy consumption is two-fold. On the one hand, it can reduce the number of charging times and prolong the device's service life. On the other hand, it is friendly to the environment. In addition, SDAGFL has some unnecessary energy consumption according to our study. Therefore, reducing the energy consumption while maintaining training performance is a problem that needs to be solved to promote the application of SDAGFL in IoT.

To reduce the energy consumption of the SDAGFL, we first analyze the energy consumption in SDAGFL, and we propose an event-triggered communication-based SDAGFL called event-triggered SDAGFL (ESDAGFL) to reduce the energy consumption of SDAGFL. The main contributions of this article are summarized as follows:

1. We analyze and give the energy consumption formula of SDAGFL. Then, we give the optimization objective of energy efficient SDAGFL;
2. Based on the energy consumption optimization objective, we propose an energy-efficient SDAGFL(ESDAGFL) scheme to reduce the energy consumption of SDAGFL. We evaluate the performance of ESDAGFL on two datasets, and the evaluation results show that ESDAGFL can efficiently reduce energy consumption compared to SDAGFL.

The rest of the paper is organized as follows. In Section 2, we review and compare the art of the federated learning framework based on DAG-DLT and show the advantages of SDAGFL over other frameworks. In Section 3, we formulate the energy consumption and give the optimization objective of energy efficient SDAGFL. We propose an energy efficient optimization scheme of SDAGFL in Section 4. Numerical simulation results are presented in Section 5. Finally, we summarize the work and give possible future researches.

2. Related Work

In this section, we will introduce and compare four federated learning frameworks based on the Tangle. Then, we analyze the advantages of SDAGFL compared to other works. Finally, we introduce the SDAGFL in detail.

Recently, the Tangle, which features high TPS, low energy usage, and decentralization, has received extensive attention from researchers. There are four Tangle-based federated learning frameworks that have been proposed as far as we know. Cao et al. [20] deployed a federated learning framework with Tangle called DAG-FL. DAG-FL includes three layers: the federated learning layer, the DAG layer, and the application layer, and it implements federated learning with asynchronous training and anomaly detection features. Shuo Yuan et al. [21] proposed ChainsFL, a two-layer hierarchical federated learning framework combining the Tangle and blockchain. The ChainsFL consists of a sub-ledger, the Raft [22]-based Hyperledger Fabric [23], deployed on edge nodes, and a Tangle-based master ledger. The ChainsFL has overcome the disadvantages of massive resource consumption and limited throughput in traditional blockchain-based FL. In addition, Schmid Robert et al. discussed the basic applicability of the combination of federated learning and the Tangle in Ref. [24]. They assumed a long-standing open network for continuous learning and development. In this framework, the tip selection algorithm based on Monte Carlo Markov Chain in the Tangle is used to implement model selection. The averaged model is used to update the model. The evaluation of this framework shows high training accuracy and model-agnostic resistance against random poisoning and label-flipping attacks. Beilharz et al. [25] further optimized the framework in Ref. [24]. They proposed an implicit model specialization framework for federated learning called Specializing DAG Federated Learning (SDAGFL). In SDAGFL, the implicit clustering of clients with a similar local dataset is achieved through the accuracy-biased random walk on DAG-DLT and the Fedavg algorithm. We compare these four frameworks in Table 1 from four different aspects.

Table 1. Comparison of four Federated Learning Frameworks based on Tangle DAG-DLT.

Framework	Decentralized	Robust for Poisoning Attack	Asynchronous Training	Individualized Model
ChainsFL [21]	✓	✓	✗	✗
DAG-FL [20]	✓	✓	✓	✗
Learning-Tangle [24]	✓	✓	✓	✗
SDAGFL [25]	✓	✓	✓	✓

Table 1 shows that SDAGFL is better than the three other works, and it sufficiently uses the DAG data structure's feature, and an accuracy-biased random walk for model update realized implicit clustering of the client. SDAGFL can adapt device and data heterogeneity, resist poisoning attacks, and support creating personal models. Thus, we think it is worthy of further study. However, from the prototype implementation of SDAGFL, we notice that the training process of SDAGFL is quite energy consuming because the client needs to compute the confidence and rating of the nodes selected by multiple random walks by traveling the ledger with 15–25 depth to obtain the "reference model" to judge whether or not to broadcast the newly trained model. In this paper, we will analyze the energy consumption of SDAGFL.

In particular, if not specified, the DAG-DLT in this work refers to the Tangle, and SDAGFL refers to the framework proposed in Ref. [25].

3. Optimization Objective of Energy Efficient SDAGFL

In this section, we will analysis the energy consumption of SDAGFL and give the energy optimization objective. The main notations used in this section is summarized in Table 2.

Table 2. Notations adopted for the energy consumption analysis.

Notations	Meaning
d	The depth of the Tangle ledger
c	The average number of the site node's children
f_i	The CPU-cycle frequency of the $client_i$
C_i	The CPU chipset's effective capacitance coefficient of the $client_i$
f_i^w	Number of CPU cycles required while the $client_i$ tests the model's accuracy
f_i^a	Number of CPU cycles required while the $client_i$ averages the model
f_i^t	Number of CPU cycles required while the $client_i$ trains one sample of dataset
f_i^c	Number of CPU cycles required while the $client_i$ compares the model
f_i^r	Number of CPU cycles required while the $client_i$ performs the random walk
E_i^{tip}	Energy consumption when the $client_i$ obtains two tip nodes
E_i^{agg}	Energy consumption when the $client_i$ averages the model
E_i^{train}	Energy consumption when the $client_i$ trains the model
$E_i^{reference}$	Energy consumption when the $client_i$ obtains and compares with the reference model
ω	Model parameter
$l(x_j, y_j, \omega)$	The loss function on data point (x_j, y_j)
$f(\omega)$	The global objective optimization in each clustered community
$F_n(\omega)$	The objective optimization in client n
N	The number of clients in each community
b	Size of the local batch
B	Number of the local batch
τ	Number of training epoch

3.1. Principle and the Workflow of SDAGFL

Before analysis of the energy consumption of SDAGFL, we first introduce the basic principle of SDAGFL. The basic principle of SDAGFL is shown in Figure 1. All the clients share a public DAG ledger in SDAGFL. The nodes in SDAGFL are divided into four categories, each containing a complete model parameter.

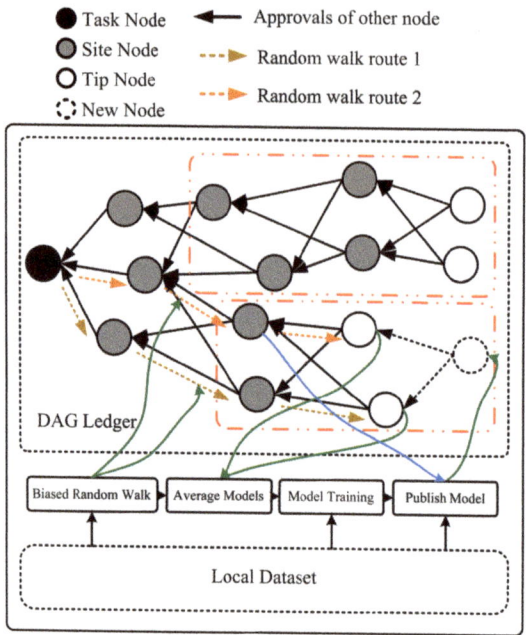

Figure 1. Basic principle of SDAGFL.

1. **Task Node:** The task node is a genesis node. It contains the basic training task and initial model parameter;
2. **Site Node:** The site node is the basic node type. It contains the model published by the clients;
3. **Tip Node:** The tip node is a special site node, which is not approved by other node;
4. **New Node:** The new node contains the model that will be published.

During the process of the SDAGFL, the clients will repeat the following five steps to update the model parameter:

1. Each client performs two accuracy-biased random walks in Algorithm 1 based on the local dataset until reaching the tip node. We show the random walk route using the yellow dashed line in Figure 1;
2. The client then averages the models in two tips obtained from Step 1;
3. Next, the client trains the average above model using its local dataset, and then obtains a new model;
4. In this step, the client will perform several accuracy-biased random walks to obtain the "reference model"(the site node approved by the blue line in Figure 1);
5. Finally, the client will compare the new model with the "reference model" to determine whether or not to publish the new model. If the test loss using the local data of new model is lower than the "reference model", the client will publish the new model to the public DAG ledger.

With the increase of the DAG ledger, the site node trained by the clients with independent identical distributed data will be clustered in the same group. As we can see, the site nodes in Figure 1 are clustered into two groups (red dashed box).

Algorithm 1: Random Walk of SDAGFL.

1 $children \leftarrow$ GetChildren(Site Node n);
2 $n = len(children)$;
3 initial $accuracy[n]$;
4 **for** $child[i]$ in children **do**
5 $\quad$ $accuracy[i] = EvaluateOnLocalData(child[i])$;
6 **end**
7 **for** $accuracy[i]$ in accuracy **do**
8 $\quad$ $normalized[i] = \frac{accuracy[i] - max(accuracy)}{max(accuracy) - min(accuracy)}$;
9 $\quad$ $weight[i] = e^{normalized} \times \alpha$;
10 **end**
   // WeightChoice is a weight-based child node selection function
11 nextnode = WeightChoice(weight);
12 RandomWalk(nextnode);

3.2. Energy Consumption Optimization Objective Analysis

In this section, we will give the optimization objective of energy efficient SDAGFL through analysis of the energy consumption and training target.

According to Section 3.1, the training process in SDAGFL can be divided into four parts: client obtains two tip nodes, Model Aggregation, Model Training, and client obtains and compares with the reference model. According to the analysis in the literature [19,26,27], we can analyze the energy consumption as follows.

Client obtains two tip nodes: The energy consumption of this process is decided by the depth of the ledger and the number of children nodes of the site node. We assumption that the each client i performs the two independently random walks on the ledger with the d depth and c average number of children. The energy consumption of the process in which the client i obtains two tip nodes is shown as follows.

$$E_i^{tip} = C_i \times 2 \times dc f_i^w \times f_i^2 \quad (1)$$

Client aggregates the model in tip node: In this process, the number of CPU cycles for the client i to aggregate the model is f_i^a. Thus, the energy consumption in model aggregation is

$$E_i^{aggregation} = C_i \times f_i^a \times f_i^2 \quad (2)$$

Client trains the model: The energy cost of model training depends on the size of the dataset. The number of CPU cycles for training one sample of the dataset is defined as f_i^t, and the training batch size and number of batch is defined as b and B. The training epoch is defined as τ. The energy consumption can be expressed as

$$E_i^{training} = C_i \times bB\tau f_i^t \times f_i^2 \quad (3)$$

Client obtains and compares with the reference model: In the prototype implementation of SDAGFL (https://github.com/osmhpi/federated-learning-dag, accessed on 6 September 2022), the client needs to compute the confidence and rating of each site node to obtain the reference model. The confidence is the number of the site node selected during the multiple random walks. The rating is the number of nodes that are directly and indirectly approved by every site node. Thus, the energy cost of this process contains two aspects: confidence computing and rating computing. Thus, the energy consumption of the client obtaining the "reference model" can be expressed as follows.

$$E_i^{reference} = C_i \times r \times d \times c \times (f_i^w + f_i^c + f_i^r) \times f_i^2 \quad (4)$$

where r is the number of random walk, which is 5 in SDAGFL.

Through the above analysis, we can get the energy consumption by the client i for one time model update as follows:

$$\begin{aligned}E_i^{total} &= E_i^{tip} + E_i^{aggregation} + E_i^{training} + E_i^{reference} \\ &= C_i \times \{2dcf_i^w + f_i^a + bB\tau f_i^t + rdc(f_i^w + f_i^c + f_i^r)\} \times f_i^2 \\ &= C_i \times \{(2+r)dcf_i^w + f_i^a + bB\tau f_i^t + rdc\{f_i^c + f_i^r\}\} \times f_i^2\end{aligned} \quad (5)$$

According to Equation (5), energy can be divided into four parts: energy consumption of $(2 + r)$ times random walk, energy consumption of model aggregation, energy consumption of model training, and energy consumption of computing reference and rating. Although reducing the walk depth d to a fixed depth of 15 to 25 transactions from the tip nodes is beneficial to reduce the energy consumption, this process must be repeatedly executed multiple times, resulting in massive energy consumption. The energy consumption of model aggregation is decided by the model's size, which is a fixed parameter in a specific federated learning task. In addition, the energy consumption of model training is connected to the training batch. To be fair and simplify the problem, we let each device execute the same batch. So, model aggregation and training energy consumption can be regarded as fixed energy. Therefore, the energy consumption should be optimized, which can be expressed in Equation (6) as follows:

$$\mathbb{E} = C_i \times \{(2+r)dcf_i^w + rdc\{f_i^c + f_i^r\}\} \times f_i^2 \quad (6)$$

In addition, like general federated aggregation algorithms such as FedAvg, we use the same suggested supervised objective function. The clients with the same data distribution will be clustered in multiple implicit communities. For every cluster, we assume that the number of clients in each community is N, that the nth client has k data points $|k| = m_n$, and that the size of the whole dataset is m. Therefore, the optimization objective of federated training is defined as:

$$\min_{\omega \in \mathbb{R}^d} f(\omega) \quad (7)$$

where

$$f(\omega) = \sum_{n=0}^{N} \frac{m_n}{m} \times F_n(\omega) \quad (8)$$

and

$$F_n(\omega) = \frac{1}{m_n} \sum_j^{m_n} l_j(x_j, y_j, \omega) \quad (9)$$

Thus, the energy consumption optimization objective of SDAGFL is formulated as follows:

$$Minimize[\mathbb{E}, f(\omega)], \quad where \quad \omega \in \mathbb{R}^d \quad (10)$$

In other words, the optimization objective can be expressed as minimizing the energy consumption of the client while minimizing the federated learning training loss function.

4. Energy Efficient Optimization Scheme of SDAGFL

Event-triggered communication mechanism is proposed for the network control systems to reduce the frequency of communication [28,29]. In an event-triggered mechanism-based system, the device will perform the predetermined operation only when it detects an event that meets the trigger conditions.

The event-triggered communication mechanism has recently been introduced into the field of parallel machine learning. In a event-triggered based machine learning scenario, the client computes the model's gradient and broadcasts the gradients that meet the trigger threshold to the neighbors [30–34].

Inspired by above researches, we utilize an event-triggered communication based on the parameter change to solve the optimized objective in Equation (10). The workflow of

ESDAGFL is shown in Figure 2. Compared to the baseline, the event-triggered communication mechanism only needs to judge the change of the model's norm, and does not need to find a reference model. Therefore, ESDAGFL can reduce the energy consumption of the client and the simulation results in Section 5 also demonstrate the advantages of the proposed scheme.

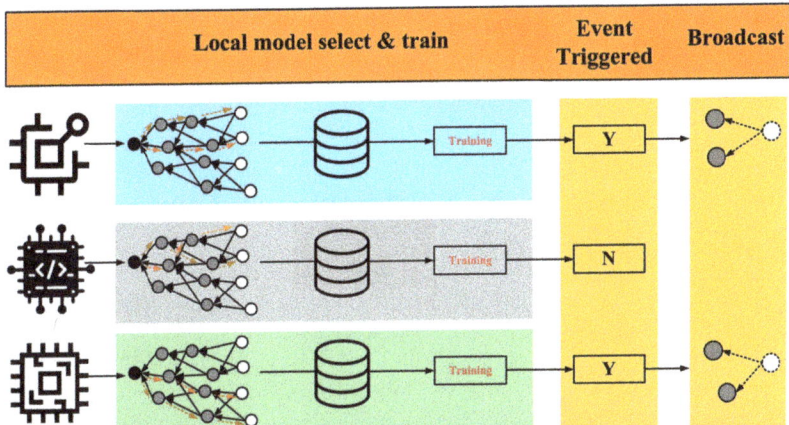

Figure 2. Workflow of the Energy Efficient SDAGFL Scheme.

The clients participating in the federated learning have the following features:

- **Data are private to the clients:** In ESDAGFL, the dataset is stored in the spatially distributed clients, and it is private to all clients. There is no data exchange among the clients;
- **Data are Non-IID for all clients:** Data is non-independent and identically distributed (Non-IID) among all clients participating in the ESDAGFL;
- **Data are IID for clients in the same cluster:** The site node published by the clients with a similar data distribution feature could be clustered into a "implicit community". So, we can regard that the data distribution among the clients clustered in the same community is an independently identical distribution;
- **Honest and Malicious client's behavior:** The honest clients always comply with the rules of the SDAGFL and use their complete local datasets to participate in the training. Once the trained model meets the broadcast condition, the client will broadcast it in time. On the contrary, the malicious clients do not comply with any rules;
- **Synchronicity:** Although the ledger is not synchronized in the vast majority of the reality cases, we think that the performance of the ESDAGFL would not be affected whether the ledger is synchronous or not. The reason is that whether the client publishes the new model or not only depends on the average model and the new trained model. Thus, we define that the ledger on each client is synchronized in real-time to analyze the performance of ESDAGFL for convenience.

In our ESDAGFL framework, each client participates in the following federated learning executed steps, as shown in Algorithm 2.

Firstly, each clients performs two random walks and obtains the aggregated model $\omega_{avg} = Avg(\omega_{tip1}, \omega_{tip2})$ with two tip nodes. (Lines 1 and 2 in Algorithm 2).

Then, the client n uses a Stochastic Gradient Descent (SGD) to compute the local gradient and perform τ local model parameter updates using the local dataset to train the new model ω_{new}. This process can be expressed as Equation (11). (Lines 3–9 in Algorithm 2).

$$\omega_{new} = \omega_{avg} - \eta \sum_{}^{\tau} \sum_{}^{B} \nabla l(\omega_{avg}) \qquad (11)$$

Next, the client computes the model parameter change rate between the trained new model and the input averaged model by Equation (12). (Lines 10 and 11 in Algorithm 2).

$$\Delta = \frac{||\Delta \omega||_2}{||\omega_{avg}||_2} \qquad (12)$$

where,

$$||\Delta \omega|| = ||\omega_{new} - \omega_{avg}|| \qquad (13)$$

Finally, the client broadcasts the trained new model ω_{new} if the model parameter change rate is equal to or greater than the trigger threshold. (Lines 12–14 in Algorithm 2)

$$\Delta \geq trigger_{threshold} \qquad (14)$$

Algorithm 2: Energy efficient SDAGFL scheme.

Input: learning rate η, local batch is B, τ is the number of local epochs
Output: New Model: ω_{new}

1 $(\omega_1, \omega_2) \leftarrow$ Random Walk of SDAGFL of client;
2 $\omega_{avg} = (\omega_1 + \omega_2)/2$;
3 $\omega_{i,j} \leftarrow \omega_{avg}$;
4 **for** local epoch $i = 0, 1, 2, ..., \tau$ **do**
5 $\quad$ **for** batch $j = 0, 1, 2, ..., B$ **do**
6 $\quad\quad$ $\omega_{i,j+1} = \omega_{i,j} - \eta \nabla l(\omega_{i,j})$;
7 $\quad$ **end**
8 **end**
9 $\omega_{new} \leftarrow \omega_{\tau,B}$;
10 $\Delta \omega = \omega_{new} - \omega_{avg}$;
11 Compute $\Delta = \frac{||\Delta \omega||_2}{||\omega_{avg}||_2}$;
12 **if** $\Delta \geq threshold_{tirgger}$ **then**
13 $\quad$ return ω_{new};
14 **end**

---

5. Experiment and Results

In this section, we evaluate ESDAGFL in the framework (https://github.com/osmhpi/federated-learning-dag, accessed on 6 September 2022) proposed by Beilharz et al. [25] with PyTorch. The parameters of the evaluation platform used in the experiment are shown in Table 3.

Table 3. Parameters of the experiment platform.

Item	Parameter
OS	Ubuntu 22.04
CPU	Intel®Core™ i7-10700 @2.9 GHz × 16
RAM	32 GB

5.1. Experiment Setting

This section introduces the experimental setting of ESDAGFL. The datasets and models used in the experiment are described in Section 5.1.1, and the fixed training hyperparameters setting is shown in Section 5.1.2.

5.1.1. Datasets and Models

We evaluated ESDAGFL on two training tasks. We evaluated a handwriting recognition task on the FMNIST-clustered dataset and a next character prediction task on the Poets dataset. The training and test datasets have a ratio of 9:1 for each client.

- **Handwriting Recognition Task:**
    - **Dataset:** The FMNIST-clustered dataset is a synthetically clustered version of Federated extended MNIST (FMNIST) built by LEAF [35]. In this dataset, the 28 × 28 pixel handwriting dataset is divided into 3 disjoint classes: [0,1,2,3], [4,5,6], [7,8,9], and the number of clients is the same in every class;
    - **Model:** For the handwriting recognition task, a Convolutional Neural Net (CNN) is used. It contains two convolution layers and two fully connected layers. The kernel size of the convolution layer is 5 with the RELU activation function. Each convolution layer is followed by a max pooling layer with pool size and stride length of 2. The two fully connected layers contain 2048 and 10 neurons, respectively.
- **Next Character Prediction Task:**
    - **Dataset:** The Poets dataset is used to evaluate the performance of ESDAGFL for the next character prediction task. The Poets dataset consists of an English dataset from William Shakespeare's works and a German dataset from Goethe's plays. The English and German datasets have an equal number of samples and are put into different clusters.
    - **Model:** For the next character prediction task, the model first maps each character to an embedding of dimension 8, calculated from the 80 character sequence. Then the model passes each character through a Long Short-Term Memory (LSTM) consisting of 2 layers with 256 units each. Finally, there is a dense layer for prediction.

5.1.2. Training Hyperparameters Setting

The training hyperparameters setting is shown in Table 4.

Table 4. Training Hyperparameters Setting of the Experiment.

Parameters	FMNIST-Clustered	Poets
Local epochs	1	1
Local batches	10	200
Batch size	10	10
Learning rate	0.05	0.8
Clients per round	10	10

5.2. Experimental Results

Different from the traditional FL, ESDAGFL is a fully asynchronous federated learning framework without any central server. It realizes the training target through each client run of the training process continuously and independently as long as its resources permit. Therefore, the concept of the rounds is introduced to enable a better demonstration of the experimental results.

In the experiment, the $threshold_{trigger}$ is set to 0.008 for the FMNIST-clustered dataset and 0.12 for the Poets datasets. The evaluation of the training accuracy and loss was done every five rounds using the test dataset with 5% of all clients randomly selected. We regard the work in Ref. [25] as the baseline, and we evaluated our ESDAGFL on three metrics: training accuracy and loss, implicit specialization, and training time cost.

5.2.1. Training Accuracy and Loss Evaluation

Training accuracy and loss are two standard metrics used to evaluate the model performance. The accuracy is the percentage of correct predictions, and the loss is the cross-entropy loss.

Firstly, we evaluated the average accuracy and loss on the test dataset for the handwriting recognition task. The evaluation results are shown as Figures 3 and 4.

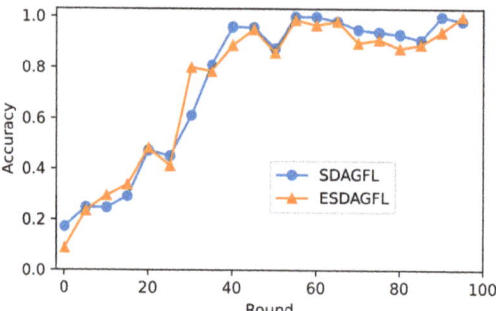

Figure 3. The average training accuracy of ESDAGFL and SDAGFL evaluated on the FMNIST-clustered dataset.

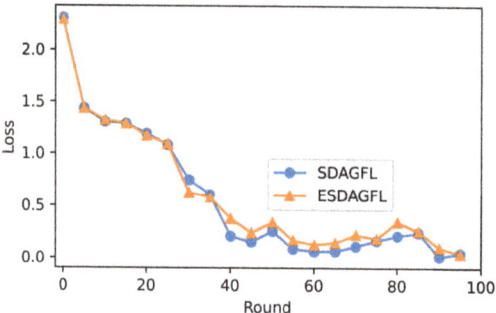

Figure 4. The average training loss of ESDAGFL and SDAGFL evaluated on the FMNIST-clustered dataset.

The evaluation result shows that the ESDAGFL achieves similar accuracy and approximate loss as the baseline. The experimental results demonstrate the applicability of ESDAGFL to the written recognition task.

Then, we evaluated the next character prediction task on the Poets dataset to further illustrate the applicability of ESDAGFL for different models. Figures 5 and 6 show ESDAGFL's average accuracy and loss in the Poets dataset. The experiment results show that ESDAGFL equally applies to the Poets dataset.

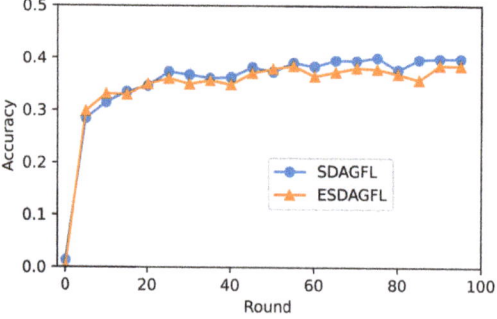

Figure 5. The average training accuracy of ESDAGFL and SDAGFL evaluated on the Poets dataset.

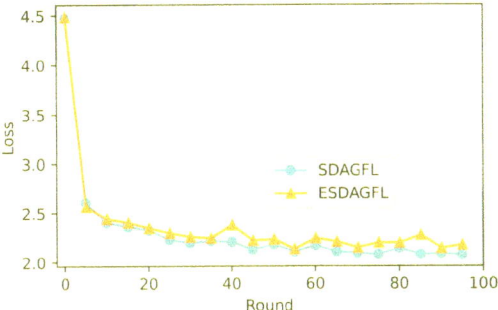

Figure 6. The average training loss of ESDAGFL and SDAGFL evaluated on the Poets dataset.

Convergence analysis of ESDAGFL: As the new models are only published if they are changed when the honest clients' training for the model is positive, the federated training will converge to an expected direction. Figures 3 and 5 show the convergence rate of ESDAGFL on the FMNIST-clustered and Poets datasets. Multiple experiment results show that the training process converges to nearly the same result. It confirmed that ESDAGFL could converge to a reasonable training accuracy.

5.2.2. Implicit Specialization Evaluation

In this part, we evaluated the implicit specialization of the ESDAGFL. There are three metrics to evaluate the implicit specialization of the ESDAGFL. The first is the modularity $m \in [-\frac{1}{2}, 1]$, which is a metric of the community segmentation in the community discovery algorithm. The more closely connected the nodes within a community and the more sparsely the connections between the communities are, the greater the modularity is. The second metric is the number of modules, which indicates the partitioning of all clients. The number of modules should be an appropriate value. The final metric is approval pureness, which measures the probability that a client approves the nodes published by other clients in the same cluster.

Firstly, we evaluated the modularity and number of modules on the FMNIST-clustered dataset. The FMNIST-clustered dataset was chosen because it has more classes and is easier to visualize.

Figures 7 and 8 show the performance of ESDAGFL in the implicit specialization. Figure 7 shows that we achieved similar modularity as the baseline on the FMNIST-clustered dataset. As shown in Figure 8, with the increase of time, we achieve the same number of modules as the baseline and almost all of the clients are clustered into the community relative to their label.

To quantify the robustness of the DAG specialized in these experiments, we test the approval pureness to quantify the "specialized" ESDAGFL on the FMNIST-clustered and Poets datasets, and the result is shown in Table 5. The base pureness refers to the approval pureness if the approvals were to be randomly spread over all clusters. Since there are three clusters in the FMNIST-clustered dataset and two clusters of Goethe and Shakespeare in Poets, their basic pureness is 0.33 and 0.5, respectively. ESDAGFL and the baseline achieves 100% pureness in the FMNIST-clustered data set, and all model approvals are from the same cluster. Although the approval pureness of ESDAGFL on the Poets dataset is lower than the FMNIST-clustered dataset, the pureness is still high. The approval pureness on the Poets dataset show the balance between specialization and generalization.

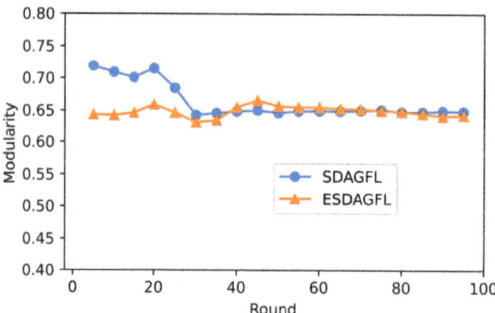

Figure 7. The modularity variation in ESDAGFL and SDAGFL evaluated on the FMNIST-clustered dataset.

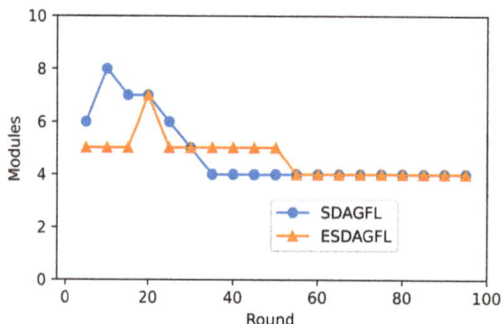

Figure 8. The modules variation in ESDAGFL and SDAGFL evaluated on the FMNIST-clustered dataset.

Table 5. Comparison of the approval pureness after 100 rounds of training between ESDAGFL and baseline.

Dataset	Base Pureness	Pureness of SDAGFL (Baseline)	Pureness of ESDAGFL (Ours)
FMNIST-clustered	0.33	1	1
Poets	0.5	0.98	0.96

The above experiments show that ESDAGFL can achieve the same implicit specialization as the Baseline (SDAGFL) for the different datasets. However, in some metrics of the implicit specialization, the performance is not as good as the baseline, which has a negligible effect for the implicit specialization.

5.2.3. Energy Consumption Evaluation

The experiments in Sections 5.2.1 and 5.2.2 show that ESDAGFL realizes the balance between the training performance and specialization on the FMNIST-clustered and Poets dataset. In this part, we will evaluate ESDAGFL's energy consumption using "pyJoules" (https://github.com/powerapi-ng/pyJoules, accessed on 6 September 2022) [36,37].

The "pyJoules" is a software-defined power meter that can measure the power consumption of the program running on a host machine. It uses the Intel "Running Average Power Limit" (RAPL) technology to monitor the energy consumption of the Intel CPU socket package, RAM, and Intel integrated GPU.

The devices that "pyJoules" can monitor match the RAPL domain, as shown in Figure 9.

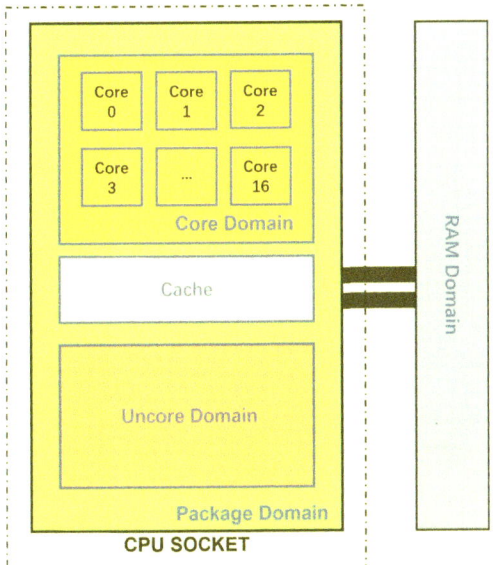

Figure 9. RAPL domain match part of our CPU socket. The energy consumption of the core measured by pyJoules is the sum of the CPU core energy consumption, the energy consumption of the packages is the wall CPU energy consumption, and the energy consumption of the uncore is the energy consumption of the integrated GPU.

In our simulation experiments, we took the statistics of the core's energy consumption per round of training multiple times and removed the bad values to get the average results.

Figures 10 and 11 show the core's energy consumption per round with 10 clients and when the client performs the random walk from the depth of 15 to 25. The energy consumption of ESDAGFL is lower than the baseline. The total CPU core's energy consumption of ESDAGFL is 61.29 KJ, and the baseline's core's energy consumption is 106.74 KJ. Our ESDAGFL scheme can reduce the 42.5% energy consumption compared with the baseline for the handwriting recognition task. As for the Poets dataset, the total CPU core's energy consumption of ESDAGFL is 2390.41 KJ, and the baseline's core's energy consumption is 4956.63 KJ. Our ESDAGFL scheme can reduce the 51.7% energy consumption for the next character prediction task.

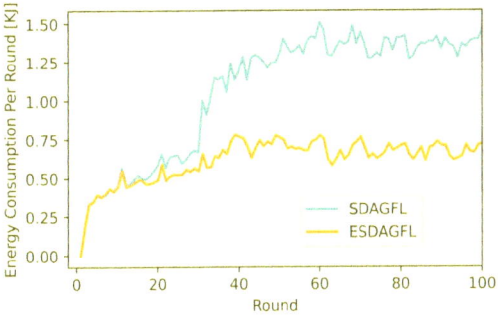

Figure 10. Energy consumption in every round evaluated on the FMNIST-clustered dataset.

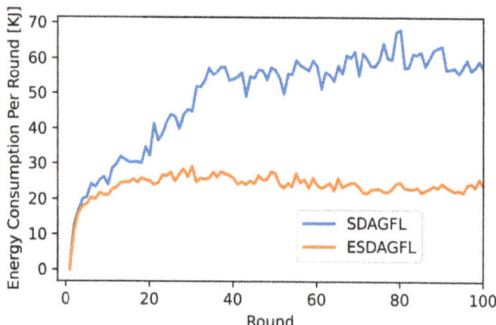

Figure 11. Energy consumption in every round evaluated on the Poets dataset.

We evaluate the energy consumption of ESDAGFL and baseline on the FMNIST-clustered dataset with a different client for each round. The result is shown in Table 6.

Table 6. Energy consumption [KJ] with a different number of clients per round evaluated on the FMNIST-clustered dataset.

Scheme	15 Clients per Round	20 Clients per Round	25 Clients per Round
SDAGFL (Baseline)	216.37	362.82	601.01
ESDAGFL (Ours)	108.35	166.21	220.42
Energy consumption reduction	47.6%	54.1%	63.3%

5.2.4. Results Analysis

We evaluate the performance of our approach on the FMNIST-clustered and Poets datasets. The experimental results show that our scheme can realize the same implicit specialization and generalization as the SDAGFL. Furthermore, compared with the SDAGFL, our scheme can reduce the energy consumption by 42.5% and 51.7% for the handwriting recognition task on the FEMNIST dataset and the character prediction task on the Poets dataset, respectively, on the condition that 10 clients participate in each round. Besides, the reduction of energy consumption increases with the number of clients.

6. Conclusions

Energy consumption is an important issue for federated learning in IoT scenarios, as many devices are battery powered. To reduce the energy consumption of the SDAGFL and make it a better federated learning scheme for IoT, the energy optimization objective of SDAGFL is constructed based on the thorough analysis of energy consumption of SDAGFL. To realize the optimization objective, we design an event-triggered communication-based SDAGFL scheme, named ESDAGFL. In ESDAGFL, the client broadcasts the trained new model only in the specific event in which the new model is significantly different form the previous one. It is necessary to search for a "reference model" by traveling the whole ledger to judge whether a new model is broadcasted or not in SDAGFL. Our simulations are performed on a platform with Intel®Core™ i7-10700 CPU. The simulation results show, that compared with SDAGFL, our scheme can reduce the energy consumption by 42.5% and 51.7% on the FMNIST-clustered dataset and the Poets dataset, respectively. Besides, the energy reduction increases along with the increase of the number of clients per round. Meanwhile, our scheme can reach an equal balance between the model's performance and specialization as SDAGFL. In the future, we plan to deploy the ESDAGFL in a practical scenario and perform an overall test. In addition, we will not only consider the energy consumption caused by computation, but also the communication.

Author Contributions: Funding acquisition, Q.L.; Investigation, Q.L. Methodology, X.X.; Resources, H.M. and Q.L.; Validation, X.X.; Writing—original draft, X.X. and H.M.; Writing—review and editing, X.X., Q.L., F.H. and A.A.A.E.-L. All authors have read and agreed to the published version of the manuscript.

Funding: This research was funded by the National Natural Science Foundation of China (grant number 62071151).

Data Availability Statement: Not applicable.

Acknowledgments: Thanks to authors of article "Implicit Model Specialization through DAG-based Decentralized Federated Learning" for the open source simulation code in https://github.com/osmhpi/federated-learning-dag (accessed on 6 September 2022).

Conflicts of Interest: The authors declare no conflict of interest.

Abbreviations

The following abbreviations are used in this manuscript:

CNN	Convolutional Neural Network
DAG	Directed Acyclic Graph
DLT	Distributed Ledger Technology
ESDAGFL	Event-Triggered Specialized DAG Federated Learning
FL	Federated Learning
FMNIST	Federated extended MNIST
IoT	Internet of Things
RAPL	Running Average Power Limit
LSTM	Long Short-Term Memory
SDAGFL	Specialized DAG Federated Learning
SGD	Stochastic Gradient Descent
TPS	Transactions Per Second

References

1. Albrecht, J.P. How the GDPR Will Change the World. *Eur. Data Prot. Law Rev.* **2016**, *2*, 287–289. [CrossRef]
2. Yi, S. Personal Information Protection: China's Path Choice. *US-China Law Rev.* **2021**, *18*, 227. [CrossRef]
3. Konečný, J.; McMahan, H.B.; Ramage, D.; Richtárik, P. Federated Optimization: Distributed Machine Learning for On-Device Intelligence. *arXiv* **2016**, arXiv:1610.02527.
4. Niknam, S.; Dhillon, H.S.; Reed, J.H. Federated Learning for Wireless Communications: Motivation, Opportunities, and Challenges. *IEEE Commun. Mag.* **2020**, *58*, 46–51. [CrossRef]
5. Chen, M.; Gündüz, D.; Huang, K.; Saad, W.; Bennis, M.; Feljan, A.V.; Poor, H.V. Distributed Learning in Wireless Networks: Recent Progress and Future Challenges. *IEEE J. Sel. Areas Commun.* **2021**, *39*, 3579–3605. [CrossRef]
6. Xu, C.; Qu, Y.; Xiang, Y.; Gao, L. Asynchronous Federated Learning on Heterogeneous Devices: A Survey. *arXiv* **2022**, arXiv:2109.04269.
7. Zhao, Y.; Li, M.; Lai, L.; Suda, N.; Civin, D.; Chandra, V. Federated Learning with Non-IID Data. *arXiv* **2018**, arXiv:1806.00582.
8. Cao, D.; Chang, S.; Lin, Z.; Liu, G.; Sun, D. Understanding Distributed Poisoning Attack in Federated Learning. In Proceedings of the 2019 IEEE 25th International Conference on Parallel and Distributed Systems ICPADS), Tianjin, China, 4–6 December 2019; pp. 233–239. [CrossRef]
9. Kim, H.; Park, J.; Bennis, M.; Kim, S.L. Blockchained On-Device Federated Learning. *IEEE Commun. Lett.* **2019**, *24*, 1279–1283. [CrossRef]
10. Wang, P.; Zhao, Y.; Obaidat, M.S.; Wei, Z.; Qi, H.; Lin, C.; Xiao, Y.; Zhang, Q. Blockchain-Enhanced Federated Learning Market with Social Internet of Things. *IEEE J. Sel. Areas Commun.* **2022**, 3213314. [CrossRef]
11. Miri Rostami, S.; Samet, S.; Kobti, Z. A Study of Blockchain-Based Federated Learning. In *Federated and Transfer Learning*; Razavi-Far, R., Wang, B., Taylor, M.E., Yang, Q., Eds.; Adaptation, Learning, and Optimization; Springer: Cham, Switzerland, 2023; pp. 139–165. [CrossRef]
12. Chen, S.; Wang, X.; Zhou, P.; Wu, W.; Lin, W.; Wang, Z. Heterogeneous Semi-Asynchronous Federated Learning in Internet of Things: A Multi-Armed Bandit Approach. *IEEE Trans. Emerg. Top. Comput. Intell.* **2022**, *6*, 1113–1124. [CrossRef]
13. Xu, X.; Duan, S.; Zhang, J.; Luo, Y.; Zhang, D. Optimizing Federated Learning on Device Heterogeneity with A Sampling Strategy. In Proceedings of the 2021 IEEE/ACM 29th International Symposium on Quality of Service (IWQOS), Tokyo, Japan, 25–28 June 2021; pp. 1–10. [CrossRef]

14. Cao, J.; Lian, Z.; Liu, W.; Zhu, Z.; Ji, C. HADFL: Heterogeneity-aware Decentralized Federated Learning Framework. In Proceedings of the 2021 58th ACM/IEEE Design Automation Conference (DAC), San Francisco, CA, USA, 5–9 December 2021; pp. 1–6. [CrossRef]
15. Li, G.; Hu, Y.; Zhang, M.; Liu, J.; Yin, Q.; Peng, Y.; Dou, D. FedHiSyn: A Hierarchical Synchronous Federated Learning Framework for Resource and Data Heterogeneity. *arXiv* **2022**, arXiv:2206.10546.
16. Huang, W.; Ye, M.; Du, B. Learn from Others and Be Yourself in Heterogeneous Federated Learning. In Proceedings of the 2022 IEEE/CVF Conference on Computer Vision and Pattern Recognition (CVPR), New Orleans, LA, USA, 18–24 June 2022; pp. 10133–10143. [CrossRef]
17. Popov, S. The Tangle. White Paper. 2018. Version 1.4.3. Available online: http://www.descryptions.com/Iota.pdf (accessed on 6 September 2022).
18. Wright, D.C.S. Bitcoin: A Peer-to-Peer Electronic Cash System. 2008. Available online: https://papers.ssrn.com/sol3/papers.cfm?abstract_id=3440802 (accessed on 6 September 2022).
19. dos Anjos, J.C.S.; Gross, J.L.G.; Matteussi, K.J.; González, G.V.; Leithardt, V.R.Q.; Geyer, C.F.R. An Algorithm to Minimize Energy Consumption and Elapsed Time for IoT Workloads in a Hybrid Architecture. *Sensors* **2021**, *21*, 2914. [CrossRef]
20. Cao, M.; Zhang, L.; Cao, B. Toward On-Device Federated Learning: A Direct Acyclic Graph-Based Blockchain Approach. *IEEE Trans. Neural Netw. Learn. Syst.* **2021**, 1–15. [CrossRef]
21. Yuan, S.; Cao, B.; Peng, M.; Sun, Y. ChainsFL: Blockchain-driven Federated Learning from Design to Realization. In Proceedings of the 2021 IEEE Wireless Communications and Networking Conference (WCNC), Nanjing, China, 29 March–1 April 2021; pp. 1–6. [CrossRef]
22. Ongaro, D.; Ousterhout, J. In Search of an Understandable Consensus Algorithm. In Proceedings of the 2014 USENIX Annual Technical Conference (USENIX ATC 14), Philadelphia, PA, USA, 19–20 June 2014; pp. 305–319.
23. Androulaki, E.; Barger, A.; Bortnikov, V.; Cachin, C.; Christidis, K.; De Caro, A.; Enyeart, D.; Ferris, C.; Laventman, G.; Manevich, Y.; et al. Hyperledger Fabric: A Distributed Operating System for Permissioned Blockchains. In Proceedings of the 13th EuroSys Conference, EuroSys'18, Porto, Portugal, 23–26 April 2018; Association for Computing Machinery: New York, NY, USA, 2018; pp. 1–15. [CrossRef]
24. Schmid, R.; Pfitzner, B.; Beilharz, J.; Arnrich, B.; Polze, A. Tangle Ledger for Decentralized Learning. In Proceedings of the 2020 IEEE International Parallel and Distributed Processing Symposium Workshops (IPDPSW), New Orleans, LA, USA, 8–22 May 2020; pp. 852–859. [CrossRef]
25. Beilharz, J.; Pfitzner, B.; Schmid, R.; Geppert, P.; Arnrich, B.; Polze, A. Implicit Model Specialization through Dag-Based Decentralized Federated Learning. In Proceedings of the 22nd International Middleware Conference, Middleware'21, Québec City, QC, Canada, 6–10 December 2021; Association for Computing Machinery: New York, NY, USA, 2021; pp. 310–322. [CrossRef]
26. Tran, N.H.; Bao, W.; Zomaya, A.; Nguyen, M.N.H.; Hong, C.S. Federated Learning over Wireless Networks: Optimization Model Design and Analysis. In Proceedings of the IEEE INFOCOM 2019—IEEE Conference on Computer Communications, Paris, France, 29 April–2 May 2019; pp. 1387–1395. [CrossRef]
27. Lu, Y.; Huang, X.; Zhang, K.; Maharjan, S.; Zhang, Y. Communication-Efficient Federated Learning and Permissioned Blockchain for Digital Twin Edge Networks. *IEEE Internet Things J.* **2021**, *8*, 2276–2288. [CrossRef]
28. Dimarogonas, D.V.; Frazzoli, E.; Johansson, K.H. Distributed Event-Triggered Control for Multi-Agent Systems. *IEEE Trans. Autom. Control* **2012**, *57*, 1291–1297. [CrossRef]
29. Lemmon, M. Event-Triggered Feedback in Control, Estimation, and Optimization. In *Networked Control Systems*; Bemporad, A., Heemels, M., Johansson, M., Eds.; Lecture Notes in Control and Information Sciences; Springer: London, UK, 2010; pp. 293–358. [CrossRef]
30. Ghosh, S.; Aquino, B.; Gupta, V. EventGraD: Event-triggered Communication in Parallel Machine Learning. *Neurocomputing* **2022**, *483*, 474–487. [CrossRef]
31. Nguyen, N.; Han, S. AET-SGD: Asynchronous Event-triggered Stochastic Gradient Descent. *arXiv* **2021**, arXiv:2112.13935.
32. George, J.; Gurram, P. Distributed Stochastic Gradient Descent with Event-Triggered Communication. *Proc. AAAI Conf. Artif. Intell.* **2020**, *34*, 7169–7178. [CrossRef]
33. George, J.; Gurram, P. Distributed Deep Learning with Event-Triggered Communication. *arXiv* **2019**, arXiv:1909.05020.
34. Kajiyama, Y.; Hayashi, N.; Takai, S. Distributed Subgradient Method With Edge-Based Event-Triggered Communication. *IEEE Trans. Autom. Control* **2018**, *63*, 2248–2255. [CrossRef]
35. Caldas, S.; Duddu, S.M.K.; Wu, P.; Li, T.; Konečný, J.; McMahan, H.B.; Smith, V.; Talwalkar, A. LEAF: A Benchmark for Federated Settings. *arXiv* **2019**, arXiv:1812.01097.
36. Bourdon, A.; Noureddine, A.; Rouvoy, R.; Seinturier, L. PowerAPI: A Software Library to Monitor the Energy Consumed at the Process-Level. *ERCIM News* **2013**, *92*, 43–44.
37. Georgiou, S.; Rizou, S.; Spinellis, D. Software Development Lifecycle for Energy Efficiency: Techniques and Tools. *ACM Comput. Surv.* **2019**, *52*, 1–33. [CrossRef]

Article

Machine Learning-Based Prediction Models of Acute Respiratory Failure in Patients with Acute Pesticide Poisoning

Yeongmin Kim [1,†], Minsu Chae [2,†], Namjun Cho [3], Hyowook Gil [3] and Hwamin Lee [2,*]

1. Department of Computer Software Engineering, Soonchunhyang University, Asan 31538, Republic of Korea
2. Department of Medical Informatics, College of Medicine, Korea University, Seoul 02841, Republic of Korea
3. Department of Internal Medicine, Soonchunhyang University Cheonan Hospital, Cheonan 31151, Republic of Korea
* Correspondence: hwamin@korea.ac.kr; Tel.: +82-2-3407-2099
† These authors contributed equally to this work.

Abstract: The prognosis of patients with acute pesticide poisoning depends on their acute respiratory condition. Here, we propose machine learning models to predict acute respiratory failure in patients with acute pesticide poisoning using a decision tree, logistic regression, and random forests, support vector machine, adaptive boosting, gradient boosting, multi-layer boosting, recurrent neural network, long short-term memory, and gated recurrent gate. We collected medical records of patients with acute pesticide poisoning at the Soonchunhyang University Cheonan Hospital from 1 January 2016 to 31 December 2020. We applied the k-Nearest Neighbor Imputer algorithm, MissForest Impuer and average imputation method to handle the problems of missing values and outliers in electronic medical records. In addition, we used the min–max scaling method for feature scaling. Using the most recent medical research, p-values, tree-based feature selection, and recursive feature reduction, we selected 17 out of 81 features. We applied a sliding window of 3 h to every patient's medical record within 24 h. As the prevalence of acute respiratory failure in our dataset was 8%, we employed oversampling. We assessed the performance of our models in predicting acute respiratory failure. The proposed long short-term memory demonstrated a positive predictive value of 98.42%, a sensitivity of 97.91%, and an F1 score of 0.9816.

Keywords: machine learning; respiratory failure; acute pesticide poisoning; logistic regression; random forests; long short-term memory

MSC: 68T07; 9A16; 4008

1. Introduction

Pesticide toxicosis is caused by the ingestion of or exposure to pesticides [1]. In the Republic of Korea, the death toll from toxicosis is 2702 people and 1675 of the 2702 people (61.99%) had toxicosis caused by the ingestion of pesticides [2]. In addition, 71% of patients with pesticide poisoning have been reported to die within 6–24 h [2]. The most common reason for the ingestion of pesticides is suicide [3]. Each year, 110,000 individuals die from pesticide poisoning [3], which accounts for 13.7% of all suicides [3]. In the Republic of Korea, some regions (Chungcheong-do, Gangwon-do, Jeolla-do) have a higher death rate from pesticide poisoning than that the capital area (Seoul, Incheon, and Gyeonggi-do) [2]. Pesticide toxicosis is easily accessible, especially in these regions [2]. Neurological, respiratory, and cardiovascular symptoms have been reported in cases of pesticide toxicosis [1]. The prognosis of pesticide toxicosis depends on the extent of respiratory failure [1]. Respiratory failure is associated with a high death rate in hospitals. Current respiratory failure treatment options can be ineffective [4]. Preventing the failure of multiple organs is crucial in reducing the rate of mortality from respiratory failure [5]. Therefore, the prediction of respiratory failure is important for patient prognosis.

Recent predictions of respiratory failure include predicting respiratory failure based on semi-supervised learning [4]; predicting respiratory failure with clinical data [5]; predicting respiratory failure in patients with coronavirus disease-2019 (COVID-19) [6]; predicting respiratory failure in the intensive care unit (ICU) [7,8]; predicting respiratory failure in pesticide intoxication [9]; and predicting respiratory failure with simple patient trajectories [10].

Machine learning algorithms such as semi-recurrent neural networks (RNNs), extreme gradient boosting, logistic regression (LR), random forest (RF), and long short-term memory (LSTM) have been used to predict respiratory failure in previous studies. In the case of respiratory failure prediction based on semi-RNNs, the positive predictive value (PPV) was 3.3% and the sensitivity was 78.0% [4]. In the case of respiratory failure prediction with clinical data, the sensitivity was 71% [5]. In the case of respiratory failure prediction in patients with COVID-19, the PPV was 74% and the sensitivity was 78% [6]. In the case of respiratory failure prediction in the ICU, the PPV was 42% and the sensitivity was 80% [7]. In the case of respiratory failure prediction in pesticide intoxication, the PPV was 83.3% and the sensitivity was 60.6% [9]. In the case of respiratory failure prediction with simple patient trajectories, the PPV was 22.6% and the sensitivity was 88.1% [10]. Therefore, the performance of algorithms for predicting acute respiratory failure is low.

Our goal is to predict the prognosis for patients with acute pesticide poisoning. However, it is difficult to predict the prognosis because of the various causes of acute pesticide poisoning. We predict acute respiratory failure, an important prognostic factor for patients with acute pesticide poisoning. We predict acute respiratory failure within 24 h using machine learning and three-hour electronic medical records (EMRs) for patients. We perform EMR preprocessing as follows: (1) solve human errors; (2) solve missing values; (3) sliding window; (4) feature selection; (5) data scaling; and (6) solve the imbalance. Data preprocessing is important to improve performance [11–14]. Using current patient data to fill in the gaps, we imputed missing values using the k-Nearest Neighbor (KNN) imputer algorithm from scikit-learn [15], the MissForest Imputer, or the data average imputation technique. We used a sliding window dataset based on 3 h data. We performed feature selection based on the current medical knowledge and p-values and oversampling. We performed data scaling using MinMaxScaler provided by scikit-learn [15]. For predicting acute respiratory failure in acute pesticide poisoning, we utilize shallow learning such as decision tree (DT), random forest (RF), logistic regression (LR), support vector machine (SVM), adaptive boosting (AB), gradient boosting (GB), and deep learning such as multilayer perceptron (MLP), recurrent neural network (RNN), long short-term memory (LSTM), and gated recurrent unit (GRU).

2. Materials

2.1. Data

This retrospective cohort study consisted of patients admitted to Soonchunhyang University Cheonan Hospital in the Republic of Korea between January 2016 and December 2020. The patients were over 19 years of age. The patients with pesticide poisoning and respiratory failure within 1 h of admission were excluded. The number of patients was 707. The pesticide categories included glyphosate, glufosinate, paraquat, organophosphate, pyrethroid, and carbamate. After replacing missing data, we performed sliding window data preprocessing, feature selection, and oversampling on the medical records.

When the data preprocessing process was completed, the total data consisted of 11,526 data with 17 features for 3 h. We split the data into the training dataset and test dataset at a 7:3 ratio. The training dataset was then divided with a 7:3 ratio into a training dataset and a test dataset. The training dataset was then divided with a 7:3 ratio into a training dataset and a holdout fold. The training dataset was then divided with a 7:3 ratio into a training dataset and a validation dataset. Using the training dataset, machine learning methods were constructed and evaluated using the holdout fold. The number of respiratory failures was only 909. Oversampling or undersampling can be used to solve the imbalance in the datasets. We used oversampling algorithms such as the synthetic minority oversampling technique (SMOTE), borderline-SMOTE, and adaptive synthetic (ADASYN) given the limited cases of respiratory failure. Figure 1 shows the processing of patient selection. The number of training datasets was 7291, the number of validation set was 1695, and the number of holdout fold was 2421, and the number of test datasets was 3458.

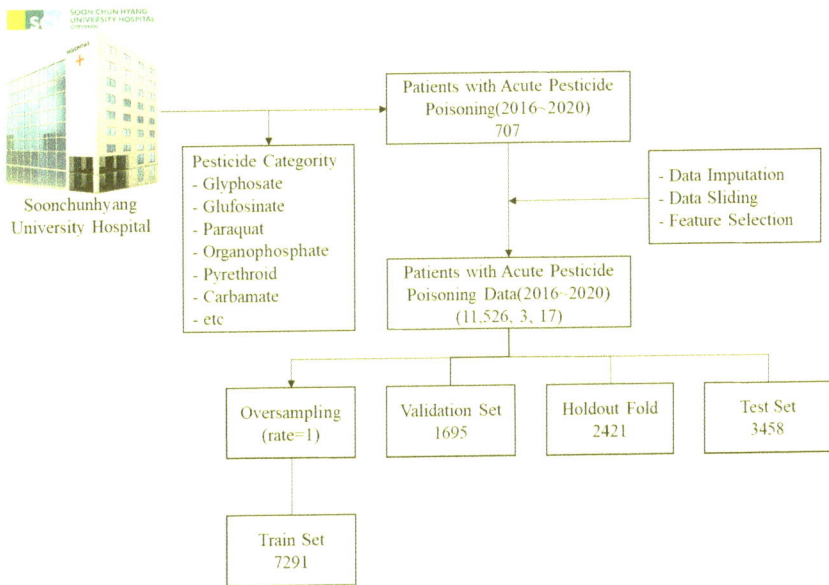

Figure 1. Patient selection, data preprocessing, and dataset. The training dataset number was 7291, and the validation set number was 1695, and the holdout fold number was 2421, and the test dataset number was 3458.

2.2. Replacement Missing Value

Medical records are not free from missing values. The patient may be absent or there may be problems with noise or human errors. To improve machine learning algorithms, missing values need to be solved [11,12]. We solved missing values with the following three steps: (1) replace missing values with the recent data imputer (RDI); (2) apply the KNN imputation algorithm of scikit-learn with highly relevant features; and (3) replace other features through average imputation.

2.2.1. Recent Data Imputer

Time-series data have continuous values over time. The RDI can be used to replace the missing values of each patient with recent data. Figure 2 shows the RDI algorithm. The average imputer, the maximum imputer, and the minimum imputer are non-consistent time-series characteristics. The RDI has time-series characteristics.

2.2.2. k-Nearest Neighbor (KNN) Imputer

There may be missing values after using the RDI. The KNN imputer replaces missing values through distance functions for highly correlated features [13]. Improved performance may be achieved with the KNN compared with that of other imputers, such as the average imputer, maximum imputer, and minimum imputer [13]. We performed KNN imputation with highly relevant features twice as follows: (1) total CO_2, pH, HCO_3 standard, base excess, and lactate features; (2) pCO_2 and pO_2 features.

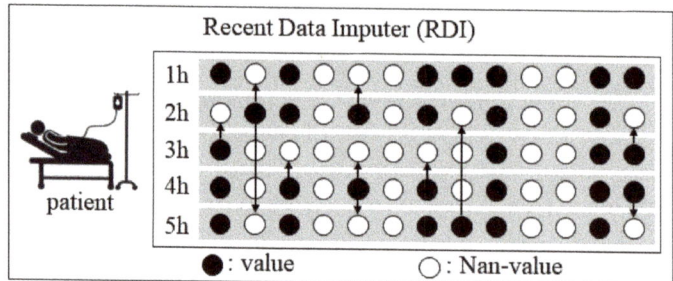

Figure 2. Processing by the recent data imputer (RDI). The black nodes indicate measured values and the white nodes indicate missing values.

2.2.3. MissForest Imputer

The MissForest is an imputer based on RF [16]. Figure 3 shows the processing of MissForest. The MissForest replaces missing values to the median and trains on the dataset to interpolate missing values into prediction results.

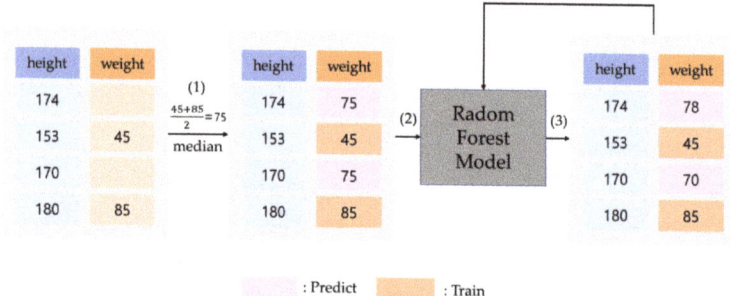

Figure 3. Processing by the MissForest. (1) Interpolate missing values into the median. (2) Train via dataset. (3) Interpolate missing values into the prediction results.

We interpolate total CO_2, pH, HCO_3 standard, base excess lactate, pCO_2, and pO_2 after performing the RDI algorithm.

2.3. Feature Selection

Feature selection is important to improve the machine learning algorithm [14]. We perform according to a feature selection. (1) We calculate p-values to confirm unrelated features. (2) We determine features based on current medical knowledge. (3) We analyze the importance of features using RF and GB. (4) Using recursive feature elimination, we analyze both high- and low-ranking features. (5) In low-ranking features, we compare performance results exclusive to each feature.

First, we calculate the p-value for each feature and respiratory failure using ordinary least squares. To calculate the p-value, we utilize the OLS method provided by statsmodels [17]. We ignored features with p-value above 0.05 because they are uncorrelated factors. Table 1 shows the p-values for each feature.

We confirm that the features of p-values above 0.05 are as follows: smoking; alcohol; cardiovascular disease; SBP max; DBP max; RR max; Hb; glucose; BUN; creatinine; pCO_2; HCO_3 standard; BE; and troponin.

Table 1. The respiratory failure was correlated to *p*-values of features.

Features	*p*-Value	Features	*p*-Value
Pesticide dose	0.000	WBC	0.000
Sex	0.000	PLT	0.000
Age	0.000	Albumin	0.000
BMI	0.023	Glucose	0.070
Smoking	0.326	BUN	0.622
Alcohol	0.313	Creatinine	0.100
Diabetes disease	0.000	Total CO_2	0.000
Respiratory disease	0.000	C-reactive protein 1	0.000
Cardiovascular disease	0.995	pH	0.000
GCS	0.000	pCO_2	0.199
SBP max	0.088	pO_2	0.000
DBP max	0.897	O_2 saturation	0.000
HR max	0.000	HCO_3 standard	0.356
RR max	0.174	BE	0.120
BT max	0.000	Troponin	0.555
Hb	0.221	Lactate	0.000

BMI: body mass index; GCS: Glasgow Coma Scale; SBP: systolic blood pressure; DBP: diastolic blood pressure; HR: heart rate; RR: respiratory rate; BT: body temperature; max: maximum; Hb: hemoglobin; WBC: white blood cell; PLT: platelet; BUN: blood urea nitrogen; BE: base excess.

Second, we perform feature selection based on current medical knowledge. The following features are determined based on current medical knowledge: pesticide category; pesticide dose; sex; age; GCS; SBP max; HR max; BT max; WBC; PLT; albumin; total CO_2; CRP1; pH; pO_2; O_2 saturation; and lactate.

Third, we analyze tree-based feature selection methods. Table 2 shows the performance results of RF and GB. We confirm that sex is an unimportant feature.

Table 2. Performance result of tree-based feature selection method.

Features	Tree-Based Feature Selection		Features	Tree-Based Feature Selection	
	RF	GB		RF	GB
Pesticide category	0.056	0.073	PLT	0.032	0.013
Pesticide dose	0.032	0.017	Albumin	0.034	0.008
Sex	0.001	0.0003	total_CO_2	0.072	0.025
Age	0.039	0.019	C-reactive_protein_1	0.053	0.057
GCS	0.053	0.068	pH	0.175	0.247
SBP_max	0.010	0.003	pO_2	0.018	0.003
HR_max	0.087	0.097	O_2_saturation	0.062	0.017
BT_max	0.033	0.029	Lactate	0.017	0.006
WBC	0.224	0.316			

RF: Random Forest; GB: Gradient Boost; GCS: Glasgow Coma Scale; SBP: systolic blood pressure; max: maximum; HR: heart rate; BT: body temperature; WBC: white blood cell; PLT: platelet.

Fourth, using recursive feature elimination, we analyze high- and low-ranking features. We perform recursive feature elimination based on SVM with linear, LR, RF, DT, and GB. We perform recursive feature elimination provided by scikit-learn. Table 3 shows the performance results of recursive feature elimination based on SVM with linear, LR, DT, and GB. We confirm that sex, albumin, and PLT are unimportant factors.

Fifth, we compare the performance results of each feature based on Tables 2 and 3. Table 4 shows the performance results of each feature using RF, GB, and MLP. The feature of the second stage is the highest performance in Table 4. We separate our dataset into train and test datasets. In the case of RF and GB, we train algorithms using stratified k-folds by train folds after separating train datasets into train folds and holdout folds. We train MLP algorithms using train folds and early stop using validation folds after separating train datasets into the train, validation, and holdout folds. In addition, we compare the performance of each feature via each algorithm using holdout folds.

Table 3. Performance result of recursive feature elimination of each algorithm. Low-rank features are albumin, platelet, and sex. High-rank feature is pH.

Machine Learning Algorithm	Low-Rank Feature	High-Rank Feature
SVM with linear	Albumin	pH
LR	PLT	pH
RF	Sex	pH
DT	Albumin	pH
GB	Sex	pH

SVM: support vector machine; LR: logistic regression; RF: random forest; DT: decision tree; GB: gradient boost; PLT: platelet.

Table 4. Performance result of feature selection based on RF, GB, and MLP.

Feature	Algorithm	PPV	Sensitivity	F1 Score	AUC
Reference	RF	98.92%	96.34%	0.9761	0.9812
	GB	96.81%	95.29%	0.9604	0.9751
	MLP	96.81%	95.29%	0.9604	0.9751
Reference exclude sex	RF	98.92%	95.81%	0.9734	0.9786
	GB	92.78%	94.24%	0.9351	0.9681
	MLP	92.78%	94.24%	0.9351	0.9681
Reference exclude PLT	RF	99.46%	95.81%	0.9760	0.9788
	GB	98.89%	93.19%	0.9596	0.9655
	MLP	92.78%	94.24%	0.9351	0.9681
Reference exclude albumin	RF	98.39%	95.81%	0.9708	0.9784
	GB	96.70%	92.15%	0.9437	0.9594
	MLP	96.70%	92.15%	0.9437	0.9594

PPV: positive predictive value; RF: random forest; GB: gradient boosting; MLP: multi-layer perceptron; PLT: platelet.

For the performance evaluation of each feature, we perform steps one through five. The first and second steps exhibit the highest performance when each feature's performance is evaluated. Therefore, we used the following features: pesticide category; pesticide dose; sex; age; GCS; SBP max; HR max; BT max; WBC; PLT; albumin; total CO_2; CRP1; pH; pO_2; O_2 saturation; and lactate.

2.4. Hour Sliding Window in 24 H

In this study, our objective was to predict respiratory failure within 24 h using 3 h data. Figure 4 shows the sliding window to time-series data in 24 h based on 3 h data.

2.5. MinMaxScaler

In machine learning, each feature unit is different; thus, the results can be biased. To solve this, it is necessary to use the same data range. In this study, we used the MinMaxScaler provided by scikit-learn [15]. It expresses a value between 0 and 1 through Equation (1).

$$X = \frac{X - \min}{\max - \min} \quad (1)$$

The X variable represents a feature in the dataset. The min variable represents the minimum value of each feature. The max variable represents the maximum value of each feature.

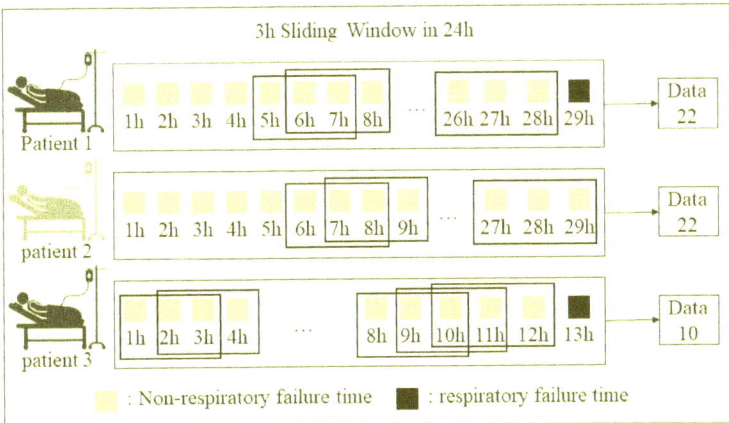

Figure 4. Processing by sliding window.

2.6. Oversampling

In this study, the prevalence of respiratory failure was 8%, which suggests an imbalance. There are two methods of solving the imbalance: (1) oversampling and (2) undersampling. In this study, the number of cases of respiratory failure was limited and oversampling rather than undersampling tends to be better for improving performance [18]. Therefore, we performed oversampling. We applied SMOTE, borderline-SMOTE, and ADADYN to solve the data imbalance [19].

2.6.1. Synthetic Minority Oversampling Technique (SMOTE)

The SMOTE algorithm was proposed by Chawla et al. [20]. The SMOTE algorithm has three steps as follows: (1) apply the KNN algorithm in the minority class after randomized minority data selection [20]; (2) choose randomized minority data in the nearest data [20]; and (3) locate generated data between one-step and two-step data. Figure 5 shows the processing of the SMOTE algorithm.

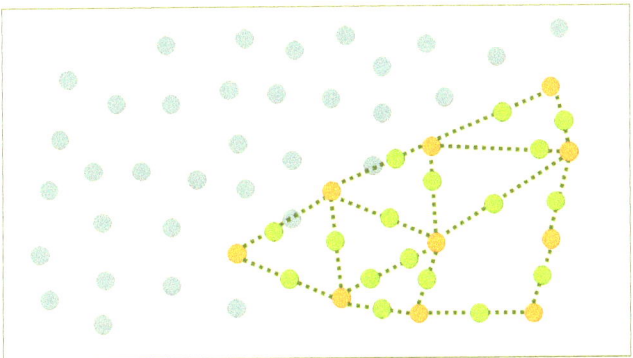

Figure 5. SMOTE processing. Blue indicates the majority dataset (non-respiratory failure). Orange indicates the minority dataset (respiratory failure). Green indicates the minority dataset (respiratory failure), which is generated by SMOTE.

2.6.2. Borderline-SMOTE

The borderline-SMOTE algorithm is based on the original SMOTE [21]. Notably, minority data may be far from the minority data-majority data boundary [21]. The borderline-SMOTE is applied at the boundary between minority data and majority data. After the

KNN algorithm is applied to the minority data, the borderline-SMOTE algorithm considers data comprising more than half of the data to be borderline. The borderline-SMOTE applies the SMOTE algorithm to minority data at the borderline. Figure 6 shows data processing by the borderline-SMOTE, which generates minority data at the borderline and achieves classifier efficiency [19,21].

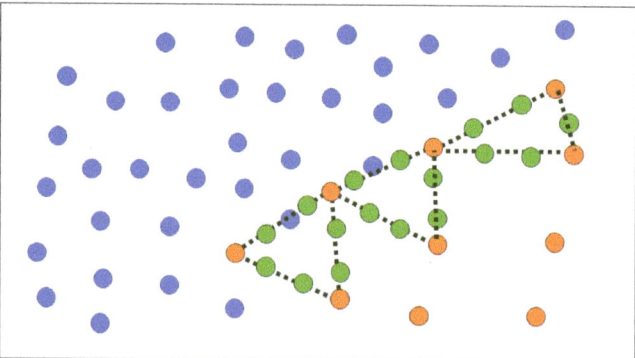

Figure 6. Borderline-SMOTE processing. Blue indicates the majority dataset (non-respiratory failure). Orange indicates the minority dataset (respiratory failure). Green indicates the minority dataset (respiratory failure) generated by Borderline-SMOTE.

2.6.3. Adaptive Synthetic (ADASYN)

The ADASYN algorithm applies the KNN algorithm to minority data to generate minority data if there is a huge amount of majority data [19]. The ADASYN algorithm consists of four steps as follows [19]: (1) KNN algorithm calculates the ratio of minority data to majority data for each minority datum [19]; (2) the sum of the ratios of majority data is divided by each ratio of majority data [19]; (3) it is calculated to repeat through Equation (2) [19]; and (4) generate minority data as much as a repeat on each minority dataset [19]. Figure 7 shows data processing by the ADASYN algorithm.

$$\text{repeat} = \text{second step result} \times (\text{number of majority} - \text{number of minority}) \quad (2)$$

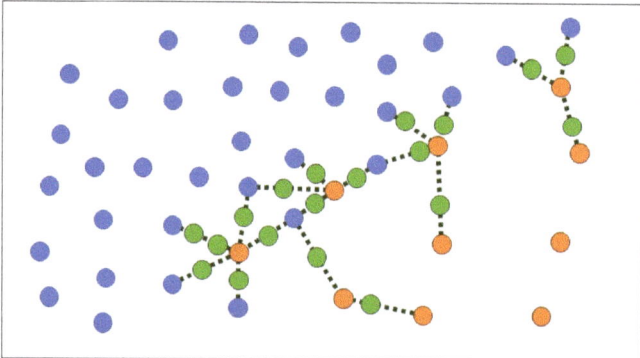

Figure 7. ADASYN processing. Blue indicates the majority dataset (non-respiratory failure). Orange indicates the minority dataset (respiratory failure). Green indicates the minority dataset (respiratory failure) generated by ADASYN. When there are more majority data surrounding minority data, more minority data are produced.

3. Methods

3.1. Shallow Learning

A time-series dataset has three dimensions: number of datasets, amount of time, and number of features, e.g., (11,148, 3, 17). However, the machine learning algorithms provided by scikit-learn [15] use only two dimensions. We used both time-series and non-time-series features. The time-series features were the maximum systolic blood pressure (SBP) in 1 h, maximum heart rate (HR) in 1 h, maximum body temperature (BT) in 1 h, white blood cell (WBC), platelet (PLT), albumin, total CO_2, cysteine-rich protein 1 (CRP1), pH, pO_2, O_2 saturation, and lactate. The non-time-series features were pesticide category, pesticide dose, sex, age, and Glasgow Coma Scale (GCS). We expressed the dataset structure as the number of datasets and number of features, e.g., (11,148, 41).

3.1.1. Logistic Regression (LR)

LR classification is based on the sigmoid function. LR calculates the weight and bias in the training dataset [22]. In this study, LR was used to calculate z with 41 features through Equation (3) [22].

$$Z = \sum_{i=1}^{41} w_i a_i + b \quad (3)$$

LR was used to calculate the sigmoid function with z through Equation (4) [22].

$$F(z) = \frac{1}{1 + e^{-z}} \quad (4)$$

LR classification was performed by f(z) through Equation (5) [22]. If f(z) is greater than 0.5, it is classified as respiratory failure; otherwise, it indicates non-respiratory failure. We utilized the LR algorithm provided by scikit-learn [15].

$$\text{predict} = \begin{cases} 0 & \text{if } f(z) \leq 0.5 \\ 1 & \text{if } f(z) > 0.5 \end{cases} \quad (5)$$

3.1.2. Decision Tree (DT)

DT utilizes a binary tree structure, which can classify highly related features of respiratory failure [3]. DT calculates impurity and branches until the leaf node impurity is 0. There are two methods of calculating impurity: (1) the Gini coefficient and (2) the entropy coefficient. In this study, we calculated impurity using the Gini coefficient through Equation (6) [15].

$$\text{Gini} = \sum_{i=1}^{n} (R_i (1 - \sum_{k=1}^{m} P_k^2)) \quad (6)$$

The n variable represents the number of nodes and the m variable represents the number of outcomes. In this study, m is 2. The R_i variable represents the sample ratio of each branch. The P_k variable represents the class ratio. DT utilizes pruning to solve overfit. Figure 8 shows the DT algorithms. The 1 h prefix indicates the medical record after 1 h of measurement. The 2 h prefix indicates the medical record after 2 h of measurement. The 3 h prefix indicates the medical record after 3 h of measurement. We used the DT model provided by scikit-learn [15].

3.1.3. Random Forest (RF)

RF is an ensemble model that uses many DTs and bootstrap aggregation [23]. Figure 9 shows the processing of RF. After each DT in the RF predicts respiratory failure, the RF classifier votes on the prediction. In this study, we used a max depth hyperparameter of 6 and a n_estimators hyperparameter of 100 in the RandomForestClassifier provided by scikit-learn [15].

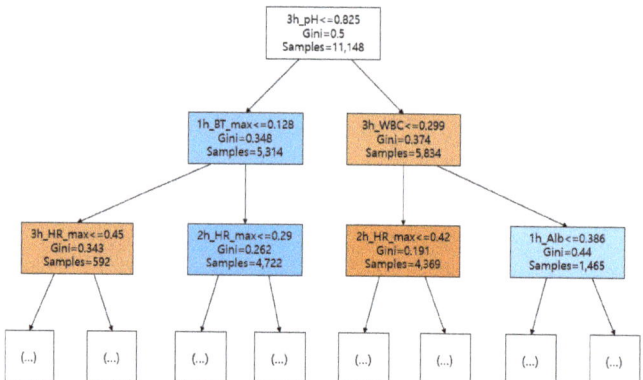

Figure 8. Decision tree. Orange indicates the prediction of respiratory failure in each node. Blue indicates non-respiratory failure. Darker nodes contain more data.

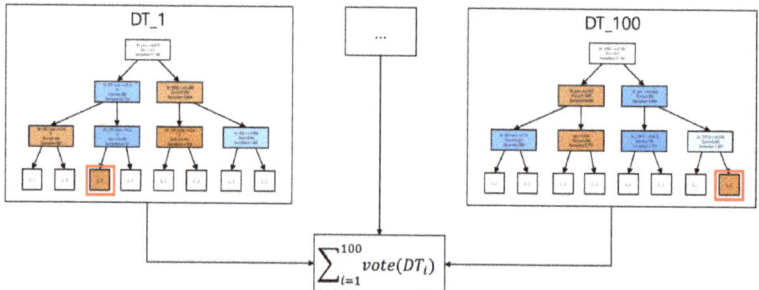

Figure 9. Processing of random forests. The red box in the decision tree indicates the predicted outcome. The random forest classifier is based on decision tree outcomes. Random Forest performs bootstrap aggregation on many decision trees and uses voting to determine the prediction results of many decision trees.

3.1.4. Support Vector Machine (SVM)

SVM is a support vector base classifier and not a decision boundary [24]. There is a problem with overfitting when other labels are near to the decision boundary. A support vector is a calculated boundary that minimizes classification error. Figure 10 shows classification based on the support vector of SVM.

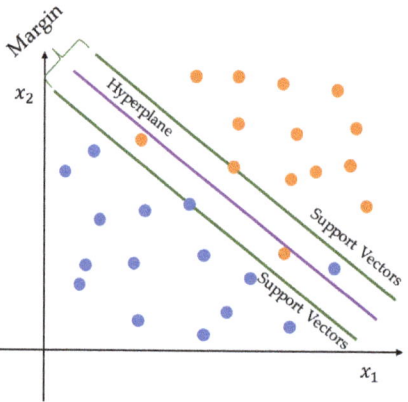

Figure 10. It shows the classification of SVM.

3.1.5. Adaptive Boost (AB)

AB has two leaf nodes named as stump [25]. AB reduces the error by providing the next estimator with the weight of the incorrectly predicted respiratory dataset through the stump [25]. Figure 11 shows the AB weight calculation process.

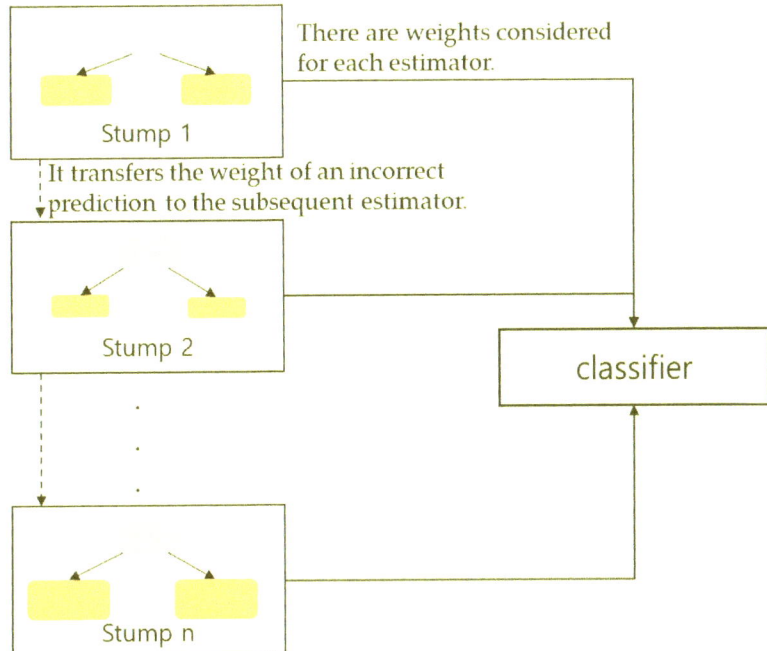

Figure 11. Processing of adaptive boost. Each estimator transfers the weight of an incorrect prediction after training via the dataset.

3.1.6. Gradient Boost (GB)

The GB calculates the residual error for solving an incorrect prediction result by subtracting the measured value from the prediction value. GB calculates the residual error based on the tree using actual respiratory failure and prediction results.

3.2. Deep Learning Algorithms

We organize datasets in three dimensions because RNN-based models support time-series data. We utilize TensorFlow to perform deep learning [26].

3.2.1. Multi-Layer Perceptron (MLP)

The perceptron calculates the weight of many features such as EMRs, for prediction. However, single perceptrons suffer with non-linear datasets. In order to solve the nonlinear problem, the perceptron organizes multiple layers, which is referred to as MLP. The MLP does not support three dimensions, so we organize datasets in two dimensions. We implement MLP utilizing dense layers provided by TensorFlow [26]. Figure 12 shows the structure of MLP.

3.2.2. Recurrent Neural Network (RNN)

Without considering the order of the time series, machine learning recognizes the features of time-series as other features [27]. An RNN model has been proposed for considering the order of the time series. The vanilla RNN model calculates the current weights using the weights from the previous EMRs and the current EMRs. We implement RNN utilizing a simple RNN layer provided by TensorFlow. Figure 13 shows the structure of RNN.

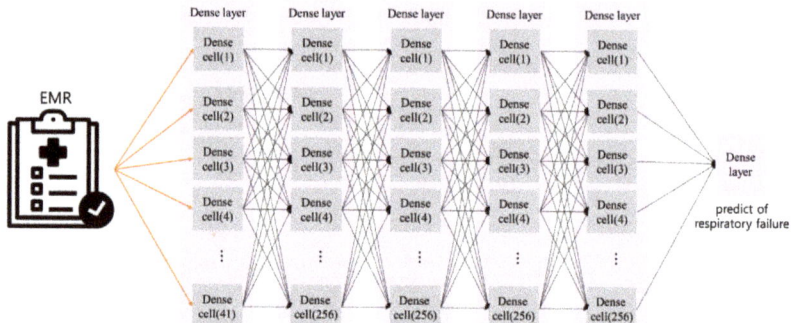

Figure 12. Structure of MLP in this paper. The number of units in hidden layer is 256.

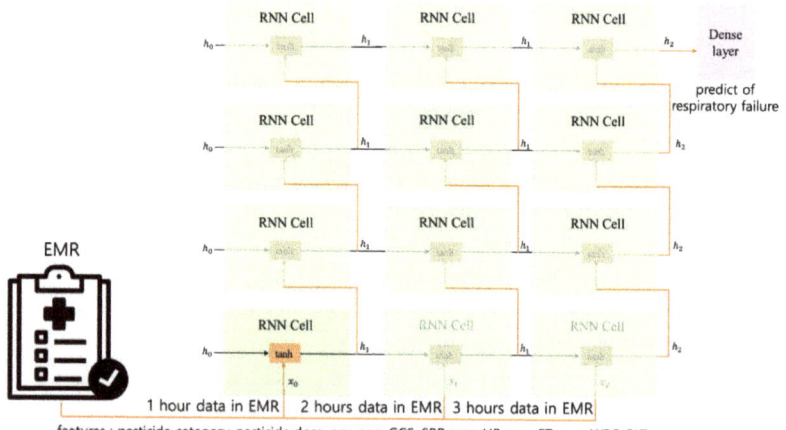

Figure 13. Structure of the RNN model in this paper. The number of units in the RNN layer is 128.

3.2.3. Long Short-Term Memory (LSTM)

The vanilla RNN has the problem of considering only previous and current EMRs [28]. LSTM organizes forget gate, input gate, and output gate to solve short-term dependent problems [28]. The input gate calculates reminder information for long-term memory using the weight of previous short-term memory and current EMRs. The forget gate calculates the removal information for long-term memory using the weight of previous short-term memory and current EMRs. The long-term memory updates using the results of the forget gate and input gate. The output gate calculates weight using previous short-term memory and current EMRs and current long-term memory. We utilize the LSTM layer provided by TensorFlow. Figure 14 shows the structure of LSTM.

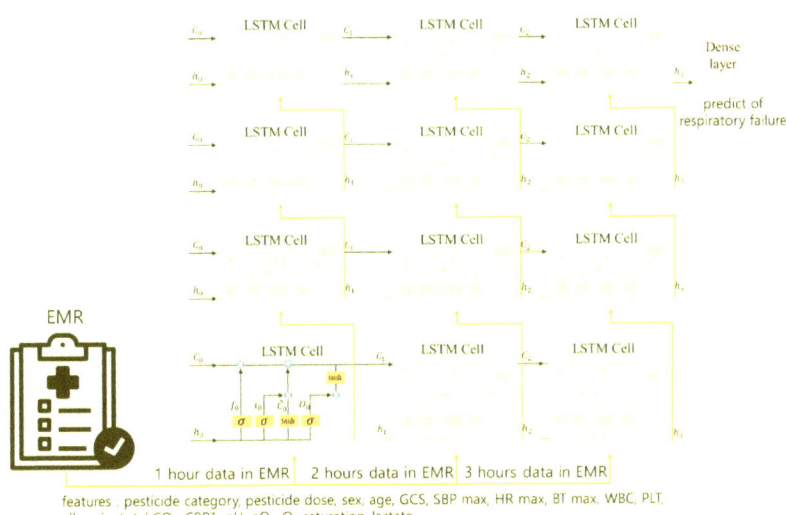

Figure 14. Structure of the LSTM model in this paper. The number of units in the LSTM layer is 128.

3.2.4. Gated Recurrent Unit (GRU)

To solve long-term dependency in vanilla RNN, the GRU organizes update gates and reset gates [29]. The weight for the prediction of acute respiratory failure is calculated through previous memory and current EMRs. The update gates determine whether to use the previous memory or the current weight. The reset gates calculate the removal information for memory. We utilize the GRU layer provided by TensorFlow. Figure 15 shows the structure of GRU.

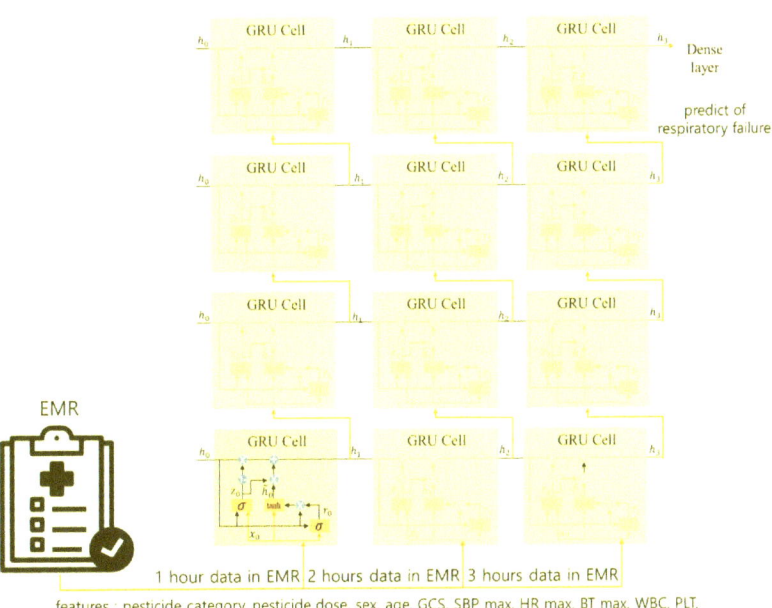

Figure 15. Structure of the GRU model in this paper. The number of units in the GRU layer is 128.

3.3. Stratified-k-Fold

Cross-validation involves the splitting of the training dataset into a training fold and a test fold [30]. The training fold is used for learning in the training step [30]. The test fold is used for validation after the machine learning training step [30]. This technique can be used to solve overfitting [31]. The stratified k-fold is a cross-validation method that separates the data into k-training folds and test folds based on the class ratio (Figure 16).

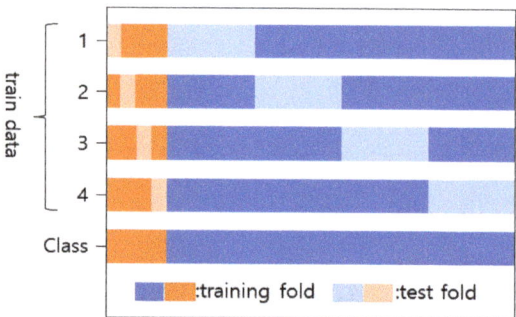

Figure 16. Stratified k-fold processing (in this case, k is 4).

4. Results

4.1. Evaluation Methods

In machine learning, binary classification is mainly evaluated by a confusion matrix. Table 5 shows the confusion matrix. In this study, the patient dataset was imbalanced. Therefore, the PPV and sensitivity are more important than accuracy. The PPV indicates the rate of actual respiratory failure patients among patients with predicted respiratory failure, which can be calculated through Equation (7).

$$\text{PPV} = \frac{\text{TP}}{\text{TP} + \text{FP}} \tag{7}$$

Table 5. Confusion matrix.

		Actual	
		Respiratory failure	Non-respiratory failure
Precision	Respiratory failure	True positive (TP)	False positive (FP)
	Non-respiratory failure	False negative (FN)	True negative (TN)

The sensitivity indicates the rate of patients with predicted respiratory failure among actual patients with respiratory failure, which can be calculated through Equation (8).

$$\text{Sensitivity} = \frac{\text{TP}}{\text{TP} + \text{FN}} \tag{8}$$

The F1 score considers both the PPV and sensitivity, which can be calculated through Equation (9).

$$\text{F1 score} = 2 \times \frac{\text{PPV} \times \text{Sensitivity}}{\text{PPV} + \text{Sensitivity}} \tag{9}$$

4.2. Characteristics of Study

We used 17 features to predict respiratory failure. Table 6 shows the mean and standard deviation of each feature in this study. We excluded the "pesticide category" from the features because there were several pesticide categories. We construct time series characteristics comprising 3 h measurements. To avoid overfitting, we oversampled the training dataset.

Table 6. The characteristics of our dataset.

	Training Data (n = 8068)	Test Data (n = 3458)
Pesticide dose	171.87 ± 138.07	177.18 ± 144.12
Sex, male	4994, 61.90%	2210, 63.91%
Age	61.07 ± 16.37	61.40 ± 16.48
GCS	13.90 ± 2.10	13.85 ± 2.19
1h_SBP_max	124.00 ± 18.03	123.89 ± 18.16
1h_HR_max	76.40 ± 14.37	76.30 ± 14.36
1h_BT_max	36.55 ± 0.36	36.55 ± 0.36
1h_WBC	7.53 ± 3.17	7.54 ± 3.15
1h_PLT	144.52 ± 64.19	142.90 ± 65.09
1h_albumin	3.59 ± 0.48	3.57 ± 0.49
1h_total_CO_2	24.14 ± 3.15	24.23 ± 3.15
1h_C-reactive_protein_1	26.97 ± 50.49	28.73 ± 51.61
1h_pH	7.43 ± 0.06	7.43 ± 0.06
1h_pO_2	93.29 ± 26.62	93.02 ± 24.74
1h_O_2_saturation	96.02 ± 3.67	96.03 ± 3.62
1h_lactate	2.43 ± 2.02	2.48 ± 2.12
2h_SBP_max	123.92 ± 17.90	123.93 ± 18.32
2h_HR_max	76.40 ± 14.38	76.38 ± 14.43
2h_BT_max	36.55 ± 0.35	36.56 ± 0.37
2h_WBC	7.53 ± 3.17	7.54 ± 3.15
2h_PLT	144.12 ± 63.80	142.47 ± 64.86
2h_albumin	3.58 ± 0.48	3.57 ± 0.49
2h_total_CO_2	24.14 ± 3.15	24.23 ± 3.15
2h_C-reactive_protein_1	26.98 ± 50.49	28.74 ± 51.61
2h_pH	7.43 ± 0.06	7.43 ± 0.07
2h_pO_2	93.41 ± 26.20	92.98 ± 24.72
2h_O_2_saturation	95.99 ± 3.79	96.00 ± 3.72
2h_lactate	2.44 ± 2.03	2.48 ± 2.13
3h_SBP_max	123.84 ± 18.01	123.79 ± 17.99
3h_HR_max	76.48 ± 14.52	76.34 ± 14.36
3h_BT_max	36.56 ± 0.35	36.56 ± 0.36
3h_WBC	7.53 ± 3.17	7.53 ± 3.15
3h_PLT	143.94 ± 63.64	142.42 ± 64.84
3h_albumin	3.58 ± 0.48	3.57 ± 0.49
3h_total_CO_2	24.14 ± 3.16	24.24 ± 3.15
3h_C-reactive_protein_1	26.99 ± 50.48	28.74 ± 51.61
3h_pH	7.43 ± 0.07	7.43 ± 0.07
3h_pO_2	93.44 ± 25.88	92.98 ± 24.74
3h_O_2_saturation	95.99 ± 3.81	95.97 ± 3.88
3h_lactate	2.44 ± 2.04	2.49 ± 2.15

GCS: Glasgow Coma Scale; SBP: systolic blood pressure; HR: heart rate; BT: body temperature; WBC: white blood cell; PLT: platelet.

4.3. Evaluation of Imputation

We compared the performances of KNN and missForest after interpolating with RDI. We separate our dataset into train and test datasets. In the case of RF and GB, we train algorithms using stratified k-folds by train folds after separating train datasets into train folds and holdout folds. We train MLP algorithms using train folds and early stop using validation folds after separating train datasets into train, validation, and holdout folds. In addition, we compare the performance of the imputer via each algorithm using holdout folds. The results of KNN and missForest imputation using validation datasets based on RF, GB, and MLP are shown in Table 7.

Table 7. Performance comparison of KNN and missForest based on RF, GB, and MLP.

Imputation	Algorithm	PPV	Sensitivity	F1 Score	AUC
KNN imputer	RF	98.92%	96.34%	0.9761	0.9812
	GB	97.80%	93.19%	0.9544	0.9651
	MLP	96.81%	95.29%	0.9604	0.9751
MissForest imputer	RF	98.92%	96.34%	0.9761	0.9812
	GB	97.74%	90.58%	0.9402	0.9520
	MLP	96.17%	92.15%	0.9412	0.9592

PPV: positive predictive value; RF: random forest; GB: gradient boosting; MLP: multi-layer perceptron.

In this paper, we confirm KNN imputer outperforms the MissForest imputer. We perform the KNN imputer after interpolating through RDI.

4.4. Evaluation of Hyperparameter Tuning

To improve the performance of each algorithm, we perform hyperparameter turning. We separate our dataset into train and test datasets. In the case of RF and GB, we train algorithms using stratified k-folds by train folds after separating train datasets into train folds and holdout folds. We train MLP algorithms using train folds and early stop using validation folds after separating train datasets into train, validation, and holdout folds. In addition, we compare the performance of hyperparameter tuning via each algorithm using holdout folds. On DT and RF, we tune the hyperparameter for the information gain method and max depth. We use SVM to turn hyperparameters for the penalty of square l2 and kernels with radial basis functions (RBF), linear, and polynomial. On AB, we tune the hyperparameters for the number of estimators and learning rate. On GB, we tune hyperparameters for the number of estimators and max depth. On MLP, RNN, LSTM, and GRU, we tune hyperparameters for the unit size and dropout rate. The performance results of each algorithm according to their hyperparameters are shown in Table 8.

Table 8. Performance comparison of hyperparameter turning.

Algorithm	Hyperparameter				PPV	Sensitivity	F1 Score	AUC
DT	Function of computational complexity	Gini	Max depth	8	95.76%	82.72%	0.8876	0.9120
				9	96.49%	86.39%	0.9116	0.9306
				10	98.17%	84.29%	0.9070	0.9208
		Entropy		8	97.27%	93.19%	0.9519	0.9648
				9	98.35%	93.72%	0.9598	0.9679
				10	98.31%	91.62%	0.9485	0.9574
RF	Function of computational complexity	Gini	Max depth	8	98.66%	76.96%	0.8647	0.8844
				9	98.76%	83.25%	0.9034	0.9158
				10	98.74%	82.20%	0.8971	0.9105
		Entropy		8	98.86%	91.10%	0.9482	0.9550
				9	98.92%	96.34%	0.9761	0.9812
				10	98.91%	95.29%	0.9707	0.9760

Table 8. *Cont.*

Algorithm	Hyperparameter				PPV	Sensitivity	F1 Score	AUC
SVM	Regularization parameter	5.5	kernel	RBF	99.32%	76.96%	0.8673	0.8846
				Linear	77.36%	42.92%	0.5522	0.7093
				Poly	97.43%	79.58%	0.8761	0.8970
		6.0		RBF	99.33%	77.49%	0.8706	0.8872
				Linear	76.58%	44.50%	0.5629	0.7167
				Poly	96.84%	80.10%	0.8768	0.8994
		6.5		RBF	99.34%	79.06%	0.8805	0.8950
				Linear	75.68%	43.98%	0.5563	0.7138
				Poly	86.84%	80.10%	0.8768	0.8994
AB	Number of estimators	140	Learning rate	0.5	94.08%	74.87%	0.8338	0.8723
				1.0	95.43%	87.43%	0.9126	0.9354
				2.0	6.91%	84.82%	0.1277	0.4344
		150		0.5	94.34%	78.53%	0.8571	0.8907
				1.0	96.07%	89.53%	0.9268	0.9461
				2.0	6.91%	84.82%	0.1277	0.4344
		160		0.5	94.97%	79.06%	0.8629	0.8935
				1.0	94.97%	89.01%	0.9189	0.9430
				2.0	6.91%	84.82%	0.1277	0.4344
GB	Number of estimators	110	Max depth	3	95.73%	82.99%	0.8845	0.9094
				4	97.77%	91.62%	0.9459	0.9572
				5	97.80%	93.19%	0.9544	0.9651
		120		3	96.45%	85.34%	0.9056	0.9254
				4	97.80%	93.19%	0.9544	0.9651
				5	97.80%	93.19%	0.9544	0.9651
		130		3	96.45%	85.34%	0.9056	0.9254
				4	97.78%	92.15%	0.9488	0.9598
				5	96.55%	87.96%	0.9205	0.9384
MLP	Unit size	64	Dropout rate	20%	97.13%	88.48%	0.9260	0.9413
				30%	96.63%	90.05%	0.9322	0.9489
				40%	95.71%	81.68%	0.8814	0.9068
		128		20%	97.71%	89.53%	0.9344	0.9467
				30%	98.24%	87.43%	0.9252	0.9365
				40%	94.97%	79.06%	0.8629	0.8935
		256		20%	96.81%	95.29%	0.9604	0.9751
				30%	97.80%	93.19%	0.9544	0.9651
				40%	92.73%	80.10%	0.8596	0.8798

Table 8. Cont.

Algorithm	Hyperparameter					PPV	Sensitivity	F1 Score	AUC
RNN	Unit size	64	Dropout rate		20%	76.28	62.30%	0.6859	0.8032
					30%	86.26%	82.20%	0.8418	0.9054
					40%	71.34%	66.49%	0.6883	0.8210
		128			20%	98.32%	92.15%	0.9514	0.9601
					30%	97.85%	95.29%	0.9655	0.9755
					40%	78.08%	59.69%	0.6766	0.7913
		256			20%	96.65%	90.58%	0.9351	0.9515
					30%	64.88%	69.63%	0.6717	0.8320
					40%	98.22%	86.91%	0.9222	0.9339
LSTM	Unit size	64	Dropout rate		20%	95.72%	93.72$	0.9471	0.9668
					30%	98.29%	90.05%	0.9399	0.9496
					40%	72.15%	59.69%	0.6533	0.7886
		128			20%	96.15%	91.62%	0.9383	0.9565
					30%	98.29%	90.05%	0.9399	0.9459
					40%	98.88%	92.67%	0.9568	0.9629
		256			20%	98.87%	91.62%	0.9511	0.9577
					30%	97.77%	91.62%	0.9459	0.9572
					40%	98.86%	91.10%	0.9482	0.9550
GRU	Unit size	64	Dropout rate		20%	97.75%	91.10%	0.9431	0.9546
					30%	98.82%	87.43%	0.9278	0.9367
					40%	98.32%	92.15%	0.9514	0.9601
		128			20%	98.88%	92.15%	0.9539	0.9603
					30%	97.16%	89.53%	0.9319	0.9465
					40%	97.30%	94.24%	0.9574	0.9701
		256			20%	97.78%	92.15%	0.9488	0.9598
					30%	98.31%	91.10%	0.9457	0.9548
					40%	98.34%	93.19%	0.9570	0.9653

PPV: positive predictive value; DT: decision tree; RF: random forest; SVM: support vector machine; RBF: radial basis function; poly: polynomial; AB: adaptive boost; GB: gradient boost; MLP: multi-layer perceptron; RNN: recurrent neural network; LSTM: long short-term memory; GRU: gated recurrent unit.

DT had the highest F1 score when the information gain method and max depth were set to entropy and 9, respectively. RF had the highest F1 score when the information gain method and max depth were set to entropy and 9, respectively. SVM had the highest F1 score when the penalty of square l2 and kernel was set to 6.5 and radial basis function, respectively. AB had the highest F1 score when the number of estimators and learning rate were set to 150 and 1.0, respectively. GB had the highest F1 score when the number of estimators and max depth was set to 120 and 4, respectively. MLP had the highest F1 score when the unit size and dropout rate were set to 256 and 20%, respectively. RNN had the highest F1 score when the unit size and dropout rate were set to 128 and 30%, respectively. LSTM had the highest F1 score when the unit size and dropout rate were set to 128 and 40%, respectively. GRU had the highest F1 score when the unit size and dropout rate were set to 128 and 40%, respectively.

4.5. Evaluation of Oversampling Algorithms

We separate our dataset into train and test datasets. In the case of shallow machine learning, we train algorithms using stratified k-folds by train folds after separating train datasets into train folds and holdout folds. We train deep learning using train folds and early stop using validation folds after separating train datasets into train, validation, and holdout folds. In addition, we compare the performance of each oversampling via each algorithm using holdout folds. Table 9 shows the performance of each machine learning algorithm and oversampling algorithm. Table 9 showed the best performance at GRU with ADASYN.

Table 9. Prediction performance. GRU with ADASYN demonstrated the highest performance.

Machine Learning Algorithm	Oversampling	PPV	Sensitivity	F1 Score	AUC
LR	SMOTE	42.42%	86.39%	0.5690	0.8817
	Borderline SMOTE	43.24%	85.34%	0.5739	0.8787
	ADASYN	44.24%	86.39%	0.5851	0.8853
DT	SMOTE	86.27%	92.15%	0.8911	0.9545
	Borderline SMOTE	92.22%	93.19%	0.9271	0.9626
	ADASYN	86.12%	94.24%	0.9000	0.9647
RF	SMOTE	96.43%	98.95%	0.9767	0.9932
	Borderline SMOTE	95.43%	98.43%	0.9691	0.9901
	ADASYN	96.43%	98.90%	0.9765	0.9929
SVM	SMOTE	91.09%	96.34%	0.9364	0.9776
	Borderline SMOTE	90.53%	90.05%	0.9029	0.9462
	ADASYN	90.40%	93.72%	0.9203	0.9643
AB	SMOTE	83.11%	95.29%	0.8878	0.9681
	Borderline SMOTE	85.57%	90.05%	0.8776	0.9438
	ADASYN	83.49%	92.67%	0.8784	0.9555
GB	SMOTE	95.96%	99.48%	0.9769	0.9956
	Borderline SMOTE	95.90%	97.91%	0.9689	0.9877
	ADASYN	98.45%	99.48%	0.9896	0.9967
MLP	SMOTE	98.94%	97.91%	0.9842	0.9891
	Borderline SMOTE	96.84%	96.34%	0.9659	0.9803
	ADASYN	98.38%	95.29%	0.9681	0.9758
RNN	SMOTE	97.33%	95.29%	0.9630	0.9753
	Borderline SMOTE	98.35%	93.72%	0.9598	0.9679
	ADASYN	96.35%	96.86%	0.9661	0.9827

Table 9. Cont.

Machine Learning Algorithm	Oversampling	PPV	Sensitivity	F1 Score	AUC
LSTM	SMOTE	97.85%	95.29%	0.9655	0.9755
	Borderline SMOTE	98.38%	95.29%	0.9681	0.9758
	ADASYN	98.93%	96.86%	0.9788	0.9838
GRU	SMOTE	98.3–8%	95.29%	0.9681	0.9758
	Borderline SMOTE	98.38%	95.29%	0.9681	0.9758
	ADASYN	98.42%	97.91%	0.9816	0.9889

PPV: positive predictive value; LR: logistic regression; DT: decision tree; RF: random forest; SVM: support vector machine; AB: adaptive boost; GB: gradient boost; MLP: multi-layer perceptron; RNN: recurrent neural network; LSTM: long short-term memory; GRU: gated recurrent unit.

4.6. Evaluation of Machine Learning Algorithms

We separate our dataset into train and test datasets. In the case of shallow machine learning, we train algorithms using stratified k-folds by train folds. We train deep learning using train folds and early stop using validation folds after separating train datasets into train and validation folds. In addition, we compare the performance of each algorithm using test datasets. Table 10 shows the highest performance achieved for each machine learning algorithm. In the case of PPV, GRU shows the highest performance. In the case of sensitivity, GB shows the highest performance. In the case of the F1 score, LSTM shows the highest performance.

Table 10. Highest prediction performance of each machine learning algorithm.

Machine Learning Algorithms	Oversampling	PPV	Sensitivity	F1 Score	AUC
LR	ADSYN	44.47%	83.88%	0.5812	0.8745
DT	Borderline SMOTE	92.11%	94.14%	0.9312	0.9672
RF	SMOTE	96.09%	98.90%	0.9747	0.9928
SVM	SMOTE	89.49%	96.70%	0.9296	0.9786
AB	SMOTE	86.33%	94.87%	0.9040	0.9679
GB	ADASYN	96.10%	99.27%	0.9766	0.9946
MLP	SMOTE	97.48%	99.27%	0.9837	0.9952
RNN	ADASYN	95.67%	97.07%	0.9636	0.9835
LSTM	ADASYN	98.18%	98.90%	0.9854	0.9937
GRU	ADASYN	98.53%	98.17%	0.9835	0.9902

PPV: positive predictive value; LR: logistic regression; DT: decision tree; RF: random forest; SVM: support vector machine; AB: adaptive boost; GB: gradient boost; MLP: multi-layer perceptron; RNN: recurrent neural network; LSTM: long short-term memory; GRU: gated recurrent unit.

4.7. Comparison of the Importance of Disease Prediction

We guess that the most important techniques in a machine learning algorithm are imputation, oversampling, and feature selection, in that order. We evaluate the performance of an RDI-based imputer and a mean-based imputer. In the RDI-based imputer, the PPV and sensitivity improved by more than 20% and 10%, respectively. The datasets in the medical field are unbalanced. In the case of diseases such as respiratory failure, they are classified according to their occurrence. The overfitting problem of prediction occurs when machine learning comprised a majority of the data in datasets. Oversampling or undersampling must be performed to solve imbalanced datasets. Table 11 shows the performance results of replacing missing values with the mean or not performing oversampling. In the case of not performing oversampling, the sensitivity is lower than if oversampling had been performed.

Table 11. Performance result of each scenario.

Scenario	Algorithm	PPV	Sensitivity	F1 Score	AUC
Reference	LR	44.47%	83.88%	0.5812	0.8745
	DT	92.11%	94.14%	0.9312	0.9672
	RF	96.09%	98.90%	0.9747	0.9928
	SVM	89.49%	96.70%	0.9296	0.9786
	AB	86.33%	94.87%	0.9040	0.9679
	GB	96.10%	99.27%	0.9766	0.9946
	MLP	97.48%	99.27%	0.9837	0.9952
	RNN	95.67%	97.07%	0.9636	0.9835
	LSTM	98.18%	98.90%	0.9854	0.9937
	GRU	98.53%	98.17%	0.9835	0.9902
Replace missing values via average	LR	22.26%	70.70%	0.3386	0.7477
	DT	63.06%	83.15%	0.7172	0.8949
	RF	53.24%	87.18%	0.6611	0.9031
	SVM	41.05%	83.15%	0.5496	0.8646
	AB	58.21%	73.99%	0.6516	0.8472
	GB	75.40%	86.45%	0.8055	0.9201
	MLP	65.49%	81.32%	0.7255	0.8882
	RNN	62.10%	78.02%	0.6916	0.8697
	LSTM	41.63%	80.22%	0.5482	0.8529
	GRU	65.73%	85.71%	0.7440	0.9094
Does not perform oversampling	LR	81.46%	45.05%	0.5802	0.7209
	DT	98.05%	92.31%	0.9509	0.9608
	RF	98.85%	94.14%	0.9644	0.9702
	SVM	99.06%	77.29%	0.8683	0.8861
	AB	94.80%	86.81%	0.9063	0.9320
	GB	98.38%	89.01%	0.9346	0.9444
	MLP	95.67%	80.95%	0.8770	0.9032
	RNN	90.87%	83.88%	0.8724	0.9158
	LSTM	77.73%	62.64%	0.6937	0.8055
	GRU	91.09%	86.08%	0.8851	0.9268

PPV: positive predictive value; LR: logistic regression; DT: decision tree; RF: random forest; SVM: support vector machine; AB: adaptive boost; GB: gradient boost; MLP: multi-layer perceptron; RNN: recurrent neural network; LSTM: long short-term memory; GRU: gated recurrent unit.

Table 4 shows the variance in performance according to the features. We consider the features with the highest performance in Table 4. In the case of machine learning algorithms, the performance depends on the dataset. The sensitivity of machine learning algorithms in Table 10 is more than 80%. However, the PPV of logistic regression is less than 50%. The deep learning models such as RNN, LSTM, and GRU all performed at comparable levels.

5. Discussion

We proposed the prediction model for acute respiratory failure, an important prognostic factor in acute pesticide poisoning patients. The effects of respiratory failure include loss of consciousness, arrhythmias, and death. Table 12 shows the performance of each algorithm for the prediction of respiratory failure. In recent years, respiratory failure prediction models have been developed to predict respiratory failure in COVID-19 patients based on deep learning with semi-supervised learning [4], respiratory failure based on XGBoost using clinical data [5], respiratory failure in COVID-19 patients based on LR [6], respiratory failure in ICU patients based on LightGBM [7], respiratory failure in patients with pesticide poisoning due to intentional pesticide ingestion based on LR [9], cardiac arrest and respiratory failure in ICU patients based on LSTM [10], and respiratory failure in ICU patients based on gradient boosting [8]. In the case of prediction based on semi-supervised learning [4], the PPV was 0.033 and the sensitivity was 0.78. In the case of prediction based on XGBoost using clinical data [5], the sensitivity was 0.71. In the case of prediction of respiratory failure with COVID-19 based on LR [6], the PPV was 0.74 and the sensitivity was 0.72. In the case of prediction based on LightGBM [7], the PPV was 0.42 and the sensitivity was 0.80. In the case of prediction of respiratory failure in patients with pesticide poisoning based on LR [9], the PPV was 0.833 and the sensitivity was 0.606. In the case of prediction based on LSTM [10], the PPV was 0.226 and the sensitivity was 0.881. However, these respiratory failure prediction algorithms are characterized by a large measurement interval, low performance, or large number of features [4–10]. Our proposed algorithm demonstrated improved respiratory failure prediction within 24 h with higher PPV and sensitivity compared with those of other models.

Table 12. Comparison of the performance between the proposed algorithm and algorithms in other studies.

	Algorithms	Features	Patient Data Range	Sensitivity	PPV	AUC
[4]	Semi-supervised learning	25	32 h	0.78	0.023	0.78
[5]	XGBoost	24	-	0.71	-	-
[6]	LR	26	-	0.72	0.74	0.89
[7]	LightGBM	25	-	0.80	-	0.746
[8]	GradientBoosting	106	6 h	0.534	0.643	0.769
[9]	LR	7	-	0.606	0.833	0.912
[10]	LSTM	8	2 h	0.881	0.226	0.886
Our algorithm	LSTM	17	3 h	0.9817	0.9890	0.9937

PPV: positive predictive value; LR: logistic regression; LSTM: long short-term memory; RF: random forest.

Our proposed algorithm demonstrated improved respiratory failure prediction within 24 h with higher PPV and sensitivity compared with those of other models. We guess that the pesticide category, white blood cell (WBC), pH, heart rate (HR), and C-reactive protein 1 are important predictors of respiratory failure in acute pesticide poisoning. The performance results of important features based on RF and GB confirmed that the highest-scoring features are the above features. These are the limitations of this paper: First, we conducted a single cohort at the Soonchunhyang University Cheonan Hospital. Second, we proceeded with retrospective research. We have not yet confirmed the validity from an external source. Third, our algorithm required three-hour EMRs. Our algorithm is not applicable to high-risk patients hospitalized for less than three hours. Fourth, our algorithm is difficult to use in hospitals. Our algorithm predicts whether respiratory failure has happened within 24 h. It does not estimate the time or risk score that a patient should experience respiratory failure. Follow-up research is required to decrease the prediction range for respiratory failure or score the patient's risk or provide information such as the estimated time of respiratory failure.

6. Conclusions

We predicted respiratory failure in patients with pesticide poisoning at Soonchunhyang University Cheonan Hospital. We analyzed the 3 h medical records of individuals with pesticide poisoning to predict respiratory failure within 24 h. In consideration of time-series properties, sliding windows, feature selection, and oversampling were used to replace missing values. Enhanced performance was achieved with the use of LSTM. Moreover, our machine learning technique algorithm could improve the prognosis of patients with pesticide poisoning. In addition, we will enhance the algorithm for predicting respiratory failure within 24 h such that it can predict respiratory failure within 4 or 8 h. We plan to conduct studies to predict respiratory failure in patients admitted to the general ward and ICU.

Author Contributions: Conceptualization, H.L. and H.G.; methodology, Y.K. and M.C.; software, Y.K. and M.C.; validation, N.C., H.G. and H.L.; formal analysis, Y.K.; resources, H.G. and N.C.; data curation, N.C.; writing—original draft preparation, Y.K. and M.C.; writing—review and editing, H.L. and H.G.; visualization, Y.K. and M.C.; supervision, H.L.; project administration, H.L. and H.G.; funding acquisition, H.L. and H.G. All authors have read and agreed to the published version of the manuscript.

Funding: This research was supported by the MSIT (Ministry of Science and ICT), Korea, under the ICAN (ICT Challenge and Advanced Network of HRD) program (IITP-2022-RS-2022-00156439) supervised by the IITP (Institute of Information and Communications Technology Planning and Evaluation), the Bio and Medical Technology Development Program (No. NRF-2019M3E5D1A02069073) and a Korea University Grant.

Institutional Review Board Statement: The study was conducted in accordance with the Declaration of Helsinki and approved by the Institutional Review Board of Soonchunhyang University Cheonan Hospital (IRB number: 2020-02-016).

Informed Consent Statement: Patient consent was waived because of the retrospective design of the study.

Data Availability Statement: Data sharing not applicable.

Conflicts of Interest: The authors declare no conflict of interest.

References

1. Cho, N.-J.; Park, S.; Lee, E.Y.; Gil, H.-W. Risk factors to predict acute respiratory failure in patients with acute pesticide poisoning. *J. Korean Soc. Clin. Toxicol.* **2020**, *18*, 116–122.
2. Lee, H.; Choa, M.; Han, E.; Ko, D.R.; Ko, J.; Kong, T.; Cho, J.; Chung, S.P. Causative Substance and Time of Mortality Presented to Emergency Department Following Acute Poisoning: 2014-2018 National Emergency Department Information System (NEDIS). *J. Korean Soc. Clin. Toxicol.* **2021**, *19*, 65–71.
3. Mew, E.J.; Padmanathan, P.; Konradsen, F.; Eddleston, M.; Chang, S.-S.; Phillips, M.R.; Gunnell, D. The global burden of fatal self-poisoning with pesticides 2006-15: Systematic review. *J. Affect. Disord.* **2017**, *219*, 93–104. [CrossRef] [PubMed]
4. Lam, C.; Tso, C.F.; Green-Saxena, A.; Pellegrini, E.; Iqbal, Z.; Evans, D.; Hoffman, J.; Calvert, J.; Mao, Q.; Das, R. Semisupervised Deep Learning Techniques for Predicting Acute Respiratory Distress Syndrome from Time-Series Clinical Data: Model Development and Validation Study. *JMIR Form. Res.* **2021**, *5*, e28028. [CrossRef]
5. Sinha, P.; Churpek, M.M.; Calfee, C.S. Machine learning classifier models can identify acute respiratory distress syndrome phenotypes using readily available clinical data. *Am. J. Respir. Crit. Care Med.* **2020**, *202*, 996–1004. [CrossRef]
6. Bartoletti, M.; Giannella, M.; Scudeller, L.; Tedeschi, S.; Rinaldi, M.; Bussini, L.; Fornaro, G.; Pascale, R.; Pancaldi, L.; Pasquini, Z. Development and validation of a prediction model for severe respiratory failure in hospitalized patients with SARS-CoV-2 infection: A multicentre cohort study (PREDI-CO study). *Clin. Microbiol. Infect.* **2020**, *26*, 1545–1553. [CrossRef]
7. Hüser, M.; Faltys, M.; Lyu, X.; Barber, C.; Hyland, S.L.; Merz, T.M.; Rätsch, G. Early prediction of respiratory failure in the intensive care unit. *arXiv* **2021**, arXiv:2105.05728.
8. Schwager, E.; Jansson, K.; Rahman, A.; Schiffer, S.; Chang, Y.; Boverman, G.; Gross, B.; Xu-Wilson, M.; Boehme, P.; Truebel, H. Utilizing machine learning to improve clinical trial design for acute respiratory distress syndrome. *NPJ Digit. Med.* **2021**, *4*, 133. [CrossRef]
9. Cho, N.-J.; Park, S.; Lyu, J.; Lee, H.; Hong, M.; Lee, E.-Y.; Gil, H.-W. Prediction Model of Acute Respiratory Failure in Patients with Acute Pesticide Poisoning by Intentional Ingestion: Prediction of Respiratory Failure in Pesticide Intoxication (PREP) Scores in Cohort Study. *J. Clin. Med.* **2022**, *11*, 1048. [CrossRef] [PubMed]

10. Kim, J.; Chae, M.; Chang, H.-J.; Kim, Y.-A.; Park, E. Predicting cardiac arrest and respiratory failure using feasible artificial intelligence with simple trajectories of patient data. *J. Clin. Med.* **2019**, *8*, 1336. [CrossRef]
11. Idri, A.; Benhar, H.; Fernández-Alemán, J.; Kadi, I. A systematic map of medical data preprocessing in knowledge discovery. *Comput. Methods Programs Biomed.* **2018**, *162*, 69–85. [CrossRef]
12. Benhar, H.; Idri, A.; Fernández-Alemán, J. Data preprocessing for heart disease classification: A systematic literature review. *Comput. Methods Programs Biomed.* **2020**, *195*, 105635. [CrossRef] [PubMed]
13. Jadhav, A.; Pramod, D.; Ramanathan, K. Comparison of performance of data imputation methods for numeric dataset. *Appl. Artif. Intell.* **2019**, *33*, 913–933. [CrossRef]
14. Li, J.; Cheng, K.; Wang, S.; Morstatter, F.; Trevino, R.P.; Tang, J.; Liu, H. Feature selection: A data perspective. *ACM Comput. Surv.* **2017**, *50*, 1–45. [CrossRef]
15. Pedregosa, F.; Varoquaux, G.; Gramfort, A.; Michel, V.; Thirion, B.; Grisel, O.; Blondel, M.; Prettenhofer, P.; Weiss, R.; Du-bourg, V.; et al. Scikit-learn: Machine learning in Python. *J. Mach. Learn. Res. JMLR* **2011**, *12*, 2825–2830.
16. Stekhoven, D.J.; Bühlmann, P. MissForest—Non-parametric missing value imputation for mixed-type data. *Bioinformatics* **2012**, *28*, 112–118. [CrossRef]
17. Seabold, S.; Perktold, J. Statsmodels: Econometric and statistical modeling with python. In Proceedings of the 9th Python in Science Conference, Austin, TX, USA, 28 June–3 July 2010; p. 10-25080.
18. Mohammed, R.; Rawashdeh, J.; Abdullah, M. Machine learning with oversampling and undersampling techniques: Over-view study and experimental results. In Proceedings of the 2020 11th International Conference on Information and Communication Systems (ICICS), Irbid, Jordan, 7–9 April 2020; pp. 243–248.
19. He, H.; Bai, Y.; Garcia, E.A.; Li, S. ADASYN: Adaptive synthetic sampling approach for imbalanced learning. In Proceedings of the 2008 IEEE International Joint Conference on Neural Networks (IEEE World Congress on Computational Intelligence), Hong Kong, China, 1–8 June 2008; pp. 1322–1328.
20. Chawla, N.V.; Bowyer, K.W.; Hall, L.O.; Kegelmeyer, W.P. SMOTE: Synthetic minority over-sampling technique. *J. Artif. Intell. Res.* **2002**, *16*, 321–357. [CrossRef]
21. Han, H.; Wang, W.-Y.; Mao, B.-H. Borderline-SMOTE: A new over-sampling method in imbalanced data sets learning. In Proceedings of the International Conference on Intelligent Computing, Hefei, China, 23–26 August 2005; pp. 878–887.
22. Kleinbaum, D.G.; Dietz, K.; Gail, M.; Klein, M.; Klein, M. *Logistic regression*; Springer: Berlin/Heidelberg, Germany, 2002.
23. Breiman, L. Random forests. *Mach. Learn.* **2001**, *45*, 5–32. [CrossRef]
24. Platt, J. Probabilistic outputs for support vector machines and comparisons to regularized likelihood methods. *Adv. Large Margin Classif.* **1999**, *10*, 61–74.
25. Freund, Y.; Schapire, R.E. A decision-theoretic generalization of on-line learning and an application to boosting. *J. Comput. Syst. Sci.* **1997**, *55*, 119–139. [CrossRef]
26. Abadi, M.; Barham, P.; Chen, J.; Chen, Z.; Davis, A.; Dean, J.; Devin, M.; Ghemawat, S.; Irving, G.; Isard, M.; et al. TensorFlow: A System for Large-Scale Machine Learning. In Proceedings of the 12th USENIX symposium on operating systems design and implementation (OSDI 16), Savannah, GA, USA, 2–4 November 2016; pp. 265–283.
27. Hüsken, M.; Stagge, P. Recurrent neural networks for time series classification. *Neurocomputing* **2003**, *50*, 223–235. [CrossRef]
28. Hochreiter, S.; Schmidhuber, J. Long short-term memory. *Neural Comput.* **1997**, *9*, 1735–1780. [CrossRef] [PubMed]
29. Cho, K.; Van Merriënboer, B.; Gulcehre, C.; Bahdanau, D.; Bougares, F.; Schwenk, H.; Bengio, Y. Learning phrase representations using RNN encoder-decoder for statistical machine translation. *arXiv* **2014**, arXiv:1406.1078.
30. Refaeilzadeh, P.; Tang, L.; Liu, H. Cross-validation. *Encycl. Database Syst.* **2009**, *5*, 532–538.
31. Berrar, D. Cross-Validation. *Encycl. Bioinform. Comput. Biol.* **2019**, *1*, 542–545. Available online: https://www.sciencedirect.com/science/article/pii/B978012809633820349X?via%3Dihub (accessed on 23 September 2022).

Article

A Malware Attack Enabled an Online Energy Strategy for Dynamic Wireless EVs within Transportation Systems

Fahad Alsokhiry [1,2], Andres Annuk [3,*], Toivo Kabanen [3] and Mohamed A. Mohamed [4,*]

1. Department of Electrical and Computer Engineering, Faculty of Engineering, King Abdulaziz University, Jeddah 21589, Saudi Arabia
2. K. A. CARE Energy Research and Innovation Center, King Abdulaziz University, Jeddah 21589, Saudi Arabia
3. Institute of Forestry and Engineering, Estonian University of Life Sciences, 51006 Tartu, Estonia
4. Electrical Engineering Department, Faculty of Engineering, Minia University, Minia 61519, Egypt
* Correspondence: andres.annuk@emu.ee (A.A.); dr.mohamed.abdelaziz@mu.edu.eg (M.A.M.)

Citation: Alsokhiry, F.; Annuk, A.; Kabanen, T.; Mohamed, M.A. A Malware Attack Enabled an Online Energy Strategy for Dynamic Wireless EVs within Transportation Systems. *Mathematics* 2022, 10, 4691. https://doi.org/10.3390/math10244691

Academic Editors: Jose Antonio Sáez Muñoz and José Luis Romero Béjar

Received: 16 October 2022
Accepted: 7 December 2022
Published: 10 December 2022

Publisher's Note: MDPI stays neutral with regard to jurisdictional claims in published maps and institutional affiliations.

Copyright: © 2022 by the authors. Licensee MDPI, Basel, Switzerland. This article is an open access article distributed under the terms and conditions of the Creative Commons Attribution (CC BY) license (https://creativecommons.org/licenses/by/4.0/).

Abstract: Developing transportation systems (TSs) under the structure of a wireless sensor network (WSN) along with great preponderance can be an Achilles' heel from the standpoint of cyber-attacks, which is worthy of attention. Hence, a crucial security concern facing WSNs embedded in electrical vehicles (EVs) is malware attacks. With this in mind, this paper addressed a cyber-detection method based on the offense–defense game model to ward off malware attacks on smart EVs developed by a wireless sensor for receiving data in order to control the traffic flow within TSs. This method is inspired by the integrated Nash equilibrium result in the game and can detect the probability of launching malware into the WSN-based EV technology. For effective realization, modeling the malware attacks in conformity with EVs was discussed. This type of attack can inflict untraceable detriments on TSs by moving EVs out of their optimal paths for which the EVs' power consumption tends toward ascending thanks to the increasing traffic flow density. In view of this, the present paper proposed an effective traffic-flow density-based dynamic model for EVs within transportation systems. Additionally, on account of the uncertain power consumption of EVs, an uncertainty-based UT function was presented to model its effects on the traffic flow. It was inferred from the results that there is a relationship between the power consumption and traffic flow for the existence of malware attacks. Additionally, the results revealed the importance of repressing malware attacks on TSs.

Keywords: transportation system (TS); electrical vehicles (EVs); malware attack; offense–defense game; traffic flow; wireless sensor network (WSN); uncertainty

MSC: 94-10

1. Introduction

Transportation systems (TSs) are revolutionary technologies derived from new technologies and communications designed to meet the needs of social welfare within cities [1]. Indeed, the growing employment of electrical vehicles (EVs) and metro-based city railways encourages willing researchers and industry holders to provide insightful ideas aiming to enhance energy management in terms of reliability and safety [2]. Here, the key factor that can make a profound impact on the TSs is misuse and unfair treatment in using these revolutionary technologies, such as wireless sensor networks (WSNs) embedded in EVs. These emerging communications, along with their privileges, constitute a guiding light for cyber-attacks, which need to be dealt with [3].

Data broadcasting structures utilize sensor nodes, which are located in varied areas and are connected by dint of the wireless communication named WSN. Their wide applications can be recapitulated in intelligent transportation, monitoring systems, fire prevention, and so forth [4]. Due to being vulnerable, this type of network is a meaningful target to malicious hackers and various kinds of attacks, e.g., flooding attack [5,6], wormhole attack [7], Sybil attack [8,9], selective forwarding [10,11], and malware attacks [12]. The authors in ref. [12] tried to model a malware attack for wireless sensor networks. Among the abovementioned attacks, malware attacks may be more perilous and destructive in comparison with other attacks because a malware contaminated-node will automatically broadcast malware to the other neighboring nodes. This means that the entire system would be infected if even a single node is exposed to a malware attack. Hence, having an underlying understanding of the malware's behavior and providing the defensive modus operandi against it is a must. WSNs infected by malware are comprised of malicious and authentic nodes. In this way, the authentic nodes in turn receive malware from malicious ones. Meanwhile, the intrusion detection system (IDS) has the ability to identify the malware in the authentic nodes. With this knowledge, the infected nodes do not send malware to other nodes at any time. Since the authentic node to detect attacks takes the high consumed energy, it does not carry out IDS every time in order to preserve energy. So, it results that the malicious and authentic nodes conflict with each other, similar to a game structure. So to speak, it is needed to use the game strategy to examine the defense methods between these nodes in wireless networks. Indeed, when an authentic node is overshadowed by malware attacks, it starts the process of propagating false data and malwares to neighboring nodes. The authors in ref. [13] tried to provide an effective algorithm to select an adaptive node considering the evolutionary game led to the improvement of network efficiency and lifetime. Additionally, ref. [14] explained a defense model based on the differential game strategy regarding the dynamic changes caused by cyber-attacks. As a result, by drawing on game theory policy, it is easy to identify the type of attack and provide defense frameworks for the infected WSNs. The widespread presence of WSNs in the industry has led to the reformation of conventional cities, becoming easier for use by management [15]. Malware attacks, in nature, utilize a type of brutal data dissemination to harm the host system. On the other hand, considering that in a smart city, the energy management of the transportation sector is linked with the entire system, if a malware attack were successful, the energy management of the entire system would be damaged. Likewise, on a small scale, because the wireless sensor network is installed on every vehicle and charging station, receiving or sending false data would lead to a severe anomaly in the urban energy management sector. The 'smartness' concept has become more prevalent in cities and, thereby, intelligent, interconnected, and instrumented structures have emerged [16]. A smart city is an interconnected environment that integrates and shares data through different services. Additionally, instrumented structures in smart cities, with the help of sensors and measurements, provide the needed ability to instantaneously collect data related to city life. Investigations indicate that a decision-making system based on an intelligent structure could guarantee finding the optimal decisions through data processing in optimization and modeling analyses and so forth [17]. The presence of advanced technologies within city infrastructure is a convincing reason to develop varied energy segments, including in transportation, smart homes, and the electrical grid. Smart cities have the potential to transfer energy among these segments in order to facilitate real-time energy management. As already mentioned, TS, as a significant energy segment in smart cities, is always undergoing a change in energy management for the sake of the EVs' performance. The traffic flow arising from moving EVs in non-optimal paths is one of these underlying concerns in the TSs' energy management. Hence, due to the relationship between the power consumption of EVs and the traffic flow density, it is essential to dynamically consider the EVs' behavior in energy management. Indeed, the constant daily movement of people using EVs sometimes gives rise to the traffic on a city's roads, which in turn affects the energy consumption related to battery-based hybrids or EVs [18]. Ref. [19], in response to this problem, proposed a new model of traffic flow with the presence of EVs. The authors in ref. [20] tried to dynamically provide an all-inclusive study of all factors on traffic flow. To be more precise, the weather patterns, stochastic EV behavior, and the time of travel are all factors to consider when making an impression on the traffic density, all of which were investigated in ref. [21]. Additionally, dynamic and static traffic models were compared in ref. [22]. It is worth mentioning that the remarkable estimating error between the real and simulated models defined for TSs does not arise from being aware of the uncertainty in the EVs' performances. Hence, in response to this concern, many investigations have advocated for varied models and approaches to adderss uncertainty. In recent years, the Monte Carlo method, as the most valid technique, has been recognized to solve energy management probability problems within the power system. Having high accuracy in the real world is a convincing reason for its use as a reference in order to evaluate the other methods [23]. However, the time-consuming nature of estimating the calculation is known as the greatest weakness of this technique [24]. However, recently, lots of efforts to overcome this problem have been made, among which the unscented transformation (UT) method could attract attention due to its properties [25]. The authors in ref. [26] tried to implement the UT method in the uncertain environment of a smart city. This reference used the property of UT in modeling the correlation among the energy segments in the smart city. Individually, the modeling of the uncertain performance of EVs based on vehicle-to-grid technology (V2G) was deterministically carried out, and inspired by the UT concept [27]. Given the dynamic model of energy management in the EVs proposed in the current paper, a UT-based uncertainty model was applied to the static/dynamic energy model of EVs within the transportation system. Indeed, this uncertainty model mapped all uncertain parameters related to the dynamic model, including the EV's speed and the traffic rate, and also related to the static model including the braking energy of metro stations and the number of EVs in the stations.

The highlights of this paper can be summarized as follows:

- We analyzed the energy consumption of EVs using on a traffic flow-based online dynamic energy model and developed it by WSNs to control the traffic flow within a transportation system.
- We modeled a malware attack and implemented it into the WSN-based EVs with the aim of infecting the wireless sensor nodes and also proposing an effective offense–defense game strategy to deal with that.
- We presented a UT-based uncertainty method to model the high-risk energy consumption of the EVs considering the varied density rates of the traffic flow.

The rest of the paper is structured as follows: Section 2 presents the traffic model considering the energy consumption model of EVs. Section 3 presents the malware attack and the offense–defense strategy for WSNs. Section 4 presents the uncertainty model. Section 5 presents and evaluates the simulation results. Finally, Section 6 provides the conclusions.

2. The Traffic Flow Density-Based Online Dynamic Model of the Transportation System
2.1. Definition of the Traffic Flow Model

On account of the ease of arriving at the destination (for instance, the city center), the drivers of EVs are inclined to move into a few areas (usually the main road) and can cause traffic density on the main roads within the city [28]. However, this is not the sole reason for traffic flow density. The abrupt emergence of physical obstacles on the roads, the flawed performance of some drivers, who fail to respect city rules, and also natural stumbling blocks are well-known as strong contributing factors to the city traffic crisis. For these reasons, it is rather unlikely that the EVs follow the same behavior in the traffic flow. Accordingly, the traffic flow density would be altered, with an emphasis on the position and time. On the basis of the city observations, the definition of the traffic model is based on two key parameters, including macro and micro parameters. From the macro standpoint, the traffic flow is defined as concerning the global indexes. However, the EVs' performances and their effects on the traffic flow density shed light on the micro parameter. Relying on this knowledge, the macro parameters include (1) flow rate, (2) density, and (3) speed. On the contrary, the time and local intervals and the vehicle's speed are the micro parameters. The volume parameter expresses how many vehicles are passing across the road based on a specific length at a certain period. Additionally, the moving rate of the vehicle at a distance is defined by the speed parameter. The density parameter is calculated based on the volume of traffic not considering the time, which is the most vital parameter in the traffic model. Table 1 indicates the classification of the parameters of the traffic model based on the macro/micro standpoints.

$$V_{x,t} = \frac{\Delta N^{EV}}{\Delta t} \qquad (1)$$

$$D_{x,t} = \frac{V_{x,t}}{S_{x,t}} \qquad (2)$$

Table 1. Definition of the micro and macro parameters.

Parameter Type	Macro Parameter	Micro Parameter
	Volume or flow rate	The speed of each vehicle
Parameters of traffic flow model	Density	Time interval
	Speed	Local interval

The traffic flow rate and density are respectively defined by Equations (1) and (2), in which speed (S) and speed changes are involved. As can be seen in Figure 1, the vehicle's length and distance between two vehicles in a row when moving on a road are significant criteria for computing density into the traffic flow. Keeping this in mind, density in a traffic line is defined by:

$$D = \frac{5280}{L_d} \qquad (3)$$

wherein D and L_d are density and the distance (m) for vehicles in a row in a line. As already mentioned, the density of traffic flow on the road has a relationship to the distance among EVs involved in the traffic. Hence, the parameter of the distance was used in Equation (3) for calculating the density. Additionally, the flow rate (V) was obtained based on the time interval (T_d) among the consecutive vehicles moving on a road, as indicated in Equation (4). Indeed, the time interval (T_d) among EVs on the road can affect the EV's speeds when moving on the road. It is worth noting that the values 5280 and 3600 are constant and were determined for all types of vehicles on average.

$$V = \frac{3600}{T_d}. \qquad (4)$$

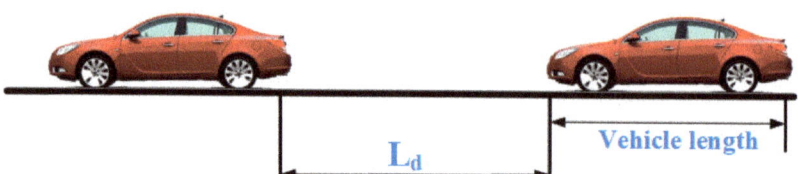

Figure 1. Illustration of the local interval for the vehicles in traffic.

Finality, the average speed in the traffic flow was calculated based on the time and local interval as follows:

$$S = 0.68 \frac{L_d}{T_d}. \qquad (5)$$

2.2. The Proposed Online Dynamic Energy Management for TSs

To realize dynamic energy management within TS, modeling the speed rate of EVs can affect the energy consumption involved in the traffic flow rate. Indeed, the speed rate obtained in the previous section was considered as a criterion to calculate the consumed energy in the EVs moving on the road. In this section, this paper tries to present an online dynamic energy model for advanced transport employed in a smart city. Indeed, the goal was to indicate the effects of traffic flow density on the dynamic energy consumption of EVs, which, in turn, will alter the energy interactions with the charging stations. The congestion of EVs on the road or the traffic flow causes inhibits the movement of vehicles from start to end. This results in EVs being driven at non-optimal velocities, in turn resulting in declined energy efficiency. To be more precise, there is a need to describe the energy consumption model for EVs concerning the speed and technical and environmental conditions that EVs on the roads are exposed to. In this model, the vehicle-specific power (VSP) and the average driving speed were considered the main criteria. Indeed, the energy efficiency and the output power in EVs were computed based on the VSP that each EV is defined on. Additionally, we assumed that all EVs take various speeds on the road at any time (second). By keeping this in mind, the traction energy of EVs being on the move is defined as follows:

$$\begin{aligned} E_t^{EV} &= E_t^{roll} + E_t^{air} + E_t^{grad} \\ &= \chi_r \times a_g \times \cos a_t^{gr} + \tfrac{1}{2}\varepsilon_a \times \tfrac{\chi_a \times \gamma}{w}(S_t)^2 + a_g \times \sin a_t^{gr}. \end{aligned} \quad (6)$$

This equation expresses three terms of energy, all of which affect the traction energy for an EV, including (1) the force of rolling resistance, (2) air resistance, and (3) gradient resistance. Indeed, this equation indicates that the consumed energy of an EV is overshadowed by the depreciation energy arising from the EV wheel and the atmosphere condition, which results in the EV's speed reduction. Hence, the first and third terms are related to the wheel energy involved by the gravity force to coefficients a_{gr}, χ_r and a_g, including the road gradient, the rolling resistance, and the gravity acceleration. As for speed, these coefficients describe the depreciation condition between the vehicle's wheels and the roads, which is a nonlinear function based on $cos(.)$ and $sin(.)$. It is worth noting that the first/third terms of Equation (6) are solvable due to the constant coefficient a_{gr}. Air resistance is another criterion that led to the EV's speed reduction once moving on the road. The second term shows the relationship between the traction energy and speed. Additionally, this relation considers the air conditions with an emphasis on coefficients ε_a = air density, χ_a = air drag, γ = front surface area of the EV, S = EV speed (m/s), and w = EV weight (kg). On the basis of city traffic knowledge, the number of EVs passing on the road is important from the standpoint of a reference point meant for the traffic flow. This definition is different from the density of the traffic. In density definition, the road length involved by EVs is considered. As a result, the traffic flow model indicates the speed variation due to the time and local intervals. Therefore, it seems the traction energy mentioned by Equation (6) needs to be updated based on the acceleration resistance force (E_{Acc}) arising from the effect of the traffic flow on the traction energy of EVs. Hence, the traction force F_{TE}^s in (7) and (8) expresses how acceleration in the vehicle's speed for the sake of traffic causes increased power consumption [29].

$$\begin{aligned} E_t^{force} &= E_t^{roll} + E_t^{air} + E_t^{grad} \\ &= \chi_r \times a_g \times \cos a_t^{gr} + \tfrac{1}{2}\varepsilon_a \times \tfrac{\chi_a \times \gamma}{w}(S_t)^2 + a_g \times \sin a_t^{gr} + AR_a \times w \times \tfrac{dS_t}{dt}, \end{aligned} \quad (7)$$

where the last term is related to the acceleration resistance with coefficient AR_a. It is essential to say that the acceleration term can affect the traction force of EVs if it becomes involved in the traffic flow. Indeed, the effects of traffic flow on the EVs' speed reduction are modeled by Equation (7) in the last term. Now, it we need to convert the traction force to the power consumption, taking into account the EV's speed when moving. Hence, finally, the power consumption considering the updated traction force is calculated as follows:

$$P_t^{force} = E_t^{force} \times S_t \quad (8)$$

Now, we model the dynamic energy management in TSs, bearing in mind that EVs, to fill up their transient power consumption, can utilize the social facilities of chargers located in parking lots on the roadside. It is essential to say we, in this paper defined, two types of charging stations, including the metro station and the grid station, as indicated in Figure 2. There are significant differences in the characteristics of these two types of charging stations (metro and grid station). 1—The metro charging station increases the uncertainty of the system by injecting unpredictable power. 2—The capacity of the metro station is lower than the grid station. 3—The oscillation of the power grid impresses the grid stations. Needless to say, we can optimally assess the energy management by taking into account the objective function's satisfaction based on the economic policy. Specifically, having a linking structure in TSs can carry out ways in which EVs have the needed ability to minimize their consumed energy cost. On the other hand, dynamically modeling the energy framework for EVs can help to make optimal decisions in the path selection when moving and controlling the traffic flow within TSs. Hence, the proposed dynamic energy framework is defined here. As aforementioned, the EVs can participate in the buying or selling of energy from/to the metro or grid stations with an emphasis on priority. Based on this, the EVs follow the objective function defined by Equation (9). The objective function is considered based on the energy cost involved by the EVs in the static/dynamic states, which defines whether the EVs are moving or not. Given that the EVs are able to exchange the power to the two grid/metro stations, the energy cost of EVs comprises two terms, C_{V2G}, C_{V2M}, which are indicated in the first/second terms of Equation (9). The last term expresses the degradation cost related to the discharge cycle of the EV's battery. Each term of this function is separately described in Equations (10)–(12). As can be seen, when the EVs exchange the energy with the grid, it is needed to calculate the energy cost based on the grid price defined by the decision-making center as shown in Equation (10). This description is correct for the metro station (see Equation (11)). The battery degradation for the hybrid or especially electrical vehicles is essential to consider in energy management because it can help with precision (see Equation (12)).

$$C_{EV} = C_{V2G} + C_{V2M} - \sum_v C_v^{deg} \tag{9}$$

$$C_{V2G} = \sum_{t,v} (C_{v,t}{}^{V2G} \times P_{v,t}^{V2G}) \tag{10}$$

$$C_{V2M} = \sum_{t,v} (C_{v,t}{}^{V2M} \times P_{v,t}^{V2M}) \tag{11}$$

$$C_{deg} = C_v^{deg} \times \sum_{s,t} R(P_{v,t}^{V2G} + P_{s,v,t}^{V2M}) \tag{12}$$

However, other key factors which should be considered when dynamically modeling energy are the technical limitations of EVs n that arise due to their dynamic energy consumption. Hence, dynamically computing the energy rate of the EV's battery concerning the consumed power (P_t^{force}) in the traffic flow can be achieved, as shown in Equation (18). As can be seen, the battery energy is dependent on the energy wasted when the EV is in operation for any time t and the energy rate in the past tense $t - 1$. Once the EV needs to visit the charging/discharging power stations, the energy rate of the battery is assessed based on the relation shown in Equation (19). In addition, EVs must adhere to the technical limitations of charging/discharging, all of which are defined in Equations (13) to (17). It is worth noting that the EVs can connect to the charging stations for two charging/discharging cycles. So, the EVs' batteries in the stations are defined based on the charging model or the discharging model at each time t. Hence, this difference can be modeled by Equation (13). It is essential to consider the charging/discharging powers in the metro stations with an emphasis on the charging/discharging limitations as shown in Equations (14) and (15). This is correct for the grid stations based on Equation (16). The power consumption arising from the traction force of EVs affects the battery energy, which is modeled by Equation (17).

$$u_{s,p,v,t}^{ch} + u_{s,p,v,t}^{dis} = u_{s,p,v,t} \tag{13}$$

$$u_{s,p,v,t}^{ch} P_v^{min\_ch} \leq P_{s,p,v,t}^{V2M\_ch} \leq u_{s,p,v,t}^{ch} P_v^{max\_ch} \tag{14}$$

$$u_{s,p,v,t}^{dis} P_v^{min\_dis} \leq P_{s,p,v,t}^{V2M\_dis} \leq u_{s,p,v,t}^{dis} P_v^{max\_dis} \tag{15}$$

$$u_{s,p,v,t}^{ch} P_v^{min\_ch} \leq P_{s,p,v,t}^{V2G\_ch} \leq u_{s,p,v,t}^{ch} P_v^{max\_ch}$$
$$u_{s,p,v,t}^{dis} P_v^{min\_dis} \leq P_{s,p,v,t}^{V2G\_dis} \leq u_{s,p,v,t}^{dis} P_v^{max\_dis} \tag{16}$$

$$E_v^{min} \leq E_{P,v,T} \leq E_v^{max} \tag{17}$$

$$E_{s,p,v,t} = E_{s,p,v,t-1} - P_t^{force} \tag{18}$$

$$E_{p,v,t,initial} = E_{p,v,t-1,final} - \sum_s (P_{s,p,v,t}^{V2G^{ch}} - P_{s,p,v,t}^{V2G^{dis}} + P_{s,p,v,t}^{V2M^{ch}} - P_{s,p,v,t}^{V2M^{dis}}) \tag{19}$$

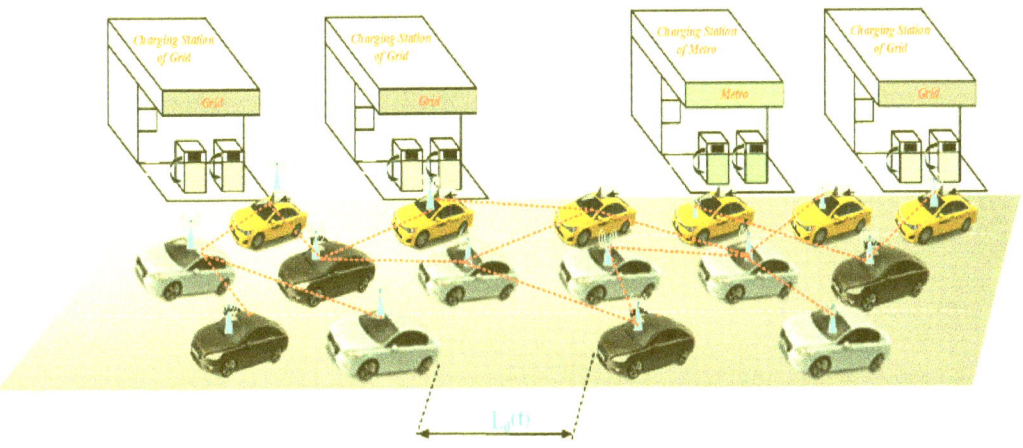

Figure 2. The illustration of moving vehicles on the road considering the traffic density.

3. The Malware Attack Analysis in the Transportation System

3.1. Malware Attack Model

In smart cities, developing the transportation segment to become more controllable, flexible, and optimal for the benefit of social welfare has been the focus of research attention in recent years [30]. For instance, equipping the transportation sector, especially with hybrid or electric vehicles with a communicable bed for the awareness-raising, may be the best and most-far sighted solution to the problem of traffic control in smart cities. Based on the literature, the advancement of communication has resulted in one of the most widely-used applications called wireless sensor networks (WSNs) [31]. Indeed, the structure of these networks, which are installed in EVs, is made up of nodes which handle data processing and preparation within the network. However, when launching a malware attack on WSNs, some nodes are infected by malware and are designated as malicious nodes. Malware attacks would be very effective in systems with wireless sensor networks, as each node has an intrusion detection system. The detection system possesses an internal battery whose energy can be its weakness when confronted with a malware attack. Dispatching the consecutive fake data leads to preoccupying the IDS part for detection. So, the manipulation of the whole system will be realized in this way [32]. In this situation, the network is made up of authentic and malicious nodes, and the malware is broadcasted from the malicious nodes to the authentic nodes. However, IDS embedded into the nodes can detect the information that is infected by malware. It is essential that the infected nodes, in addition to malware, send normal data to others to blot out their attacks. The IDS-based defense scheme causes remarkable energy consumption for nodes when receiving new information at any time. Hence, a node will be removed from the system if it has no energy for propagating data in WSN [33]. This type of node is named a dead node. In this way, to thwart the effect of dead nodes on the network's performance, we carry out the process of adding new nodes to the network with an emphasis on keeping the total number of nodes unchanged. Note that considering the self-mending process for the infected nodes is an ideal assumption. Therefore, assume that these malicious nodes will be repaired by authentic ones when identifying the attack's behavior arising from the infected nodes within the network. Figure 3 indicates which state each node is transferred to. Wherein, the authentic, malicious, and dead nodes take symptoms A, m, and D, respectively. When running a WSN, the probability of an authentic node being infected is κ, while the probability of repairing it is μ, and γ_1 and γ_2 are defined as the removing probabilities for both authentic and malicious nodes, respectively. In addition, the network would add new nodes to the rate of λN due to having no energy and wiping out the related nodes. N is defined as the total number of nodes, among which N_m is the number of malicious nodes with the proportion of m and N_A is the number of authentic nodes with the proportion of A. When running an attack, we assume that each malicious node can involve an authentic node infected at any time. Hence, Equations (20)–(23) show how the system is evaluated.

$$\frac{dN_A}{dt} = \alpha A N_m + \beta A N_m - \gamma_1 N_A + \lambda N \tag{20}$$

$$\frac{dN_m}{dt} = \alpha A N_m - \beta A N_m - \gamma_2 N_m \tag{21}$$

$$\lambda N = \gamma_1 N_A + \gamma_2 N_m \tag{22}$$

$$N = N_A + N_m \tag{23}$$

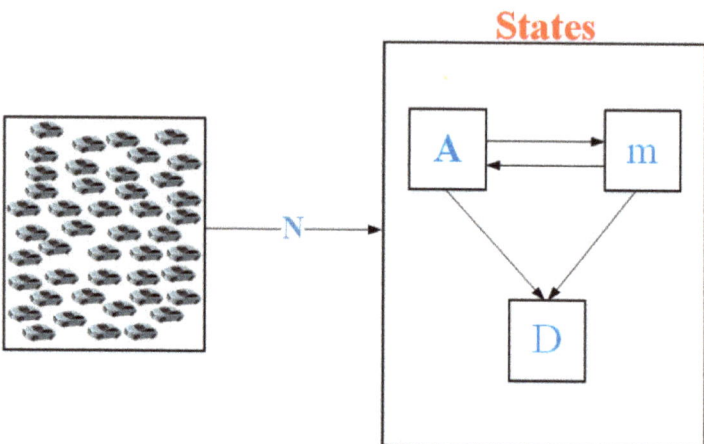

Figure 3. The states of malicious, authentic, and dead nodes.

Dividing by N:

$$\frac{dA}{dt} = \alpha A m + \beta A m - \gamma_1 A + \lambda \tag{24}$$

$$\frac{dm}{dt} = \alpha A m - \beta A m - \gamma_2 m \tag{25}$$

$$\lambda = \gamma_1 A + \gamma_2 m \tag{26}$$

$$A + m = 1 \tag{27}$$

Simplifying Equation (24), we can obtain m_t as follows:

$$\frac{dm}{m(1-\gamma_2/(\kappa-\mu)-m)} = (\kappa-\mu)dt$$
$$\text{when } t \to 0 \quad m_t = \frac{(1-(\gamma_2/(\kappa-\mu)))R\, e^{(\kappa-\mu-\gamma_2)t}}{1+R\, e^{(\kappa-\mu-\gamma_2)t}} \quad (28)$$
$$R = \frac{m_0}{1-(\gamma_2/(\kappa-\mu))-m_0}.$$

As already mentioned, the aim is to compute the number of malicious nodes for $t = \infty$. Based on Equation (28), for $t = \infty$, it is $\kappa > \mu + \gamma_2$, and $m_{t\to\infty} = 1 - (\gamma_2/(\kappa-\mu))$. Indeed, the changes in the number of malicious nodes can be obtained in line with the removing probability of those nodes. Looking over Equation (28), it is deducted that the changes in the number of malicious nodes can be zero if the gap between κ and μ is removed. This should be carried out during the time of the attack. It is worth noting the maximum number of malicious nodes is the maximum for $t = 0$, while, $m_{t\to\infty}$ will be zero if $\kappa \leq \mu + \gamma_2$.

Therefore, the proportion value for malicious nodes is obtained by the following:

$$\text{when } t \to \infty \quad m_t = \begin{cases} 1 - \frac{\gamma_2}{\kappa-\mu} & \kappa > \mu + \gamma_2 \\ 0 & \kappa \leq \mu + \gamma_2. \end{cases} \quad (29)$$

3.2. The Malware Attack-Based Offense–Defense Strategy

As mentioned, nodes in WSNs need to be supplied using batteries. Therefore, when these nodes (authentic nodes) enforce the detection program for new information, they will quickly lose their saving energy. Accordingly, the authentic nodes, concerning the consumed energy cost and the attack probability, should decide whether the detection is carried out or not. With the knowledge of the infected WSNs, the mutually opposed behavior between the authentic and malicious nodes is similar to a game process, in which the best strategy is picked by the offenders and defenders. Providing and designing a defense game algorithm can be analogous to how a malware attack is broadcasted with a network. For instance, the strategies used by defenders and hackers can change the process of the malware's propagation. Indeed, an authentic node attacked by the malware takes certain defensive measures, resulting in the success rate of generating malicious nodes. It is worth noting that the game strategies proposed in the literature are not able to prevent malware's propagation within a network. Therefore, modeling a game strategy based on the Nash equilibrium solution to provide malware propagation in WSNs is needed. Hence, this paper tried to apply a game theory-based method to deal with the strategy of offenders. The proposed game-based detection method seemed to have advantages related to modeling the Nash equilibrium solution based on malware propagation compared with other methods. The defined offense–defense game model in WSNs is based on a triple $\Xi = [N, \{S_i\}, \{u_i\}]$ wherein N and S_i were defined as sets of players and strategies in the game process. In this model, if i is an authentic node, the participants can pick out the strategy: $S_{\text{legitimate}} = \{\text{check, not check}\}$, otherwise, the strategy will be $S_{\text{malicious}} = \{\text{offense, not offense}\}$ if i is a malicious node. Here, the strategy of the non-offense means that the malicious node not only sends no false information, but can also send normal information. The benefit resulting from the game for adoption of a strategy is u_i for the participants. Additionally, the reward strategies defined for players in the game model are indicated in Table 2.

Table 2. The attack and detection strategies based on the benefit function.

Node Type	Game Strategies		
	Not-Attack	Attack	
Authentic node	$r + C_A$, $-C_m$	g_d-C_A-r, $-g_d$-C_m	Detected
Malicious node	$-r$, $-C_m$	$-g_a$-r, g_a-C_m	Not detected

Based on Table 2, two strategies for analysis were considered: (1) attack and (2) not-attack. On the other hand, the performance of the authentic and malicious nodes against these strategies was examined based on the energy cost involved the nodes. Hence, during the offense–defense conflict, a malicious node forwards the malware or normal information with the consumed energy cost of C_m while it is $r + C_A$ for an authentic node once it wants to detect or receive information. Meanwhile, an authentic node will take the benefit of a, if detected quickly when launching a malware attack by a malicious node. Conversely, the penalty of g_d is considered for a malicious node due to its identity detection. The malicious node can earn the benefit of g_a when its attack is successful. In the situation of a successful attack, the authentic node receives a penalty of b due to indiscrimination. To have a g_d meaningful game process, two conditions were defined for the model for satisfying: $g_d > C_A + r$ and $g_a > C_m$. Additionally, sending nodes to sleep in order to save energy is an underlying concern in WSNs. However, the game model to overcome this problem does not need to consider the additional incentives for two reasons: (1) receiving the normal information is itself an incentive for humans controlling the authentic node and (2) the additional incentives have no impact on the final game (Nash equilibrium solution) even if an additional incentive to authentic nodes is considered. Based on Table 2, an authentic node will carry out its detection process when a malware attack is launched by a malicious node. A malicious node stops launching an attack if node one starts the detection system; otherwise, it can launch its attack. Afterward, the behavior of players in the game model based on the attacking and defensive states is considered concerning the integrated equilibrium approach. Finality, keeping these in mind, all nodes set their strategies based on the Nash equilibrium strategy in order to detect or attack in WSNs. For instance, the malicious nodes with the probability of $P_A = C_A/(g_d + g_a)$ launch their attacks while the authentic nodes can detect them with a probability of $P_m = g_a/(g_d + g_a)$. On the other hand, if an authentic node has no detection, it will be infected by the malicious node with a probability of $\kappa = P_A(1 - P_m)$. Likewise, recovery of the malicious nodes is carried out at the probability of $\mu = P_A P_m$. Replacing these equations into Equations (24)–(27) results in Equation (30), considering the following parameters:

$$\text{when } t \to \infty \ m_t = \begin{cases} 1 - \frac{\gamma_2 (g_d+g_a)^2}{(g_d-g_a)C_A} & g_d - g_a > \frac{(g_d+g_a)^2 \gamma_2}{C_A} \\ 0 & a - b \leq \frac{(g_d+g_a)^2 \gamma_2}{C_A} \end{cases} \tag{30}$$

4. UT-Based Uncertainty Model

Having no unerring estimation for the energy consumption in TSs is an underlying concern that is worthy of attention. Hence, identifying and modeling the uncertain parameters which affect energy management in a non-optimal way is necessary. In the literature, the uncertainty methods have been provided, one of which is the UT model, well known as a useful tool regarding the ability to map the correlation environment [34]. This method takes advantage of coding and low computing time. Indeed, the UT-based uncertainty using a probability distribution function (pdf) approximates a nonlinear function [35]. To this end, in this section, we used the UT-based uncertainty concept to estimate uncertain parameters, such as the consumed power of EVs and their travel time. To briefly recapitulate, this method was executed based on the randomly generated points from a probability area into a discreet one. The estimated parameters saved in vector X are defined based on the nonlinear function $T = \hat{f}(X)$ with parameters p, mean value z, and covariance matrix Y_{aa}. all steps related to the process of the UT method are as below:

Step 1: Calculate all points with Equations (31)–(33).

$$X^0 = z \tag{31}$$

$$X^k = z + \left(\sqrt{\frac{p}{1-W^0}} Y_{aa}\right)_k \quad k = 1, 2, \ldots, p \tag{32}$$

$$X^{k+c} = z - \left(\sqrt{\frac{p}{1-W^0}} Y_{aa}\right)_k \quad k = 1, 2, \ldots, p \tag{33}$$

where Y_{aa} is the covariance matrix and $\overline{R} = z$.

Step 2: calculate the point weight with Equation (34):

$$W^k = \frac{1-W^0}{2p} \quad k = 1, 2, \ldots, 2p \tag{34}$$

The sum of the weights is equal to 1.

Step 3: by using these points, the model outcome can be obtained by inserting these points into $\hat{f}(X^k)$. In this step, we need to define the output T considering points p based on the objective function. The weighting coefficients for points are applied to the output vector and calculated in the previous step. Then, vector T is computed by Equation (35) considering all points. Following that, to obtain the mean value and covariance matrix of the output, the UT method carries out Equation (36) based on the points p.

$$\overline{T} = \sum_{k=0}^{2p} W^k T^k \tag{35}$$

$$C_{TT} = \sum_{k=1}^{2p} W^k \left(T^k - \overline{T}\right) \left(T^k - \overline{T}\right)^R \tag{36}$$

5. Emulation and Evaluation

In this section, we consider the simulation of the studied transportation system in the dynamic mode in which EVs can decide to voluntarily choose their paths, with emphasis on the traffic information for any path. This information is distributed by the wireless sensor networks embedded in EVs and upon receiving them, with which path drivers can make an optimal decision to move on. Indeed, this can help the decision-making center of the TSs to control the traffic flow during overloaded times within smart cities. In this simulation, we used a transportation system based on the EV fleets validated by ref. [36]. In this case, we considered six EV fleets, each of which includes 30 hybrid or electrical vehicles moving on the varied paths for special hours [36]. The authors in ref. [36] showed that the number of 180 EVs in the transportation system can cover the varied challenges facing the considered traffic flow within the transportation system. Here, it is assumed that some paths are involved in the traffic flow density for any reason, thanks to which EVs will consume the surplus power in overloaded times. In addition, to make reparations, two types of charging stations including the metro charging parking lots and the grid charging parking lots are proposed for all paths, whereby which EVs that have incurred energy losses as a consequence of traffic can use them. From the viewpoint of security, we simulated and launched a malware attack on a wireless-enabled EV network equipped with the proposed attack–defense game scheme. It is worth mentioning that we used the GAMS software to implement the studied transportation system and the relevant dynamic analysis; the malware attack model along with the proposed defense approach were also implemented on MATLAB software linked with GAMS software on a PC with 32 GB of RAM. For better realization, we provide the simulation consequences in three varied forms in the following.

5.1. The Numerical and Dynamic Simulation of TSs

This section is dedicated to dynamically assessing the EVs' behaviors in the power consumption caused by traffic flow density based on the proposed traffic model-based online energy management. Spreading out information related to the path traffic density among EVs by WSN can practically remedy the controlling patterns of traffic in the smart city. To do so, we first need to have a close look at dynamically consuming the surplus energy created EVs owing to traffic. Hence, we carried out the studied case for all EV fleets moving on paths considering their traffic density. The consequences related to the dynamic analysis with the help of the proposed online model are illustrated in Figures 4–7. Varying the vehicle's speed on a high scale in traffic can cause significant acceleration in the vehicle's wheels, leading to excess power consumption (see Formula (7)). Keeping this in mind, the proposed analysis outcome for EV fleet 6 during an hour (0 to 3600 s) from 11 to 12 can be seen in Figure 4. As shown, the EVs consumed various amounts of power at any given time due to environmental conditions, traffic flow, and so forth. Indeed, this shows that the EVs use a mean consumed power of less than 400 w from 0 to 1000 s, owing to the vehicle's amortization arising from the moving wheels. Assuming that a human accident occurred in the overload time, we carried out the determination of traffic flow on the road during 1100 to 1700 s following this. As can be seen, in this situation, the EVs had to consume power exceeding 400 w and even 900 w due to the speed changes which arose from the braking energy in traffic. It is clear the energy consumption pattern of EVs, on average, did not exceed 300 kW if assuming limited traffic density on the road. However, this consumption pattern would increase in frequency when the traffic flow increases for any reason.

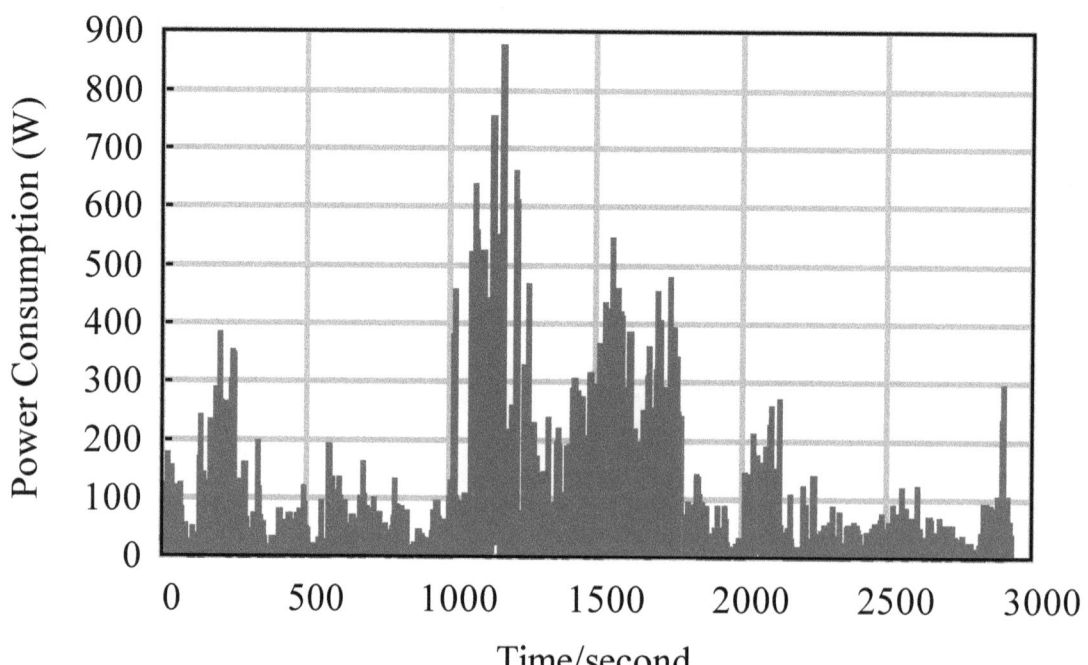

Figure 4. The EVs' energy consumptions considering the traffic.

However, the discharging/charging patterns related to EV fleets in the parking lots underwent a change because the EVs consumed energy in excess of that which was expected during the scheduled travel time. To be more precise, based on Equations (22) and (23), balancing the energy balance of the EVs' batteries would be dependent on the traffic energy consumption, charging, and discharging energies at any time. In other words, it is expected that as energy consumption on movement increases, so would the charging rate of energy for the EVs when stopping at metro or grid stations. Hence, the simulation results related to the fleets' performances in the charging parking lots considering the traffic flow are indicated in Figure 5. It can be observed that all EV fleets during the traffic flow (see Figure 4 and traffic time 11 to 12) had to to link up the parking lots and increase their charging rates due to being exceeding the consumed power in the traffic flow when moving on the road. However, the absence of traffic and saving the initial energy may be convincing reasons for the high discharging rate of EV fleets at $t = 1, 8, 14$, and 18. To realize the proposed online model's effectiveness for dynamic analysis, we provided additional results to describe the effects of the traffic density rate while accounting for the proportion of rates 10% (down) and 40% (up) to the normal rate on the charging/discharging rate for all EV fleets. These results are indicated in Figures 6 and 7. It is clear that varying traffic rates cause the charging/discharging rates to increase or decrease for some EV fleets, while the charging/discharging times experience no change when not altering the traffic time or overload time.

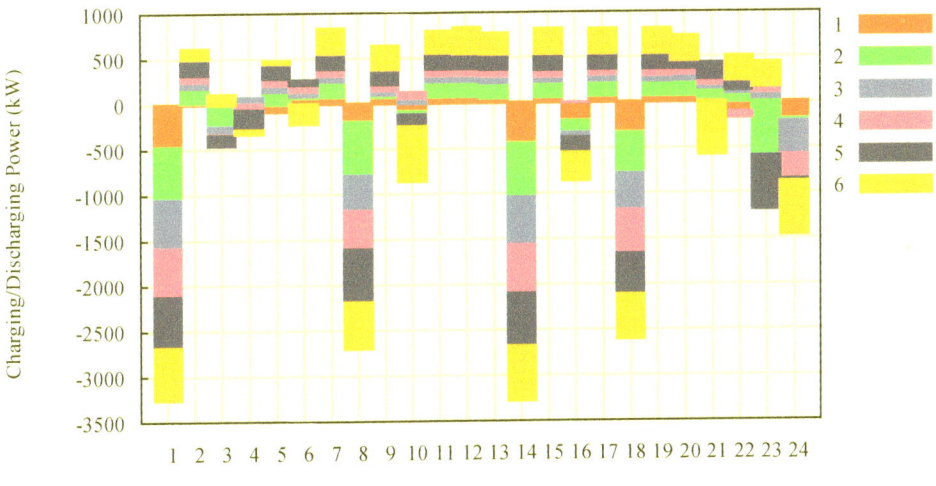

Figure 5. The EVs' performances at the charging station at a normal rate.

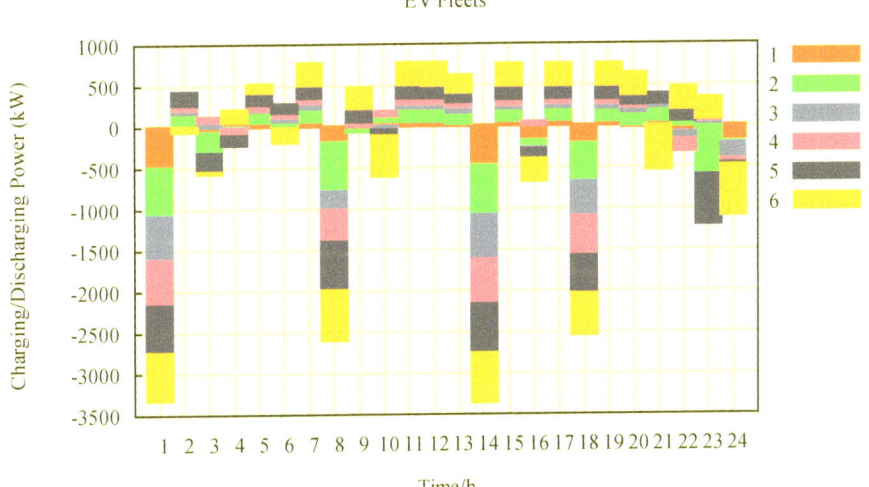

Figure 6. The EVs' performances at the charging station at a rate of 40%.

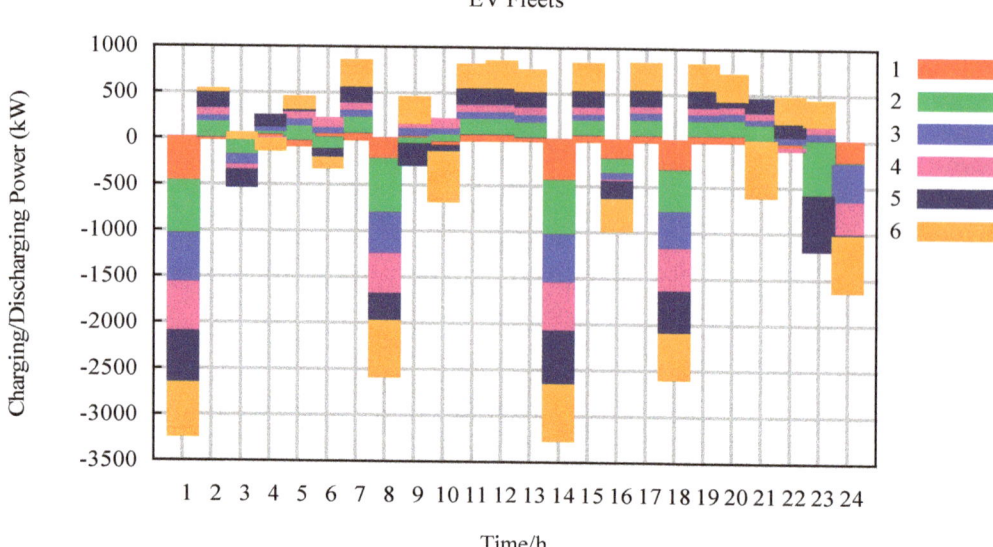

Figure 7. The EVs' performances at the charging station at a rate of 10%.

5.2. Analyzing Malware Model and Offense–Defense Strategy

In this section, we concentrate on simulating a malware attack in the studied transportation system equipped with the WSNs in order to facilitate the control of the traffic flow. Indeed, each node was made up of a wireless-enabled EV that connected to others through the main wireless network. In this case, the number of EV fleets was six, each of which included 30 EVs moving on the WSN-based road, and, in turn, the total number of nodes was 180. Here, we assumed that some nodes or EVs were randomly infected by malware and they propagated malware along with the normal data to the authentic nodes or EVs based on the proposed model through the wireless network. In this simulation, based on ref. [14], due to the memory and the energy level of nodes, the initial malware attack can affect the number of nodes visibly infected by malware. We initially considered 5% of the total number of nodes as the malware nodes and, in turn, 95% of nodes were authentic nodes. These nodes, considering their consumed energies, try to broadcast malware with the aim of infecting the authentic nodes connected to them. By doing this, the authentic nodes infected by malware will continue the propagation process if they consume enough energy. However, the authentic nodes, to discern the normal data from the malware, were equipped with IDS to detect the malware after receiving information. Hence, we assumed that the success probability of an attack by the malicious nodes and the probability of detection by the other nodes are obtained based on the proposed game strategy with the authentic and malicious gains 2 and 4. The energy cost for receiving normal data was considered as 1 while it was 0.5 when propagating malware. When the malware attack was launched, the number of malicious nodes or EVs could be different at any time for the sake of changing the roles of nodes in the offense–defense game strategy. In essence, setting the gains related to the game strategy affects the proportion of the infected wireless-based EVs during a given time period. Indeed, the malware's propagation speed in WSN was controllable thanks to the offense–defense game strategy. To prove the truth of the matter, the simulation results of the proposed strategy in the studied transportation system are indicated in Figures 8 and 9. It can be seen in Figure 8 that the propagation speed of malicious nodes decreased when setting the detection gain as 3. However, by setting the gain as 6, the majority of the nodes were infected by malware very quickly, as shown in Figure 8. In addition, the effect of attack gain on the propagation speed of malicious nodes was evaluated for varied values in Figure 9. It can be seen that if the attack gain takes a value greater than 3, the offense–defense game strategy could provide better effective performance during a given time period when launching a malware attack. Additionally, the expected benefits for the authentic and malicious nodes are dependent on the success probability in playing roles based on the game strategy. Looking over Figure 10, when the attack probability was >0.5, the benefit of the authentic nodes tended toward ascending and, in turn, it adopted the detection strategy. On the contrary, the authentic node selected the non-detection strategy if the attack probability was less than 0.5, considering an ascending trend for the detection probability. Figure 11 indicates how the malicious nodes can adopt the non-attack and attack strategies with an emphasis on their expected benefit based on the proposed offense–defense game strategy. It can be seen that when the attack probability was >0.5, the infected nodes adopted the attack strategy when playing in the game due to the positive benefit. However, the non-attack strategy was imposed on that node for attack probabilities <0.5.

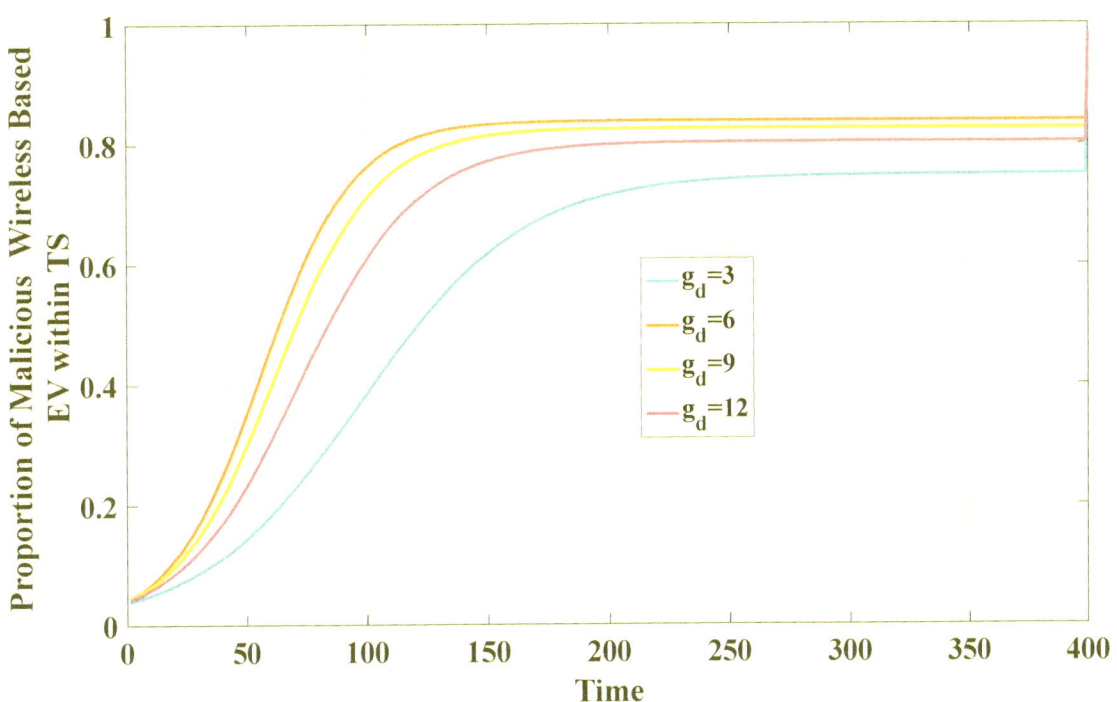

Figure 8. The fraction of malicious wireless-based EVs based on the varied detection gains.

5.3. The Simulation of the Uncertainty Model Based on UT

Modeling and evaluating the irrefutable fluctuations in forecasting some parameters have salutary effects on the precise scheduling of energy in the system. Hence, the transportation systems, owing to not being certain of the energy consumption and travel time related to EVs, need an effective uncertainty model to deal with this concern in energy management. Indeed, modeling the uncertain parameters in transportation systems increases the chance of having accurate online data in order to optimally control the traffic flow. In this section, we simulate the studied case by using the UT concept and obtained the simulation outcomes, as shown in Table 3. For better realization, the results are indicated for the varied traffic rates of 10%, 40%, and normal in the forms of determinacy and uncertainty. As can be seen, the total energy cost underwent remarkable changes of 12%, 2.9%, and 11.39% for normal, 10%, and 40% rates, respectively. In a nutshell, having uncertainty awareness for dynamic energy management can help the decision-making center of TS to provide flexible and controllable traffic management.

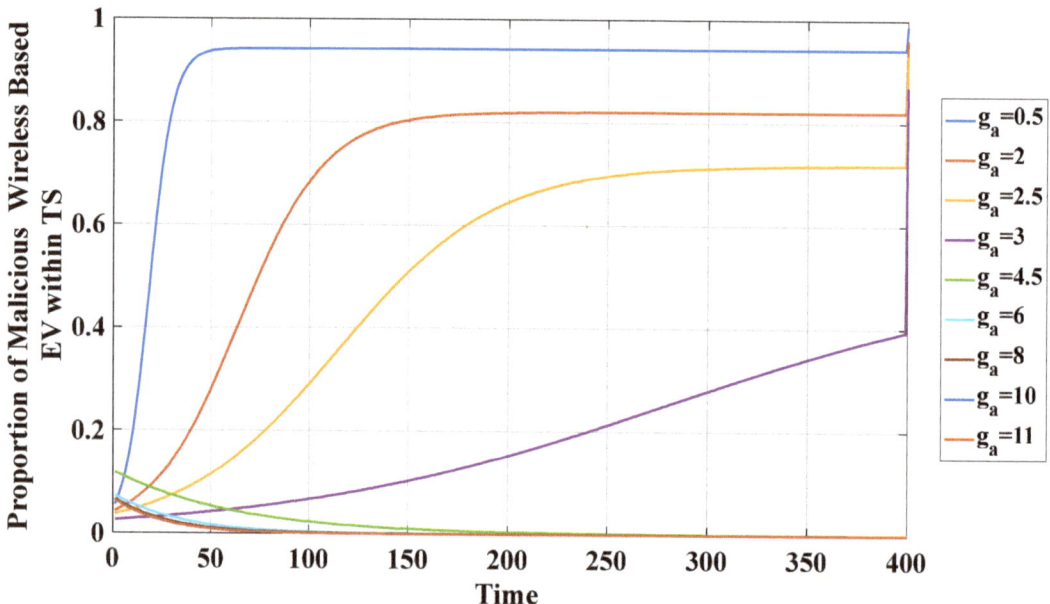

Figure 9. The fraction of malicious wireless-based EVs based on the varied attack gains.

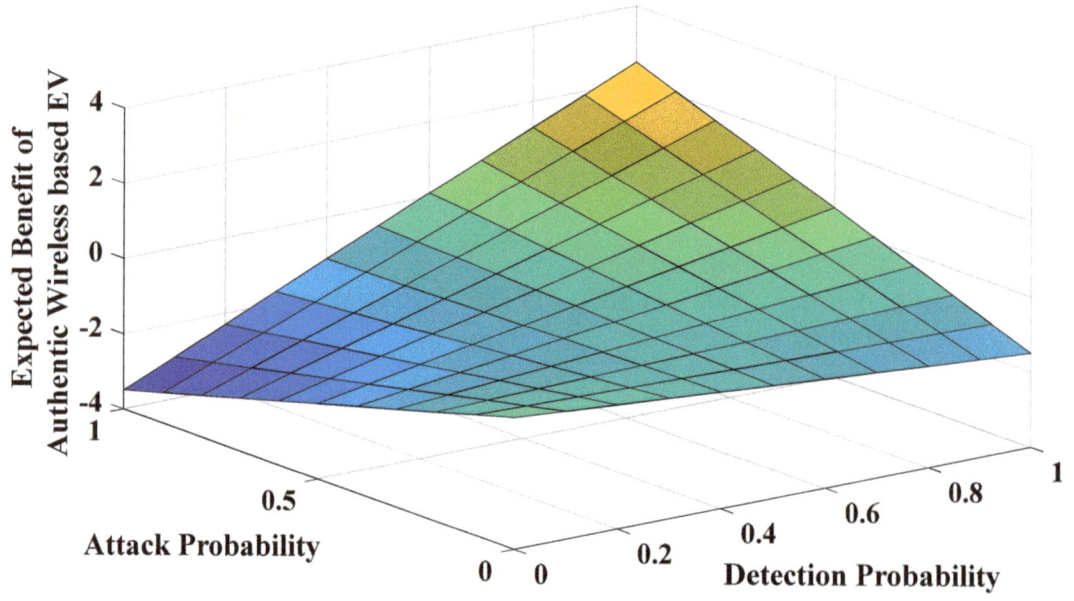

Figure 10. The phase diagram of the expected benefit of the authentic wireless-based EVs.

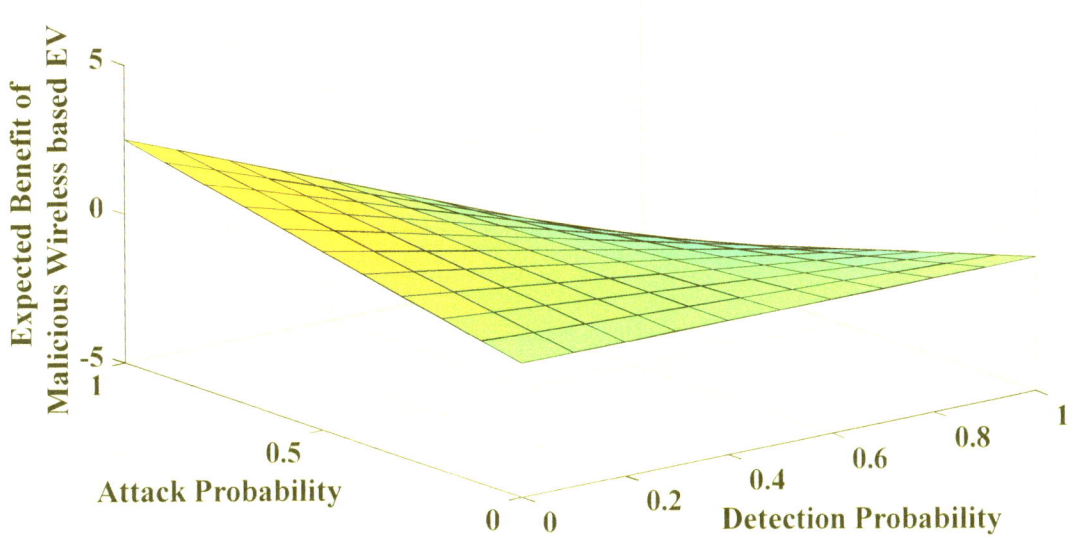

Figure 11. The phase diagram of the expected benefit of the malicious wireless-based EVs.

Table 3. The total cost under the deterministic and uncertain conditions.

Studied Cases	Total Energy Cost (¢)		
	Traffic Rate (Normal)	Traffic Rate (10%)	Traffic Rate (40%)
Determinacy	1.9497×10^5	1.8928×10^5	2.0854×10^5
Uncertainty	2.2191×10^5	1.9511×10^5	2.3536×10^5

In this investigation, by analyzing the advantages of utilizing a wireless sensor network, we were able to model the dynamic behavior of vehicle movement to implement intelligent energy management. However, deploying this infrastructure poses security risks due to its vulnerability to malware attacks for whole-system energy management at the scale of a smart city. For this reason, to address this weakness, we offered a solution based on the game theory method to neutralize a simulated malware attack.

6. Conclusions

In this article, we simulated a transportation system including 6 fleets along with 30 vehicles on varied paths. These vehicles were assumed to be equipped with sensor nodes with the aim of connecting to the WNS. Then, we initially subjected 5% of the nodes to a malware attack, which was broadcasted throughout WSN within the transportation system. Hence, we carried out the proposed offense–defense strategy in the proposed transportation network to deal with the malware infection. The relevant consequences are attested to the proposed energy strategy's effectiveness. Looking over the results, it can be realized, by dynamic analysis based on the proposed traffic model, that the EVs facing the traffic flow on the road would suffer an increase of 40% in dynamic energy consumption. Indeed, propagating false data by malware attacks among the wireless sensor-based EVs can help address the high traffic flow density in smart cities. This problem was considered out for the varied density rates of 10% and 40% in the dynamic analysis. Based on the proposed detection model, once the attack probability was >0.5, the benefit of the authentic nodes was increased such that they selected the detection strategy. However, the authentic nodes selected the non-detection strategy for low probabilities of attack based on the proposed strategy. Additionally, evaluating the offense–defense strategy demonstrated that precisely balancing the attack and detection gains can assist the authentic nodes to adopt the detection strategy very quickly considering the energy consumption. Analyzing the uncertainty model indicated remarkable changes of 12%, 2.9%, and 11.39% in the energy cost for the varied traffic rates.

Author Contributions: Conceptualization, F.A., A.A., T.K. and M.A.M.; Data curation, M.A.M.; Formal analysis, M.A.M.; Funding acquisition, F.A., A.A. and T.K.; Investigation, F.A., T.K. and M.A.M.; Methodology, M.A.M.; Resources, A.A.; Software, A.A. and M.A.M.; Supervision, M.A.M.; Validation, F.A. and T.K.; Visualization, M.A.M.; Writing – original draft, F.A., A.A., T.K. and M.A.M.; Writing – review & editing, F.A. and M.A.M. All authors have read and agreed to the published version of the manuscript.

Funding: This research received no external funding.

Institutional Review Board Statement: Not applicable.

Informed Consent Statement: Informed consent was obtained from all subjects involved in the study.

Data Availability Statement: The data supporting reported results are available in the manuscript.

Acknowledgments: The Authors acknowledge the support provided by King Abdullah City for Atomic and Renewable Energy (K.A.CARE) under K.A.CARE-King Abdulaziz University Collaboration Program. The authors are also thankful to Deanship of Scientific Research, King Abdulaziz University for providing financial support vide grant number (RG-49-135-42).

Conflicts of Interest: The authors declare no conflict of interest.

Nomenclature

Sets/Indices

$\Omega p/p$	Set/index of roads.
$\Omega s/s$	Set/index of stations.
$\Omega t/t$	Set/index of time.
$\Omega v/v$	Set/index of EVs.

Constants

$V_{x,t}$	The velocity for the special values of x and t.
ΔN^{EV}	The change in number of EVs.
$D_{x,t}$	The density of the traffic flow.
$S_{x,t}$	The EVs' speed.
L_d	The distance parameter.
T_d	The time interval.
$P_v^{min\_ch}, P_v^{max\_ch}$	Min/max charging/discharging powers of EV.
E_v^{min}, E_v^{max}	Min/max EV battery energy.
N_A, N_m, N	The number of authentic, malicious nodes and the total number of nodes.
$\alpha A, \beta A$	The proportions of the authentic and malicious nodes.
r	Cost/benefit for the detection.
C_A, C_m	The energy costs for the malicious and authentic nodes.
g_a, g_d	The benefit for the successful attack, and the penalty for the attack failed.
C_e^d	The degradation cost of EVs battery.
η_c, η_d	Charging and discharging efficiencies, respectively.

Variables

$E_t^{EV}, E_t^{roll}, E_t^{air}, E_t^{grad}$	The consumed total energy of the EV, the EV's wheel energy, the involved energy of air resistance, the involved energy of gradient resistance.
P_t^{force}, F_t^{force}	The power consumption and the traction force.
$C_{EV}, C_{V2G}, C_{V2M}, C_v^{deg}$	EVs, V2G and V2M energy transaction costs and degradation costss
$u_{s,p,v,t}^{ch}, u_{s,p,v,t}^{dis}, u_{s,p,v,t}$	Binary variables related to the charging, and discharging of the EVs.
$P_{s,p,v,t}^{V2M\_ch}, P_{s,p,v,t}^{V2M\_dis}, P_{s,p,v,t}^{V2G\_ch}, P_{s,p,v,t}^{V2G\_dis}$	Charging/discharging powers during V2M and V2G, respectively.

$E_{P,v,T}$	The EV battery capacity.
γ_2, γ_1	The removing probabilities of γ_1 and γ_2 for both authentic and malicious nodes.
κ, μ	The infected probability, the repairing probability.
$P_A P_m$	The successful probabilities of authentic and malicious nodes.
X^0	The generated points.
z	The mean value.
Y_{aa}	The covariance matrix.
C_{TT}	The output points of the uncertainty model.

References

1. Wang, B.; Ma, H.; Wang, F.; Dampage, U.; Al-Dhaifallah, M.; Ali, Z.M.; Mohamed, M.A. An IoT-Enabled Stochastic Operation Management Framework for Smart Grids. *IEEE Trans. Intell. Transp. Syst.* **2022**, 1–10. [CrossRef]
2. Avatefipour, O.; Al-Sumaiti, A.S.; El-Sherbeeny, A.M.; Awwad, E.M.; Elmeligy, M.A.; Mohamed, M.A.; Malik, H. An intelligent secured framework for cyberattack detection in electric vehicles' CAN bus using machine learning. *IEEE Access* **2019**, *7*, 127580–127592. [CrossRef]
3. Batista, F.K.; del Rey, A.M.; Queiruga-Dios, A. A new individual-based model to simulate malware propagation in wireless sensor networks. *Mathematics* **2022**, *8*, 410. [CrossRef]
4. Al-Mousawi, A.J.; AL-Hassani, H.K. A survey in wireless sensor network for explosives detection. *Comput. Electr. Eng.* **2019**, *72*, 682–701. [CrossRef]
5. Farjamnia, G.; Gasimov, Y.; Kazimov, C. Review of the techniques against the wormhole attacks on wireless sensor networks. *Wirel. Pers. Commun.* **2019**, *105*, 1561–1584. [CrossRef]
6. Jamshidi, M.; Zangeneh, E.; Esnaashari, M.; Darwesh, A.M.; Meybodi, M.R. A novel model of sybil attack in cluster-based wireless sensor networks and propose a distributed algorithm to defend it. *Wirel. Pers. Commun.* **2019**, *105*, 145–173. [CrossRef]
7. Yuan, Y.; Huo, L.; Wang, Z.; Hogrefe, D. Secure APIT localization scheme against sybil attacks in distributed wireless sensor networks. *IEEE Access* **2018**, *6*, 27629–27636. [CrossRef]
8. Dung, T.N.; Choe, J.; Thang, L.D.; Tai, L.D.; Zalyubovskiy, V.V.; Choo, H. Delaysensitive flooding based on expected path quality in low duty-cycled wireless sensor networks. *Int. J. Distrib. Sens. Netw.* **2016**, *12*, 15501477166642548.
9. Zhang, Q.; Zhang, W. Accurate detection of selective forwarding attack in wireless sensor networks. *Int. J. Distrib. Sens. Netw.* **2019**, *15*, 15501477188240081. [CrossRef]
10. Seto, K.C.; Güneralp, B.; Hutyra, L.R. Global forecasts of urban expansion to 2030 and direct impacts on biodiversity and carbon pools *Proc. Natl. Acad. Sci. USA* **2012**, *109*, 16083–16088. [CrossRef]
11. Bridge, G.; Bouzarovski, S.; Bradshaw, M.; Eyre, N. Geographies of energy transition: Space, place and the low-carbon economy. *Energy Policy* **2013**, *53*, 331–340. [CrossRef]
12. Zhou, H.; Shen, S.; Liu, J. Malware propagation model in wireless sensor networks under attack–defense confrontation. *Comput. Commun.* **2020**, *162*, 51–58. [CrossRef]
13. Liu, J.; Yue, G.; Shen, S.; Shang, H.; Li, H. A game-theoretic response strategy for coordinator attack in wireless sensor networks. *Sci. World J.* **2014**, *2014*, 950618. [CrossRef] [PubMed]
14. Shen, S.; Li, H.; Han, R.; Vasilakos, A.V.; Wang, Y.; Cao, Q. Differential game-based strategies for preventing malware propagation in wireless sensor networks. *IEEE Trans. Inf. Forensics Secur.* **2014**, *9*, 1962–1973. [CrossRef]
15. Harrison, C.; Eckman, B.; Hamilton, R.; Hartswick, P.; Kalagnanam, J.; Paraszczak, J.; Williams, P. Foundations for smarter cities. *IBM J. Res. Dev.* **2010**, *54*, 1–16. [CrossRef]
16. Mohamed, M.A. A relaxed consensus plus innovation based effective negotiation approach for energy cooperation between smart grid and microgrid. *Energy* **2022**, *252*, 123996. [CrossRef]
17. Yang, W.; Liu, W.; Chung, C.Y.; Wen, F. Joint planning of EV fast charging stations and power distribution systems with balanced traffic flow assignment. *IEEE Trans. Ind. Inform.* **2020**, *17*, 1795–1809. [CrossRef]
18. Qian, Z.; Yi, Z.; Zhong, W.; Yue, H.; Yaojia, S. Siting and sizing of electric vehicle fast-charging station based on quasi-dynamic traffic flow. *IET Renew. Power Gener.* **2020**, *14*, 4204–4214. [CrossRef]
19. Zhang, C.; Chen, X. Stochastic nonlinear complementarity problem and applications to traffic equilibrium under uncertainty. *J. Optim. Theory Appl.* **2018**, *137*, 277–295. [CrossRef]
20. Lippi, M.; Bertini, M.; Frasconi, P. Short-term traffic flow forecasting: An experimental comparison of time-series analysis and supervised learning. *IEEE Trans. Intell. Transp. Syst.* **2013**, *14*, 871–882. [CrossRef]
21. Roustaei, M.; Niknam, T.; Salari, S.; Chabok, H.; Sheikh, M.; Kavousi-Fard, A.; Aghaei, J. A scenario-based approach for the design of Smart Energy and Water Hub. *Energy* **2020**, *195*, 116931. [CrossRef]
22. Chabok, H.; Aghaei, J.; Sheikh, M.; Roustaei, M.; Zare, M.; Niknam, T.; Lehtonen, M.; Shafi-khah, M.; Catalão, J.P. Transmission-constrained optimal allocation of price-maker wind-storage units in electricity markets. *Appl. Energy* **2021**, *310*, 118542. [CrossRef]
23. Korayem, A.H.; Khajepour, A.; Fidan, B. Vehicle-trailer lateral velocity estimation using constrained unscented transformation. *Veh. Syst. Dyn.* **2022**, *60*, 1048–1075. [CrossRef]
24. Zhang, L.; Cheng, L.; Alsokhiry, F.; Mohamed, M.A. A Novel Stochastic Blockchain-Based Energy Management in Smart Cities Using V2S and V2G. *IEEE Trans. Intell. Transp. Syst* **2022**, 1–8. [CrossRef]

25. Ding, S.; Cao, Y.; Vosoogh, M.; Sheikh, M.; Almagrabi, A. A directed acyclic graph based architecture for optimal operation and management of reconfigurable distribution systems with PEVs. *IEEE Trans. Ind. Appl.* **2020**, 1. [CrossRef]
26. Zou, H.; Tao, J.; Elsayed, S.K.; Elattar, E.E.; Almalaq, A.; Mohamed, M.A. Stochastic multi-carrier energy management in the smart islands using reinforcement learning and unscented transform. *Int. J. Electr. Power Energy Syst.* **2021**, *130*, 106988. [CrossRef]
27. Gong, X.; Dong, F.; Mohamed, A.M.; Abdalla, O.M.; Ali, Z.M. A secured energy management architecture for smart hybrid microgrids considering PEM-fuel cell and electric vehicles. *IEEE Access* **2020**, *8*, 47807–47823. [CrossRef]
28. Wang, P.; Wang, D.; Zhu, C.; Yang, Y.; Abdullah, H.M.; Mohamed, M.A. Stochastic management of hybrid AC/DC microgrids considering electric vehicles charging demands. *Energy Rep.* **2020**, *6*, 1338–1352. [CrossRef]
29. Almalaq, A.; Albadran, S.; Mohamed, M.A. Deep machine learning model-based cyber-attacks detection in smart power systems. *Mathematics* **2022**, *10*, 2574. [CrossRef]
30. Chen, J.; Mohamed, M.A.; Dampage, U.; Rezaei, M.; Salmen, S.H.; Obaid, S.A.; Annuk, A. A multi-layer security scheme for mitigating smart grid vulnerability against faults and cyber-attacks. *Appl. Sci.* **2021**, *11*, 9972. [CrossRef]
31. Mohamed, M.A.; Abdullah, H.M.; El-Meligy, M.A.; Sharaf, M.; Soliman, A.T.; Hajjiah, A. A novel fuzzy cloud stochastic framework for energy management of renewable microgrids based on maximum deployment of electric vehicles. *Int. J. Electr. Power Energy Syst.* **2021**, *129*, 106845. [CrossRef]
32. Chaudhary, R.; Jindal, A.; Aujla, G.S.; Aggarwal, S.; Kumar, N.; Choo, K.K.R. BEST: Blockchain-based secure energy trading in SDN-enabled intelligent transportation system. *Comput. Secur.* **2019**, *85*, 288–299. [CrossRef]
33. Al-Saud, M.; Eltamaly, A.M.; Mohamed, M.A.; Kavousi-Fard, A. An intelligent data-driven model to secure intravehicle communications based on machine learning. *IEEE Trans. Ind. Electron.* **2019**, *67*, 5112–5119. [CrossRef]
34. Tan, H.; Yan, W.; Ren, Z.; Wang, Q.; Mohamed, M.A. A robust dispatch model for integrated electricity and heat networks considering price-based integrated demand response. *Energy* **2022**, *239*, 121875. [CrossRef]
35. Norouzi, M.; Aghaei, J.; Pirouzi, S.; Niknam, T.; Fotuhi-Firuzabad, M.; Shafie-khah, M. Hybrid stochastic/robust flexible and reliable scheduling of secure networked microgrids with electric springs and electric vehicles. *Appl. Energy* **2021**, *300*, 117395. [CrossRef]
36. Lan, T.; Jermsittiparsert, K.; Alrashood, S.T.; Rezaei, M.; Al-Ghussain, L.; Mohamed, M.A. An advanced machine learning based energy management of renewable microgrids considering hybrid electric vehicles' charging demand. *Energies* **2021**, *14*, 569. [CrossRef]

Article

A Dynamic Spatio-Temporal Stochastic Modeling Approach of Emergency Calls in an Urban Context

David Payares-Garcia [1,*], Javier Platero [2] and Jorge Mateu [2]

1 ITC Faculty Geo-Information Science and Earth Observation, University of Twente, 7522 NB Enschede, The Netherlands
2 Department of Mathematics, University Jaume I, 12006 Castellon, Spain
* Correspondence: d.e.payaresgarcia@utwente.nl

Abstract: Emergency calls are defined by an ever-expanding utilisation of information and sensing technology, leading to extensive volumes of spatio-temporal high-resolution data. The spatial and temporal character of the emergency calls is leveraged by authorities to allocate resources and infrastructure for an effective response, to identify high-risk event areas, and to develop contingency strategies. In this context, the spatio-temporal analysis of emergency calls is crucial to understanding and mitigating distress situations. However, modelling and predicting crime-related emergency calls remain challenging due to their heterogeneous and dynamic nature with complex underlying processes. In this context, we propose a modelling strategy that accounts for the intrinsic complex space–time dynamics of some crime data on cities by handling complex advection, diffusion, relocation, and volatility processes. This study presents a predictive framework capable of assimilating data and providing confidence estimates on the predictions. By analysing the dynamics of the weekly number of emergency calls in Valencia, Spain, for ten years (2010–2020), we aim to understand and forecast the spatio-temporal behaviour of emergency calls in an urban environment. We include putative geographical variables, as well as distances to relevant city landmarks, into the spatio-temporal point process modelling framework to measure the effect deterministic components exert on the intensity of emergency calls in Valencia. Our results show how landmarks attract or repel offenders and act as proxies to identify areas with high or low emergency calls. We are also able to estimate the weekly average growth and decay in space and time of the emergency calls. Our proposal is intended to guide mitigation strategies and policy.

Keywords: Cox processes; crime data; diffusion; emergency calls; spatio-temporal point processes; stochastic integro-differential equations; volatility

MSC: 60-01

Citation: Payares-Garcia, D.; Platero, J.; Mateu, J. A Dynamic Spatio-Temporal Stochastic Modeling Approach of Emergency Calls in an Urban Context. *Mathematics* 2023, 11, 1052. https://doi.org/10.3390/math11041052

Academic Editors: José Luis Romero Béjar and Jose Antonio Sáez Muñoz

Received: 26 January 2023
Revised: 14 February 2023
Accepted: 16 February 2023
Published: 19 February 2023

Copyright: © 2023 by the authors. Licensee MDPI, Basel, Switzerland. This article is an open access article distributed under the terms and conditions of the Creative Commons Attribution (CC BY) license (https://creativecommons.org/licenses/by/4.0/).

1. Introduction

Emergency calls are considered a crucial tool to respond to incidents that require immediate attention, including accidents, wildfires, crimes, and medical emergencies. The information provided in these calls typically encompasses not only the description of the incident, but also its location and time, which are vital elements for a prompt response [1]. In order to ensure an effective response, authorities analyse the spatial and temporal characteristics of emergency calls to allocate resources and infrastructure, identify areas of high risk, and formulate contingency plans. In this context, emergency calls' spatial and temporal analysis is crucial to understanding and mitigating distress situations.

The spatio-temporal analysis of emergency calls is relatively modest. Some papers employ GIS techniques to explore the spatial and temporal dynamics of emergency calls to further improve emergency services [2,3]. Most works use statistical methods to forecast future events and to determine emergency call driving factors [4], spatial and temporal clusters [5–8], or approximating population sizes [9]. In particular, Heaton et al. [10] and Li et al. [11] apply spatial and spatio-temporal point processes, a discipline within spatial statistics, to model the spatio-temporal characteristics of emergency calls. As emergency calls often mirror crimes, these authors adapted popular point process methodologies for crime data to analyse distress signals. For example, Li et al. [11] analysed emergency calls in Montgomery County, Pennsylvania (2016–2017) using a non-parametric spatio-temporal self-exciting point process model previously employed for crime data. The model captured the clustering features in emergency calls and distinguished the areas with intrinsically high emergency calls and those temporal intervals with higher peaks in the calls to direct emergency interventions. For a neater exposition, a summarised version of previous works can be found in Appendix A.

Spatial and spatio-temporal point processes define a suitable mathematical framework to model location-based data in various scientific disciplines, including ecology, epidemiology, and criminology. In particular, the pattern formed by the spatio-temporal coordinates of a crime in a region or a city can be represented by a stochastic point process that could be augmented with additional spatial or temporal covariate information. Point processes to analyse crime data are frequent in the literature given the natural and context-dependent tendency of crime to cluster [12]. Notably, many studies use Hawkes-type point processes or log-Gaussian Cox processes (LGCPs, also called doubly stochastic Poisson processes) for modelling crime event data since both techniques account for spatio-temporal dependencies, covariate inclusion, and clustering phenomena [13,14].

Cox processes serve as suitable models for point process phenomena that are stimulated by environmental factors. However, they are not as apt for phenomena that are predominantly instigated by interactions among the points themselves. A particular property in LGCPs is that the logarithm of the intensity surface is a Gaussian process. As noted by Mohler et al. [12] and Diggle et al. [15], this produces a range of advantageous features that simplify the estimation, interpretation, and simulation of the model. Additionally, the stochastic nature of the intensity process enables the capture of spatio-temporal dependencies [16]. In light of this, it can be challenging, or even unfeasible, to differentiate empirically between processes that represent stochastic, independent fluctuations in a heterogeneous environment, and those that represent stochastic interactions in a homogeneous environment [15]. In the same line, Hawkes point processes, being a type of self-exciting processes, can model the space–time structure of events conditional on the history through the specification of a triggering function.

Emergency calls are defined by an ever-expanding utilisation of information and sensing technology, leading to extensive volumes of high-resolution data. These data are typically heterogeneous and dynamic, characterised by intricate underlying processes in conflict processes, such as diffusion, heterogeneous escalation, and volatility. Consequently, the temporal dynamics of crime data recorded from emergency calls can not be trivially handled by the triggering function in Hawkes processes or by classical formulations of LGCPs. We note that the latter type of processes depend on a space–time covariance structure which is difficult to handle against large datasets and complex non-separable structures. This poses both a mathematical and a computational problem. In this line, Hawkes processes cannot easily address complex dispersion processes such as advection and diffusion [17,18] and these are basic characteristics associated with emergency calls, which have to be prudently introduced into the modelling framework. To account for a system's complex temporal dynamics and to reinforce the discrete-time series definition in LGCPs, Zammit-Mangion et al. [19] introduced stochastic integro-difference equations (SIDEs) as a way to fill the existing gap in modelling complicated latent factors.

As many sorts of emergency calls (such as different types of crimes) share patterns and trends in space and time as armed conflicts, and events are generally registered in discrete-time format, we build upon the reasoning of Zammit-Mangion et al. [19] to study the spatio-temporal dynamics of several emergency calls in Valencia, Spain, for ten years (2010–2020). Our ambition is to model and predict the spatio-temporal behaviour of emergency calls in an urban environment to support criminal activity mitigation strategies and policy responses. Our strategy differs from the existing literature on LGCPs and Hawkes processes in the way we consider the transition from one time to the next one, as rather than depending on covariance structures or triggering functions, we consider integro-difference equations that are able to mimic complex latent processes, and, in turn, are able to model fractional growth or decay. This paper indeed improves on Zammit-Mangion et al. [19]'s proposal in several aspects. We consider crimes in much smaller regions, such as the street network of cities, that make the spatio-temporal interaction rather different from much larger regions. This is indeed a step forward as LGCPs are very scarce for network data. We then consider an analytical procedure for parameter selection rather than a more empirical-based approach. We finally make a finer and more friendly implementation of the methodology. Our modelling strategy is able to account for the intrinsic complex space–time dynamics of some crime data on cities by handling complex advection and diffusion processes.

The structure of this paper is as follows. Section 2 presents the data and the motivating problem and details the spatio-temporal point processes and SIDE methodology. Section 3 presents the data analysis, and Section 4 is devoted to conclusions.

2. Materials and Methods

2.1. Data

The dataset consists of the geocoded locations and times of the calls reported to the 112 emergency telephone number of the Valencian agency for security and emergency with headquarters in the city of Valencia (Spain). The data relate to individual crime activities from 1 January 2010 to 31 October 2020, in Valencia, Spain. Valencia, located on the southeastern coast of Spain, is one of the most important cities in the Valencian community of Spain. According to the national statistical records of Spain in 2018 (https://www.ine.es), (accessed on 30 November 2022) the city has a population of around 0.8 million. We use data from 2010 to 2019 to fit our model while holding out the first ten months of 2020 for prediction purposes and model validation. We note that we also have access to data until mid-2020. However, data from 1 March onwards cannot be used to validate the model due to, first, the quarantine, and then the severe restrictions imposed in Spain due to the COVID-19 epidemiological situation, which changed the natural effects and structure of the calls and the criminal behaviour itself.

During these ten years, 83,379 calls were registered by the emergency number in Valencia. These calls can be divided into four subgroups due to the nature and reason of the call. Among them, we have 51,533 calls due to assaults, 23,282 due to robberies to individuals in the streets, 388 due to aggression against women, and the remaining 8176 calls were due to a cause other than the previous ones but without specifying the reason for the call.

Figure 1 shows weekly data from 2010 to 2020. We note that the first weeks of 2020 show a similar trend to previous years. However, from the ninth week (first week of March), just when the mandatory quarantine for Spanish citizens was implemented, the number of calls decreased considerably to reach levels well below the rest of the years. For this reason, we work only with data until the end of February 2020 (so we only consider the first eight weeks of 2020). Figure 1 highlights a slight upward trend and similar behaviour each year. We note a rise in the intermediate weeks of the year corresponding to the summer holiday period, which is somehow expected for a touristic city. Additionally, we observe a constant peak around weeks 10–11 each year, corresponding to the weeks when the city's local festivities (well-known all over Spain) call for many tourists.

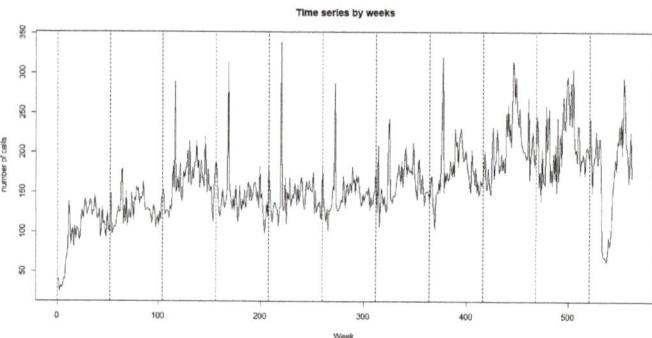

Figure 1. Time series by weeks of the 83,379 emergency calls in Valencia (2010–2019).

The city has been divided into 81 neighbourhoods or districts, and the spatial distribution of the calls per district is given in Figure 2. We note a higher number of calls in central districts compared to others located outdoors of the city.

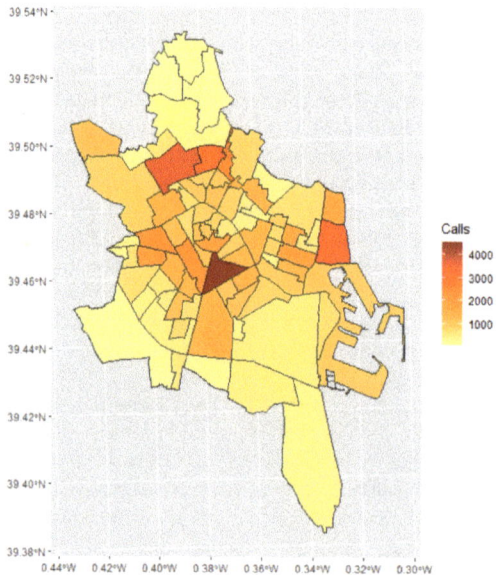

Figure 2. Map of total number of calls per district in Valencia.

With the idea of showing some underlying aspects of the number of calls, we now show in Figures 3–5 the weekly number of calls in six selected districts that we will use in the rest of the paper to show our analytical and prediction results. We show two neighbourhoods in the centre of the city (with a higher number of calls), Russafa and Sant Francesc, two neighbourhoods in the maritime east of the city, Cabanyal-Canyamelar and Malva-Rosa, and two neighbourhoods located on the outskirts of the city, one in the north, Benicalap, and one in the south, La Torre. Maritime districts highlight the effect of the arrival of tourists in summer with an increase in the number of calls. In addition, the central districts show a clear peak of calls in March due to the local festivities whose activities gather people in downtown Valencia.

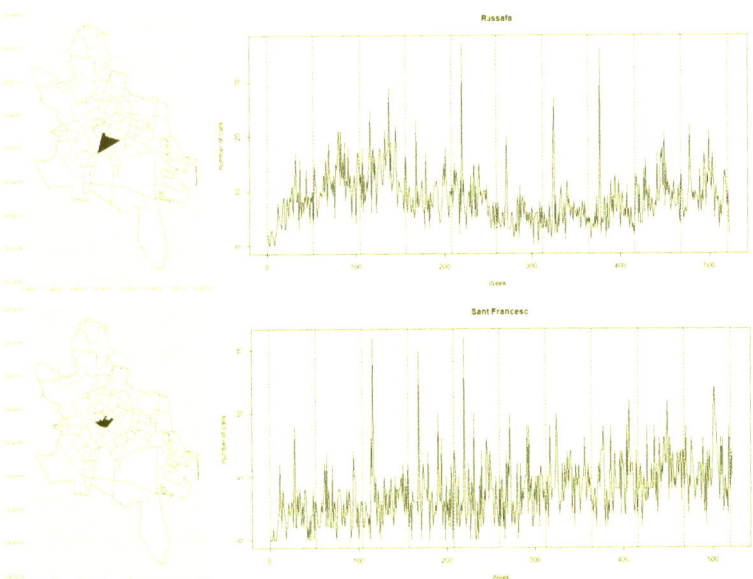

Figure 3. Weekly calls distribution over time at central districts in Valencia, Russafa (top row), and Sant Francesc (bottom row).

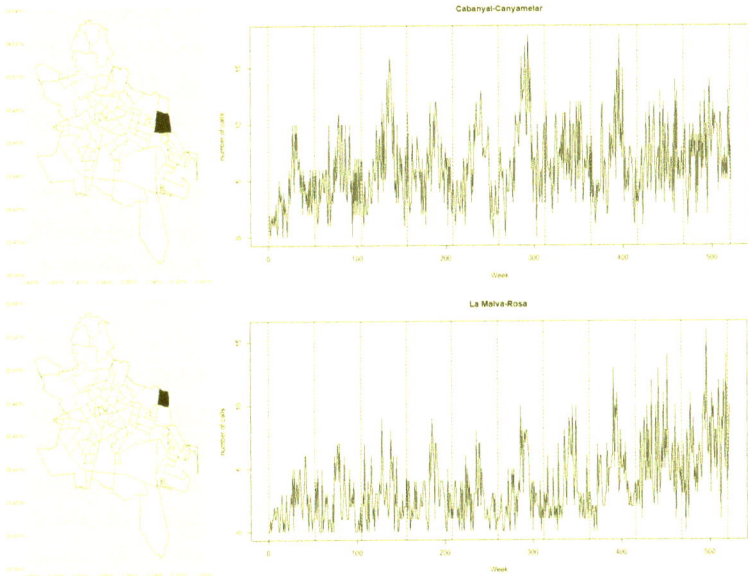

Figure 4. Weekly calls distribution over time at maritime districts in Valencia, Cabanyal-Canyamelar (top row), and Malva-Rosa (bottom row).

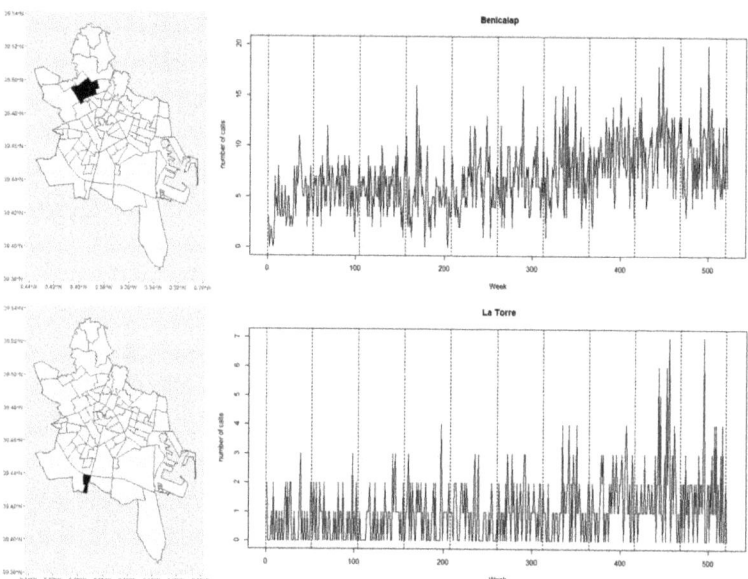

Figure 5. Weekly calls distribution over time at Benicalap, northern Valencia (top row), and at La Torre, southern Valencia (bottom row).

To have a graphical idea of the spatial pattern of the calls due to the selected crimes, we represent such point patterns for 2012 (see Figure 6). Here, we observe an increase in calls in March and during the summer months. In addition, we can see that the increase in March happens in central districts, while the increase in calls in the summertime is more concentrated in maritime districts. For comparative purposes and to graphically analyse the space–time interaction of the calls, we also show in Figure 7 the month of January over the ten years. This graphical output shows an increasing number of calls per year for January, giving light to such dynamics and interaction in space and time.

A final piece of information we have associated with the space–time locations of the crime-related calls is based on several covariates that are minimum distances from an event (a call) to a set of selected landmarks in the city. This is highly important as different landmarks increase or decrease the number of calls, which means criminal activity can be related to being closer or further from a particular landmark. In this line, we considered the following selected landmarks or points of interest: ATMs, banks, bars, coffee shops, industries, markets, nightclubs, police stations, pubs, restaurants, and taxi stops. Furthermore, we have measured the nearest distance from a call to such landmarks. We only show some distributions and corresponding patterns for some graphical testing for brevity. Thus, on the one hand, we see how the minimum distances to restaurants or pubs (Figure 8) are dominant in the lower values of the distribution of the distances. On the other hand, in the cases of police stations or ATMs (Figure 9), the minimum distances are not that short, indicating that crimes are happening much farther from these landmarks. Finally, in the industries or markets (Figure 10), the distances seem equal or constant, indicating these landmarks have no or little effect on crimes.

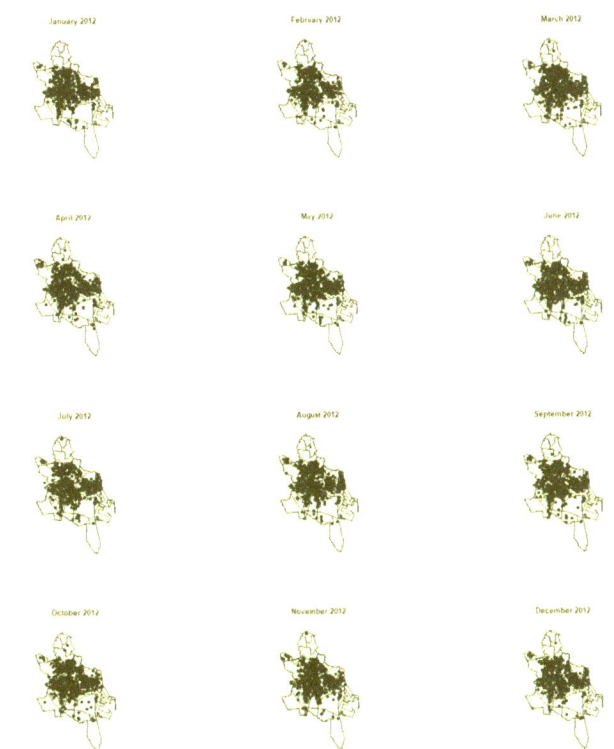

Figure 6. Spatial point patterns of the emergency calls in Valencia for each month in 2012.

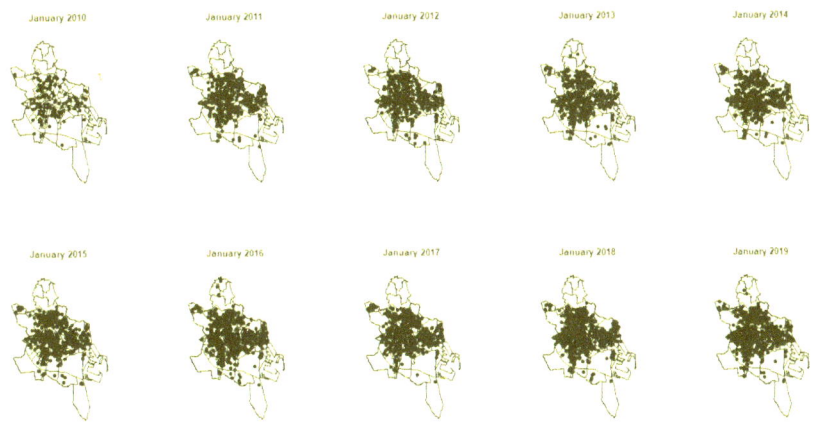

Figure 7. Spatial point patterns of the emergency calls for the month of January of each year between 2010 and 2019.

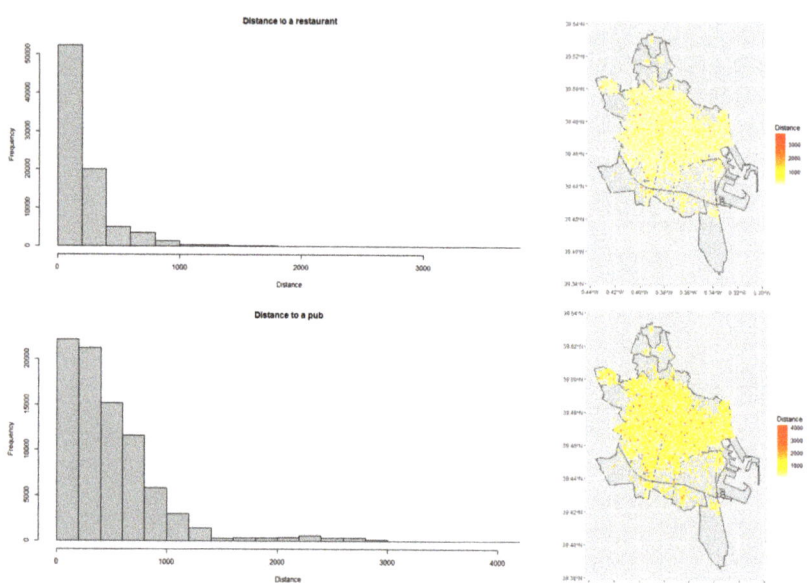

Figure 8. Histogram and corresponding point patterns coloured by their minimum distances to restaurants (top row) and to pubs (bottom row).

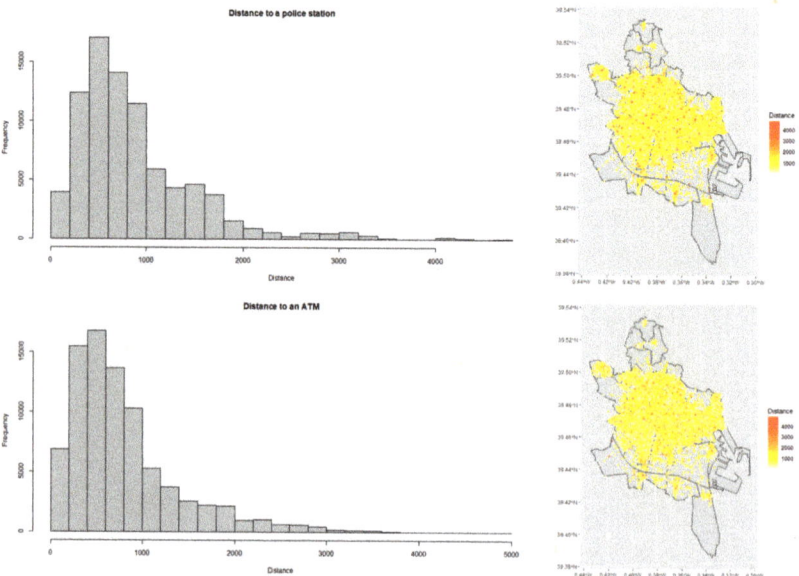

Figure 9. Histogram and corresponding point patterns coloured by their minimum distances to police stations (top row) and to ATMs (bottom row).

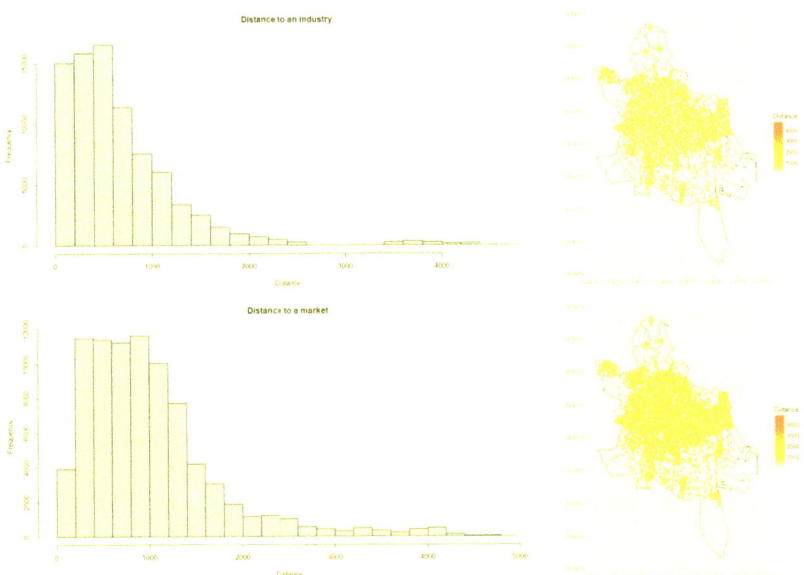

Figure 10. Histogram and corresponding point patterns coloured by their minimum distances to industries (top row) and to markets (bottom row).

2.2. Methodology

We follow the logic of Zammit-Mangion et al. [19] to propose a suite of dynamic spatio-temporal modelling tools to identify complex underlying processes of crime data coming from calls to the emergency phone in an urban environment. Our modelling approach sits in the family of LGCP as it provides great flexibility over the more simple inhomogeneous Poisson process. Poisson point processes focus on modelling event-based data by assuming a Poisson distribution to the probability of observing a particular number of events within a defined area, D. The mean of this distribution is determined by the integral over D of an intensity function, denoted as $\lambda(\mathbf{s})$, which is dependent on the location vector $\mathbf{s}$, belonging to D. A doubly stochastic or Cox process is defined if the intensity function is assumed to be a random function. In the case of a LGCP, the logarithm of the intensity is assumed to be a Gaussian process (GP), defined as a latent structure.

Let us assume we have a discrete time series of continuous spatial LGCPs, since the temporal range is discrete. Formally, let $k \in \mathcal{K}$, $\mathcal{K} = \{1, \ldots, K\}$ denote a discrete time index set, and $\{z_k(\mathbf{s})\}$, $z_k(\mathbf{s}) \sim \mathrm{GP}\big(\mu_k(\mathbf{s}), \sigma_k^2 \Psi_k(\mathbf{s}, \mathbf{r})\big)$, a set of temporally correlated spatial Gaussian processes, each with mean $\mu_k(\mathbf{s})$ and covariance function $\sigma_k^2 \Psi_k(\mathbf{s}, \mathbf{r})$. The intensity function of the point process is represented by a exponential function of $z_k(\mathbf{s})$ for each k, i.e., $\lambda_k(\mathbf{s}) = \exp(z_k(\mathbf{s}))$. To mitigate prediction uncertainty, the mean function of $z_k(\mathbf{s})$ can be linked to explanatory variables. In this scenario, a vector of spatially referenced covariates denoted as $\mathbf{d}(\mathbf{s})$ and its corresponding regression coefficients represented by $\mathbf{b}^T$ can be employed. As such, the intensity of the LGCP at time k is given by the exponential of the sum of the regression coefficients and the mean function, i.e., $\lambda_k(\mathbf{s}) = \exp\big(\mathbf{b}^T \mathbf{d}(\mathbf{s}) + z_k(\mathbf{s})\big)$.

In order to account for the intricate temporal dynamics of the intensity functions through $z_k(\mathbf{s})$, we adopt the stochastic integro-difference equation (SIDE) framework. This flexible modelling approach is capable of capturing dynamic temporal effects such as diffusion and dispersal. Specifically, the SIDE relates the spatio-temporal dependent variable $z_k(\mathbf{s})$ to $z_{k+1}(\mathbf{s})$ through the following integral equation:

$$z_{k+1}(\mathbf{s}) = \int_D k_M(\mathbf{s}, \mathbf{r}) f(z_k(\mathbf{r})) \mathrm{d}\mathbf{r} + e_k(\mathbf{s}), \tag{1}$$

where $k_M(\mathbf{s}, \mathbf{r})$ is the mixing kernel in the integral, and $e_k(\mathbf{s}) \sim \text{GP}(\mu_Q(\mathbf{s}), k_Q(\mathbf{s}, \mathbf{r}))$ is an added disturbance, modelled as a Gaussian field with mean $\mu_Q(\mathbf{s})$ and covariance function $k_Q(\mathbf{s}, \mathbf{r})$, and D is the spatial domain under study. The non-linear mapping $f(\cdot)$ distorts the field in the sedentary stage; in the absence of a priori knowledge, the identity map $f(z_k(\mathbf{r})) = z_k(\mathbf{r})$ can be adopted.

2.2.1. Correlation Analysis

We need to measure the correlation between the crime events within the same and across subsequent time frames. In point process statistics, these are quantified through the pair correlation function. Indeed, the pair auto-correlation function (PACF) $g_{k,k}(\mathbf{s}, \mathbf{r})$, and the pair cross-correlation function (PCCF) $g_{k,k+1}(\mathbf{s}, \mathbf{r})$ quantify the probability of finding an event at location $\mathbf{r}$ given that an event has occurred at $\mathbf{s}$ within the same time frame k or at previous time frame $k - 1$. These functions are given by

$$g_{k,k}(\mathbf{s}, \mathbf{r}) = \frac{\lambda_{k,k}^{(2)}(\mathbf{s}, \mathbf{r})}{\lambda_k^{(1)}(\mathbf{s}) \lambda_k^{(1)}(\mathbf{r})},$$

$$g_{k,k+1}(\mathbf{s}, \mathbf{r}) = \frac{\lambda_{k,k+1}^{(2)}(\mathbf{s}, \mathbf{r})}{\lambda_k^{(1)}(\mathbf{s}) \lambda_{k+1}^{(1)}(\mathbf{r})},$$

where $\lambda_k^{(1)}(\mathbf{s}) = \mathrm{E}[\lambda_k(\mathbf{s})]$ and $\lambda_{k,k}^{(2)}(\mathbf{s}, \mathbf{r}) = \mathrm{E}[\lambda_k(\mathbf{s})\lambda_k(\mathbf{r})]$ are real and positive. The PACF determines qualitative characteristics of the events; if $g_{k,k}(\mathbf{s}, \mathbf{r}) = 1$, no spatial pattern can be inferred from the data; $g_{k,k}(\mathbf{s}, \mathbf{r}) > 1$ and $g_{k,k}(\mathbf{s}, \mathbf{r}) < 1$ indicate event clustering and inhibition, respectively.

Given a spatial point pattern at time k, $\mathscr{P}_k$, a realisation of a particular point process model, a standard non-parametric kernel estimator of the first-order intensity $\lambda_k^{(1)}(\mathbf{s})$ is given by

$$\hat{\lambda}_k^{(1)}(\mathbf{s}) = \sum_{\mathbf{s}_i \in \mathscr{P}_k} \frac{k_b(\|\mathbf{s} - \mathbf{s}_i\|)}{c_{D,b}(\mathbf{s}_i)},$$

where $\|\cdot\|$ denotes the Euclidean distance on D, $c_{D,b}(\mathbf{s}_i)$ is an edge-correction factor, and $k_b(s)$ is here representing the Epanechnikov kernel.

Consequently, non-parametric estimators of PACF and PCCF for $v = \|\mathbf{s} - \mathbf{r}\|$ are given by

$$\hat{g}_{k,k}(v) = \frac{1}{2\pi v s.|D|} \sum_{\mathbf{s}_i, \mathbf{s}_j \in \mathscr{P}_k}^{\neq} \frac{k_b(\|\mathbf{s}_i - \mathbf{s}_j\| - v)}{\hat{\lambda}_k^{(1)}(\mathbf{s}_i) \hat{\lambda}_k^{(1)}(\mathbf{s}_j) w(\mathbf{s}_i, \mathbf{s}_j)},$$

and

$$\hat{g}_{k,k+1}(v) = \frac{1}{2\pi v s.|D|} \sum_{\mathbf{s}_i \in \mathscr{P}_k, \mathbf{s}_j \in \mathscr{P}_{k+1}}^{\neq} \frac{k_b(\|\mathbf{s}_i - \mathbf{s}_j\| - v)}{\hat{\lambda}_k^{(1)}(\mathbf{s}_i) \hat{\lambda}_{k+1}^{(1)}(\mathbf{s}_j) w(\mathbf{s}_i, \mathbf{s}_j)},$$

where $w(\mathbf{s}_i, \mathbf{s}_j)$ is the fraction of the circle (in two dimensions) with centre $\mathbf{s}_i$ and radius $\|\mathbf{s}_i - \mathbf{s}_j\|$ lying in D.

If the processes are taken to be second-order stationary also in time, to smooth out the non-parametric estimates an average over all K time steps may be taken so that $\tilde{g}_{k,k}(v) = (1/K) \sum_{k=1}^{K} \hat{g}_{k,k}(v)$ and $\tilde{g}_{k,k+1}(v) = (1/(K-1)) \sum_{k=1}^{K-1} \hat{g}_{k,k+1}(v)$.

Finally, note that $\ln g_{k,k+1}(v) = [k_M * \ln g_{k,k}](v)$, with $*$ being the convolution operator. The above indicates that the kernel k_M may be acquired by performing a deconvolution on the previous equation, and conventional image processing methods such as direct inverse filtering may be applied to achieve this. Furthermore, one can show that $k_Q(v) = \ln g_{k+1,k+1}(v) - [k_M * k_M * \ln g_{k,k}](v)$. Note that if temporally averaged PACF/PCCFs are used, the inverse filter is given as $\hat{k}_M(v) = \mathscr{F}^{-1}\left(\frac{\mathscr{F}(\ln \tilde{g}_{k,k+1}(v))}{\mathscr{F}(\ln \tilde{g}_{k,k}(v))}\right)$.

2.2.2. Dimensionality Reduction

A computationally convenient truncated basis function representation of the spatio-temporal field is considered to develop an inferential approach. The choice of basis functions relies on the non-parametric estimation of the PACF. Specifically, we recall here two results. In the fundamental lemma of LGCPs, the log PACF equals the field auto-correlation function,

$$g_{k,k}(\mathbf{s},\mathbf{r}) = \exp(\sigma_k^2 \Psi_k(\mathbf{s},\mathbf{r})),$$

The second result represents the auto-correlation theorem which indicates that the spectrum of the signal is the Fourier transform of the auto-correlation function. This connection between the frequency content of the point process and the PACF is employed to pick a suitable collection of basis functions. Having the basis functions, the kernel, the mean disturbance and the field can be decomposed as

$$z_k(\mathbf{s}) = \boldsymbol{\phi}(\mathbf{s})^T \mathbf{x}_k,$$
$$\mu_Q(\mathbf{s}) = \boldsymbol{\phi}(\mathbf{s})^T \boldsymbol{\vartheta},$$
$$k_M(\mathbf{s},\mathbf{r}) = \boldsymbol{\phi}(\mathbf{s})^T \Sigma_M \boldsymbol{\phi}(\mathbf{r}),$$
$$k_Q(\mathbf{s},\mathbf{r}) = \boldsymbol{\phi}(\mathbf{s})^T \Sigma_Q \boldsymbol{\phi}(\mathbf{r}),$$

where $\boldsymbol{\phi}(\mathbf{s}) \in \mathbb{R}^n$ is the vector of basis functions, $\mathbf{x}_k \in \mathbb{R}^n$ and $\boldsymbol{\vartheta} \in \mathbb{R}^n$ are weights which reconstruct the spatio-temporal field and the disturbance mean, respectively. Furthermore, $\Sigma_M \in \mathbb{R}^{n \times n}$ and $\Sigma_Q \in \mathbb{R}^{n \times n}$ reconstruct the kernel covariance function and the disturbance covariance function, respectively. In this context, the SIDE of Equation (1) can be rewritten as

$$\mathbf{x}_{k+1} = \mathbf{A}(\Sigma_I)\mathbf{x}_k + \mathbf{w}_k(\boldsymbol{\vartheta}, \Sigma_Q), \tag{2}$$

In this equation, $\mathbf{A}(\Sigma_I) \in \mathbb{R}^{n \times n}$ and $\mathbf{w}_k \in \mathbb{R}^n$ represents a Gaussian coloured noise term with a mean of $E[\mathbf{w}_k] = \boldsymbol{\vartheta}$ and a covariance of $\text{cov}[\mathbf{w}_k] = \Sigma_Q$. The objective is to estimate the unknown parameters $\theta = \boldsymbol{\vartheta}, \Sigma M, \Sigma Q^{-1}$ and the states $\mathscr{X}K = \mathbf{x}0 : K = \mathbf{x}kk = 0^K$ using the data $\mathscr{Y}_K = \mathbf{y}kk = 1^K$, where each $\mathbf{y}_k$ is a set of coordinates of the logged events at the k-th time point.

Basis Function Selection

Based on the link between the Fourier transform of the PACF and the signal spectrum, we select the collection of basis functions using a frequency-based approach. The Fourier transform of the average PACF is computed, and a cut-off frequency of ν_c cycles/unit is selected from it. Then, localised reconstruction kernels are placed at small regular intervals throughout the spatial domain. The centres of the basis functions $\{\zeta_i\}_{i=1}^n$ equate the sequence of vectors defining the regular partition of length Δ_s over the spatial domain D, so that

$$\Delta_s < \frac{1}{2\nu_c} = \frac{1}{2\alpha_0\nu_c}$$

where α_0 is an oversampling parameter. Gaussianity makes mathematical development easier while still producing flexible close forms. Thus, if basis functions are defined as Gaussian radial basis functions (GRBFs) with functional form $\phi(s) = \exp(-s^2/2\sigma_b^2)$, their Fourier transforms are also Gaussian radial functions

$$\phi(\nu) = \mathscr{F}\{\phi(s)\} = \sqrt{2\pi\sigma_b^2}\exp\left(-2\pi^2\sigma_b^2\nu^2\right)$$

so that the spatial and frequency variances relate through the mappings

$$\sigma_\nu^2 \leftarrow \frac{1}{4\pi^2\sigma_b^2}, \sigma_b^2 \leftarrow \frac{1}{4\pi^2\sigma_\nu^2}$$

To ensure appropriate reconstruction, the frequency and spatial range of the basis functions needs to exceed that of the field. This condition is addressed when $\sigma_\nu = \frac{1}{\sqrt{2}}\nu_c$.

Given σ_ν, the previously stated relation between σ_ν and σ_b can be used to find the width of the desired Gaussian radial basis functions (GRBFs), obtaining that $\sigma_b^2 = (2\nu_c^2 \pi^2)^{-1}$. The resulting basis functions are a set of GRBFs with parameter σ_b placed in the spatial domain D centred on the coordinates $\{\zeta_i\}_{i=1}^n$. However, since the GRBFs are not of compact support, a compact GRBF (CGRBF) is used instead, which takes the form

$$\phi(s) = \begin{cases} \frac{(2\pi - \tau\|s\|)(1 + (\cos\tau\|s\|/2) + 1.5\sin(\tau\|s\|))}{3\pi}, & \|\tau s\| < 2\pi \\ 0, & \text{otherwise} \end{cases} \quad (3)$$

for $\tau > 0$, and where $\|\cdot\|$ denotes the Euclidean distance on D. The CGRBF is similar to the usual GRBF with $\phi(s) = \exp(-\tau^2\|s\|^2/2\pi)$ but it is of compact support. The CGRBF parameter τ can be estimated having the GRBF parameter σ_b, or a cutoff frequency ν_c, as follows:

$$\tau = \sqrt{\pi}/\sigma_b = \sqrt{2\nu_c^2 \pi^3}$$

The cutoff frequency of ν_c is selected from the Fourier transform of the average PACF. The value corresponds to where the average PACF function decays abruptly towards zero.

2.2.3. Variational Bayesian Inference

We use the following likelihood function for inference:

$$p(\mathbf{y}_k | \lambda_k(\mathbf{s})) = \prod_{\mathbf{s}_j \in \mathbf{y}_k} \lambda_k(\mathbf{s}_j) \exp\left(-\int_D \lambda_k(\mathbf{s}) d\mathbf{s}\right),$$

where each $\lambda_k(\mathbf{s})$ is approximated using the same basis representation

$$\lambda_k(\mathbf{s}) = \exp(\mathbf{b}^T \mathbf{d}(\mathbf{s}) + z_k(\mathbf{s})) \approx \exp(\mathbf{b}^T \mathbf{d}(\mathbf{s}) + \boldsymbol{\phi}(\mathbf{s})^T \mathbf{x}_k).$$

The full posterior distribution is approximated through the variational Bayes method, and takes the form

$$p(\mathcal{X}_K, \boldsymbol{\theta}, \mathbf{b} \mid \mathcal{Y}_K) = p(\mathcal{X}_K, \boldsymbol{\vartheta}, \boldsymbol{\Sigma}_M, \boldsymbol{\Sigma}_Q^{-1}, \mathbf{b} \mid \mathcal{Y}_K) \approx \tilde{p}(\mathcal{X}_K)\tilde{p}(\boldsymbol{\vartheta})\tilde{p}(\boldsymbol{\Sigma}_M)\tilde{p}(\boldsymbol{\Sigma}_Q^{-1})\tilde{p}(\mathbf{b})$$

with $\tilde{p}(\cdot)$ being the variational marginals. The variational marginals inform about important properties of the crime events' progression. $\mathcal{X}_K$ reconstructs the spatio-temporal field at every time point, $\boldsymbol{\vartheta}$ shows the spatially varying escalation in events, $\boldsymbol{\Sigma}_M$ displays the extent of the spatial dynamics, and $\boldsymbol{\Sigma}_Q^{-1}$ informs about the volatility of the event occurrences. The number of unknown parameters in the reduced model scales as $D(n^2)$, where n is the number of basis functions retained.

The variational Bayes marginals for the unknown states $\mathcal{X}_K$ and parameters $\boldsymbol{\theta} = \left(\boldsymbol{\vartheta}, \boldsymbol{\Sigma}_Q^{-1}\right)$ and $\mathbf{b} = [b_1, b_2, \ldots, b_d]$, with d denoting the number of covariates, can be estimated by finding the lower on the marginal likelihood. We then have

$$\tilde{p}(\mathcal{X}_K) \propto \exp\left(\mathbb{E}_{\tilde{p}(\boldsymbol{\theta})\tilde{p}(\mathbf{b})}[\ln p(\mathcal{Y}_K, \mathcal{X}_K, \boldsymbol{\theta}, \mathbf{b})]\right)$$

$$\tilde{p}(\boldsymbol{\vartheta}) \propto \exp\left(\mathbb{E}_{\tilde{p}(\mathcal{X}_K)\tilde{p}(\boldsymbol{\theta}/\boldsymbol{\vartheta})\tilde{p}(\mathbf{b})}[\ln p(\mathcal{Y}_K, \mathcal{X}_K, \boldsymbol{\theta}, \mathbf{b})]\right)$$

$$\tilde{p}\left(\boldsymbol{\Sigma}_Q^{-1}\right) \propto \exp\left(\mathbb{E}_{\tilde{p}(\mathcal{X}_K)\tilde{p}\left(\boldsymbol{\theta}/\boldsymbol{\Sigma}_Q^{-1}\right)\tilde{p}(\mathbf{b})}[\ln p(\mathcal{Y}_K, \mathcal{X}_K, \boldsymbol{\theta}, \mathbf{b})]\right)$$

$$\tilde{p}(b_i) \propto \exp\left(\mathbb{E}_{\tilde{p}(x_k)\tilde{p}(\theta)\tilde{p}(b^{/b_i})}\right)[\ln p(\mathcal{Y}_K, \mathcal{X}_K, \theta, b)]), i = 1\ldots d$$

where $\theta^{/\theta}$ denotes the set of variables θ without θ and $\mathbb{E}_{\tilde{p}(\cdot)}[\cdot]$ is employed to compute expectations with respect to the distribution in question.

In the upcoming section, we present the method utilised to deduce the unknown variables. It should be noted that the notation $i|j$ denotes the estimate at time i based on the data observed until time j. For the sake of clarity, we have reformulated the model as follows: $\mathbf{x}k + 1 = \mathbf{x}k + \boldsymbol{\vartheta} + \tilde{\mathbf{w}}k$, where $\tilde{\mathbf{w}}k$ has a zero mean.

Parameter Estimation

Starting with the state inference, the distribution $\tilde{p}(\mathcal{X}_K)$ can be computed by an approximate variational Kalman smoother. Let $\mathbf{x}_0 \sim \mathcal{N}_{x0}(\boldsymbol{\mu}_0, \boldsymbol{\Sigma}_0)$. Considering the variational forward $\tilde{\alpha}(\mathbf{x}_k) = \tilde{p}(\mathbf{x}_k \mid \mathbf{y}_{1:k})$ and backward $\tilde{\beta}(\mathbf{x}_k) = \tilde{p}(\mathbf{y}_{k+1:K} \mid \mathbf{x}_k)$ messages, and using the Laplace method approximation, we can further write $\tilde{\alpha}(\mathbf{x}_k) \to \mathcal{N}_{\mathbf{x}_k}\left(\hat{\mathbf{x}}_{k|k}, \boldsymbol{\Sigma}_{k|k}\right)$ and $\tilde{\beta}(\mathbf{x}_k) \to \mathcal{N}_{\mathbf{x}_k}\left(\hat{\mathbf{x}}_{k|k+1:K}, \boldsymbol{\Sigma}_{k|k+1:K}\right)$.

The two messages are then combined to give the smoothed estimate

$$\tilde{p}(\mathbf{x}_k \mid \mathbf{y}_{1:K}) \propto \tilde{p}(\mathbf{x}_k \mid \mathbf{y}_{1:k})\tilde{p}(\mathbf{y}_{k+1:K} \mid \mathbf{x}_k) = \tilde{\alpha}(\mathbf{x}_k)\tilde{\beta}(\mathbf{x}_k)$$
$$= \mathcal{N}_{\mathbf{x}_k}\left(\hat{\mathbf{x}}_{k|K}, \boldsymbol{\Sigma}_{k|K}\right)$$

In relation to escalation inference, and considering the prior $p(\boldsymbol{\vartheta}) \sim \mathcal{N}_{\boldsymbol{\vartheta}}(\hat{\boldsymbol{\vartheta}}_p, \boldsymbol{\Sigma}_{\vartheta,p})$, its posterior $\tilde{p}(\boldsymbol{\vartheta})$ can be written as

$$\tilde{p}(\boldsymbol{\vartheta}) \propto p(\boldsymbol{\vartheta}) \exp\left(-\frac{1}{2}\mathbb{E}_{\tilde{p}(\mathcal{X}_K)\tilde{p}(\boldsymbol{\Sigma}_Q^{-1})} \left[\sum_{k=0}^{K-1}(\mathbf{x}_{k+1} - \mathbf{x}_k - \boldsymbol{\vartheta})^T \times \boldsymbol{\Sigma}_Q^{-1}(\mathbf{x}_{k+1} - \mathbf{x}_k - \boldsymbol{\vartheta})\right]\right)$$

Considering now volatility inference, let the prior $p\left(\boldsymbol{\Sigma}_Q^{-1}\right) = \mathcal{W}i_{\boldsymbol{\Sigma}_Q^{-1}}(V_p, d_p)$ where $\mathcal{W}i_{\boldsymbol{\Sigma}_Q^{-1}}(V, d)$ denotes a Wishart distribution with V a positive definite, symmetric scale matrix and d degrees of freedom. The variational posterior can be then written as

$$\tilde{p}\left(\boldsymbol{\Sigma}_Q^{-1}\right) \propto p\left(\boldsymbol{\Sigma}_Q^{-1}\right) \exp\left(\frac{K}{2}\ln\left|\boldsymbol{\Sigma}_Q^{-1}\right| - \frac{1}{2}\text{tr}\left(\boldsymbol{\Gamma}\boldsymbol{\Sigma}_Q^{-1}\right)\right)$$

where the evaluation of $\boldsymbol{\Gamma}$ requires evaluation of the cross-covariance matrix in addition to the usual posterior covariance matrices. The computation of the cross-covariance also requires Laplace approximations.

Finally, in terms of the regression parameters, under the variational Bayes approach, we let $\tilde{p}(\mathbf{b}) = \prod_{i=1}^d \tilde{p}(b_i)$, and set the prior $p(b_i) \sim \mathcal{N}_{b_i}\left(\hat{b}_{i,p}, \sigma_{b_i,p}^2\right)$. Its variational posterior $\tilde{p}(b_i)$ is then given by

$$\tilde{p}(b_i) \propto p(b_i) \prod_{k \in \mathcal{K}} \left\{ \left[\prod_{s_j \in \mathbf{y}_k} \exp\left(\mathbb{E}_{\tilde{p}(\mathcal{X}_K)\tilde{p}(\mathbf{b}^{/b_i})}\left[\mathbf{b}^T d(s_j)\right.\right.\right.\right.$$
$$\left.\left.\left.\left. + \boldsymbol{\phi}(s_j)^T \mathbf{x}_k\right]\right)\right] \exp\left(\mathbb{E}_{\tilde{p}(\mathcal{X}_K)\tilde{p}(\mathbf{b}^{/b_i})}\left[-\int_D \exp\left(\mathbf{b}^T d(s)\right.\right.\right.\right.$$
$$\left.\left.\left.\left. + \boldsymbol{\phi}^T(s)\mathbf{x}_k\right)\right]\right) ds \right\} \xrightarrow{\text{Laplace}} \mathcal{N}_{b_i}\left(\hat{b}_i, \sigma_{b_i}^2\right), i = 1\ldots d$$

Note, finally, that the estimation of $\tilde{p}(\mathcal{X}_K)$, $\tilde{p}\left(\boldsymbol{\Sigma}_Q^{-1}\right)$, $\tilde{p}(b_i)$ requires the Laplace approximation. We refer to Zammit-Mangion et al. [19] in their supplementary information for further technical details.

2.2.4. Prediction

Assuming a linear relationship from t to $t+1$, the prediction of number of events $\hat{Y}_{i,t+1}$ for $i = 1, \ldots, L$ neighbourhoods and target time $t+1$ immediately posterior to time t is given by

$$\hat{Y}_{i,t+1} = \frac{N_{i,t+1}}{N_{i,t}} Y_{i,t} \qquad (4)$$

where $N_{i,t+1}$ and $N_{i,t}$ are the estimated number of events derived from the predictive Algorithm 1, and $Y_{i,t}$ corresponds to the number of reported events for time t and neighbourhood i.

Algorithm 1 Prediction algorithm for time $t+1$

Number of iterations is set to L

Monte Carlo estimation of the intensity
for iteration $N = 1$ to L **do**
  1: Sample the trajectory z_k through $\tilde{p}(\mathcal{X}_K)$ in t
  2: Forward simulate each trajectory for $t+1$ using the generative model with parameters ϑ, Σ_Q and $\mathbf{b}$, set to $\mathbb{E}_{\tilde{p}(\vartheta)}[\vartheta]$, $\left(\mathbb{E}_{\Sigma_Q^{-1}}\left[\Sigma_Q^{-1}\right]\right)^{-1}$ and $\mathbb{E}_{\tilde{p}(\mathbf{b})}[\mathbf{b}]$, respectively.
  3: Integrate the interpolated sample over each i neighbourhood to obtain $\hat{z}_{k,i}$.
  4: Estimate the intensity $\hat{\lambda}_{k,i}$, and average over fixed predefined intervals to obtain $\hat{\lambda}_{t,i}$ and $\hat{\lambda}_{t+1,i}$.
  5: Generate two samples $N_{i,t+1}$ and $N_{i,t}$ from Poisson random variables with intensity parameters $\hat{\lambda}_{t,i}$ and $\hat{\lambda}_{t+1,i}$.
  6: Predict $\hat{Y}_{i,t+1}$ using Equation (4)
end for

Computation of statistics
Calculate mean, median, and standard deviations out of the L estimations of $\hat{Y}_{i,t+1}$.

2.2.5. Experimental Set-Up

The VB algorithm (Algorithm 2) was assumed to have converged when the change in ϑ and $\mathbf{b}_i$, $i = 2, 3, \ldots, 5$ in subsequent iterations was less than 0.005, and when all diagonal elements in $\mathbb{E}\left[\Sigma_Q^{-1}\right] = \hat{a}\hat{V}$ changed by less than 1%. Note that the prior scale matrix V_p and the background rate $\mathbf{b}_1$ arise from the observed data itself. V_p was chosen such that its mean is $16I$; this value is the squared reciprocal of the standard deviation of the week with the highest variance in the Levene's test for homoscedasticity. In particular, $\mathbf{b}_1$ was set to -4.0, indicating the expected weekly events per year. The coefficients of the covariates $\mathbf{b}_i$ and their variances were initialised in 0. We set the VB algorithm to run for 200 interactions; however, it usually converged between the 50th and 65th iterations.

The exploratory analysis, VB algorithm, and predictions were implemented entirely in MATLAB R2020a. We employed parallel computing, statistics and machine learning, optimisation and mapping toolboxes, and customised functions for most of the above-mentioned methods. Further details about functions and parameters can be found in the code documentation at https://github.com/DavidPayares/ValenciaCallsSIDE (accessed on 17 January 2022). The VB algorithm was trained in a Windows CPU equipped with 16 GB RAM and 11th Gen Intel(R) Core(TM) i7 processor. Overall, the training time of the VB algorithm was 8 h and 32 min. Other methods' computational time was negligible.

Algorithm 2 VB-Laplace smoothers (adapted from Zammit-Mangion et al. [19]).

Time interval $\Delta_t = 1$ is assumed throughout.
Expectations are taken with respect to the relevant distributions
Input: Data set $\mathcal{Y}_K$, parameters b, μ_0, Σ_0 and parameter distributions $\tilde{p}(\vartheta)$, $\tilde{p}\left(\Sigma_Q^{-1}\right) = \tilde{p}(Q)$.

Forward message
Set $\hat{x}_{0|0} = \mu_0$ and $\Sigma_{0|0} = \Sigma_0$
for $k = 1$ to K **do**
$\quad \Sigma_{k-1}^* = \left(\Sigma_{k-1|k-1}^{-1} + \mathbb{E}[Q]\right)^{-1}$
$\quad \tilde{\Sigma}_k = \left(\mathbb{E}[Q] - \mathbb{E}[Q]\Sigma_{k-1}^*\mathbb{E}[Q]\right)^{-1}$
$\quad \tilde{x}_k = \tilde{\Sigma}_k\left[\mathbb{E}[Q]\Sigma_{k-1}^*\left(\Sigma_{k-1|k-1}^{-1}\hat{x}_{k-1|k-1} - \mathbb{E}[Q]\mathbb{E}[\theta]\right) + \mathbb{E}[Q]\mathbb{E}[\vartheta]\right]$
$\quad \hat{x}_{k|k} = \arg\max_{x_k} \sum_{s_j \in y_k}\left(\mathbb{E}\left[b^T d(s_j)\right] + \phi(s_j)^T x_k\right) - \int_D \mathbb{E}\left[\exp\left(b^T d(s)\right)\right]\exp(\phi^T(s)x_k)ds - \frac{1}{2}(x_k - \tilde{x}_k)^T\tilde{\Sigma}^{-1}(x_k - \tilde{x}_k)$
$\quad \Sigma_{k|k} = \left(\tilde{\Sigma}_k^{-1} + \int_D^x \phi(s)\phi(s)^T \exp\left(\phi(s)^T x_{k|k}\right)\mathbb{E}\left[\exp\left(b^T d(s)\right)\right]ds\right)^{-1}$
end for

Backward message
Set $\Sigma_{K|K+1:K}^{-1} = 0$ (ignore estimate of end condition)
for $K = (K-1)$ down to 0 **do**
$\quad x'_{k+1} = \arg\max_{x_{k+1}} \sum_{s_j \in y_{k+1}}\left(\mathbb{E}\left[b^T d(s_j)\right] + \phi(s_j)^T x_{k+1}\right) - \int_D \mathbb{E}\left[\exp\left(b^T d(s)\right)\right]\exp(\phi^T(s)x_{k+1})ds - \frac{1}{2}\left(x_{k+1} - \tilde{x}_{k+1|K+2:K}\right)^T\tilde{\Sigma}^{-1}\left(x_{k+1} - \tilde{x}_{k+1|K+2:K}\right)$
$\quad \Sigma'_{k+1} = \left(\Sigma_{k+1|K+2:K}^{-1} + \int_D \phi(s)\phi(s)^T \exp\left(\phi(s)^T x'_{k+1}\right)\mathbb{E}\left[\exp\left(b^T d(s_j)\right)\right]ds\right)^{-1}$
$\quad \Sigma_{k|k+1:K} = \left(\mathbb{E}[Q] - \mathbb{E}[Q]\left(\Sigma'^{-1}_{k+1} + \mathbb{E}[Q]\right)^{-1}\mathbb{E}[Q]\right)^{-1}$
$\quad x_{k|k+1:K} = \Sigma_{k|k+1:K}\left(-\mathbb{E}[Q]\mathbb{E}[\theta] + \mathbb{E}[Q]\left(\Sigma'^{-1}_{k+1} + \mathbb{E}[Q]\right)^{-1}\left(\Sigma'^{-1}_{k+1}x'_{k+1} + \mathbb{E}[Q]\mathbb{E}[\vartheta]\right)\right)$
end for

Smoothed estimate
for k = 0 to K **do**
$\quad \Sigma_{k|K} = \left(\Sigma_{k|k}^{-1} + \Sigma_{k|k+1:K}^{-1}\right)^{-1}$
$\quad \hat{x}_{k|K} = \Sigma_{k|K}\left[\Sigma_{k|k}^{-1}\hat{x}_{k|k} + \Sigma_{k|k+1:K}^{-1}\hat{x}_{k|k+1:K}\right]$
end for

Computation of cross-covariance $\{M_k\}_{k=1}^K$
for $K = (K-1)$ down to 0 **do**
$\quad M_{k|K} = \Sigma_{k-1}^*\mathbb{E}[Q][\Sigma_{k|k+1:K}^{-1} + \mathbb{E}[Q] + \int_D \phi(s)\phi(s)^T \exp(\phi(s)^T x_k \mid K)\mathbb{E}\left[\exp\left(b^T d(s)\right)\right]ds$
$\quad -\mathbb{E}[Q]\Sigma_{k-1}^*\mathbb{E}[Q]^{-1}$.
end for

Output: $\left\{\hat{x}_{k|K}, \Sigma_{k|K}\right\}_{k=0}^K, \left\{M_{k|K}\right\}_{k=1}^K \ldots$

3. Results

3.1. Temporal Analysis

One of the main premises of the SIDE-driven LGCPs methodology to model the temporal dynamics is that the increments between two consecutive times are normally distributed, and thus, the system can be expressed as a Geometric Brownian motion (GBM). The GBM is redefined in terms the mean $\mu_Q(\mathbf{s})$ and covariance function $k_Q(\mathbf{r},\mathbf{s})$ as in Equation (1). One obtains the random walk model in Equation (2) by decomposing the field $z_{k+1}(s)$. Indeed, considering that the intensity of the LGCP at time k is given by $\lambda_k(\mathbf{s}) = \exp\left(\mathbf{b}^T \mathbf{d}(\mathbf{s}) + z_k(\mathbf{s})\right)$, we have $d\lambda_k(\mathbf{s}) = R(\mathbf{s})\lambda_k(\mathbf{s})dk + \lambda_k(\mathbf{s})dW_k(\mathbf{s})$, with the increment $dW_k(\mathbf{s})$ a Gaussian process with zero mean and covariance function $k_Q(\mathbf{r},\mathbf{s})$, and $R(\mathbf{s})$ a spatially varying drift.

We analysed the weekly number of emergency calls in Valencia between January 2010 and December 2019. As mentioned in the data description section, the weekly behaviour is similar yearly with a general increasing trend throughout the studied period. When we examined the increments between week N and $N+1$, we found that the data's 2.9% (12 weeks) were outliers. After removing these data, we confirmed that the increment rates between adjacent weeks were normally distributed. Figure 11 shows the distribution of the increments as well as their normal probability plot, suggesting normality in the temporal increments.

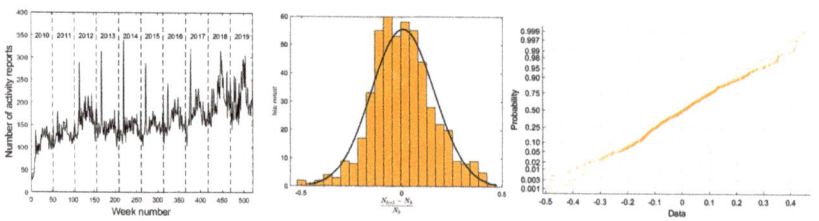

Figure 11. Analysis of the temporal increments. The left panel displays the temporal distribution of the emergency calls over the study period (2010–2019); the central panel shows the normally-distributed weekly increments; the right panel is the normal probability plot corroborating normality in the weekly increments.

3.2. Basis Function Selection

In order to select a set of basis functions that describe the spatio-temporal field of the LGCP, non-parametric estimations of both the PACF and PCCF were conducted (see Section 3.1). Furthermore, the relationship between the PACF and PCCF is essential to determine the mixing kernel $k_M(\mathbf{s},\mathbf{r})$ and the noise kernel $k_Q(\mathbf{s},\mathbf{r})$ (see Algorithm 3).

Algorithm 3 Analysis for dynamic, homogeneous, isotropic spatiotemporal point processes

---

1: Estimate $\lambda_k^{(1)}(s) \forall k$ using $\lambda_k^{(1)} = \frac{N_k}{|D|}$ for stationary systems or simple regression where clear trends are evident. N_k is the cardinality of a spatial point process $\mathscr{P}_k$ at point k.
2: Estimate $\hat{g}_{k,k}(v), \hat{g}_{k,k+1}(v) \forall k$.
3: Estimate $k_M(v)$ using $\bar{g}_{k,k}(v)$ and $\bar{g}_{k,k+1}(v)$
4: Estimate $\hat{k}_Q(v)$ using k_M and $\bar{g}_{k,k}(v)$.

Figure 12a shows the non-parametric estimations of both the average of $\ln g_{k,k}(\mathbf{s},\mathbf{r})$ (PACF) and the average of $\ln g_{k,k+1}(\mathbf{s},\mathbf{r})$ (PCCF) in terms of $v = \|\mathbf{s} - \mathbf{r}\|$ using the expressions in Section 3.1. Note that v corresponds to approximately 270 m. Both functions are symmetric concerning zero ($v = 0$).

We also note that the behaviour of $\ln g_{k,k}(\mathbf{s},\mathbf{r})$ and $\ln g_{k,k+1}(\mathbf{s},\mathbf{r})$ are almost identical; property also noticed in Zammit-Mangion et al. [19]. Once $\ln g_{k,k}(\mathbf{s},\mathbf{r})$ and $\ln g_{k,k+1}(\mathbf{s},\mathbf{r})$ have been estimated, the mixing kernel $k_M(\mathbf{s},\mathbf{r})$ and the noise kernel $k_Q(\mathbf{s},\mathbf{r})$ can be inferred using the exact inverse filter. Figure 12b shows the estimated mixing kernel of the process. This kernel should closely resemble the true underlying kernel.

Figure 12c displays a positive cross-section of $\ln g_{k,k}(\mathbf{s},\mathbf{r})$ and its corresponding confidence interval and that of the isotropic basis function selected for the modelling. $\phi(\nu)$ directly comes from Equation (3) with the cut-off value ν_c obtained from the average PACF. We chose a cut-off frequency of $\nu_c = 0.22$ cycles/units giving a basis parameter $\tau \approx 1.7325$ and an oversampling parameter of $\alpha_0 = 1.2$ for the placement of the CGRBFs across the study area.

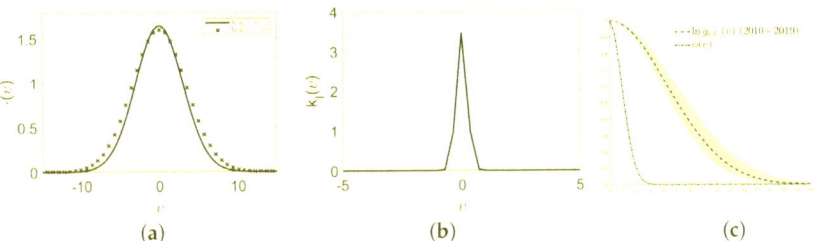

Figure 12. Average natural logarithm PACF and average natural logarithm PCCF ((**a**), left plot), $k_M(v)$ ((**b**), centre plot), cross-section on the positive real line of natural logarithm PACF, and corresponding chosen basis function ((**c**), right plot).

As in Zammit-Mangion et al. [19]'s scheme, 256 basis functions were placed on a 16 × 16 grid covering Valencia. The centre of the basis functions was separated by $\Delta_s = 1.9$ grid units. The functions were then filtered out to remove non-representative areas having sparse events. Only basis functions whose intensity was above a constant background event rate of $b_1 = -4$ (approximately five reported events per year with a distance of 450 m from its centre) and whose centre does not exceed 300 m beyond Valencia's boundary were chosen. Figure 13 shows the distribution of reported emergency calls across Valencia and the basis functions selected for the study. In total, 129 basis functions meet the criteria; they cover the areas with high rates of emergency calls. Some areas in the northern and southern areas of the city do not have a basis function representation, given their low event rates. Nonetheless, we assigned to these areas the baseline background rate b_1 to ensure identifiability.

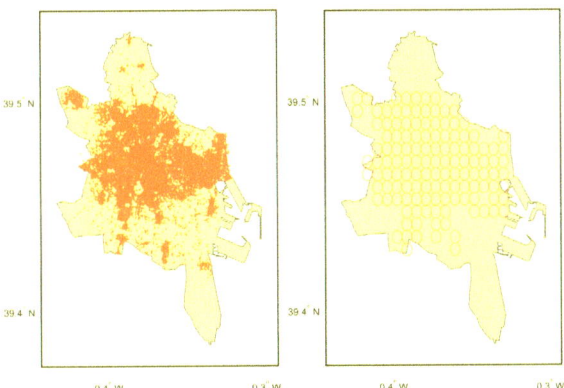

Figure 13. Spatial locations of logged events (2010–2019), and 118 basis functions placement in the city of Valencia.

3.3. Fixed Effects

Crimes vary significantly in space and time within a region due to many demographic and economic factors, and while it is known that the combined effect of these factors favours criminal behaviour, their geographical character defines crime locations. Typically, crimes occur in business sectors within low and middle-income neighbourhoods. These sectors concentrate most of the facilities (e.g., banks, ATMs, restaurants) that compose neighbourhoods' economic activity. Offenders target victims close to locations that can maximise their profit and reduce the risk of apprehension [20,21]. In this context, we have included putative geographical variables into the spatio-temporal point process modelling framework to measure the effect deterministic components exert on the intensity of emergency calls in Valencia.

We introduced ten covariates in the form of distances to relevant landmarks in Valencia. The landmarks included financial facilities such as banks and ATMs, and leisure places such as bars, pubs, restaurants, cafes, and nightclubs. We also included industrial areas, taxi stop areas, and markets. We measured the degree of correlation between the landmarks' distances and the intensity of the 112 calls. Half of the covariates displayed evidence of association with the emergency calls: distance to banks, ATMs, bars, cafes, and restaurants.

Figures 14 and 15 display the spatial distribution of distances to landmarks within Valencia and their relationship to the 112 calls' intensity values. Overall, short distances to landmarks occur primarily in the city centre and densely populated neighbourhoods; facilities, shops, and venues are located strategically to provide location advantages (e.g., access to services and amenities) for residents and tourists. Distances to restaurants are short throughout the city except in the northern and part of the southwestern neighbourhoods.

Emergency calls, which in our case reflect criminal behaviour, are known to occur frequently near facilities with features that offenders find attractive [20]; for example, places with multiple or desired targets, and our results corroborate this assumption; while the correlation between the emergency calls log intensity and the distances vary differently for each landmark, the behaviour of these relationships is relatively similar. One would expect many emergency calls in locations close to banks, ATMs, bars, cafes, and restaurants. Figure 14 shows that high intensities concentrate in a radius of approximately 1 km to 1.5 km centred in the facilities; beyond these distances, the emergency calls' intensity decreases. We notice that the standard deviations of the correlation functions (red dotted lines) widen with distances exceeding 2 km. This occurs given the sparse number of emergency calls in low crime incidence areas.

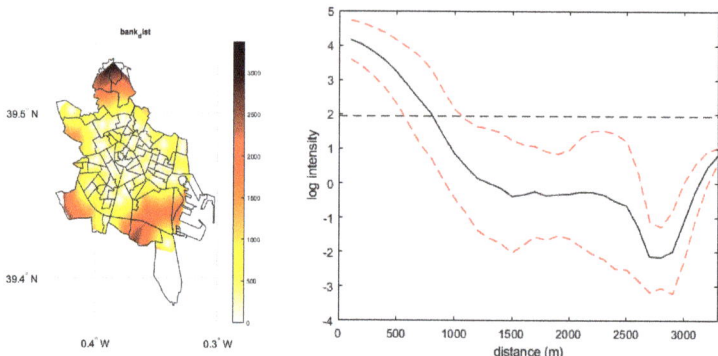

Figure 14. Maps of fixed effects and empirical relationship between distance-based fixed effects (banks) and the log spatial intensity.

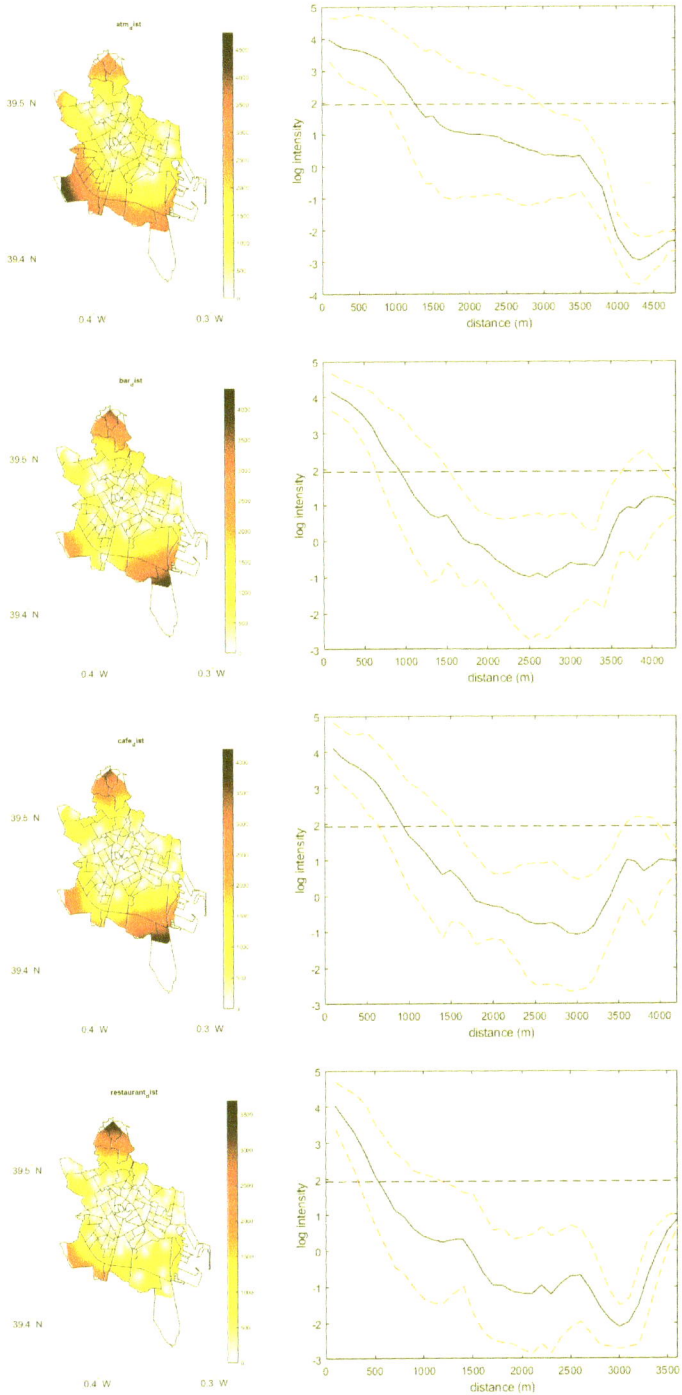

Figure 15. Maps of fixed effects and empirical relationship between distance-based fixed effects (atms, bars, cafes and restaurants) and the log spatial intensity.

In order to account for the factors contributing to the spatial intensity of emergency calls in Valencia, the distances to various amenities, including banks, ATMs, bars, industrial areas, markets, taxi stops, cafes, pubs, nightclubs, police stations, and restaurants were considered as the deterministic component of the intensity function $\lambda_k(\mathbf{s})$. This was due to the observed strong association between these distances and the average intensity of emergency calls in the region, while we introduced eleven covariates as fixed effects in the intensity function, six presented a regression coefficient of zero. Our model could identify covariates with no quantitative influence over the emergency calls intensity. Following Algorithm 2 for inference, we found the regression coefficients for only five of the distance covariates we introduced in the intensity function; the distance to banks and ATMs parameters confidence intervals were $-1.9 \times 10^{-2} \pm 1.2 \times 10^{-11}$ and $-3.4 \times 10^{-2} \pm 6.8 \times 10^{-11}$, respectively. The results indicate that emergency calls tend to occur in proximity to economic facilities. This proximity suggests that potential perpetrators may identify victims in these areas, potentially leading to some financial gain. The regression parameters and the confidence interval for distances to bars and cafes were $0.5 \times 10^{-2} \pm 7.3 \times 10^{-11}$ and $1.3 \times 10^{-2} \pm 7.7 \times 10^{-11}$, respectively. In this case, emergency calls are located far from these places. It is important to note that while these types of landmarks attract a large number of potential victims, they also increase the likelihood of apprehension by law enforcement. For example, bars and cafes often have enhanced security systems due to their elevated risk of crimes, such as robbery and assault Weisburd et al. [21]. The coefficients for the distance to restaurants were found to be negative, with a magnitude of $4.0 \times 10^{-2} \pm 6.2 \times 10^{-11}$. This suggests that emergency calls tend to occur in proximity to restaurants. This is particularly worrying, as restaurants are often targeted by criminals for robbery, burglary, and theft, due to the accumulation of significant amounts of cash during daily operations.

These results show how landmarks, to some extent, attract or repel offenders and act as proxies to identify areas with high or low emergency calls.

3.4. Heterogeneous Growth and Decay

A spatio-temporal analysis of events is concerned with identifying high-intensity spots and their evolution over space and time, and while traditional cluster analysis excels in determining hotspots, it cannot portray the temporal characteristics that govern advection and diffusion processes. Our methodology allows us not only to locate hotspots but also to determine their behaviour over time.

Figure 16 presents the weekly average fractional growth and decay of emergency calls in Valencia. As anticipated, the majority of the city has experienced a rise in the frequency of emergency calls. However, Sau Pau, Malilla, and La Torre neighbourhoods display the highest increments over the years. For example, Sau Pau is a neighbourhood prone to robbery; by 2015, it had accumulated approximately 9% of all robberies in Valencia Las Provincias [22]. Despite the fact that the central neighbourhoods of the city have the highest incidence of events, they did not exhibit a marked increase in the number of emergency calls. Conversely, areas with fewer events, such as the Quatre Carreres district, have experienced an increase in emergency call hotspots over the course of the study period. The spatial intensity of emergency calls has shown a decrease in certain areas throughout Valencia. An interesting finding is the decay in the Benicalp neighbourhood. It is considered one of the most notorious neighbourhoods in Valencia due to high poverty levels, illegal occupation, and drug traffic, and while criminal activity has grown consistently in this neighbourhood over the last few years, the number of reported emergency calls has decreased. A possible explanation is that victims or witnesses do not report crimes as they are afraid of repercussions by organised crime. Other neighbourhoods, such as La Punta and El Castellar, also display a reduction in hotspots. However, volatility (Figure 17) suggests that the data in these areas are not very reliable.

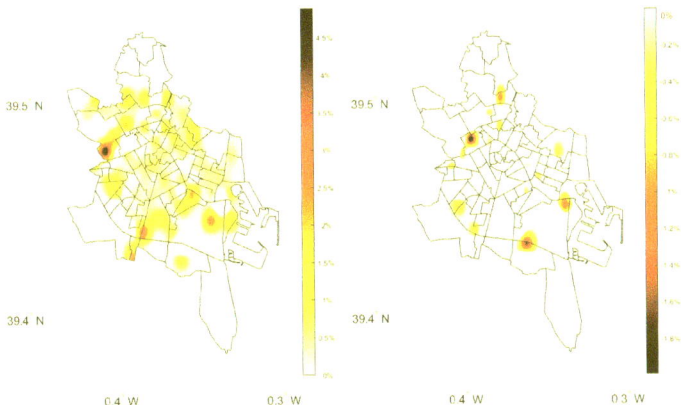

Figure 16. Posterior mean fractional growth (left panel) and (right panel) decay of emergency calls per week in Valencia (2010–2019).

3.5. Volatility

The volatility (Σ_Q) in the SIDE-driven LGCP allows us to measure the accuracy of future intensity estimations. The lower the value in the diagonal of Σ_Q, the more accurate the predictions are, and conversely. Figure 17 shows the volatility map for emergency calls in Valencia. High volatility is present in the neighbourhoods of La Punta, Benimamet, La Llum, and Castellar-L'Oliveral. In both Banimamet and La Llum, the volatility is high as the neighbourhoods reported zero emergency calls in most of the weeks of the study. This produces a volatile temporal trend fluctuating between zero and the number of logged events (e.g., La Llum reported a maximum of four events per week) that the model cannot capture, while La Punta and Castellar-L'Oliveral also report many weeks with zero observations, high volatility results from scattered events in both time and space. For example, emergency calls in La Punta are primarily located close to the boundary with adjacent neighbourhoods. The majority of the city displays low volatility values, particularly areas with generous data, such as the city centre and surrounding neighbourhoods. The volatility map suggests that we will obtain less accurate predictions as we move from the city centre towards the suburbs.

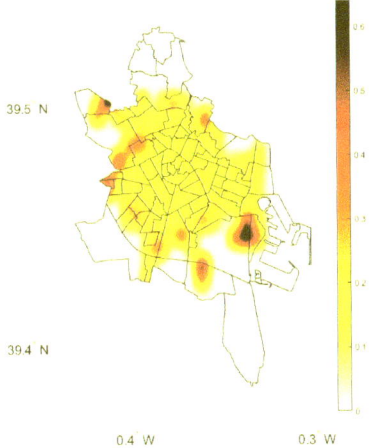

Figure 17. Volatility map in emergency calls in Valencia (2010 to 2019).

3.6. Model Fitting

We perform model parameter estimation through Bayesian inference as detailed in Section 2.2.3, and using Algorithm 2. In particular, we estimate the intensity quartiles through the posterior smoothed estimate, the smoothed covariance matrix, and the effect of the covariates. Figure 18 shows the fitted model for five different neighbourhoods in Valencia. Overall, the model fits the data well. Real values are essentially contained in the 90% confidence intervals except in weeks when the number of reported emergency calls drops down to zero, as in La Malva-Rosa neighbourhood. The model also captures the temporal trend of the events, in some cases, even when abrupt spikes occur. An example is the neighbourhood of Arrancanpins, whose events count shot up from 9 calls in week 376 to 66 in week 377. The model accurately imitates this peak.

How accurately the model fits the data varies according to the weekly changes and the overall temporal trend. The larger the shifts from week to week are, the narrower the bandwidths become. Nonetheless, generally, the model fairly reconstructs the spatial and temporal character of the reported emergency calls in Valencia.

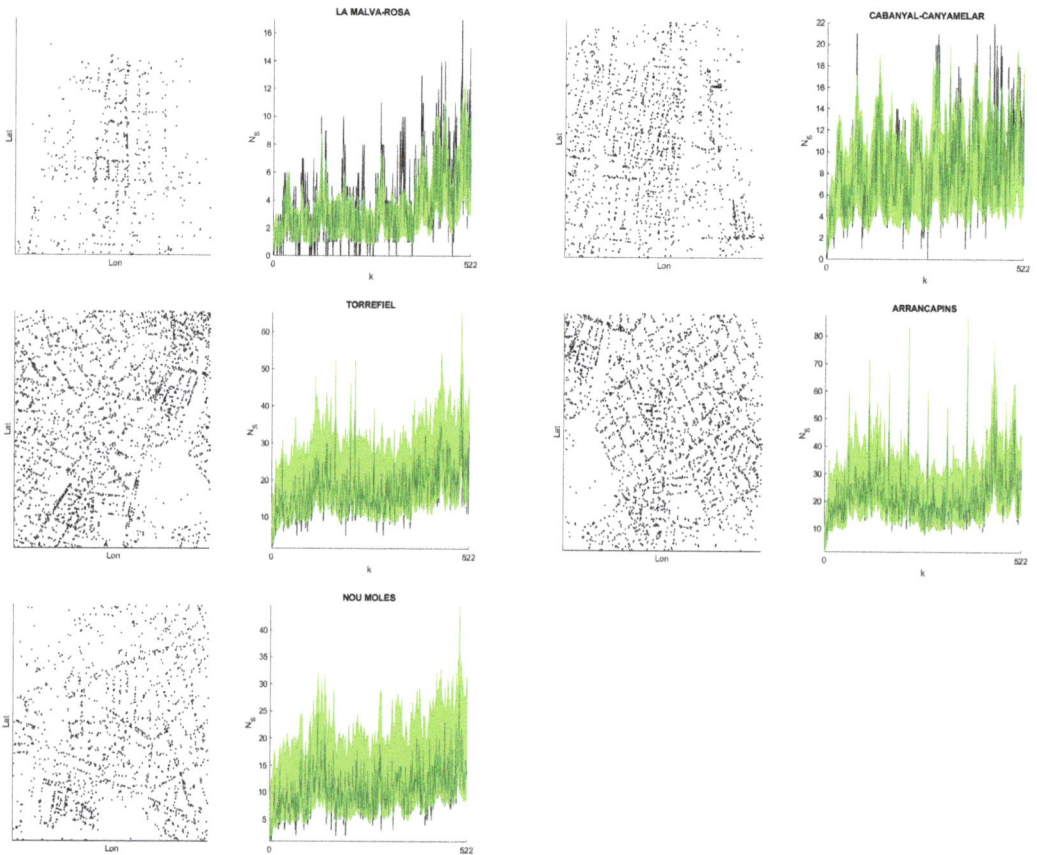

Figure 18. Model fitting for five different neighbourhoods in Valencia. For each neighbourhood, the left panel shows the spatial distribution of the emergency calls with a buffer of 100 m, and the right panel displays the weekly counts (black) with their corresponding 90% fitted confidence intervals (green).

3.7. Prediction

One of the key strengths of our methodology is its ability to make predictions. Given that we have accurately modelled the spatio-temporal dynamics of emergency calls in Valencia, we can now forecast their future behaviour. To evaluate the predictive capability of the model, we used Algorithm 1 to estimate the number of emergency calls in Valencia for the first 40 weeks of 2020 (Figure 19). We selected this period aiming to assess our model's predictive robustness. The emergency calls recorded in 2020 contrast with the training data (i.e., calls from 2010 to 2019) due to the implementation of containment measures in response to the COVID-19 pandemic. As seen in Figure 19, there was a significant decrease in emergency calls between weeks 532 and 543, which coincided with the quarantine and isolation policies imposed by the Spanish government. However, there was an increase in calls as measures were relaxed. We expect our method to be robust enough to effectively model these variations based on past data, covariates, and the interaction of space and time.

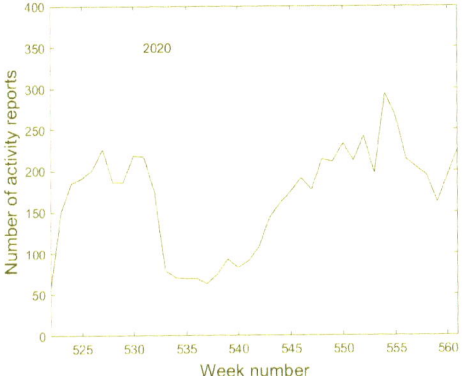

Figure 19. Time series of the weekly calls of the first 40 weeks of 2020 in Valencia.

To stabilise the variance, we first transformed the counts into logarithms. Table 1 displays the Pearson correlation coefficients between the predicted and actual counts and log counts. These coefficients demonstrate a strong correlation between the predicted and actual values (0.87 for counts and 0.89 for log counts), indicating the model's strong predictive ability.

Table 1. Pearson correlation coefficients between the count and log predictions of the SIDE model and the actual values for the first 40 weeks of 2020.

Prediction	ρ	p-Value
counts vs. predicted counts	0.8724	<0.001
log counts vs. predicted log counts	0.8994	<0.001

The scatter plot depicted in Figure 20 presents a comparison between the logarithmic median prediction of the model and the logarithmic reported cases for the year 2020. The concentration of points around the ideal prediction line demonstrates the high level of correspondence between the predicted data and the observed data. However, it should be noted that the error bars display significant uncertainty in the median estimates for some neighbourhoods, such as Castellar-L'Oliveral and Ciutat De Les Arts I de Les Ciences. This higher level of uncertainty is likely due to the presence of multiple zero observations in these areas.

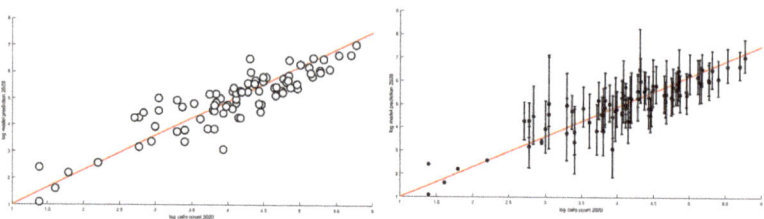

Figure 20. Scatter plot and error bar plot between the log median predictions and log real values for 2020. The circles in the scatter plot represent individual neighbourhoods in Valencia, and the red line refers to the ideal prediction. The error bars represent the 99% confidence intervals for the predicted values.

We cannot only predict the number of emergency calls per week but also estimate their growth and decay over space and time. Figure 21 shows the histogram of the percentage of growth and decay of emergency calls for 2020. We can see that the predictions of the growth/decay rates are remarkably accurate in La Vega Baixa, Favara, La Carrasca, Ciutat Fallera, Sani Isidre, L'Hort de Senabre, and La LLum; the estimated percentage is almost identical to the observed one. These neighbourhoods are characterised by low volatility values that range from 0.01 to 0.12. The model predicts the changes with higher uncertainty in those areas with excessive zero counts (e.g., Carpesa, Exposicio, Castellar LÓliveral, and Cami Real) and, naturally, with more considerable volatility.

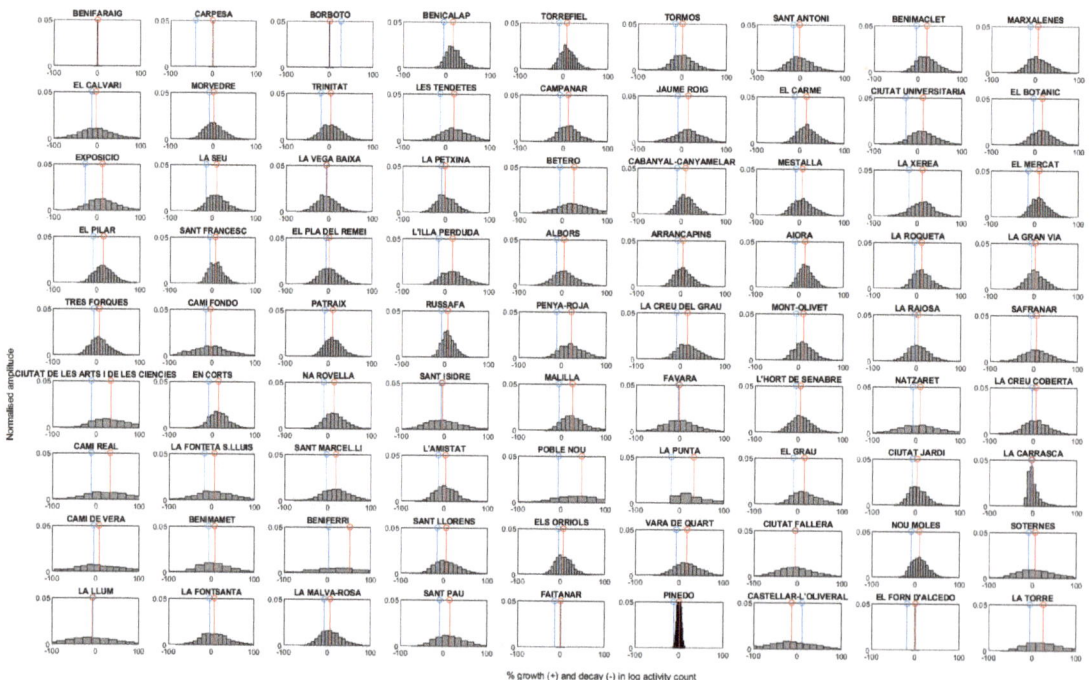

Figure 21. Normalised histograms of log counts in 2020 per province as obtained from MC simulations with true change (circle in blue) and sample median (circle in red). The closer the blue circle is to the red circle, the most accurate is the median prediction.

While the predicted counts and change rates may vary, and in some cases, differ considerably, from the ground truth values, the model is rather accurate as (i) the predicted values are always contained in the 99% confidence intervals, (ii) the predicted log counts exhibit high correlation with the actual log counts, (iii) the distributions between the observed and predicted changes are virtually analogous, and (iv) predicted values come with a measurement of uncertainty.

Comparison with Alternative Benchmark Models

In order to evaluate the predictive capability of our model, a comparison was conducted between using its logarithmic predictions and those generated by three conventional and known point process modelling methods:

(i) A spatial point process with a deterministic intensity function $\lambda_k(\mathbf{s}) = \exp(\mathbf{b}^T \mathbf{d}(\mathbf{s}))$, where $\mathbf{d}(\mathbf{s})$ is a vector of spatial covariates and $\mathbf{b}^T$ is the vector of corresponding regression parameters.

(ii) A spatio-temporal point process with a separable and deterministic intensity function $\lambda(\mathbf{s})\mu(t)$, where $\lambda(\mathbf{s})$ follows the above-mentioned structure and $\mu(t)$ is a log-linear regression model in the form

$$log(\mu(t)) = \alpha_1 \cos(\omega t) + \beta_1 \sin(\omega t) + \alpha_2 \cos(2\omega t) + \beta_2 \sin(2\omega t) + \gamma_t$$

with $\alpha_1, \beta_1, \alpha_2$ and β_2 as regression parameters, $\omega = 2*\pi/52$ corresponding to annual periodicity, and γ_t the slope parameter overall trend.

(iii) A spatio-temporal log-Gaussian Cox processes (LGCPs) with intensity function $\lambda(\mathbf{s})\mu(t)\exp\{\mathcal{Y}(s,t)\}$, where $\lambda(\mathbf{s})$ follows the specification in (i), $\mu(t)$ is defined as in (ii), and $\mathcal{Y}(s,t)$ is a second-order stationary Gaussian process with a minimally-parametrised exponential covariance function.

The results of this comparison provides valuable insights into the strengths and limitations of our model, as well as its overall performance in the prediction of point process data. Further details on the models are described in [15,23,24].

We assessed the four models using the mean squared prediction error (MSPE) and the Pearson correlation (ρ) between the log predicted counts and the log real counts. The results are presented in Table 2. Note that the spatio-temporal LGCP coupled with the SIDE framework displays the lowest MSPE and the highest ρ. As expected, the purely spatial point process model displayed the poorest assessment metrics, due to the absence of the temporal component in the intensity function modelling. The spatio-temporal point process model and the spatio-temporal LGCP model showed similar metrics, with the latter being a superior alternative. Despite the relatively good performance of the benchmark models, they were unable to compete with our approach.

Table 2. MSPE and Pearson correlation coefficients between the log real counts and log predictions of benchmark models and our method for the first 40 weeks of 2020.

Method	MSPE	ρ
Spatial point process	102.56	0.32
Spatio-temporal point process	99.86	0.53
Spatio-temporal LGCP	93.75	0.64
Spatio-temporal LGCP + SIDE	62.25	0.89

The computation of models (i) and (ii) was straightforward as did not entail the use of Monte Carlo simulations. The estimation of model (iii), on the other hand, was performed using the `lgcp` R package, with a processing time of 3 h and 45 min. Despite the prolonged processing time of our approach compared to other available models, the results exhibit a substantial increase in prediction accuracy. This trade-off between processing time and accuracy is justified, as the primary objective of point process models is to accurately forecasts future events in space and time.

4. Conclusions and Discussion

Nowadays, technology makes enormous amounts of data readily available to researchers, and being able to handle such large quantities of information provides a step forward in many societal problems. One such problem is considered here as emergency calls in an urban context. Authorities must allocate resources and infrastructure for an effective response, identify high-risk event areas, and develop contingency strategies. We highlight that such emergency calls' spatial and temporal analysis is crucial to understanding and mitigating distress situations.

We have proposed a modelling framework to handle heterogeneous, dynamic, and complex underlying processes to observe crime-related emergency calls. Our strategy can account for the intrinsic complex space–time dynamics by handling complex advection, diffusion, relocation, and volatility processes. This is shown by analysing the dynamics of some emergency calls in Valencia, Spain, for ten years (2010–2029), and we believe this is a case study that can be an excellent example for many similar emergency calls in other urban contexts.

For example, other more complex scenarios can be idealised in our framework by considering SIDE with a non-linear transformation of the random field at time k in the integrand of our SIDE equation. This would complicate the model while providing a more flexible case. A combination with some deep learning methods could be imagined here. This can be the motivation for further research.

There are, however, some limitations due to the baseline assumptions imposed in the modelling approach. Note that the SIDE-driven LGCPs methodology sits on the assumption that the increments between two consecutive times are normally distributed, and thus, the system can be expressed as a Geometric Brownian motion. This must be verified a priori, and thus we can not expect that all types of data behave this way. Another sort of computational disadvantage is that the procedure involves a large number of matrix inverse calculations that can bring trouble in cases where the determinants are close to zero.

The code used in this paper has been made publicly available on the authors' Github repository (https://github.com/DavidPayares/ValenciaCallsSIDE) (accesed date on 17 January 2022). The authors have translated the LGCP + SIDE methodology into easily understandable and executable scripts, which not only replicate the results reported in this paper but also facilitate its implementation with similar datasets. The availability of the code enhances the reproducibility and transparency of the research findings.

Author Contributions: Conceptualization, J.M.; Methodology, D.P.-G. and J.M.; Software, D.P.-G.; Validation, D.P.-G. and J.M.; Formal analysis, D.P.-G. and J.M.; Data curation, J.P.; Writing—original draft, D.P.-G. and J.P.; Writing—review & editing, D.P.-G. and J.M.; Visualization, J.P.; Supervision, J.M. All authors have read and agreed to the published version of the manuscript.

Funding: This research received no external funding.

Institutional Review Board Statement: Not applicable.

Informed Consent Statement: Not applicable.

Data Availability Statement: Data sharing is not applicable to this article.

Conflicts of Interest: The authors declare no conflict of interest.

Appendix A. Literature Review Table

Table A1. Literature review of spatio-temporal models for analysing and predicting emergency calls.

Authors	Method	Contribution
Hashtarkhani et al. [2]	GIS	Spatial and temporal analysis of emergency calls in Iran
Sabet et al. [3]	GIS and Kernel estimation	Spatio-temporal insights of emergency response by fire departments in Canada
Towers et al. [4]	Linear regression	Forecast of future emergency events and determination of driving factors in Chicago, USA
Cramer et al. [5]	Hotspot analysis and Geographically weighted regression	Spatio-temporal analysis of 911 calls in Oregon, USA
Chohlas-Wood et al. [6]	Rolling Forecast Prediction Model	Temporal analysis and forecasting of 911 calls in New York City
Marco et al. [7]	Poisson log-linear mixed model	Spatio-temporal mapping of suicide-related emergency calls
Robles et al. [8]	Bivariate–Gaussian kernel, decision tree learning, random forest, and logistic regression	Spatio-temporal relations to predict 911 events in Cuenca, Ecuador
Zhou et al. [9]	Geographically weighted regression	Estimation of the spatial distribution of urban populations based on first-aid calls based in Shanghai, China
Heaton et al. [10]	Inhomogeneous Poisson point process	Inference of number and type of 911 calls in Houston, USA
Li et al. [11]	Non-parametric self-exciting point processes	Spatio-temporal modelling of emergency calls in Pennsylvania, USA

References

1. Dunnett, S.; Leigh, J.; Jackson, L. Optimising police dispatch for incident response in real time. *J. Oper. Res. Soc.* **2019**, *70*, 269–279.
2. Hashtarkhani, S.; Kiani, B.; Mohammadi, A.; MohammadEbrahimi, S.; Eslami, S.; Tara, M.; Matthews, S.A. One-year spatiotemporal database of Emergency Medical Service (EMS) calls in Mashhad, Iran: data on 224,355 EMS calls. *BMC Res. Notes* **2022**, *15*, 8.
3. Sabet, M.S.; Asgary, A.; Solis, A.O. Emergency calls during the 2013 southern Ontario ice storm: case study of Vaughan. *Int. J. Emerg. Serv.* **2019**, *8*, 292–314.
4. Towers, S.; Chen, S.; Malik, A.; Ebert, D. Factors influencing temporal patterns in crime in a large American city: A predictive analytics perspective. *PLoS ONE* **2018**, *13*, e0205151.
5. Cramer, D.; Brown, A.A.; Hu, G. Predicting 911 Calls Using Spatial Analysis. *Softw. Eng. Res. Manag. Appl.* **2012**, *2011*, 15–26.
6. Chohlas-Wood, A.; Merali, A.; Reed, W.; Damoulas, T. Mining 911 Calls in New York City: Temporal Patterns, Detection, and Forecasting. In Proceedings of the AAAI Workshop: AI for Cities, Austin, TX, USA, 25–26 January 2015.
7. Marco, M.; Lopez-Quilez, A.; Conesa, D.; Gracia, E.; Lila, M. Spatio-Temporal Analysis of Suicide-Related Emergency Calls. *Int. J. Environ. Res. Public Health* **2017**, *14*, 735.
8. Robles, P.; Tello, A.; Zuniga-Prieto, M.; Solano-Quinde, L. Modeling 911 Emergency Events in Cuenca-Ecuador Using Geo-Spatial Data. In Proceedings of the 4th International Conference on Technology Trends (CITT), Babahoyo, Ecuador, 29–31 August 2018.
9. Zhou, Y.; Zhu, Q.Z.; Luo, L. Simulation of the spatial distribution of urban populations based on first-aid call data. *Geospat. Health* **2020**, *15*, 267–273.

10. Heaton, M.J.; Sain, S.R.; Monaghan, A.J.; Wilhelmi, O.V.; Hayden, M.H. An Analysis of an Incomplete Marked Point Pattern of Heat-Related 911 Calls. *J. Am. Stat. Assoc.* **2015**, *110*, 123–135.
11. Li, C.; Song, Z.; Wang, X. Nonparametric Method for Modeling Clustering Phenomena in Emergency Calls Under Spatial-Temporal Self-Exciting Point Processes. *IEEE Acces* **2019**, *7*, 24865–24876.
12. Mohler, G.O.; Short, M.B.; Brantingham, P.J.; Schoenberg, F.P.; Tita, G.E. Self-Exciting Point Process Modeling of Crime. *J. Am. Stat. Assoc.* **2011**, *106*, 100–108.
13. Reinhart, A. A Review of Self-Exciting Spatio-Temporal Point Processes and Their Applications. *Stat. Sci.* **2018**, *33*, 299–318.
14. Shirota, S.; Gelfand, A.E. Space Furthermore, Circular Time Log Gaussian Cox Processes with Application To Crime Event Data. *Ann. Appl. Stat.* **2017**, *11*, 481–503.
15. Diggle, P.J.; Moraga, P.; Rowlingson, B.; Taylor, B.M. Spatial and spatio-temporal log-gaussian cox processes: Extending the geostatistical paradigm. *Stat. Sci.* **2013**, *28*, 542–563.
16. Shirota, S.; Banerjee, S. Scalable inference for space-time Gaussian Cox processes. *J. Time Ser. Anal.* **2019**, *40*, 269–287.
17. Tang, Y.; Zhu, X.; Guo, W.; Ye, X.; Hu, T.; Fan, Y.; Zhang, F. Non-Homogeneous Diffusion of Residential Crime in Urban China. *Sustainability* **2017**, *9*, 934.
18. Short, M.B.; D'Orsogna, M.R.; Pasour, V.B.; Tita, G.E.; Brantingham, P.J.; Bertozzi, A.L.; Chayes, L.B. A statistical model of criminal behavior. *Math. Model. Methods Appl. Sci.* **2008**, *18*, 1249–1267.
19. Zammit-Mangion, A.; Dewar, M.; Kadirkamanathan, V.; Sanguinetti, G. Point process modelling of the Afghan War Diary. *Proc. Natl. Acad. Sci. USA* **2012**, *109*, 12414–12419.
20. Eck, J.E.; Clarke, R.V.; Guerette, R.T. Risky Facilities: Crime Concentration in Homogeneous Sets of Establishments and Facilities. In *Imagination for Crime Prevention: Essays in Honour of Ken*; Lynne Rienner Publishers: Boulder, CO, USA, 2007; Volume 21, pp. 225–264.
21. Weisburd, D.; Groff, E.R.; Yang, S.M. *The Criminology of Place: Street Segments and Our Understanding of the Crime Problem*; Oxford University Press: Oxford, UK, 2013.
22. Las Provincias. Valencia Neighborhoods With Highest Thefts Incidence. *Las Prov.* **2017**.
23. Diggle, P.; Rowlingson, B.; Su, T.L. Point process methodology for on-line spatio-temporal disease surveillance. *Off. J. Int. Environmetr. Soc.* **2005**, *16*, 423–434.
24. Taylor, B.M.; Davies, T.M.; Rowlingson, B.S.; Diggle, P.J. Lgcp: Inference with spatial and spatio-temporal log-gaussian cox processes in R. *J. Stat. Softw.* **2013**, *52*, 1–40.

Disclaimer/Publisher's Note: The statements, opinions and data contained in all publications are solely those of the individual author(s) and contributor(s) and not of MDPI and/or the editor(s). MDPI and/or the editor(s) disclaim responsibility for any injury to people or property resulting from any ideas, methods, instructions or products referred to in the content.

Article

Validity Evidence for the Internal Structure of the Maslach Burnout Inventory-Student Survey: A Comparison between Classical CFA Model and the ESEM and the Bifactor Models

Raimundo Aguayo-Estremera [1], Gustavo R. Cañadas [2,*], Elena Ortega-Campos [3], Tania Ariza [4] and Emilia Inmaculada De la Fuente-Solana [5]

1. Departamento de Psicobiología y Metodología de las Ciencias del Comportamiento, Facultad de Psicología, Universidad Complutense de Madrid, Campus de Somosaguas, Ctra. De Húmera, s/n, Pozuelo de Alarcón, 28223 Madrid, Spain; raaguayo@ucm.es
2. Department of Didactic of Mathematics, Faculty of Education Science, University of Granada, 18071 Granada, Spain
3. CEINSA-UAL, Universidad de Almería, Carretera de Sacramento s/n, La Cañada de San Urbano, 04120 Almería, Spain; elenaortega@ual.es
4. Department of Educational Psychology and Psychobiology, Faculty of Education, Universidad Internacional de la Rioja, Av. De la Paz, 137, 26006 Logroño, Spain; tania.ariza.castilla@unir.net
5. Brain, Mind and Behaviour Research Center (CIMCYC), University of Granada, 18071 Granada, Spain; edfuente@ugr.es
* Correspondence: grcanadas@ugr.es

Citation: Aguayo-Estremera, R.; Cañadas, G.R.; Ortega-Campos, E.; Ariza, T.; De la Fuente-Solana, E.I. Validity Evidence for the Internal Structure of the Maslach Burnout Inventory-Student Survey: A Comparison between Classical CFA Model and the ESEM and the Bifactor Models. *Mathematics* 2023, *11*, 1515. https://doi.org/10.3390/math11061515

Academic Editor: Jie Wen

Received: 9 February 2023
Revised: 10 March 2023
Accepted: 18 March 2023
Published: 21 March 2023

Copyright: © 2023 by the authors. Licensee MDPI, Basel, Switzerland. This article is an open access article distributed under the terms and conditions of the Creative Commons Attribution (CC BY) license (https://creativecommons.org/licenses/by/4.0/).

Abstract: Academic burnout is a psychological problem characterized by three dimensions: emotional exhaustion, depersonalization, and personal accomplishment. This paper studies the internal structure of the MBI-SS, the most widely used instrument to assess burnout in students. The bifactor model and the ESEM approach have been proposed as alternatives, capable of overcoming the classical techniques of CFA to address this issue. Our study considers the internal structure of the MBI-SS by testing the models most frequently referenced in the literature, along with the bifactor model and the ESEM. After determining which model best fits the data, we calculate the most appropriate reliability index. In addition, we examined the validity evidence using other variables, namely the concurrent relationships with depression, anxiety, neuroticism, and conscientiousness, and the discriminant relationships with the dimensions of engagement, extraversion, and agreeableness. The results obtained indicate that the internal structure of the MBI-SS is well reflected by the three-factor congeneric oblique model, reaching good values of reliability and convergent and discriminant validity. Therefore, when the scale is used in applied contexts, we recommend considering the total scores obtained for each of the dimensions. Finally, we recommend using the omega coefficient and not the alpha coefficient as an estimator of reliability.

Keywords: academic burnout syndrome; MBI-SS; internal structure; reliability and validity; ESSEM and bifactor model

MSC: 62-11

1. Introduction

Academic burnout has traditionally been defined as a psychological problem arising from continual exposure to stressors related to the educational institution and to study activities. The syndrome is usually characterized by three dimensions: emotional exhaustion, depersonalization, and low personal accomplishment [1–3]. In the academic context, emotional exhaustion (EE) refers to feelings of stress related to the educational center, particularly chronic fatigue. Depersonalization (D) is manifested as an indifferent or distant attitude towards academic tasks and a view of studies as meaningless. Low personal accomplishment (PA) alludes to perceived inefficacy when studying, lack of academic success, and scant benefit obtained from the study course [4].

Academic burnout can have serious physical and psychological consequences for students' health [5], for example, provoking sleeplessness, depression, low self-esteem, poor academic performance, absenteeism, and dropout. With a prevalence of 2–41% [5,6], the syndrome continues to be a significant social problem, calling for further study and better understanding.

1.1. The Maslach Burnout Inventory: Internal Structure Validity Evidence and Reliability

Various measurement instruments have been developed to assess burnout syndrome among the working population, among which the Maslach Burnout Inventory (MBI) [7] is the most widely used [8–10]. Currently, three versions are available: MBI-Human Services Survey (HSS), MBI-Educators Survey (ES), and MBI-General Survey (GS). However, few such instruments specifically target the university population. To our knowledge, only the MBI-Student Survey [4], the Granada Burnout Questionnaire for university students (CBG-US) [5,11], and the Student Burnout Inventory [12] have been developed. The MBI-SS was created as an adaptation of the MBI-GS [13], and measures the three dimensions of burnout established for the MBI (EE, D, and PA). This instrument is the most widely used to assess burnout in students [14,15], and has been adapted for use in numerous other linguistic populations, namely Portuguese, Dutch, Spanish [4], Brazilian [14], Italian [16], French [17], Chinese [18], Iranian [19], Turkish [20], Colombian-Spanish [21], Serbian [22], Hungarian [23], and Sri Lankan [15].

Most studies of the psychometric properties of the MBI-SS have focused on its internal structure and reliability. However, those seeking evidence of internal structure validity have obtained mixed results. In some cases, the three-dimensional structure of the original scale has been replicated [14,20,22], but in others, the solution obtained is unsatisfactory [17] and/or different from the original due to modifications in the specification of the model, for example, changes in the number of factors, the elimination of items with psychometric problems, or the specification of correlations between item error variances in order to improve the overall fit, and specification of orthogonal factors [15,16,18,19].

In addition to the three-dimensional model, other models have been proposed, based on the results of empirical investigations, mainly performed on the MBI-HSS and the MBI-GS. Thus, [10] presented a hierarchical model with three first-order factors (EE, D, and PA) and a second-order general factor (burnout). This model, motivated especially by the strong correlation observed between the three factors, can be integrated into the original theoretical proposal of [2]. In addition, a two-factor model has been proposed, excluding PA, which is viewed as a non-nuclear component of the syndrome [14,24,25]. In another two-factor model, EE and D form a single factor (the burnout core component), with PA as the second factor [26–28]. Finally, models with four [29] and even five factors [30] have been proposed for the MBI-HSS, but not for the MBI-GS or the MBI-SS, since these have different numbers of questionnaire items.

As concerns reliability evidence for the MBI-SS, the vast majority of studies use the alpha coefficient [31], which requires compliance with assumptions such as unidimensionality, tau (τ)-equivalence, and normality of the distribution of the items. These assumptions are rarely verified and, when they are, seldom satisfied [32–34]. When the alpha coefficient assumptions are not met, this index tends to underestimate the true reliability of the scale. For this reason, to deal with congeneric scales (which do not satisfy the assumption of τ-equivalence), the omega coefficient is usually recommended [35]. Furthermore, the correct coefficient must be calculated since there are different types of omega coefficients (total, hierarchical, and subscale, among others), the choice of which depends on the type of model proposed (e.g., single or multi-factor). In this context, none of the studies previously conducted to estimate the validity of the MBI-SS have tested τ-equivalent models or have used omega as a reliability estimator.

1.2. The Bifactor Model

The bifactor model was proposed by [36] and has been discussed in detail in various papers [37,38]. In this model, each item depends on two or more orthogonal factors: a general factor and one or more group factors that characterize a specific subset of items.

Formally, the factor analysis model (whether exploratory or confirmatory) can be represented with the generic matrix expression [39]:

$$Y = \Lambda X + \Psi E \qquad (1)$$

where Y is an n × 1 random vector of observed random variables (responses to items); Λ is an n × r factor-pattern matrix (factor loadings); X is an r × 1 random vector of latent common factors (factor scores); Ψ is an n × n diagonal matrix of unique-factor-pattern loadings (residual variances or uniquenesses); and E is an n × 1 random vector of latent unique-factor variables (residual scores). In confirmatory factor analysis (CFA), not all items are forced to load on all factors; residual variances may be correlated, and restrictions can be made for items, for example, all items may be forced to load equally on the same factor. Equation 1 can also be expressed as follows [40]:

$$Y = \Lambda_y \eta + \varepsilon \qquad (2)$$

where Y is an observed variable, Λy are the coefficients describing the effects of the latent variables on the observed variables, η is a latent factor score, and ε is the measurement error (uniqueness) that can be decomposed into two terms, such as ε = s + e, where s represents the specific variance associated with each variable and e is the remaining random component in Y.

From Equation (2), specific models such as the three oblique factors and the bifactor can be represented. Thus, in the case of the MBI factors (EE, D, and PA), the first of these would be expressed as:

$$Y = \lambda_{EE}\eta_{EE} + \lambda_D\eta_D + \lambda_{PA}\eta_{PA} + \varepsilon \qquad (3)$$

in which COV(ηj, ηj) = ψ, and where the bifactor model (with the addition of a general factor, G) is expressed as follows [38]:

$$Y = \lambda_G\eta_G + \lambda_{EE}\eta_{EE} + \lambda_D\eta_D + \lambda_{PA}\eta_{PA} + \varepsilon \qquad (4)$$

where COV(ηj, ηj) = 0. For Equations (3) and (4), η are latent factor scores, and λ are standard factor loadings.

With the bifactor model, we can determine whether the responses obtained by a measurement instrument are essentially unidimensional. The term essential unidimensionality refers to structures in which the general factor (i.e., the variance element that is common to all items) dominates in the presence of a certain degree of multidimensionality reflected by the group factors [38]. This is a great advantage in contexts in which it is unlikely to find models that are purely unidimensional or strictly multidimensional, that is, where there are no correlations between the factors.

The bifactor model is a suitable means of representing multidimensionality due to the construct-relevant multidimensionality of instruments that measure general constructs where different content domains coexist [38]. According to [41], there are at least two sources of construct-relevant psychometric multidimensionality: one refers to the hierarchical nature of the construct and the other reflects the fallible nature of the indicators. In consequence, the bifactor model is appropriate for assessing the hierarchical nature of the constructs [41].

In other words, the value of the bifactor model lies in its ability to determine unidimensionality in the presence of multidimensionality and, moreover, to detect relevant (or irrelevant) group factors in the presence of essential unidimensionality. These two potentialities cannot be addressed through classical CFA models, such as unifactorial or correlated factor models [42]. For example, the CFA correlated factor model is subject to significant cross-loads, which reflects the fact that, to a certain extent, multidimensionality is not perfect or unequivocal, and, therefore, cannot be directly evaluated by interpreting the correlation between factors. Although these questions can be addressed via modification indices, these do not allow us to consider essential unidimensionality; furthermore, the danger exists that atheoretical re-specifications may be introduced into the models to improve the fit [43].

1.3. Exploratory Structural Equation Modelling

The exploratory structural equation modelling (ESEM) technique, proposed by [44] as an alternative to classical CFA, specifies that all items load on all factors (unrestricted model), as would be done in an exploratory factor analysis (EFA), but with a confirmatory technique as in CFA. Formally, ESEM can be represented with the following equations [44]:

$$Y = \nu + \Lambda\eta + KX + \varepsilon \tag{5}$$

$$\eta = \alpha + B\eta + \Gamma X + \zeta \tag{6}$$

in which there are p dependent variables $Y = (Y_1, \ldots, Y_p)$, q independent variables $X = (X_1, \ldots, X_q)$, and m latent variables $\eta = (\eta_1, \ldots, \eta_m)$. The standard assumptions of this model are that the ε and ζ residuals are normally distributed with mean 0 and variance covariance matrix θ and ψ, respectively. Equation (5) represents the measurement model where ν is a vector of intercepts, Λ is a factor loading matrix, η is a vector of continuous latent variables, K is a matrix of Y on X regression coefficients, and ε is a vector of residuals for Y. Equation (6) represents the latent variable model where α is a vector of latent intercepts, B is a matrix of η times η regression coefficients, Γ is a matrix of η times X regression coefficients, and ζ is a vector of latent variable residuals.

ESEM was proposed as a means of overcoming the problems encountered with classical CFA models, which often fit the data poorly [44], meaning that models generated with EFA cannot be confirmed using CFA [45]. This problem is partly due to the fact that the classical CFA specification, in which all cross-loadings are set to zero, is unrealistic [44,45]. In general, the measurement instruments used in this context do not have pure items with a single construct [41,45], but present cross-loadings with other constructs or latent variables [44,45]. When zero loadings are misspecified by classical CFA, this can produce distorted factors, and often leads to overestimated factor correlations [44].

ESEM provides a modelling framework that can be considered a generalization of EFA. Both approaches specify unrestricted factor models that can test whether an item loads on the hypothesized factor, using target rotation, and can check the fit of the model to the data, using the chi-square test and fit indices [44]. However, in addition, ESEM has greater modelling flexibility because, among other attributes, it provides local measures of parameter fit, characterizes correlated residuals and enables structural and measurement invariance to be tested. Moreover, it can be incorporated into larger structural models, or into models with method factors, covariates and direct effects, among other features [44,45].

ESEM can also be considered a generalization of CFA in that it specifies an unrestricted model in which all cross-loads are estimated, while the latter specifies a restricted model in which all or most cross-loads are set to zero. In fact, formal tests can be performed to compare the two models [44,45]. Furthermore, despite the loss of parsimony (presenting fewer degrees of freedom and with more parameters to be estimated), ESEM is capable of accurately recovering the factorial structure of population models made up of independent clusters, such as the oblique multifactorial solutions that are typical of classical CFAs [41].

The advantage of ESEM is that it can model one of the two sources of construct-relevant psychometric multidimensionality, namely that which is due to the fallible nature of the items [41], i.e., the fact that the items are rarely pure indicators of the construct to be measured. On the one hand, they contain a degree of measurement error, which is modelled by the error variances in classical CFA models. On the other hand, they present a systematic association with other constructs, which is usually apparent in the form of cross-loadings. ESEM incorporates these cross-loadings, thus making the model constraints more realistic and achieving unbiased factor loadings and factor correlations.

1.4. Limitations of Classical CFA Applied to the MBI and Advantages of Bifactor and ESEM Models

Since the MBI was first presented, numerous studies have examined the internal structure of its different versions, mainly using EFA and CFA to do so. In their systematic review, [10] identified 35 applications of EFA and 28 of CFA. Given the current popularity of structural equation modelling, there are now probably many more applications of CFA than of EFA. Regardless of the technique used, studies have yielded conflicting solutions. According to [10], most EFA applications obtain a three-factor solution, but 25% do not. Regarding CFA applications, 90% replicate the three-factor oblique model. However, of these, 58% introduce some form of re-specification into the model, seeking to improve the global fit indices (for example, by eliminating items or by specifying correlated error variances and cross-loadings). An important consideration is that using modification indices and other ad hoc strategies to respecify the model can produce results that are misleading (for example, confirming a structure that had not previously been hypothesized) or simply incorrect (for example, removing items that are necessary to properly represent the construct) [46]. Furthermore, in some studies, models are retained in accordance with criteria that fail to meet the minimum requirements for deeming the fit to be acceptable [47]; this shortcoming was again observed in a later study focused on the MBI-SS [17]. In short, many studies fail to replicate the original structure of the MBI when classical CFA is applied.

These results are in line with the conclusions drawn in previous reviews of the literature on factor analysis, which have observed that it is fairly common to find factor structures that are not repeated in a subsequent CFA, because the specification of these models is usually unrealistic, especially when multidimensional instruments are involved [41,48]. Furthermore, many studies conclude that the tested model fits the data well (and is therefore retained), despite the fact that its global fit indices do not meet the minimum criteria established for an acceptable fit [45,48].

In view of the debates that have arisen on the structure of the MBI and acknowledging the difficulties encountered with classical CFA models in achieving an acceptable fit, especially with multidimensional instruments, we believe that both the bifactor model and the ESEM can be considered useful methodological tools with which to clarify some of the questions posed regarding the internal structure of the MBI. The authors of [49] were among the first to apply the bifactor model to the MBI-HSS, finding it to obtain the best fit of the options considered. Subsequently, other researchers have tested the bifactor model, either with the MBI-HSS [50,51] or with the MBI-ES [52,53]. However, to our knowledge, none have used the bifactor model to address the MBI-SS. Neither have any such studies used ESEM to study the internal structure of any version of the MBI. To date, the only analysis conducted in this area has been that of Biachi et al., who used ESEM to study the overlap between burnout, depression, and anxiety [54–56].

1.5. Objectives

Due to the above-mentioned disparities in empirical results, no consensus has yet been reached on the internal structure of the MBI-SS. The techniques commonly used to address this question, which in many cases is that of classical CFA, are subject to limitations in determining the possible reasons for a repeated failure to obtain a good fit (such as cross-loadings or the importance of a general factor). The bifactor model and the ESEM approach have been proposed as alternatives, capable of overcoming these limitations. For example, the bifactor method has helped clarify the internal structure of both the MBI-HSS and the MBI-ES. However, neither of these models has yet been used to target the MBI-SS. Our study, therefore, considers the internal structure of the MBI-SS in a sample of Spanish undergraduates by testing the models most frequently referenced in the literature, together with the bifactor model and ESEM. After determining which model best fits the data, we then calculate the most appropriate index of reliability. In addition, we examine the evidence of validity using other variables, specifically concurrent relationships with depression, anxiety, neuroticism, and conscientiousness, and discriminant ones with the dimensions of engagement, extraversion, and agreeableness.

2. Method

2.1. Participants

The study sample was composed of 1162 students recruited at various Spanish universities by non-probabilistic sampling. Of these participants, 64.7% were female and, overall, their mean age was 20.9 years (SD = 1.92).

2.2. Procedure

The study data were collected, using the same procedure in every case, during the second quarter of 2018. The process took place in the classroom and during academic hours, with the approval of the university staff involved. All students gave prior informed consent to participate and were assured confidentiality and anonymity.

2.3. Instruments

All participants completed an ad hoc sociodemographic data questionnaire, including their age and sex. The following measuring instruments were administered.

- The Maslach Burnout Inventory-Student Survey (MBI-SS), adapted for Spanish speakers [57]. This questionnaire contains 15 items scored on a seven-point response scale, to measure the three dimensions of the syndrome stipulated in the original proposal by [2]; emotional exhaustion (depletion of psychological and emotional resources); depersonalization (feelings of cynicism and detachment); and scant personal accomplishment (feelings of ineffectiveness and inadequate performance).
- The Utrecht Work Engagement Scale (UWES) [4]: composed of 24 items scored on a seven-point response scale, to measure the three dimensions of engagement: absorption (full concentration and placid immersion in one's own tasks), dedication (commitment to one's own tasks, recognition of their importance, and enthusiasm) and vigor (energy and mental resilience).
- Four of the five dimensions in the Spanish version of the NEO Five Factor Inventory (NEO-FFI) [58]: neuroticism, extraversion, conscientiousness, and agreeableness. Each scale consists of 12 items, scored on a five-point Likert response format.
- The depression and anxiety dimensions of the Educational-Clinical Questionnaire: Anxiety and Depression (CECAD) [59]. This questionnaire consists of 50 items with a five-point Likert-type response format. It produces a global evaluation of emotional disorders, based on the scores obtained for six dimensions: depression, anxiety, uselessness, irritability, problematic thoughts, and psychophysiological symptoms.

2.4. Data Analysis

All statistical analyses were performed with R 4.2.1. (R Core Team, Vienna, Austria, 2022), using the lavaan package [60] for factor analyses and based on the unbiased variance-covariance matrix, since this sample statistic has better statistical properties than its biased version. The parameter estimation method used was robust maximum likelihood (MLR), in view of the number of response categories established, the multivariate normality test performed and the asymmetry and kurtosis indices obtained. Target rotation was used in ESEM. Missing values were dealt with by the full information maximum likelihood (FIML) procedure.

The internal structure of the MBI-SS was assessed using the following models: (a) one congeneric factor: a one-dimensional model in which all items freely load on a single general burnout factor; (b) two congeneric factors: a model in which the items load freely on a factor composed of EE and D, where PA is the other factor; (c) two congeneric factors without PA: a model in which the items load freely in the EE and D dimensions, but from which the items of the PA dimension are excluded; (d) three congeneric factors: the model proposed by Maslach et al. [7,13]; (e) a hierarchical model with three first-order factors (EE, D and PA) plus a second-order general factor (burnout); (f) a bifactor model in which the items load freely on each of the three group factors (EE, D and PA) and also on a general factor (burnout); (g) ESEM: a model in which all the items load freely on the three factors, using a Target rotation in accordance with the model proposed by Maslach et al. [7,13].

Additionally, a τ-equivalent specification was tested in the three-factor model, that is, imposing the restriction that the items for each of the factors should have the same factor loading. This requirement was made for two reasons: (a) the three-factor model is the original form and the one most commonly employed; (b) the alpha coefficient is the index that is normally used to estimate reliability, and this measure requires compliance with the τ-equivalence assumption.

The measurement models were assessed using the χ^2 statistic and the following global fit indices [61,62]: Tucker–Lewis index (TLI), comparative fit index (CFI), root mean square error of approximation (RMSEA) and standardized root mean squared residual (SRMR). For CFI and TLI, values above 0.90 or 0.95 are considered adequate, whereas for SRMR, values below 0.08 are acceptable [63]. RMSEA values below 0.06 are reasonable [63] while those below 0.05 are considered evidence of a satisfactory fit [64]. For all the measurement models that produced adequate fit indices, likelihood-ratio tests were performed to determine whether the difference between the log-likelihoods was statistically significant, and to calculate the difference between these and the Akaike Information Criteria (AIC). When the fit of a more complex model was significantly better than that of a simpler model and, at the same time, the estimated factor loadings were high enough (above 0.30, according to [61]), the more complex model was considered to better represent the internal structure of the MBI-SS. In contrast, when the difference in fit was not statistically significant, the simpler factor structure was retained.

Once it was decided which measurement model best represented the internal structure of the MBI-SS, an appropriate reliability index was chosen and computed. Formulas and references for the reliability indices considered are detailed in Table S2 of the supplementary material. The alpha coefficient was considered for the one-factor τ-equivalent model, and appropriate versions of the omega coefficient for the other measurement models. In the one-factor congeneric model, the total omega, ωt, was used for the whole scale, while each of the corresponding subscales was used with the three-factor model. When a bifactor model was fitted, the hierarchical omega, ωh, was the reliability index that best accounted for the general factor, while the omega subscale, ωs, was used to account for the group factors (EE, D and PA).

In order to assume essential unidimensionality (i.e., the presence of a strong general factor) or essential multidimensionality (the presence of a strong group factor), explained common variance (ECV) values greater than 0.60–0.70 and ωh values greater than 0.70 [65–67] are recommended.

A correlation analysis was performed between the mean scores for each factor (and the total), the depression and anxiety scales (CECAD) and the personality factors (NEO-FFI). To estimate the sample size required for this study, an a priori analysis was performed using online software [68]. In this calculation, the minimum expected effect size was $r = 0.15$, with $\alpha = 0.05$ and a level of statistical power $1 - \beta = 0.9$, and with a maximum of 4 latent variables and 15 observed variables. The minimum sample size needed to detect the proposed effect was N = 799, which is smaller than the sample size used.

3. Results

3.1. Validity Evidence Based on Internal Structure

Descriptive statistics for the MBI-SS items, together with the inter-item correlations, are shown in Table S1 of the Supplementary Material. Separately, Table 1 shows the results of the fit obtained for each congeneric model tested. The only models that achieved acceptable fits were the bifactor and the ESEM. Except for the SRMR index, the three-factor and hierarchical models did not produce acceptable fit indices. Neither the one- nor the two-factor models (EE + D and PA) achieved an acceptable fit to the data. However, the two-factor model without PA achieved an acceptable fit for all indices except RMSEA. Although the three-factor congeneric model did not achieve an acceptable fit, that of the τ-equivalent version of the model is presented. This model did not achieve acceptable fit indices (see Table 1) and the comparison with the congeneric model showed that the latter fitted the data better (χ^2diff(12) = 151.28, $p < 0.001$). In the chi-square difference tests, the three-factor model was compared with the bifactor and the latter with the ESEM. In this comparison, the bifactor was better than the three-factor model (χ^2diff(12) = 217.26, $p < 0.001$) while the ESEM outperformed the bifactor (χ^2diff(12) = 69.68, $p < 0.001$).

Table 1. Factor analysis results.

	χ^2	df	p	CFI	TLI	RMSEA	90% CI	SRMR	AIC
One-factor	1677.12	90	<0.001	0.640	0.580	0.135	[0.130–0.141]	0.109	61,810.51
Two factors									
(EE and D), PA	941.42	89	<0.001	0.809	0.775	0.099	[0.093–0.105]	0.075	60,909.10
EE, D	214.21	26	<0.001	0.928	0.901	0.088	[0.077–0.099]	0.053	38,103.89
Three factors									
Congeneric	551.29	87	<0.001	0.896	0.875	0.074	[0.068–0.080]	0.062	60,447.20
τ-equivalent	701.25	99	<0.001	0.866	0.858	0.079	[0.073–0.084]	0.079	60,598.93
Hierarchic	551.29	87	<0.001	0.896	0.875	0.074	[0.068–0.080]	0.062	60,447.20
Bifactor	264.50	75	<0.001	0.958	0.941	0.051	[0.044–0.057]	0.037	60,127.88
ESEM	198.95	63	<0.001	0.969	0.949	0.047	[0.040–0.055]	0.023	60,080.58

df—Degrees of freedom; CFI—Comparative fit index; TLI—Tucker-Lewis index; RMSEA—Root mean square error of approximation; SRMR—Standardised root mean squared residual; AIC—Akaike information criteria; EE—Emotional exhaustion; D—Depersonalisation; PA—Personal accomplishment; ESEM—Exploratory structural equation modelling.

Table 2 details the results obtained by the three-factor oblique congeneric, the three-factor oblique τ-equivalent, the bifactor, and the ESEM models. The path diagrams for these models are shown in Figures 1–4. In the three-factor congeneric, three-factor τ-equivalent and ESEM models, the factor loadings were adequate, i.e., above 0.30 [61]. This finding is especially significant for the ESEM, since the items can load on any factor. In fact, in this model, the factor loading for item RP6 was higher in D than in PA. In the bifactor model, adequate loadings were obtained in all the EE items (although those for EE1 and EE5 were higher on the general factor) and for PA (although the loading of PA6 was higher on the general factor). In D, the loadings of three of the four items were low. In the general factor, low loads were obtained on EE2 and on PA1, PA2, PA3, and PA5.

Evaluation of essential unidimensionality showed that the ECV and the hierarchical omega indices, ωh, were below the recommended values for the general factor. Regarding dimensionality, the EE and PA factors obtained satisfactory indices of ECV but not of ωs. For D, none of the indices obtained a good value. In both the three-factor congeneric and the τ-equivalent models, factor correlations were high between D and EE and between D and PA, and moderate between PA and EE. Correlations were lower (low-moderate), however, in ESEM. In the bifactor model, the correlations were set to zero in the specification.

3.2. Reliability

The reliability estimates for all the factors in the three-factor oblique congeneric, three-factor oblique τ-equivalent, bifactor, and ESEM models are shown in Table 2. Different estimators were considered according to the model selected. For example, the alpha coefficient, α, was used for the three-factor τ-equivalent model, but not for the others, since they are congeneric. Total omega, ωt, was calculated for all models, including the τ-equivalent model, since α and ωt are equivalent if the assumptions of the former are met [32,35]. Hierarchical omega, ωh, was used with the bifactor model as an index to estimate the internal consistency of the total test score that is exclusive of the overall factor;

the omega subscale, ω_s, was applied to the subscales for the same purpose. The formulas for each estimator are listed in Table S2 of the Supplementary Material.

Table 2. Factor loadings, correlations between factors, and reliability indices.

Item	Three Congeneric Factors			Three τ-Equivalent Factors			Bifactor				ESEM		
	EE	D	PA	EE	D	PA	EE	D	PA	General	EE	D	PA
EE1	0.648	0.000	0.00	0.661	0.000	0.000	0.382	0.000	0.00	0.503	0.472	0.248	0.070
EE2	0.593	0.000	0.00	0.613	0.000	0.000	0.480	0.000	0.00	0.373	0.588	0.057	0.058
EE3	0.579	0.000	0.00	0.618	0.000	0.000	0.491	0.000	0.00	0.343	0.580	0.011	0.125
EE4	0.560	0.000	0.00	0.623	0.000	0.000	0.662	0.000	0.00	0.201	0.701	0.069	0.078
EE5	0.770	0.000	0.00	0.676	0.000	0.000	0.503	0.000	0.00	0.550	0.591	0.239	0.096
D1	0.000	0.566	0.00	0.000	0.679	0.000	0.000	0.465	0.00	0.663	0.262	0.402	0.068
D2	0.000	0.551	0.00	0.000	0.653	0.000	0.000	0.083	0.00	0.561	0.157	0.476	0.004
D3	0.000	0.800	0.00	0.000	0.743	0.000	0.000	0.001	0.00	0.783	0.002	0.822	0.022
D4	0.000	0.840	0.00	0.000	0.742	0.000	0.000	0.001	0.00	0.830	0.068	0.759	0.085
PA1	0.000	0.000	0.528	0.000	0.000	0.586	0.000	0.000	0.475	0.247	0.051	0.031	0.547
PA2	0.000	0.000	0.574	0.000	0.000	0.528	0.000	0.000	0.581	0.206	0.032	0.134	0.667
PA3	0.000	0.000	0.687	0.000	0.000	0.656	0.000	0.000	0.640	0.294	0.030	0.035	0.717
PA4	0.000	0.000	0.722	0.000	0.000	0.688	0.000	0.000	0.574	0.401	0.072	0.142	0.637
PA5	0.000	0.000	0.579	0.000	0.000	0.596	0.000	0.000	0.566	0.235	0.097	0.144	0.673
PA6	0.000	0.000	0.578	0.000	0.000	0.622	0.000	0.000	0.319	0.523	0.080	0.419	0.355
α	-	-	-	0.772	0.781	0.778	-	-	-	-	-	-	-
ωh	-	-	-	-	-	-	0.485	0.001	0.579	0.649	-	-	-
ωt	0.768	0.790	0.777	0.772	0.794	0.777	0.632	0.074	0.701	0.767	0.725	0.719	0.775
ECV	-	-	-	-	-	-	0.605	0.138	0.717	0.516	-	-	-
rCE	1			1			1			0.000	1		
rD	0.651	1		0.669	1		0.000	1		0.000	0.376	1	
rPA	−0.287	−0.528	1	−0.277	−0.538	1	0.000	0.000	1	0.000	−0.094	−0.438	1

Note. ωh refers to ωs when it is applied to the subscales (EE, D and PA). ECV—Explained common variance.

In general, all of the models, except the bifactor, presented good reliability, above the usual recommended value of 0.70 [69]. For EE and PA, the reliability estimates (whether using alpha or omega) of the τ-equivalent model were practically the same as those of the congeneric model. For D, the reliability estimate of the τ-equivalent model was somewhat lower than that of the congeneric model. The ESEM solution produced slightly lower coefficients. With the bifactor model, only PA and the general factor obtained an ωt value above 0.70. As mentioned above, the values for ωh and ωs were below the recommended minimum.

3.3. Validity Evidence Based on Relations to Other Variables

Table 3 shows the descriptive statistics for the total scores of the variables included in the study, together with the Pearson bivariate correlation in each case. In accordance with the results described above and with the theory, we expected to obtain evidence of convergent and discriminant validity with different variables, depending on the burnout dimension considered. Considering the EE and D dimensions, the correlations with depression, anxiety, and neuroticism were positive, moderate, and statistically significant. The correlations with the dimensions of engagement (vigor, dedication, and absorption), conscientiousness, agreeableness, and extraversion were negative and of low or moderate magnitude. The correlations between PA and the dimensions of engagement (vigor, dedication, and absorption), conscientiousness, agreeableness, and extraversion were positive, strong, and statistically significant. In addition, low or moderate negative correlations were obtained with depression, anxiety, and neuroticism.

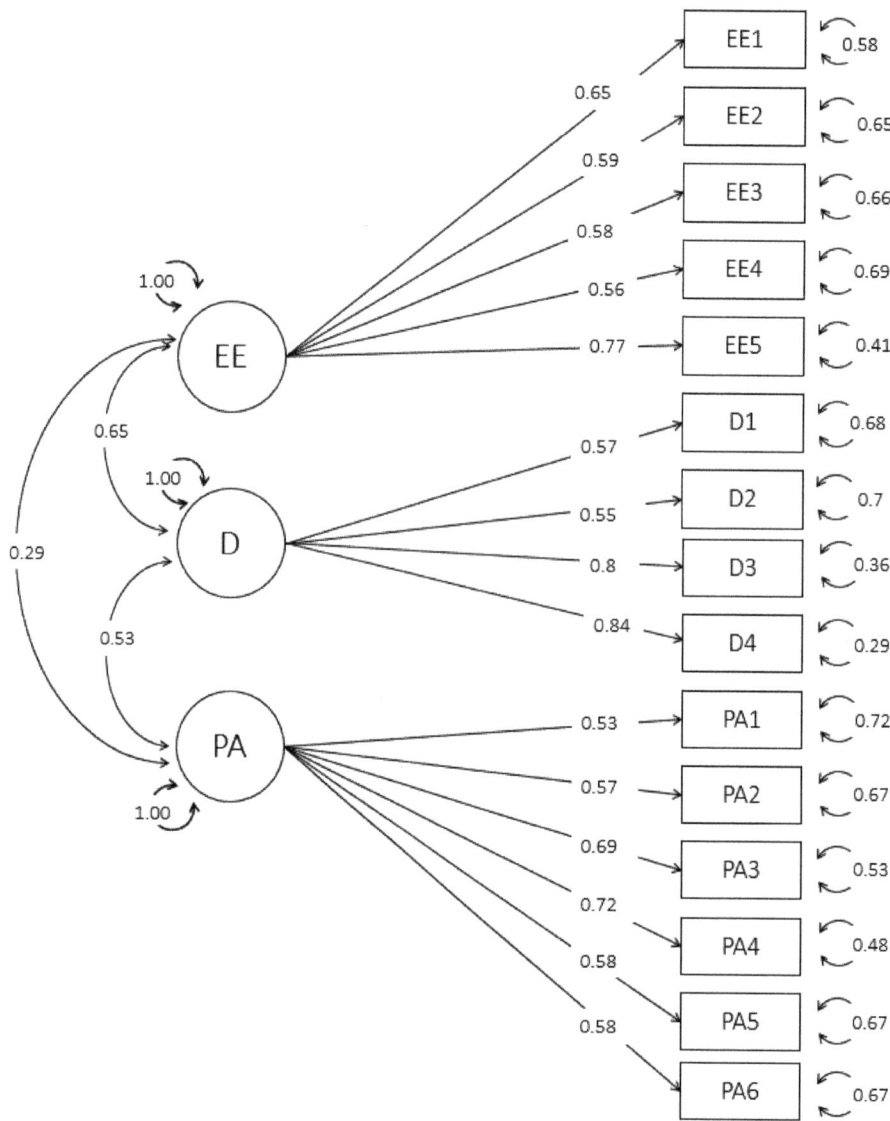

Figure 1. Three-factor congeneric model.

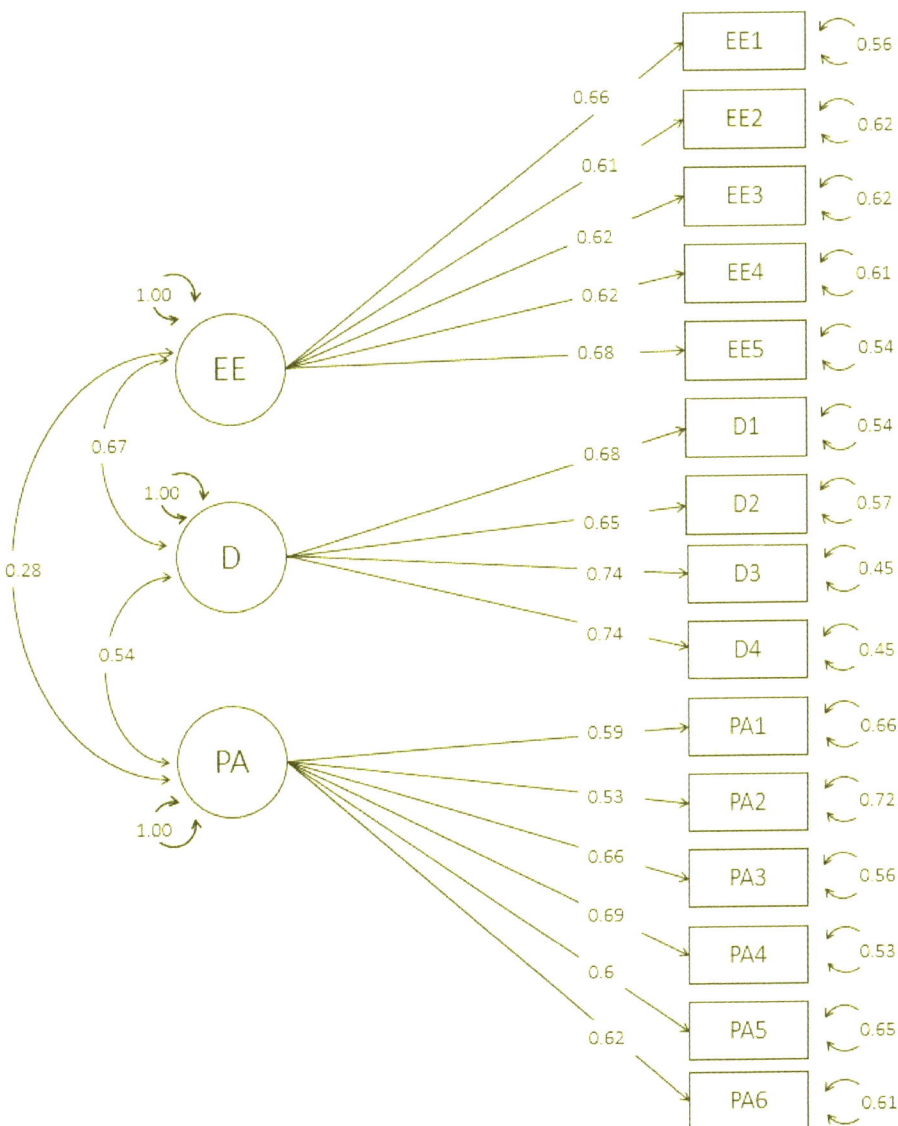

Figure 2. Three-factor τ-equivalent model.

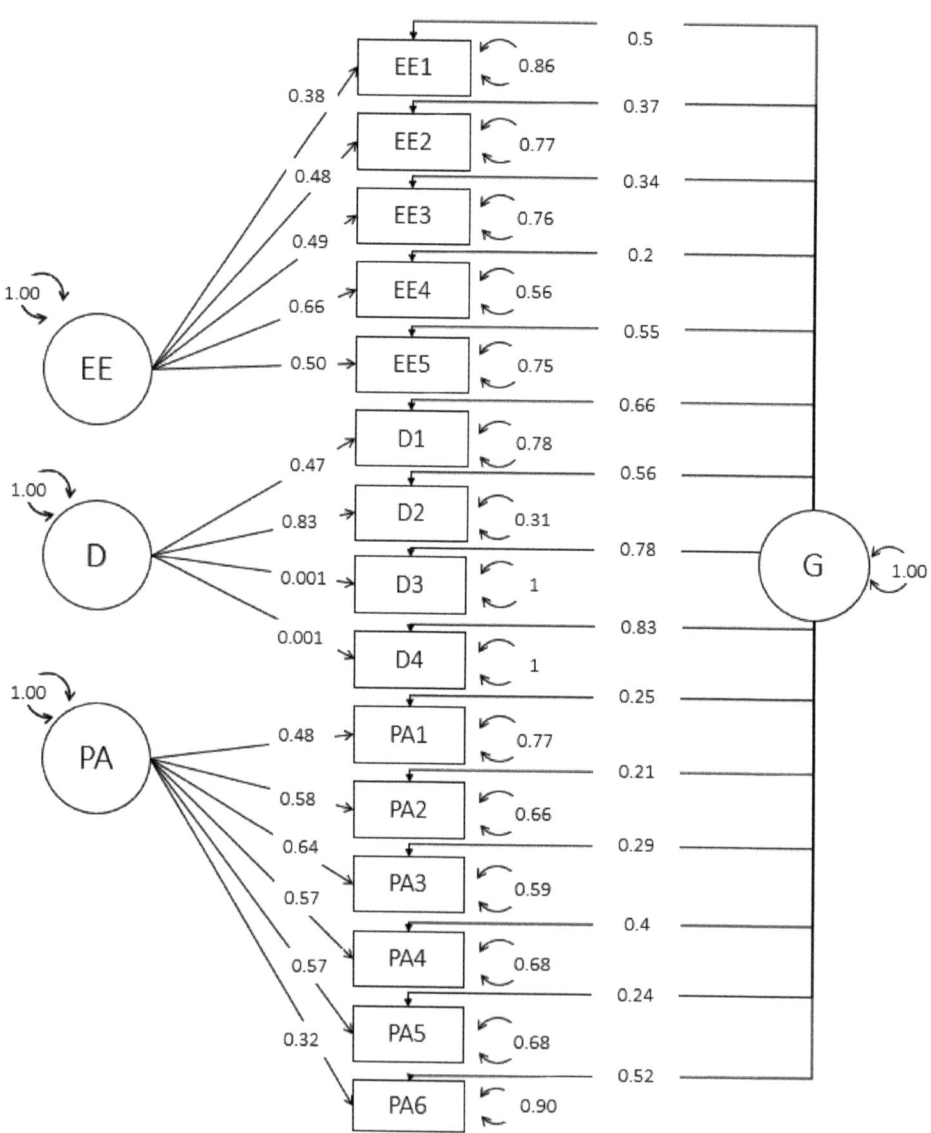

Figure 3. Bifactor model.

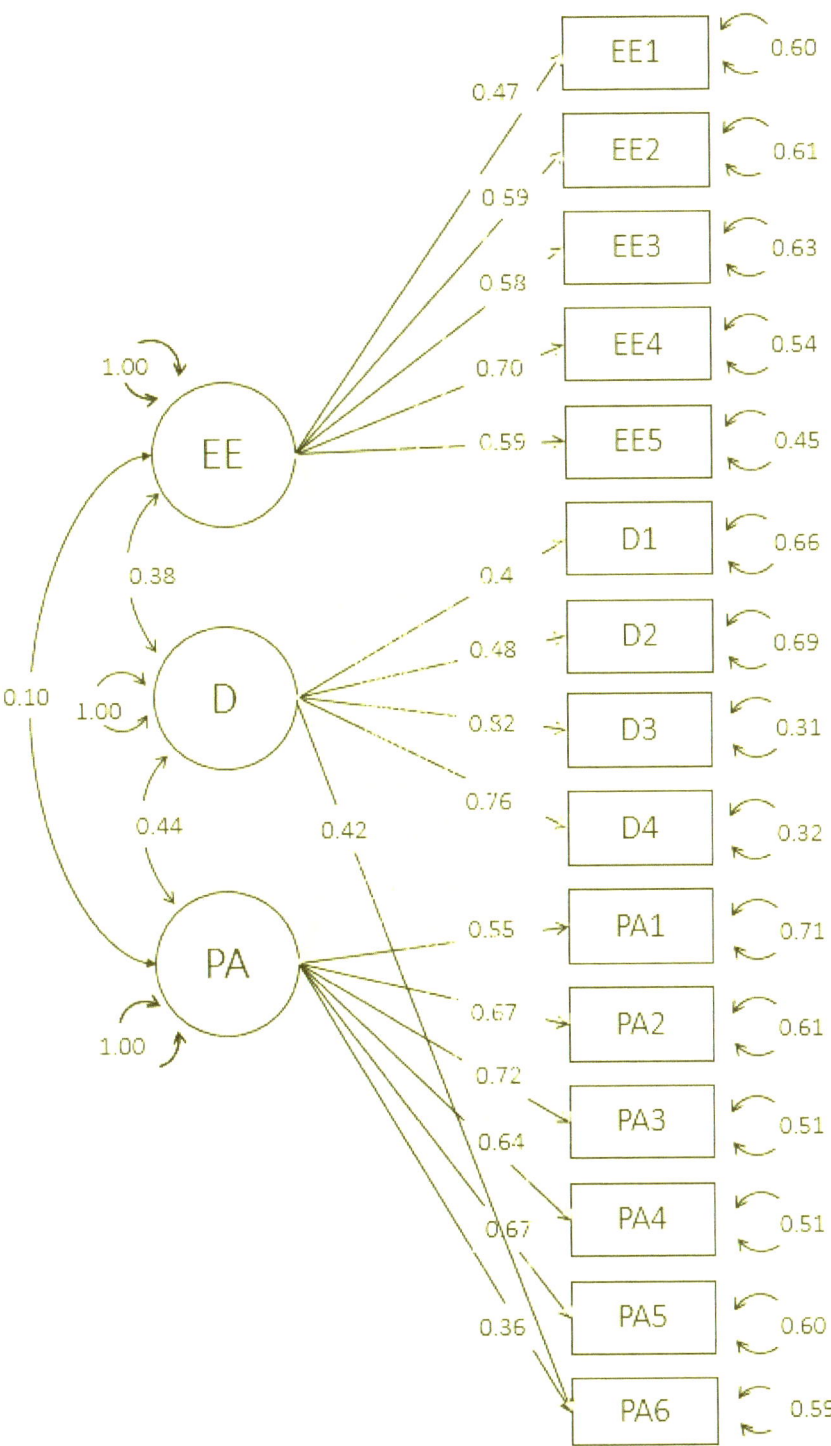

Figure 4. ESEM model.

Table 3. Descriptive statistics and correlations between the study variables.

Item	M	SD	1	2	3	4	5	6	7	8	9	10	11	12
1. EE	13.74	6.25	0.77											
2. D	7.19	5.50	0.52 ***	0.79										
3. PA	26.98	5.72	−0.21 ***	−0.41 ***	0.78									
4. VI	31.95	7.81	−0.24 ***	−0.32 ***	0.64 ***	0.73								
5. DE	37.22	9.25	−0.27 ***	−0.63 ***	0.73 ***	0.65 ***	0.92							
6. AB	25.59	7.00	−0.12 ***	−0.29 ***	0.65 ***	0.71 ***	0.63 ***	0.74						
7. DP	46.39	13.70	0.43 ***	0.26 ***	−0.29 ***	−0.22 ***	−0.19 ***	−0.13 ***	0.94					
8. AN	40.74	11.11	0.51 ***	0.30 ***	−0.22 ***	−0.16 ***	−0.18 ***	−0.06 **	0.74 ***	0.89				
9. NE	32.30	7.70	0.38 ***	0.20 ***	−0.25 ***	−0.22 ***	−0.15 ***	−0.12 ***	0.72 ***	0.60 ***	0.81			
10. CO	43.00	5.52	−0.06 **	−0.24 ***	0.49 ***	0.39 ***	0.40 ***	0.45 ***	−0.25 ***	−0.13 ***	−0.19 ***	0.80		
11. AG	43.19	5.87	−0.17 ***	−0.24 ***	0.27 ***	0.25 ***	0.27 ***	0.23 ***	−0.24 ***	−0.16 ***	−0.25 ***	0.21 ***	0.69	
12. EX	45.51	7.14	−0.19 ***	−0.27 ***	0.40 ***	0.32 ***	0.36 ***	0.29 ***	−0.40 ***	−0.32 ***	−0.33 ***	0.32 ***	0.29 ***	0.83

Note. EE = Emotional exhaustion; D = Depersonalisation; PA = Personal accomplishment; VI = Vigour; DE = Dedication; AB = Absorption; DP = Depression; AN = Anxiety; NE = Neuroticism; CO = Conscientiousness; AG = Agreeableness; EX = Extraversion. The diagonal contains Omega Total coefficients. *** = $p < 0.001$; ** = $p < 0.01$.

4. Discussion

The Maslach Burnout Inventory-Student Survey (MBI-SS; [13]) is probably the instrument most widely used to assess academic burnout syndrome [44,45], and has been adapted for use by numerous language groups (e.g., Spanish, Italian, Chinese, and German). According to the original conceptual proposal [7,13], the MBI-SS evaluates three dimensions of burnout: emotional exhaustion (EE), depersonalization (D), and personal accomplishment (PA). The most commonly evaluated psychometric properties of the MBI-SS are its internal structure and reliability. The evidence obtained for its validity, based on the internal structure, has produced mixed results: while some studies have replicated the original structure of three oblique factors [14,20,22], others have obtained different solutions, due to model re-specification [15,16,18,19] or have not achieved an acceptable fit [17]. As concerns the reliability of the instrument, most studies report the alpha coefficient, but this index can underestimate the true reliability when the assumptions of τ-equivalence, unidimensionality and normality in the distribution of the items are not met, as it is often the case with psychological assessment instruments [32–34]. A discussion on reliability can be consulted elsewhere [70–74].

To date, evidence of validity has been obtained using EFA or classical CFA (e.g., one factor, three factors or hierarchical), but these present various limitations. Firstly, EFA does not allow the introduction of restrictions in the models, which means that τ-equivalent solutions or those with correlated errors cannot be tested. Moreover, it does not provide local measures of parameter fit, nor can it be incorporated into larger structural models [44]. The classical CFA specification, on the other hand, is often unrealistic in multidimensional instruments since it sets the cross-loadings to zero [41,48]. This type of specification error tends to result in distorted factors, and often produces overestimated factor correlations [44].

Some new approaches to studying the internal structure of measurement instruments have recently been proposed, such as the bifactor model and ESEM. Both of these methods seek to overcome the limitations of classical CFA. The main advantage is that both can explain more sources of variability in the item scores (psychometric multidimensionality), and, therefore, are more realistic, generating less biased results. The authors of [41] highlighted two sources of construct-relevant psychometric multidimensionality: the first concerns the hierarchical nature of the construct, i.e., the expectation that all the items considered will present a significant level of association with their own subscales (for example, each of the items of which EE is composed with the factor as a whole), as well as with hierarchically superior constructs (such as burnout). The second source arises from the fallible nature of the indicators typically used to measure psychological constructs, most of which assess conceptually related and partially overlapping constructs (e.g., burnout and depression).

Despite the advantages offered by the bifactor model and ESEM, to our knowledge, no studies have yet been conducted using either of these approaches to evaluate the internal structure of the MBI-SS. To address this gap, the present study examines the internal structure of the MBI-SS in a sample of Spanish undergraduates by jointly testing the models most frequently cited in the literature, together with the bifactor model and ESEM. Specifically, we analyzed the results obtained by the following models: (a) one congeneric factor: a one-dimensional model in which all items freely load on a single general burnout factor; (b) two congeneric factors: a model in which the items load freely on a factor composed of the EE and D items, with PA being the other factor; (c) two congeneric factors without PA: a model in which the items load freely in the EE and D dimensions, but the items of the PA dimension are excluded from the model specification; (d) three congeneric factors: the model proposed by Maslach et al. [7,13]; (e) hierarchical: a model with three first-order factors (EE, D, and PA) together with a second-order general factor (burnout); (f) bifactor: a model in which the items load freely on each of the three group factors (EE, D, and PA) and on a general factor (burnout); (g) ESEM: a model in which all the items load freely on the three factors, using a Target rotation in accordance with the model presented by Maslach et al. [7,13]. Since the alpha coefficient is the most widely used means of estimating the reliability of the MBI-SS, we also tested the three-factor oblique τ-equivalent model, which requires compliance with the assumption of τ-equivalence.

4.1. Internal Structure of the MBI-SS

According to the results obtained, the only models that achieved an acceptable fit in all of the evaluation indices were the bifactor model and ESEM. In the chi-square difference tests, ESEM was statistically better than the bifactor model. τ-equivalent models do not usually fit the data well unless the congeneric model does so too [61]. However, we tested the fit for the model with three oblique τ-equivalent factors, since this was the only one that satisfied the assumptions for using the alpha coefficient. As expected, the model did not achieve acceptable fit indices, and proved to be statistically worse than the three-factor congeneric model. This result suggests that the alpha coefficient should not be used to estimate the reliability of the MBI-SS.

Regarding the factor loadings of the items, in all models except a few cases with the bifactor model, the recommended level was exceeded. The loadings of the oblique three-factor model were higher than those of the bifactor model and ESEM. This result was expected given that in the three-factor model, the only source of item variability is its factor; however, in the bifactor model, the variability is divided between the general factor and the group factors, while in ESEM, it is divided among the three factors. The fact that acceptable factor loadings are obtained in ESEM indicates that this model adequately explains the variability of cross-loadings while maintaining the structure of three oblique factors. Nevertheless, future studies could further investigate the reasons for the high fluctuations of the loadings in the D. The loadings of the τ-equivalent model were more homogeneous than those of the other models due to the restriction imposed that all the items of the same factor must have the same factor loading. With the bifactor model, the EE and PA items achieved good factor loadings; however, in the general factor and in D, the loadings of some items were less than 0.30 (see Table 2).

The factor loadings obtained with the bifactor model are in line with the results from the evaluation of the essential unidimensionality. Except for PA, none of the factors achieved a satisfactory ωh coefficient. As the ωh index uses factor loadings to estimate reliability (and is interpreted as the percentage of item variance that is explained by the factor, after eliminating the variance explained by the other factors), this result explains the pattern of factor loadings obtained for the different factors. Thus, several items presented significant cross-loadings in the general factor: specifically, four items (of five) in EE, all the items in D, but only two (of six) in PA. Taking the general factor as a reference, 12 items (of 15) showed relevant cross-loadings in one or more of the three group factors. The results for the ECV index are in line with these findings, from which we conclude that the values obtained for EE and, above all, PA are acceptable.

The results obtained with ωh and ECV show that the requirements for the MBI-SS to present essential unidimensionality were not met. These results also show that the most important factors generating multidimensionality (i.e., those which explain a large part of the variance of the items) are EE and PA, while D is incorporated within the general factor.

Regarding the factor correlations, in each of the three-factor models (congeneric and τ-equivalent), the values obtained are in accordance with the theory and with previous research findings [7,13]; specifically, there is a moderate positive correlation between EE and D, while the correlations between PA, EE, and D are weak-moderate and negative. In the bifactor model, the correlations were previously set to zero, that is, it was stipulated that the factors should be orthogonal. In ESEM, the correlations are weaker than with the three-factor oblique models. This result agrees with those obtained for other psychometric evaluations with ESEM and suggests that the correlations obtained with classical CFA models are overestimated [44]. Thus, according to the correlations obtained with this model, EE and D present a positive correlation of intermediate size; EE and PA have a weak negative correlation; and PA and D have a negative correlation of intermediate magnitude. These results for factor correlations may explain the poor fit of the one- and two-factor models (EE + D and PA).

4.2. Reliability of the MBI-SS

Psychometric studies of the MBI-SS have estimated its reliability using the alpha coefficient. In the present research, the alpha value results for the three dimensions of the MBI were above the recommended values [69]. However, this coefficient is not a good estimator of reliability when conditions such as τ-equivalence between the items, the unidimensionality of the scale, and the normal distribution of the item scores are not met [32–34]. As discussed above, the tau-equivalent three-factor model did not achieve an acceptable fit, which suggests that the alpha coefficient should not be used for MBI-SS applications. Since this result is common in psychological scales, the omega coefficient is normally recommended as a reliability estimator [35]. An additional advantage of omega over alpha is that it is calculated from the model that achieves a good fit to the data. If alpha was used instead, this measure could suggest a level of model reliability that, ultimately, is not maintained.

The present study is the first to focus on the psychometric properties of the MBI-SS, using omega as a reliability estimator, after fitting by CFA. According to the results obtained, the ESEM model best fitted the data. In this model, the MBI-SS achieved acceptable reliability values for all three factors [69]. The bifactor model also achieved good global indices of fit, but the reliability of the MBI dimensions in this model was not acceptable, except for PA. This result is explained by the low factor loadings of some items on their respective factors and by high ones on the general factor.

Of the oblique three-factor models, neither the congeneric nor the τ-equivalent achieved an acceptable fit to the data. However, since psychometric theory suggests that the reliability of a scale is underestimated by the alpha coefficient when the assumptions are not met, we calculated the reliability of the factors in both models. In this respect, similar values were obtained for EE and PA using the alpha coefficient for the τ-equivalent model and omega for the congeneric model. However, for D, the alpha value was somewhat lower than the omega result. These findings are consistent with the psychometric literature [33]. For PA, the factor loadings in all the models were high, except for item PA6 in the bifactor model and ESEM. As these values are high and similar, the differences between the alpha and omega coefficients are less. This is reflected in the fact that the value of the ECV index for this factor suggested that PA was a relevant group factor. On the other hand, the opposite case was observed for D; in this case, the items obtained very different factor loadings in various models (for example, very low ones in the bifactor model) and the factor obtained a very low ECV index, reflecting the slight relevance of this factor.

4.3. Relation between the MBI-SS and Other Constructs

According to the results obtained, the MBI-SS dimensions showed evidence of convergent and discriminant validity with the constructs examined. Persons who scored more highly in the dimensions of EE and D tended to obtain higher scores, too, for depression, anxiety, and neuroticism. An association was also observed between the MBI-SS and the dimensions of engagement (vigor, dedication, and absorption), conscientiousness, agreeableness, and extraversion. Thus, at higher levels of EE and D, the scores in the latter constructs were lower. As expected, an inverse pattern was obtained between PA and the other variables.

5. Limitations and Future Research

This study presents some limitations. First, the sample selection was not probabilistic, and students were not recruited from all regions of the country. Future studies should confirm the results we present with those from a broader sample. Second, we did not test the τ-equivalent versions for the bifactor and ESEM models. In future consideration of one of these models, it would be interesting to perform these tests, too, to better characterize the differences between the alpha and omega coefficients. Third, the factorial invariance of the model has not been tested. Although the age variable is not relevant in this case because the population considered is very homogeneous, variables such as the sex of the participants should be included in a future analysis. Finally, in this study, the sources of psychometric variability arising from relevant constructs were evaluated separately. On the one hand, we used the bifactor model, which enabled us to determine when it is more appropriate to use a single total score (assuming essential unidimensionality) or when subscale scores should be preferred [42]. On the other hand, we used the ESEM approach to model the cross-loadings that occur due to the fallible nature of the indicators. In future studies, it would be advisable to adjust the ESEM bifactor model [41] to simultaneously explore the two sources of construct-relevant psychometric multidimensionality, that is, the hierarchical nature of the constructs and the fallible nature of the indicators.

6. Conclusions and Practical Recommendations

To our knowledge, the present study is the first to use both the bifactor model and the ESEM approach to study the internal structure of the MBI-SS and to use a reliability estimator consistent with the retained model. Furthermore, it is the first to test tau-equivalent models and compare them with congeneric ones. To sum up, the results obtained indicate that the internal structure of the MBI-SS in Spanish undergraduates is well reflected by the three-factor oblique congeneric model, achieving good values for reliability and convergent and discriminant validity.

Firstly, the results obtained from the bifactor model show that there is insufficient evidence to suggest the essential unidimensionality of the MBI-SS. Therefore, we recommend using the scores for all three dimensions and not a global burnout score. This recommendation is in line with the proposal of [7].

Secondly, according to the ESEM results, the three-factor oblique model fits the data well when cross-loadings are taken into account, that is, accounting for the variability that occurs due to the fallible nature of the indicators. Hence, the results from both models suggest that the measurement model that best represent the data is the three-factor oblique congeneric model.

Therefore, when the scale is used in applied contexts, we recommend considering the total scores obtained for each of the dimensions (emotional exhaustion, depersonalization, and personal accomplishment), and not a global burnout score. Furthermore, when using the MBI-SS for substantive research purposes, such as testing a hypothesis related to the Job Demands-Resources theory [70], the analysis should be based on the ESEM model since this facilitates control of the cross-loadings of the scale and provides unbiased factorial correlations.

Thirdly, all three dimensions of the MBI-SS scored well for reliability. In this respect, similar values were obtained with the alpha coefficient in the three-factor τ-equivalent model and with the omega coefficient in the three-factor congeneric model. However, we recommend reporting the omega values by default since this approach does not need to meet the assumptions of tau-equivalence, item normality, and unidimensionality in order to function correctly.

Finally, the study results we describe provide a new perspective from which to consider the problem of the overlap between the MBI dimensions, especially those of EE and D. Some researchers have proposed a two-factor structure based on the strong correlation between these two dimensions [10]. According to [44], the correlations between the factors obtained with classical CFA tend to be overestimated due to the cross-loading problem. Our study results corroborate this view, and also suggest that the correlation between EE and D is not strong, but rather intermediate. Therefore, combining these two factors, as in the two-factor model proposed by some authors [26–28], would not be justified.

In summary, taking all results together, we can conclude that the MBI-SS in Spanish undergraduates is well reflected by the three-factor oblique congeneric model, achieving good values for reliability and convergent and discriminant validity.

Supplementary Materials: The following supporting information can be downloaded at: https://www.mdpi.com/article/10.3390/math11061515/s1, Table S1: Descriptive statistics and correlations for the MBI-SS items (N = 1162); Table S2: Formulas for the reliability indices [31,35,75].

Author Contributions: Conceptualization and investigation, R.A.-E. and E.I.D.l.F.-S.; formal analysis, G.R.C.; methodology, T.A.; resources, R.A.-E.; software, G.R.C. and T.A.; supervision, E.O.-C. and E.I.D.l.F.-S.; validation, G.R.C.; visualization, G.R.C. and E.O.-C.; writing—original draft, R.A.-E. and T.A.; writing—review and editing, R.A.-E. and E.O.-C. All authors have read and agreed to the published version of the manuscript.

Funding: This article has been funded by FEDER/Consejería de Universidad, Investigación e Inovación de la Junta de Andalucía. Project P20-00637.

Institutional Review Board Statement: The study was conducted in accordance with the Declaration of Helsinki and approved by the Institutional Review Board (or Ethics Committee) of the University of Granada (393/CEIH2017).

Data Availability Statement: Not applicable.

Conflicts of Interest: The authors declare no conflict of interest.

References

1. Leiter, M.P.; Maslach, C. Burnout and engagement: Contributions to a new vision. *Burn. Res.* **2017**, *5*, 55–57. [CrossRef]
2. Maslach, C.; Jackson, S.E. The measurement of experienced burnout. *J. Organ. Behav.* **1981**, *2*, 99–113. [CrossRef]
3. Schaufeli, W.B.; Leiter, M.P.; Maslach, C. Burnout: 35 years of research and practice. *Career Dev. Int.* **2009**, *14*, 204–220. [CrossRef]
4. Schaufeli, W.B.; Salanova, M.; Gonzalez-Roma, V.; Bakker, A.B. The measurement of engagement and burnout: A two sample confirmatory factor analytic approach. *J. Happiness Stud.* **2002**, *3*, 71–92. [CrossRef]
5. Aguayo, R.; Cañadas, G.R.; Assbaa-Kaddouri, L.; Cañadas-De la Fuente, G.A.; Ramírez-Baena, L.; Ortega-Campos, E. A risk profile of sociodemographic factors in the onset of academic burnout syndrome in a sample of university students. *Int. J. Environ. Res. Public Health* **2019**, *16*, 707. [CrossRef] [PubMed]
6. Aguayo-Estremera, R.; Cañadas, G.R.; Albendín-García, L.; Ortega-Campos, E.; Ariza, T.; Monsalve-Reyes, C.S.; De la Fuente-Solana, E.I. Prevalence of burnout syndrome and fear of COVID-19 among adolescent university students. *Children* **2023**, *10*, 243. [CrossRef]
7. Maslach, C.; Jackson, S.E.; Leiter, M.P. *Maslach Burnout Inventory Manual*, 4th ed.; Mind Garden: Menlo Park, CA, USA, 2018.
8. Aguayo, R.; Vargas, C.; De La Fuente, E.I.; Lozano, L.M. A meta-analytic reliability generalization study of the Maslach Burnout Inventory. *Int. J. Clin. Health Psychol.* **2011**, *11*, 343–361.
9. De Beer, L.T.; Bianchi, R. Confirmatory factor analysis of the Maslach Burnout Inventory: A Bayesian structural equation modeling approach. *Eur. J. Psychol. Assess.* **2019**, *35*, 217–224. [CrossRef]
10. Worley, J.A.; Vassar, M.; Wheeler, D.L.; Barnes, L.L.B. Factor structure of scores from the Maslach Burnout Inventory: A review and meta-analysis of 45 exploratory and confirmatory factor-analytic studies. *Educ. Psychol. Meas.* **2008**, *68*, 797–823. [CrossRef]
11. De la Fuente, E.I.; Lozano, L.M.; García-Cueto, E.; Luis, C.S.; Vargas, C.; Cañadas, G.R.; Cañadas-De la Fuente, G.A.; Hambleton, R.K. Development and validation of the Granada Burnout Questionnaire in Spanish police. *Int. J. Clin. Health Psychol.* **2013**, *13*, 216–225. [CrossRef]
12. Salmela-Aro, K.; Read, S. Study engagement and burnout profiles among Finnish higher education students. *Burn. Res.* **2017**, *7*, 21–28. [CrossRef]

13. Schaufeli, W.B.; Leiter, M.P.; Maslach, C.; Jackson, S.E. Maslach burnout inventory-general survey (MBI-GS). In *Maslach Burnout Inventory-Test Manual*, 3rd ed.; Consulting Psychologists Press: Palo Alto, CA, USA, 1996.
14. Maroco, J.; Campos, J.A.D.B. Defining the student burnout construct: A structural analysis from three burnout inventories. *Psychol. Rep.* **2012**, *111*, 814–830. [CrossRef] [PubMed]
15. Wickramasinghe, N.D.; Dissanayake, D.S.; Abeywardena, G.S. Validity and reliability of the Maslach Burnout Inventory-Student Survey in Sri Lanka. *BMC Psychol.* **2018**, *6*, 52. [CrossRef] [PubMed]
16. Portoghese, I.; Leiter, M.P.; Maslach, C.; Galletta, M.; Porru, F.; D'Aloja, E.; Finco, G.; Campagna, M. Measuring burnout among university students: Factorial validity, invariance, and latent profiles of the Italian Version of the Maslach Burnout Inventory Student Survey (MBI-SS). *Front. Psychol.* **2018**, *9*, 2105. [CrossRef]
17. Faye-Dumanget, C.; Carré, J.; Le Borgne, M.; Boudoukha, P.A.H. French validation of the Maslach Burnout Inventory-Student Survey (MBI-SS). *J. Eval. Clin. Pract.* **2017**, *23*, 1247–1251. [CrossRef]
18. Hu, Q.; Schaufeli, W.B. The factorial validity of the Maslach Burnout Inventory–Student Survey in China. *Psychol. Rep.* **2009**, *105*, 394–408. [CrossRef] [PubMed]
19. Rostami, Z.; Abedi, M.R.; Schaufeli, W.B.; Ahmadi, S.A.; Sadeghi, A.H. The Psychometric Characteristics of Maslach Burnout Inventory Student Survey: A study students of Isfahan University. *Zahedan J. Res. Med. Sci.* **2013**, *16*, 55–58.
20. Yavuz, G.; Dogan, N. Maslach Burnout Inventory-Student Survey (MBI-SS): A validity study. *Procedia. Soc. Behav. Sci.* **2014**, *116*, 2453–2457. [CrossRef]
21. Hederich-Martínez, C.; Caballero-Domínguez, C. Validación del cuestionario Maslach Burnout Inventory-Student Survey (MBI-SS) en contexto académico colombiano. *CES Psicol.* **2016**, *9*, 1–15. [CrossRef]
22. Ilic, M.; Todorovic, Z.; Jovanovic, M.; Ilic, I. Burnout syndrome among medical students at one university in Serbia: Validity and reliability of the Maslach Burnout Inventory-Student Survey. *Behav. Med.* **2017**, *43*, 323–328. [CrossRef]
23. Hazag, A.; Major, J.; Ádám, S. Assessment of burnout among students. Validation of the Hungarian version of the Maslach Burnout Inventory-Student Version (MBI-SS). *Mentálhigiéné És Pszichoszomatika* **2010**, *11*, 151–168. [CrossRef]
24. Brookings, J.B.; Bolton, B.; Brown, C.E.; McEvoy, A. Self-reported job burnout among female human service professionals. *J. Organ. Behav.* **1985**, *6*, 143–150. [CrossRef]
25. Kalliath, T.J.; O'Driscoll, M.P.; Gillespie, D.F.; Bluedorn, A.C. A test of the Maslach Burnout Inventory in three samples of healthcare professionals. *Work. Stress* **2000**, *14*, 35–50. [CrossRef]
26. Halbesleben, J.R.; Demerouti, E. The construct validity of an alternative measure of burnout: Investigating the English translation of the Oldenburg Burnout Inventory. *Work. Stress* **2005**, *19*, 208–220. [CrossRef]
27. Schaufeli, W.B.; Maslach, C.; Marek, T. *Professional Burnout: Recent Developments in Theory and Research*; Routledge: Oxfordshire, UK, 2017.
28. Schaufeli, W.B.; Taris, T.W. The conceptualization and measurement of burnout: Common ground and worlds apart. *Work. Stress* **2005**, *19*, 256–262. [CrossRef]
29. Iwanicki, E.F.; Schwab, R.L. A cross validation study of the Maslach Burnout Inventory. *Educ. Psychol. Meas.* **1981**, *41*, 1167–1174. [CrossRef]
30. Densten, I.L. Re-thinking burnout. *J. Organ. Behav. Int. J. Ind. Occup. Organ. Psychol. Behav.* **2001**, *22*, 833–847. [CrossRef]
31. Cronbach, L.J. Coefficient alpha and the internal structure of tests. *Psychometrika* **1951**, *16*, 297–334. [CrossRef]
32. Flora, D.B. Your coefficient alpha is probably wrong, but which coefficient omega is right? A tutorial on using R to obtain better reliability estimates. *Adv. Methods Pract. Psychol. Sci.* **2020**, *3*, 484–501. [CrossRef]
33. Gignac, G.E. On the inappropriateness of using items to calculate total scale score reliability via coefficient alpha for multidimensional scales. *Eur. J. Psychol. Assess.* **2014**, *30*, 130–139. [CrossRef]
34. Trizano-Hermosilla, I.; Alvarado, J.M. Best alternatives to Cronbach's Alpha reliability in realistic conditions: Congeneric and asymmetrical measurements. *Front. Psychol.* **2016**, *7*, 769. [CrossRef] [PubMed]
35. McDonald, R.P. *Test Theory: A Unified Treatment*; Psychology Press: London, UK, 1999.
36. Holzinger, K.J.; Swineford, F. The bi-factor method. *Psychometrika* **1937**, *2*, 41–54. [CrossRef]
37. Chen, F.F.; West, S.; Sousa, K. A comparison of bifactor and second-order models of quality of life. *Multivar. Behav. Res.* **2006**, *41*, 189–225. [CrossRef] [PubMed]
38. Reise, S.P. The rediscovery of bifactor measurement models. *Multivar. Behav. Res.* **2012**, *47*, 667–696. [CrossRef] [PubMed]
39. Mulaik, S.A. *Foundations of Factor Analysis*; CRC Press: Boca Raton, FL, USA, 2010.
40. Bollen, K.A. *Structural Equations with Latent Variables*; John Wiley & Sons: Hoboken, NJ, USA, 1989; pp. 1–514.
41. Morin, A.J.S.; Arens, A.K.; Marsh, H.W. A bifactor exploratory structural equation modeling framework for the identification of distinct sources of construct-relevant psychometric multidimensionality. *Struct. Equ. Model. A Multidiscip. J.* **2016**, *23*, 116–139. [CrossRef]
42. Ondé, D.; Alvarado, J.M. Contribución de los modelos factoriales confirmatorios a la evaluación de estructura interna desde la perspectiva de la validez. *Rev. Iberoam. De Diagnóstico Y Evaluación E Avaliação Psicológica* **2022**, *66*, 5. [CrossRef]
43. McDonald, R.P.; Ho, M.-H.R. Principles and practice in reporting structural equation analyses. *Psychol. Methods* **2002**, *7*, 64. [CrossRef]
44. Asparouhov, T.; Muthén, B. Exploratory structural equation modeling. *Struct. Equ. Model. A Multidiscip. J.* **2009**, *16*, 397–438. [CrossRef]

45. Marsh, H.W.; Morin, A.J.S.; Parker, P.D.; Kaur, G. Exploratory structural equation modeling: An integration of the best features of exploratory and confirmatory factor analysis. *Annu. Rev. Clin. Psychol.* **2014**, *10*, 85–110. [CrossRef] [PubMed]
46. Marsh, H.W.; Muthén, B.; Asparouhov, T.; Lüdtke, O.; Robitzsch, A.; Morin, A.J.S.; Trautwein, U. Exploratory structural equation modeling, integrating CFA and EFA: Application to students' evaluations of university teaching. *Struct. Equ. Model. A Multidiscip. J.* **2009**, *16*, 439–476. [CrossRef]
47. Kokkinos, C.M. Factor structure and psychometric properties of the Maslach Burnout Inventory-Educators Survey among elementary and secondary school teachers in Cyprus. *Stress Health* **2006**, *22*, 25–33. [CrossRef]
48. Marsh, H.W.; Lüdtke, O.; Muthén, B.; Asparouhov, T.; Morin, A.J.S.; Trautwein, U.; Nagengast, B. A new look at the big five factor structure through exploratory structural equation modeling. *Psychol. Assess.* **2010**, *22*, 471–491. [CrossRef]
49. Mészáros, V.; Ádám, S.; Szabó, M.; Szigeti, R.; Urbán, R. The Bifactor Model of the Maslach Burnout Inventory–Human Services Survey (MBI-HSS)—An Alternative Measurement Model of Burnout. *Stress Health* **2014**, *30*, 82–88. [CrossRef]
50. Doherty, A.S.; Mallett, J.; Leiter, M.P.; McFadden, P. Measuring burnout in social work. *Eur. J. Psychol. Assess.* **2021**, *37*, 6–14. [CrossRef]
51. Trógolo, M.A.; Morera, L.P.; Castellano, E.; Spontón, C.; Medrano, L.A. Work engagement and burnout: Real, redundant, or both? A further examination using a bifactor modelling approach. *Eur. J. Work. Organ. Psychol.* **2020**, *29*, 922–937. [CrossRef]
52. Hawrot, A.; Koniewski, M. Factor structure of the Maslach Burnout Inventory–educators survey in a Polish-speaking sample. *J. Career Assess.* **2018**, *26*, 515–530. [CrossRef]
53. Szigeti, R.; Balázs, N.; Bikfalvi, R.; Urbán, R. Burnout and depressive symptoms in teachers: Factor structure and construct validity of the Maslach Burnout inventory-educators survey among elementary and secondary school teachers in Hungary. *Stress Health* **2017**, *33*, 530–539. [CrossRef]
54. Bianchi, R. Do burnout and depressive symptoms form a single syndrome? Confirmatory factor analysis and exploratory structural equation modeling bifactor analysis. *J. Psychosom. Res.* **2020**, *131*, 109954. [CrossRef]
55. Schonfeld, I.S.; Verkuilen, J.; Bianchi, R. An exploratory structural equation modeling bi-factor analytic approach to uncovering what burnout, depression, and anxiety scales measure. *Psychol. Assess.* **2019**, *31*, 1073–1079. [CrossRef]
56. Verkuilen, J.; Bianchi, R.; Schonfeld, I.S.; Laurent, E. Burnout–depression overlap: Exploratory structural equation modeling bifactor analysis and network analysis. *Assessment* **2021**, *28*, 1583–1600. [CrossRef]
57. Schaufeli, W.B.; Martínez, I.M.; Pinto, A.M.; Salanova, M.; Bakker, A.B. Burnout and engagement in university students: A cross-national study. *J. Cross-Cult. Psychol.* **2002**, *33*, 464–481. [CrossRef]
58. Costa, P.T.; McRae, R.R. *Inventario de Personalidad Neo Revisado (NEO-PI-R), Inventario Neo Reducido de Cinco Factores (NEO-FFI)*; TEA Ediciones: Madrid, Spain, 2002.
59. Lozano, L.; García, E.; Lozano, I.M. *Cuestionario Educativo-Clínico: Ansiedad y Depresión*; TEA: Madrid, Spain, 2007; pp. 1–14.
60. Rosseel, Y. lavaan: An R package for structural equation modeling. *J. Stat. Softw.* **2012**, *48*, 1–36. [CrossRef]
61. Brown, T.A. *Confirmatory Factor Analysis for Applied Research*; Guilford Publications: New York, NY, USA, 2015; pp. 1–462.
62. Kline, R.B. *Principles and Practice of Structural Equation Modeling*; Guilford Publications: New York, NY, USA, 2015; pp. 1–534.
63. Hu, L.T.; Bentler, P.M. Cutoff criteria for fit indexes in covariance structure analysis: Conventional criteria versus new alternatives. *Struct. Equ. Model. A Multidiscip. J.* **1999**, *6*, 1–55. [CrossRef]
64. Browne, M.W.; Cudeck, R. Alternative ways of assessing model fit. In *Testing Structural Equation Models*; Bollen, K.A., Long, J.S., Eds.; Sage: Newbury Park, CA, USA, 1993; pp. 136–162.
65. Reise, S.P.; Scheines, R.; Widaman, K.F.; Haviland, M.G. Multidimensionality and structural coefficient bias in structural equation modeling: A bifactor perspective. *Educ. Psychol. Meas.* **2013**, *73*, 5–26. [CrossRef]
66. Rodriguez, A.; Reise, S.P.; Haviland, M.G. Applying bifactor statistical indices in the evaluation of psychological measures. *J. Personal. Assess.* **2016**, *98*, 223–237. [CrossRef]
67. Rodriguez, A.; Reise, S.P.; Haviland, M.G. Evaluating bifactor models: Calculating and interpreting statistical indices. *Psychol. Methods* **2016**, *21*, 137–150. [CrossRef]
68. Soper, D.S. A-Priori Sample Size for Structural Equation Models. Available online: https://www.danielsoper.com/statcalc/calculator.aspx?id=89 (accessed on 15 November 2022).
69. Nunnally, J.C.; Bernstein, I.H. *Psychometric Testing*, 2nd ed.; McGraw-Hill: New York, NY, USA, 1978; pp. 1–701.
70. Bakker, A.B.; Demerouti, E. Job demands–resources theory: Taking stock and looking forward. *J. Occup. Health Psychol.* **2017**, *22*, 273–285. [CrossRef]
71. Mao, K.; Liu, X.; Li, S.; Wang, X. Reliability analysis for mechanical parts considering hidden cost via modified quality loss model. *Qual. Reliab. Eng. Int.* **2021**, *37*, 1373–1395. [CrossRef]
72. Wang, Y.; Liu, X.; Wang, H.; Wang, X.; Wang, X. Improved fatigue failure model for reliability analysis of mechanical parts inducing stress spectrum. *Proceedings of the Institution of Mechanical Engineers, Part O. J. Risk Reliab.* **2021**, *235*, 973–981. [CrossRef]
73. Teng, D.; Feng, Y.-W.; Chen, J.-Y.; Lu, C. Structural dynamic reliability analysis: Review and prospects. *Int. J. Struct. Integr.* **2022**, *13*, 753–783. [CrossRef]

74. Wang, M.L.; Liu, X.; Wang, Y.S.; Luo, J. Reliability Analysis and Evaluation of Key Parts for Automobiles on the Basis of Dimensional Changes during High-Speed Operation. *J. Test. Eval.* **2015**, *43*, 1464–1471. [CrossRef]
75. Reise, S.P.; Bonifay, W.E.; Haviland, M.G. Scoring and modeling Psychological Measures in the Presence of Multidimensionality. *J. Personal. Assesment* **2012**, *95*, 129–140. [CrossRef]

Disclaimer/Publisher's Note: The statements, opinions and data contained in all publications are solely those of the individual author(s) and contributor(s) and not of MDPI and/or the editor(s). MDPI and/or the editor(s) disclaim responsibility for any injury to people or property resulting from any ideas, methods, instructions or products referred to in the content.

Article

On the Quality of Synthetic Generated Tabular Data

Erica Espinosa [1] and Alvaro Figueira [2,3,*]

[1] Department of Mathematics Engineering, Politecnico di Milano, 20133 Milan, Italy
[2] Faculty of Sciences, University of Porto, 4169-007 Porto, Portugal
[3] INESCTEC, 4200-465 Porto, Portugal
* Correspondence: arfiguei@fc.up.pt

Abstract: Class imbalance is a common issue while developing classification models. In order to tackle this problem, synthetic data have recently been developed to enhance the minority class. These artificially generated samples aim to bolster the representation of the minority class. However, evaluating the suitability of such generated data is crucial to ensure their alignment with the original data distribution. Utility measures come into play here to quantify how similar the distribution of the generated data is to the original one. For tabular data, there are various evaluation methods that assess different characteristics of the generated data. In this study, we collected utility measures and categorized them based on the type of analysis they performed. We then applied these measures to synthetic data generated from two well-known datasets, Adults Income, and Liar+. We also used five well-known generative models, Borderline SMOTE, DataSynthesizer, CTGAN, CopulaGAN, and REaLTabFormer, to generate the synthetic data and evaluated its quality using the utility measures. The measurements have proven to be informative, indicating that if one synthetic dataset is superior to another in terms of utility measures, it will be more effective as an augmentation for the minority class when performing classification tasks.

Keywords: utility measures; synthetic data; class imbalance; tabular data

MSC: 68T99

Citation: Espinosa, E.; Figueira, A. On the Quality of Synthetic Generated Tabular Data. *Mathematics* 2023, *11*, 3278. https://doi.org/10.3390/math11153278

Academic Editors: Jose Antonio Sáez Muñoz and José Luis Romero Béjar

Received: 18 June 2023
Revised: 21 July 2023
Accepted: 24 July 2023
Published: 26 July 2023

Copyright: © 2023 by the authors. Licensee MDPI, Basel, Switzerland. This article is an open access article distributed under the terms and conditions of the Creative Commons Attribution (CC BY) license (https://creativecommons.org/licenses/by/4.0/).

1. Introduction

Predictions made through classification models are nowadays used in many fields, but a common challenge that arises is class imbalance. This is a well-known problem in machine learning: it occurs when one or more classes in a dataset are significantly underrepresented compared to other classes. There are several problems associated with a class imbalance in machine learning as discussed in [1,2]. The algorithms trained on imbalanced data tend to be biased towards the majority class, which leads to poor performance in predicting the minority class. For example, in the field of medicine, rare diseases often lack sufficient data for analysis causing problems in decision-making [3], as well as identifying security bugs from a bug repository [4] or traffic accidents [5]. Fraud detection is also hindered by imbalanced datasets [6,7]. Additionally, fake news detection represents a common scenario of class imbalance, necessitating addressing this issue prior to algorithms development [8,9]. In recent years, to address this issue, synthetic data augmentation has been developed to increase the representation of the minority class. Some examples of this approach include the generation of synthetic images [10,11], time series [12], and tabular data [13,14]. The aim is to increase the amount of data available for the underrepresented class, thereby enhancing the performance of machine learning models in accurately predicting this class [15–17]. One way to upsample the minority class is through the generation of synthetic data, that is, data not from actual sampling, but created by generative models that attempt to emulate the same distribution as real data.

There are several ways to generate synthetic data. One common method to generate new samples is the SMOTE algorithm [18] (and variations of it), which generates new data by interpolating the original available ones. Other very common generative models are generative adversarial networks (GANs) [19], and variational autoencoders (VAEs) [20]. However, to ensure that these synthetic data are useful and do not only add noise to our real dataset, it is important to verify and evaluate whether they are representative of the real sample. Therefore, we need objective tools to compare the synthetic data to the real data and then evaluate the differences.

In this paper, we aim to explore the utility measures for assessing the quality of synthetic data generated from real datasets, particularly in the context of tabular data. Our research aims to investigate and analyze the utility measures that can effectively quantify the difference between real and synthetic tabular data. By examining the applicability of these measures in the context of real-world datasets, we aim to provide valuable insights into the assessment of synthetic tabular data quality. Finally, a crucial aspect of our research is to evaluate the effectiveness of data augmentation using synthetic data specifically in the context of tabular data classification tasks. By incorporating synthetic data into the training process, we aim to examine its impact on improving classification performance. Through this evaluation, we not only assess the performance of the classification models but also gain insights into the generative methods employed. The ability of these methods to accurately learn and capture the underlying distribution of the data is therefore intrinsically evaluated. We consider that this evaluation provides valuable information regarding the suitability and effectiveness of the generative methods in generating synthetic data that closely aligns with the characteristics and patterns of the original dataset. By thoroughly assessing the classification performance and considering the impact of data augmentation, we can shed light on the potential of synthetic data for improving classification accuracy and expanding the capabilities of machine learning models in handling tabular data. Moreover, the evaluation of different generative methods contributes to advancing the field of generative modeling, providing insights into the strengths and limitations of these methods in learning complex data distributions.

In Section 2, we present the statistical utility measures categorized based on their ability to capture different types of information. Section 3 examines how we can determine the usefulness of the generated data for our classification task. The datasets used in this study are described in Section 4. Section 5 provides a list and description of the generative methods employed to create the synthetic data used for augmentation. Lastly, Section 6 presents the results obtained in our analysis.

2. Statistical Utility Measures

To ensure that synthetic data accurately represents the characteristics of the real data, it is essential to evaluate the similarity between their distributions. Utility measures, also known as evaluation metrics, are used to assess the performance or effectiveness of a system, model, algorithm, or any other process. This type of evaluation comes in handy for our purpose: in order to assess the goodness or truthfulness of synthetic data we want to measure their similarity to real data in terms of distribution. Utility measures are divided into three categories based on how thoroughly they investigate the distributions. In *univariate* measures, the focus is on preserving the individual distributions of each column in the original data. This is achieved by comparing the similarity between the synthetic data and the original data column-by-column. The second category, known as *bivariate* extends the measure to consider the correlation between the columns being studied. Hence, it is a pairwise study. Finally, the third category, *multivariate*, examines the joint distribution of all the columns together, translating into a comparison made between two complete datasets.

In the following sections, we revisit some of these techniques. We use the following notation to refer to both the real and synthetic datasets: let $X = (X_1, \ldots, X_d)$ be a multivariate random variable with distribution F, where each component $X_{i=1,\ldots,d}$ represents one of the d columns of the real dataset. Similarly, let $Y = (Y_1, \ldots, Y_d)$ be a random variable with distribution G, representing the synthetic dataset. We denote the l-th row of the real dataset as x_l, which represents a realization of the random variable X. The entire real dataset can be viewed as a collection of such realizations and is denoted as $X = \{x_l\}_l$. The same applies to the synthetic dataset, which is indicated as $Y = \{y_m\}_m$.

2.1. Univariate Measures

To assess the similarity between the distributions of the synthetic and real data, it is necessary to examine their univariate distributions. Specifically, we need to measure the deviation in distribution between each pair of corresponding variables X_i and Y_i for $i = 1, \ldots, N$. This distance reflects the differences in distribution between the same columns of the two datasets. These measures are the easiest to measure and interpret, but nonetheless fundamental in the analysis of synthetic data because if there is already a large difference between the actual and synthetic data at this level, the similarity is unlikely to increase by considering more variables at a time. Univariate measures yield a distinct value for each variable, which can be further analyzed through visualization techniques such as boxplots.

One of the most widely used measures for assessing the univariate distance between two distributions is the Hellinger distance [21]. It is defined as follows: let P and Q denote two probability measures on a measure space $\mathcal{X}$, the Hellinger distance between P and Q is

$$H(P, Q) = \frac{1}{\sqrt{2}} \sqrt{\int_{\mathcal{X}} \left(\sqrt{P(dx)} - \sqrt{Q(dx)} \right)^2}. \tag{1}$$

Since we have only the realizations of the probability measures F and G, for each variable i we will adapt the distance to the discrete case: $F_i = (f_1, \ldots, f_k)$ and $G_i = (g_1, \ldots, g_k)$,

$$H(F_i, G_i) = \frac{1}{\sqrt{2}} \sqrt{\sum_{j=1}^{k} \left(\sqrt{f_j} - \sqrt{g_j} \right)^2}, \tag{2}$$

where F_i and G_i are the marginal distribution of the i-th component of X and Y, respectively, and f_j and g_j are the proportion of counts of instances for the j-th bin of the interval of values for each variable i.

Another widely used measure is the Kullback–Lieber divergence [22]. Specifically, the Kullback–Leibler divergence of Q from P denoted $D_{KL}(P\|Q)$, is the measure of information lost when Q is used to approximate P. In our discrete case, it is defined as:

$$D_{KL}(F_i\|G_i) = \sum_{j=1}^{k} f_j \log \frac{f_j}{g_j}. \tag{3}$$

When comparing two probability distributions using the Kullback–Liebler divergence, it is important to note that the supports of the distributions must match, otherwise the divergence may be infinite due to the presence of zero probabilities in the G_i distribution. When the supports do not match, it means that there are some values that have non-zero probability in one distribution but zero probability in the other.

To address this issue, the Jensen–Shannon divergence [23] was proposed, which is a symmetrized version of the Kullback–Liebler divergence that avoids this problem by smoothing out the differences between the supports of the two distributions:

$$D_{JS}(F_i\|G_i) = \sqrt{\frac{1}{2} \left[D_{KL}\left(F_i\|M_i\right) + D_{KL}\left(G_i\|M_i\right) \right]} \tag{4}$$

with

$$M_i = \frac{F_i + G_i}{2}.$$

Specifically, it is calculated as the square root of the average of the Kullback–Liebler divergence between each distribution and their average. This makes it a useful measure for comparing probability distributions, even when their supports do not match. It is always non-negative and ranges from 0 (when the distributions are identical) to 1 (when they have no common support).

2.2. Bivariate Measures

To ensure that the synthetic data accurately represent the relationships among the variables in the original dataset, we need to use bivariate utility measures. While univariate measures can be useful for evaluating the overall distribution of each variable separately, they do not capture the complex relationships that may exist between multiple variables. By using bivariate measures, we can verify that the same underlying relationships between variables are maintained in the synthetic data as well. To assess this, the pairwise correlation difference (PCD) [24] measures how much the correlation between real and synthetic data differs. The PCD is defined as:

$$PCD(X,Y) = \|Corr(X) - Corr(Y)\|_F, \tag{5}$$

where X, Y are the real and synthetic datasets, respectively. $\|\cdot\|_F$ is the Frobenius norm [25], which for an $m \times n$ matrix A is defined as:

$$\|A\|_F = \sqrt{\sum_{i=1}^{m}\sum_{j=1}^{n}|a_{ij}|^2},$$

where a_{ij} denotes the (i,j)-th entry of the matrix A. The choice of this norm as a measure for the PCD is motivated by its intuitive interpretation as a matrix size measure and its similarity to the L^2 norm for vectors, as it shares similar properties with the latter.

2.3. Multivariate Measures

While univariate and bivariate measures can provide valuable insights into individual variables and their pairwise relationships, they fail to fully capture the intricate multivariate relationships that exist within both the original and synthetic datasets. To ensure that the synthetic data accurately represents the higher-order dependencies and interactions among variables in the original dataset, we need to use multivariate measures. Multivariate measures are designed to assess the joint distribution of three or more variables, providing a deeper understanding of the underlying patterns and structures that may be present in the data. In particular, they can be used to quantify the difference between the multivariate distributions of two datasets, which in our case is $|F - G|$.

One way to do this is through the Kolmogorov–Smirnov test statistics that compare the empirical cumulative distribution functions of the real and synthetic cases:

$$D_m = \max_n |\mathcal{F}_X(z_n) - \mathcal{F}_Y(z_n)| \tag{6}$$

$$D_s = \frac{1}{N}\sum_n [\mathcal{F}_X(z_n) - \mathcal{F}_Y(z_n)]^2 \tag{7}$$

where $\mathcal{F}_X$ and $\mathcal{F}_Y$ are, respectively, the empirical cumulative density functions for X and Y, z_n is a d dimensional vector representing a possible instance for the two random variables, and N is their total number. The deviation of D_m and D_s from zero represents the distance between the two distributions, the former in terms of the maximum absolute difference capturing the largest deviation, the latter in terms of mean squared difference which quantifies the overall distance between the distributions. Unfortunately, this measure encounters two primary challenges. Firstly, the curse of the dimensionality problem undermines its accuracy when applied to high-dimensional data, as estimating multivariate distributions becomes increasingly complex. Secondly, the test's sensitivity to changes in the tail of the cumulative distribution may hinder its ability to detect significant differences in other regions of the distribution [26].

The propensity score [27] evaluates the distance between the two datasets through a classification model. The original and synthetic datasets are assigned distinct labels to enable differentiation, subsequently combined, and provided as input to a classifier. The propensity score measures how well the classifier is able to distinguish the two types of data and it is defined as follows:

$$pMSE = \frac{1}{N}\sum_{n}(\hat{p}_n - 0.5)^2 \qquad (8)$$

where p_n is the probability of belonging to the synthetic class assigned by the classifier to each instance. The less the classifier is able to distinguish between the two classes (thus returning p_n values of roughly 0.5) the closer the propensity score is to zero. In their study, Snoke et al. [27] proposed the standardized pMSE (St pMSE), which quantifies the difference between the expected and observed values in terms of the estimated null standard deviation. Higher values of this measure indicate a poorer fit between the data and the underlying synthesis, while values closer to zero indicate a better fit. Rescaling by the null statistic provides a more intuitive measure of its utility, particularly when applied to synthetic data. Notably, this rescaling renders the measure independent of sample size, allowing for easier comparisons across datasets.

Another useful criterion for evaluating the quality of synthetic data is the order of variable relevance, which can be assessed by constructing decision trees on both real and synthetic datasets and comparing the resulting variable orders. Different methods can be employed to measure the distance between the two sequences, such as focusing on the first three variables or considering the majority of the order.

3. Application-Specific Measures: Classification

When generating synthetic data to perform specific tasks, it is recommended to evaluate their quality by comparing the performance of real and synthetic data to complete these jobs. In this study, our objective is to examine the utility of synthetic data in enhancing the performance of a classification model when dealing with class imbalance. To achieve this, we augment the minority class by introducing synthetic data into the training set. Subsequently, we compare the model's performance on the test set with the performance of a model trained exclusively on the original data. The higher the classifier's performance, the more the generated data are similar to the minor class and hence more useful to the classifier.

4. The Datasets

In order to assess the effectiveness of generative models in creating realistic synthetic samples, we evaluate the performance of different utility measures on two real-world application datasets: the Adults Income dataset (http://archive.ics.uci.edu/dataset/2/adult, accessed on 17 March 2023) [28] and the Liar+ dataset (https://www.cs.ucsb.edu/~william/data/liar_dataset.zip, accessed on 17 March 2023) [29]. The Adults Income dataset was chosen for its widespread use in the literature [30–32] and its relevance to socioeconomic factors that influence income. The Liar+ dataset is a well-established dataset that finds extensive usage in the domains of natural language processing and machine learning [33–35]. We selected it due to its different nature with respect to the Adults dataset and because it represents a typical example of class imbalance in the real world. Moreover, following the preprocessing steps outlined in the forthcoming sections, the dataset will feature a substantial number of covariates, presenting a challenge in generating synthetic data.

The Adult Income dataset, also known as the Census Income dataset, is a widely used dataset in machine learning and data analysis. It contains demographic and employment-related information on 48,842 individuals from the 1994 US Census database. The dataset consists of 13 attributes (six continuous and seven categorical) including the individual's age, education level, occupation, marital status, relationship, race, gender, and native country. The target variable of the dataset is the income bracket, divided into two classes: "$\leq$50 K" with 41,001 samples, and ">50 K" with 7841 cases, indicating whether an individual's income is below or above 50,000 USD per year. To reduce the complexity of the categorical variables and ensure the feasibility of the analysis, we transformed them into binary variables using the following criteria:

- For variables "race" and "native country" there was one class representing more than 85% of the cases, therefore we selected label 1 for the predominant class and 0 for all the others.
- We transformed the other categorical variables into binary variables by grouping cases that shared similar conceptual characteristics.

For example, if the original variable was "native country" and the most common category was "USA", then the new binary variable would have a value of 1 for "USA" and 0 for all other categories. The dataset was partitioned into training and test sets to enable the evaluation of classifiers based on application-specific measures. The data were randomly partitioned in a manner that allocated 80% of the samples to the training set and 20% to the test set while ensuring that the original balance between the classes was maintained. The training set consisted of 32,801 samples in Class 0 and 6273 samples in Class 1, while the test set comprised 8200 samples in Class 0 and 1568 samples in Class 1.

The Liar+ dataset is a dataset of statements made by politicians that have been labeled as "true", "mostly true", "half true", "barely true", "false", and "pants on fire". The dataset utilized in this study underwent the following preprocessing steps: only samples belonging to the "pants-fire", "false", "mostly true", and "true" classes were retained, with the first two classes merged and labeled as 1, and the latter two labeled as 0. To address the class imbalance observed in the real world, the class labeled as 1 was downsampled to a third of the size of the class labeled as 0, resulting in 4087 instances in class 0 and 1362 instances in Class 1. The remaining data processing followed the methodology outlined by Bruno Vaz in his thesis [36] which is the following: categorical variables such as "speaker", "job", "state", "party", and "subject$_i$" were subjected to target encoding, while the variables "statement", "context", and "justification", which contained essential information for predicting the target variable, underwent processing using the Doc2Vec algorithm. After preprocessing, the dataset consisted of 50 features and one target variable. Finally, the dataset was split into training and test sets with an 80/20% ratio, maintaining the proportional distribution of classes in each set. The resulting training set was composed of 3615 samples in Class 0 and 1216 samples in Class 1, while the test set consisted of 904 samples in Class 0 and 304 samples in Class 1.

5. Synthetic Data Generative Methods

We utilized several techniques for data synthesis, each serving a specific purpose. These included Borderline SMOTE [37], which is an important variation of the widely used upsampling technique SMOTE. We also employed DataSynthesizer [38] as one of the simplest of the models that attempt to sample from the distribution of real data. Additionally, CTGAN [39], a leading GAN model for tabular data, was utilized. We also explored CopulaGAN [40], as an improvement upon the CTGAN approach. Finally, we incorporated REaLTabFormer [41], which introduces a novel approach to tabular data synthesis by leveraging large language models and transformer architectures.

5.1. Description of the Generative Methods

Borderline SMOTE is a variant of the more traditional SMOTE (synthetic minority over-sampling technique) [18] algorithm that generates new data through the interpolation of existing data. It is characterized by interpolating only the minority class data that are close to the decision boundaries, allowing the algorithm to focus on examples that are more difficult to correctly classify. The approach is as follows: the nearest neighbors algorithm first identifies instances close to the decision boundaries, called borderline instances, and then the synthetic samples are generated by randomly selecting a borderline instance and interpolating between it and one or more of its k-nearest neighbors, which are also minority class instances. The main advantage of Borderline SMOTE over the original SMOTE algorithm is that it can help to reduce the problem of overfitting, which can occur when synthetic samples are generated from all minority class instances, including those that are easy to classify.

DataSynthesizer is a system that generates synthetic datasets based on private input datasets. It consists of two main modules such as DataDescriber, and DataGenerator. DataDescriber infers attribute types and domains, supporting numeric, categorical, and datetime data. It calculates value frequency distributions for categorical attributes and equi-width histograms for numeric and datetime attributes. The module handles missing values and incorporates differential privacy by adding Laplace noise. DataGenerator takes the dataset description generated by DataDescriber and generates synthetic data by sampling from the specified distributions. It offers three modes: random mode, independent attribute mode, and correlated attribute mode. In the random mode, values are generated randomly based on attribute types. In the independent attribute mode, sampling is performed from bar charts or histograms using uniform sampling. Finally, in the correlated attribute mode, values are sampled in the appropriate order from a Bayesian network. In our study, we utilize the correlated attribute mode to incorporate intra-feature relationships into consideration. This probabilistic model is constructed using a Bayesian network that captures the correlation structure between attributes. Once the Bayesian network is obtained, it determines the order in which attribute values are sampled, resulting in the generation of synthetic data.

The CTGAN (conditional tabular generative adversarial network) model is a GAN-based approach designed to model the distribution of tabular data and generate synthetic rows that follow the same distribution. The popularity of this model in the world of synthetic tabular data stems from its ability to tackle issues related to non-Gaussian and multimodal distributions, as well as imbalanced discrete columns. Each continuous variable is remodeled using a variational Gaussian mixture (VGM) model, where the number of modes in the distribution is estimated. For each instance, the nearest mode is indicated using a one-hot vector, and a scale parameter is determined. This allows for flexible representation of continuous values within their respective modes. The CTGAN model incorporates a conditional generator and training-by-sampling to address imbalanced discrete columns. The conditional generator enables the generation of synthetic rows conditioned on a specific discrete column value. The generator is penalized during training to produce an exact copy of the given condition, ensuring that the generated rows preserve the specified condition. Training-by-sampling ensures that all categories from discrete attributes are sampled evenly (but not necessarily uniformly) during the training process, facilitating the exploration of all possible discrete values. The network structure of CTGAN consists of fully connected layers in both the generator and discriminator. Overall, CTGAN provides a solution for modeling tabular data distributions, handling complex distributions, imbalanced discrete columns, and generating synthetic rows that resemble the original data distribution.

The synthetic data vault CopulaGAN model can be seen as an elaboration of the CTGAN model. It employs a two-stage approach to generate synthetic data while preserving the statistical properties of the original dataset. In the first stage, known as statistical learning, the synthesizer focuses on learning the marginal distributions of the columns in the real dataset. This involves understanding the shape and characteristics of each individual column's distribution. For instance, it might identify a column as having a Beta distribution with parameters $\alpha = 2$ and $\beta = 5$. The synthesizer normalizes the values to a Gaussian distribution using the learned distributions. In the second stage, the synthesizer employs CTGAN to train the normalized data. By combining statistical learning with GAN-based modeling, the SDV CopulaGAN synthesizer generates synthetic data that not only captures the individual column characteristics but also maintains the complex dependencies and structure present in the original dataset.

In conclusion, REaLTabFormer introduces an innovative approach to generating realistic tabular data as it is a transformer-based framework designed for generating non-relational and relational tabular data. In this study, we focus only on the generation of non-relational data. Therefore, it treats each observation as a sequence and learns the conditional distribution of columnar values in each row. This allows it to generate the next values in the sequence, effectively creating realistic non-relational data. The underlying architecture used for this purpose is GPT-2, a transformer-based autoregressive model. GPT-2 is known for its ability to capture the conditional distribution of sequential data effectively. To encode the tabular data efficiently, a fixed-set vocabulary is adopted for each column. This means that a predetermined set of possible values is defined for each column, and the model uses these values during the generation process. To address overfitting and improve the quality of the generated samples, REaLTabFormer incorporates target masking during training. Target masking involves replacing the target or label tokens with a special mask token. Hence, it forces the model to learn the masks instead of the actual values, encouraging it to generalize and generate more diverse samples. During the generation process, the model fills the masked values with probabilistically valid tokens. By leveraging the learned distribution, the model selects appropriate values from the predefined vocabulary.

This approach ensures that the generated non-relational data adhere to the patterns and distributions observed in the training data. In summary, the REaLTabFormer model uses an autoregressive approach and the GPT-2 architecture to generate realistic non-relational tabular data. By treating each observation as a sequence and learning the conditional distribution of columnar values, the model can accurately generate the next values in the sequence. The use of a fixed-set vocabulary, target masking during training, and probabilistic sampling further enhance the quality and diversity of the generated non-relational data.

5.2. Experiments on the Generative Methods

We recall that the goal of this process is to create new data using these models so that it is possible to add synthetic data to the minority class and improve the performance of the classifiers. To avoid overfitting and better evaluate the performance of our classification models, we trained the oversampling algorithms only on samples belonging the Class 1 of the training set. This allowed us to more accurately assess the generalization performance of our models and determine their ability to handle imbalanced data.

The Borderline SMOTE algorithm automatically recognizes the class with fewer samples and generates a total of synthetic samples so that the cardinality in both classes is the same. The first target class in the Adults Income dataset has 32,800 samples, while the minor class has 6273 samples; thus, the method generated 26,527 synthetic data. Instead, the difference between the two classes in the Liar+ dataset was 2399; therefore, these many samples were generated. The other models generated a different amount of samples each time, resulting in distinct samples in each run. We created a series of 1000, 5000, 10,000, and 20,000 samples for the Adults Income dataset and 500, 1000, 1500, and 2000 samples for the Liar+ dataset. The intrinsic motivation is to assess the effect of sample size on utility measures while ensuring that the sample size does exceed the number of samples in the majority class.

6. Results and Discussion

The analysis is categorized based on the used dataset. Initially, we examined the Adults Income dataset, followed by the Liar+ dataset. We assess the synthetic data using utility measures and subsequently evaluate its classification performance.

The synthesis algorithms are outlined in the following study on the Adults Income dataset: from this point forward when we write SMOTE, we mean Borderline SMOTE and

DS_{1k}, DS_{5k}, DS_{10k}, DS_{20k} stand for, respectively, the datasets made by 1000, 5000, 10,000, and 20,000 samples from the DataSynthesizer model. The same follows for the datasets generated by CTGAN, CopulaGAN, and REaLTabFormer, the last labeled as RTF. We begin by analyzing the quality of the generated data with the univariate measures, that is, we want to verify the similarity, column by column, of the generated dataset compared to the real dataset. Because univariate measures return a value for each component of the random variable that represents our dataset, i.e., our columns, we opted to represent them using boxplots. The more the values are grouped to values close to zero, the better the similarity between the real dataset and the generated one. The boxplots of the Jensen–Shannon Measures for the Adults Income dataset are shown in Figure 1.

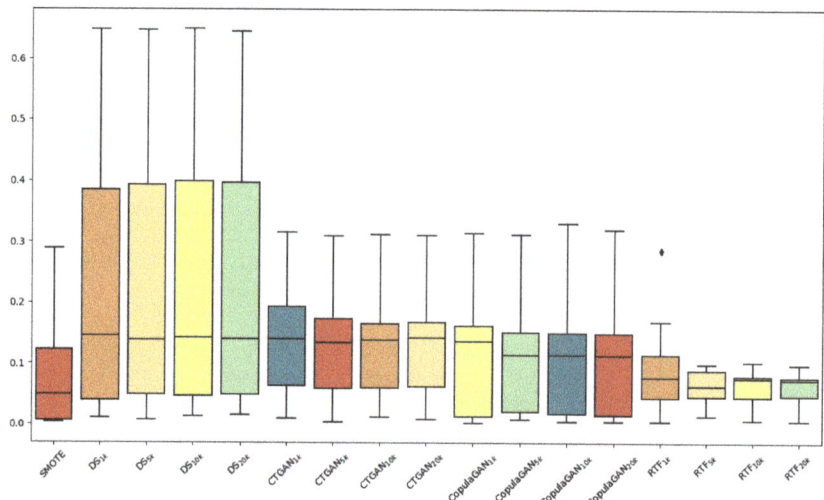

Figure 1. Boxplots of the Jensen–Shannon measures grouped by each synthesis method for the Adults Income dataset. On the y-axis there are the values of the distance measured on each column of the data frame.

The scores show little sensitivity to the sample size, suggesting that a relatively small sample of 5000 points already provides a robust representation of the variable distribution for generative models. Notably, among the models examined, REaLTabFormer shows a marginal improvement in performance. Among the methods compared, DataSynthesizer had higher median and variability scores, demonstrating a limited ability to capture the variables' distribution. REaLTabFormer and Borderline SMOTE achieved better results as the former had lower variability and the latter had lower median scores. CopulaGAN had lower median scores compared to the CTGAN but showed greater variability in scores. Table 1 presents the pairwise correlation difference (PCD) values in bivariate analysis, which evaluates the generative methods' capability to preserve the same interaction between the components as observed in the real dataset.

Table 1. Pairwise correlation difference in the Adults Income dataset for each synthesis method used.

	PCD		PCD
SMOTE	0.3387		
DS_{1k}	1.2528	$CopulaGAN_{1k}$	1.2812
DS_{5k}	1.1521	$CopulaGAN_{5k}$	1.1126
DS_{10k}	1.1113	$CopulaGAN_{10k}$	1.1325
DS_{20k}	1.1407	$CopulaGAN_{20k}$	1.1472
$CTGAN_{1k}$	1.3441	RTF_{1k}	0.3787
$CTGAN_{5k}$	1.3120	RTF_{5k}	0.2894
$CTGAN_{10k}$	1.3447	RTF_{10k}	0.2404
$CTGAN_{20k}$	1.3340	RTF_{20k}	0.2584

The value for the synthetic data reduces slightly as the sample size increases. This may be due to the fact that as datasets increase in size, the relationships among variables are more represented. Yet, in terms of maintaining interactions between variables, the Borderline SMOTE and REaLTabFormer methods appear to perform better.

To proceed to multivariate analysis, the propensity score was computed using a CART (otherwise known as decision tree) model in order to calculate the distinguishability between the original and synthetic datasets.

Table 2 shows the obtained values.

Table 2. Propensity score and standardized propensity score for the synthetic data generated from the Adults Income dataset.

	pMSE	St pMSE		pMSE	St pMSE
SMOTE	0.0433	Inf			
DS_{1k}	0.1186	351.4	$CopulaGAN_{1k}$	0.1102	597.1
DS_{5k}	0.2468	122.2	$CopulaGAN_{5k}$	0.2287	88.91
DS_{10k}	0.2369	3087.0	$CopulaGAN_{10k}$	0.2232	3524.1
DS_{20k}	0.1818	Inf	$CopulaGAN_{20k}$	0.1724	Inf
$CTGAN_{1k}$	0.2311	94.6	RTF_{1k}	0.0988	742.3
$CTGAN_{5k}$	0.2468	108.8	RTF_{5k}	0.2127	137.5
$CTGAN_{10k}$	0.2252	2203.0	RTF_{10k}	0.2080	5946.3
$CTGAN_{20k}$	0.1735	Inf	RTF_{20k}	0.1588	10,281.6

The closer the pMSE and Standardized pMSE are to 0, the better the indistinguishability between the two datasets under consideration. Because the null distribution of the pMSE is not theoretically derivable in this manner, it is estimated through resampling (details in [27]). After examining the scores of the models that were sampled with varying numbers of samples, it appears that pMSE follows a particular trend. In most cases, its value is the smallest when the sample size is low, increases to the maximum at 5000 samples, and then decreases again with 20,000 samples. In contrast, St pMSE exhibits an opposite trend to that of pMSE and even reaches infinite values with the largest sample size. In [27], the authors describe this indicator as unstable in the case of high sample sizes. Alternatively, this trend may be interpreted as follows: since pMSE employs a classifier to evaluate the quality of synthetic data, it is affected by class imbalance. With 6273 real samples, the point where the number of real and synthetic samples is most similar is when 5000 are generated by the models. Therefore, pMSE may be more significant in this case, as it scores lower for every model. A comparison of the performance of the 26,527 data generated by Borderline SMOTE with the other cases that generated 5000 samples is not meaningful in this context. Therefore, among the remaining models, the most suitable one for replicating the distribution of the real dataset is REaLTabFormer. And, since St pMSE is not affected by sample size, it can determine the optimal value in this case.

Finally, considering the application of the dataset for binary classification modeling, we evaluated and compared the performance of various algorithms—decision trees, logistic regression, random forest, and XGBoost. This comparison was made between the results from the datasets augmented with synthetic data and the baseline performances to observe any potential improvements or differences. The more the performances improve, the more the generated data help the minority class by increasing its representation. This means that the generated data comes from the same distribution as the minority class. Table 3 displays the F1 scores for Class 1 across all classifiers and synthesis methods computed on the test set. The classification models have been fitted on all the samples from Class 0 and on the real samples of Class 1, augmented with 20,000 synthetic instances to mitigate class imbalance.

Decision trees are found to be quite insensitive to the addition of new samples for Class 1, regardless of how they were generated as the F1 score improves only slightly with the augmentation. Logistic regression, on the other hand, which is a model that is very sensitive to class imbalance, shows significant performance improvement with class augmenting, especially if the data were generated using Borderline SMOTE, as the F1 score from a baseline value of 18.24 reached a value of 76.02. Figure 2 illustrates how the performance of the classifiers changes with varying numbers of added synthetic samples.

Table 3. F1 score in percentage on Class 1 for the classification on the Adults Income dataset on the test set in the case of Class 1 augmentation with 20,000 synthetic data.

	Decision Tree	Logistic Regression	Random Forest	XGBoost
Baseline	40.14	18.24	35.29	34.57
Borderline SMOTE	41.31	**76.02**	**50.36**	**49.55**
DataSyntesizer	41.11	35.59	34.67	34.50
CTGAN	41.05	61.47	37.50	34.24
CopulaGAN	**41.79**	67.47	38.24	33.86
REaLTabFormer	40.56	66.20	42.69	39.16

As previously seen, the decision tree model does not appear to be sensitive to data augmentation while the other classification models show an increase in accuracy, precision, and F1 score as the number of samples increases. There is also a decrease in the recall, potentially indicating a reduction in model bias. Also here, logistic regression is shown to be very sensitive to data augmentation, and the performances increase as the number of samples added increases.

We computed the same measures for the Liar+ dataset. Nevertheless, in this situation, $DS_{0.5k}$, DS_{1k}, $DS_{1.5k}$, and DS_{2k} correspond to samples with a cardinality of 500, 1000, 1500, and 2000, respectively, and the same goes with the other generative models. Figure 3 depicts the boxplots of the univariate Jensen–Shannon measures. In this scenario, the Borderline SMOTE algorithm and the REaLTabFormer outperform the other models showing a median that is almost half of all the other cases and very low variability. Again, the DataSynthetsizer model turns out to have poor emulation capabilities of univariate distributions.

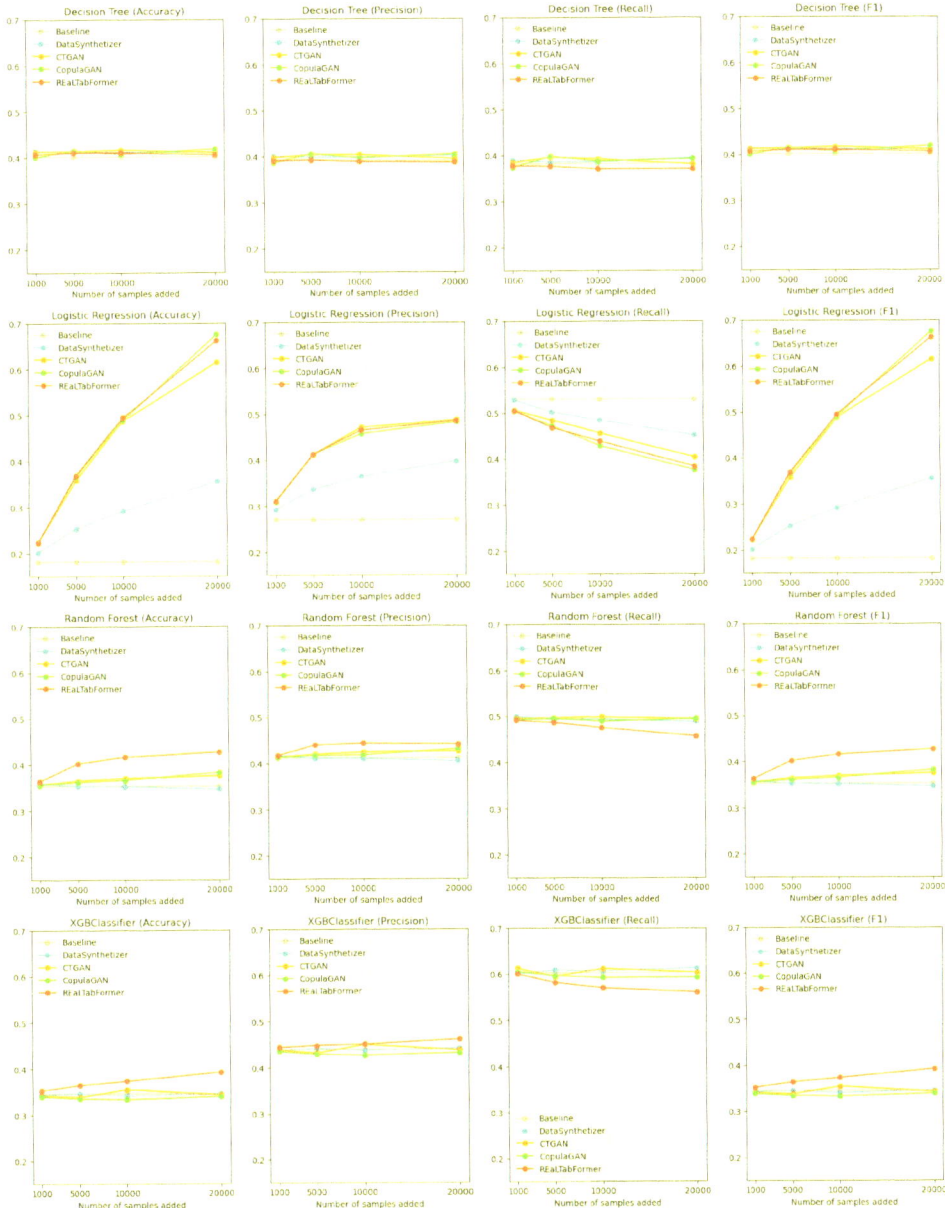

Figure 2. Comparison of the four classification model performance metrics on Class 1 on the test set for each synthesis method used for data augmentation for Adults Income dataset. On the y-axis there is the value of each score which can go from 0 to a maximum of 1.

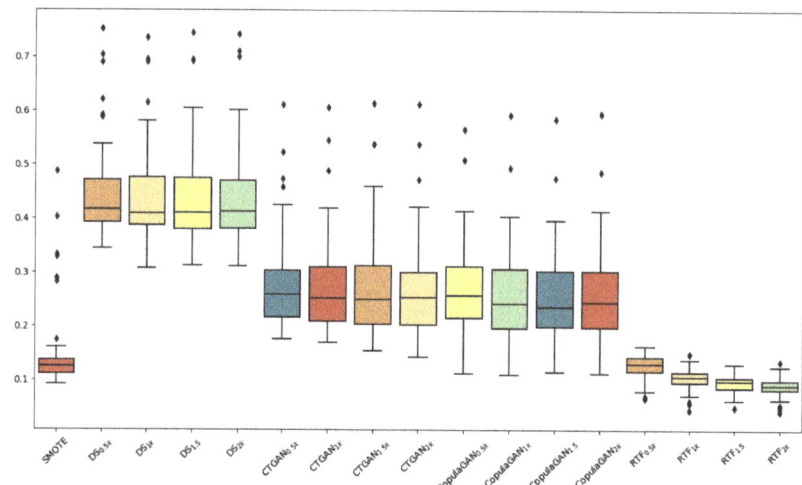

Figure 3. Boxplots of the Jensen–Shannon measures for each synthesis method for the Liar+ dataset. On the y-axis we have the measure computed on each variable of the dataframe.

The PCD follows the same pattern as the Adults Income dataset (Table 4). As the number of samples increases, the relationships among variables are better captured. Borderline SMOTE and REaLTabFormer produce the best-performing synthetic data by demonstrating the ability to capture and then represent relationships component by component. Propensity scores are collected in Table 5. Similar to the Adult Income dataset, the pMSE is highest when the generated sample size is similar to the original sample size (1216), which in this case is 1000. The values exhibit minimal fluctuations based on the generation model used, with the exception of Borderline SMOTE, which outperforms the others. In this scenario, St pMSE values remain consistent within a narrow range regardless of the volume of the synthetic data, unlike the previous case with the Adult Income dataset. This highlights the score's independence from the sample size. Furthermore, since Borderline SMOTE generated 2399 samples, we compare its score with that of the other datasets generated with 2000 samples. It emerges as the model that yields superior values. In terms of classification models, we focused on analyzing the classification models' ability to identify Class 1 data, which is more of our interest as it is the minority class. The F1 scores for Class 1 are presented in Table 6.

Table 4. Pairwise correlation difference in the Liar+ dataset for each synthesis method used.

	PCD		PCD
SMOTE	1.6030		
$DS_{0.5k}$	4.3507	$CopulaGAN_{0.5k}$	5.4978
DS_{1k}	4.0252	$CopulaGAN_{1k}$	5.3046
$DS_{1.5k}$	3.8881	$CopulaGAN_{1.5k}$	5.0944
DS_{2k}	3.8144	$CopulaGAN_{2k}$	5.0632
$CTGAN_{0.5k}$	5.5091	$RTF_{0.5k}$	2.9800
$CTGAN_{1k}$	5.1432	RTF_{1k}	2.6132
$CTGAN_{1.5k}$	5.0166	$RTF_{1.5k}$	2.3818
$CTGAN_{2k}$	5.0557	RTF_{2k}	2.3149

Table 5. Propensity score and standardized propensity score for the synthetic data generated from the Liar+ dataset.

	pMSE	St pMSE		pMSE	St pMSE
SMOTE	0.1654	3.2076			
$DS_{0.5k}$	0.1992	3.5757	$CopulaGAN_{0.5k}$	0.2041	3.6839
DS_{1k}	0.2404	3.5669	$CopulaGAN_{1k}$	0.2458	3.6514
$DS_{1.5k}$	0.2391	3.5069	$CopulaGAN_{1.5k}$	0.2462	3.6545
DS_{2k}	0.2294	3.6406	$CopulaGAN_{2k}$	0.2346	3.7681
$CTGAN_{0.5k}$	0.2034	3.6841	$RTF_{0.5k}$	0.2041	3.7280
$CTGAN_{1k}$	0.2434	3.6127	RTF_{1k}	0.2456	3.6599
$CTGAN_{1.5k}$	0.2439	3.6097	$RTF_{1.5k}$	0.2452	3.6772
$CTGAN_{2k}$	0.2315	3.7486	RTF_{2k}	0.2348	3.8048

Table 6. F1 scores in percentage of the classification on the Liar+ dataset computed on Class 1 of the test set for every classification model trained on Class 1 augmented with 2000 synthetic samples.

	Decision Tree	Logistic Regression	Random Forest	XGBoost
Baseline	54.28	25.32	49.23	53.94
Borderline SMOTE	52.85	**72.04**	**62.94**	**61.51**
DataSyntesizer	51.31	40.46	51.42	55.92
CTGAN	53.17	31.58	52.74	56.91
CopulaGAN	53.84	44.41	53.51	53.94
REaLTabFormer	**55.04**	67.11	50.55	54.61

The decision tree (Figure 4, row 1), random forest, and XGBoost exhibit superior performance only when the augmented data are derived from Borderline SMOTE. Conversely, when using data generated by other generative models, these three classifiers do not display any improvement, as we can see in the first, third, and fourth rows of Figure 4, which are the scores corresponding to the decision trees, random forest, and XGBoost models. These rows present the accuracy, precision, recall, and F1 scores from left to right, respectively. Notably, these scores remain unchanged from the baseline (gray line) and show no variation with the number of samples added. This phenomenon can be attributed to the Liar+ dataset's high dimensionality (50 features), making it more challenging to classify accurately. However, logistic regression stands out as the only model consistently showing improvements across all performance scores when class augmentation is applied, as evident from the second line in the figure. Due to its simplicity and heightened sensitivity to class imbalance, logistic regression displays a robust reliance on the number of added samples, leading to enhanced performance across all generative models.

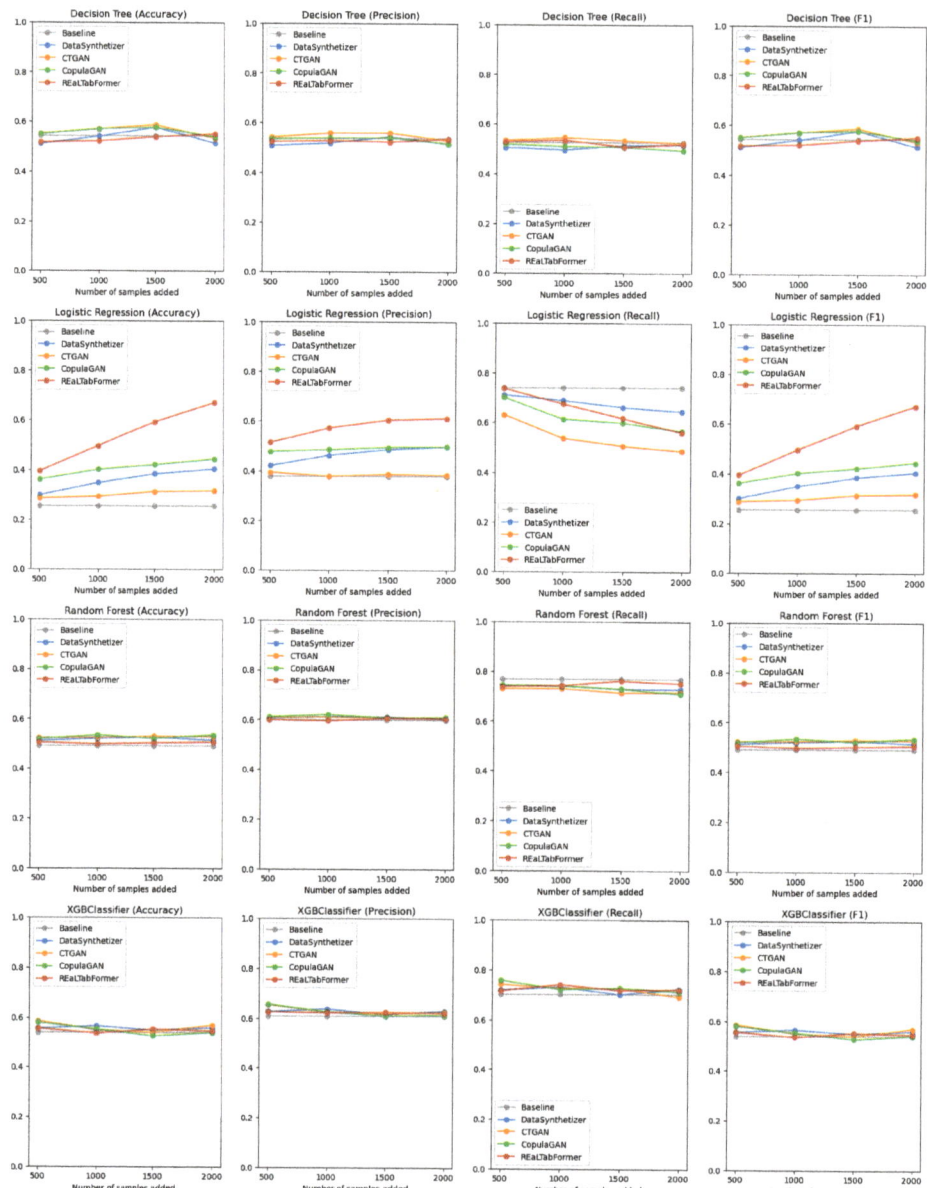

Figure 4. Performance scores on Class 1 of the classifier models on the test set of the Liar+ dataset with varying numbers of synthetic samples from different generation models. On the y-axis, there is the value of each score which can go from 0 to a maximum of 1.

In summary, for both datasets, Borderline SMOTE and REaLTabFormer stand out as generative models that exhibit superior performance across various utility measures. Notably, Borderline SMOTE consistently outperforms other models across all evaluated measures, in particular, in terms of PCD (Tables 1 and 4) and pMSE (Tables 2 and 5). The superiority of these models is also evident in the classification results, as all classification models perform better when trained on data generated by Borderline SMOTE and REaLTabFormer (Table 6). Based on the obtained results, it can be concluded that the utility measures demonstrate a general consistency among themselves. If a generative model excels in one measure, it is likely to excel in other measures as well. Analyzing the data generated by DataSynthesizer, we find that the Jensen–Shannon measure (Figures 1 and 3) indicates a poor univariate match between the synthetic data and the real data as the scores are very high compared to other models. However, the pairwise correlation difference (Tables 1 and 4) and pMSE (Tables 2 and 5) suggest that the bivariate and multivariate distributions are as good as the ones generated by the other generative models. Notably, when it comes to classification tasks, augmenting the data with DataSynthesizer-generated samples leads to relatively less pronounced performance improvement compared to other augmentation techniques, as evident from Figures 2 and 4. The blue line representing DataSynthesizer is consistently positioned below all the other lines in these figures. This indicates that if the univariate distribution of synthetic data diverges significantly from the real data, the other measures related to bivariate and multivariate distributions may have limited relevance in evaluating the quality of the synthetic data. CTGAN and CopulaGAN, with the latter being an improvement over the former, demonstrate comparable performance across most measures, except for a slight advantage of the latter in the classification task. In Figures 2 and 4, the two models correspond to the yellow and green lines, respectively. It can be seen that the green line stands above the yellow line for accuracy, precision, and F1, while below it for recall, which in this case means less bias in the model. REaLTabFormer and Borderline SMOTE emerge as the top-performing generative models across all evaluation criteria, notably surpassing other models in terms of pairwise correlation difference. Overall, the high quality of the generated data translates into superior classification performance.

7. Conclusions

In this study, we have examined various utility measures for synthetic tabular data, exploring their properties and the aspects they can evaluate. Furthermore, we assessed their applicability in the context of classification tasks with imbalanced classes.

To investigate these measures, we employed five generative algorithms: Borderline SMOTE, DataSynthesizer, CTGAN, CopulaGAN, and ReaLTabFormer. These algorithms were evaluated on two distinct datasets: Adults Income and Liar+. Our study reveals that Borderline SMOTE and REaLTabFormer exhibit superior performance among the generative models investigated. Borderline SMOTE adopts a modified approach to SMOTE, generating samples by interpolating borderline data points from the opposing class, while REaLTabFormer utilizes the GPT-2 generative model. The utility measures demonstrate consistency as the dimensionality of the compared distributions increases. Indices that perform well for univariate measures also tend to excel in bivariate and multivariate measures. However, it should be noted that good performance in bivariate and multivariate cases does not necessarily guarantee the same for univariate distributions. Importantly, these measures align with the classification results obtained after augmenting the minority class samples.

In conclusion, the utility measures analyzed in this study provide valuable insights into the quality of generated synthetic data and can serve as informative tools for data analysis. Future research directions may involve extending our findings by incorporating these measures, for example, in the training process of generative models such as CTGAN, with the aim of enhancing their performance.

Author Contributions: Conceptualization, A.F.; Software, E.E.; Validation, A.F.; Investigation, E.E.; Resources, A.F.; Writing—original draft, E.E.; Writing—review & editing, A.F.; Visualization, E.E.; Supervision, A.F. All authors have read and agreed to the published version of the manuscript.

Funding: This research received no external funding.

Data Availability Statement: No new data were created or analyzed in this study. Data sharing is not applicable to this article.

Conflicts of Interest: The authors declare no conflict of interest.

References

1. Kaur, H.; Pannu, H.S.; Malhi, A.K. A systematic review on imbalanced data challenges in machine learning: Applications and solutions. *ACM Comput. Surv.* **2019**, *52*, 1–36. [CrossRef]
2. Krawczyk, B. Learning from imbalanced data: Open challenges and future directions. *Prog. Artif. Intell.* **2016**, *5*, 221–232. [CrossRef]
3. Weng, W.H.; Deaton, J.; Natarajan, V.; Elsayed, G.F.; Liu, Y. Addressing the real-world class imbalance problem in dermatology. In Proceedings of the Machine Learning for Health, PMLR, Durham, NC, USA, 7–8 August 2020; pp. 415–429.
4. Zheng, W.; Xun, Y.; Wu, X.; Deng, Z.; Chen, X.; Sui, Y. A comparative study of class rebalancing methods for security bug report classification. *IEEE Trans. Reliab.* **2021**, *70*, 1658–1670. [CrossRef]
5. Rivera, G.; Florencia, R.; García, V.; Ruiz, A.; Sánchez-Solís, J.P. News classification for identifying traffic incident points in a Spanish-speaking country: A real-world case study of class imbalance learning. *Appl. Sci.* **2020**, *10*, 6253. [CrossRef]
6. Isangediok, M.; Gajamannage, K. Fraud Detection Using Optimized Machine Learning Tools Under Imbalance Classes. *arXiv* **2022**, arXiv:2209.01642.
7. Varmedja, D.; Karanovic, M.; Sladojevic, S.; Arsenovic, M.; Anderla, A. Credit card fraud detection-machine learning methods. In Proceedings of the 2019 18th International Symposium INFOTEH-JAHORINA (INFOTEH), East Sarajevo, Bosnia and Herzegovina, 20–22 March 2019; pp. 1–5.
8. Salah, I.; Jouini, K.; Korbaa, O. On the use of text augmentation for stance and fake news detection. *J. Inf. Telecommun.* **2023**, 1–17. [CrossRef]
9. Vaz, B.; Bernardes, V.; Figueira, Á. On Creation of Synthetic Samples from GANs for Fake News Identification Algorithms. In *Information Systems and Technologies: WorldCIST 2022*; Springer: Berlin/Heidelberg, Germany, 2022; Volume 3, pp. 316–326.
10. Frid-Adar, M.; Klang, E.; Amitai, M.; Goldberger, J.; Greenspan, H. Synthetic data augmentation using GAN for improved liver lesion classification. In Proceedings of the 2018 IEEE 15th International Symposium on Biomedical Imaging (ISBI 2018), Washington, DC, USA, 4–7 April 2018; pp. 289–293.
11. Jain, S.; Seth, G.; Paruthi, A.; Soni, U.; Kumar, G. Synthetic data augmentation for surface defect detection and classification using deep learning. *J. Intell. Manuf.* **2022**, *33*, 1007–1020. [CrossRef]
12. Fawaz, H.I.; Forestier, G.; Weber, J.; Idoumghar, L.; Muller, P.A. Data augmentation using synthetic data for time series classification with deep residual networks. *arXiv* **2018**, arXiv:1808.02455.
13. Hernandez, M.; Epelde, G.; Alberdi, A.; Cilla, R.; Rankin, D. Synthetic data generation for tabular health records: A systematic review. *Neurocomputing* **2022**, *493*, 28–45. [CrossRef]
14. Assefa, S.A.; Dervovic, D.; Mahfouz, M.; Tillman, R.E.; Reddy, P.; Veloso, M. Generating synthetic data in finance: Opportunities, challenges and pitfalls. In Proceedings of the First ACM International Conference on AI in Finance, New York, NY, USA, 15–16 October 2020; pp. 1–8.
15. Shafique, R.; Rustam, F.; Choi, G.S.; Díez, I.d.l.T.; Mahmood, A.; Lipari, V.; Velasco, C.L.R.; Ashraf, I. Breast cancer prediction using fine needle aspiration features and upsampling with supervised machine learning. *Cancers* **2023**, *15*, 681. [CrossRef]
16. Abd Al Rahman, M.; Danishvar, S.; Mousavi, A. An improved capsule network (WaferCaps) for wafer bin map classification based on DCGAN data upsampling. *IEEE Trans. Semicond. Manuf.* **2021**, *35*, 50–59.
17. Strelcenia, E.; Prakoonwit, S. Improving Classification Performance in Credit Card Fraud Detection by Using New Data Augmentation. *AI* **2023**, *4*, 172–198. [CrossRef]
18. Chawla, N.V.; Bowyer, K.W.; Hall, L.O.; Kegelmeyer, W.P. SMOTE: Synthetic minority over-sampling technique. *J. Artif. Intell. Res.* **2002**, *16*, 321–357. [CrossRef]
19. Arjovsky, M.; Chintala, S.; Bottou, L. Wasserstein generative adversarial networks. In Proceedings of the International Conference on Machine Learning, PMLR, Sydney, Australia, 6–11 August 2017; pp. 214–223.
20. Doersch, C. Tutorial on variational autoencoders. *arXiv* **2016**, arXiv:1606.05908.
21. Pardo, L. *Statistical Inference Based on Divergence Measures*; Chapman & Hall/CRC Press: Boca Raton, FL, USA, 2005.
22. Kullback, S.; Leibler, R.A. On information and sufficiency. *Ann. Math. Stat.* **1951**, *22*, 79–86. [CrossRef]
23. Lin, J. Divergence measures based on the Shannon entropy. *IEEE Trans. Inf. Theory* **1991**, *37*, 145–151. [CrossRef]
24. Goncalves, A.; Ray, P.; Soper, B.; Stevens, J.; Coyle, L.; Sales, A.P. Generation and evaluation of synthetic patient data. *BMC Med. Res. Methodol.* **2020**, *20*, 108. [CrossRef]
25. Golub, G.H.; Van Loan, C.F. *Matrix Computations*; Johns Hopkins University Press: Baltimore, MD, USA, 2013.
26. Fasano, G.; Franceschini, A. A multidimensional version of the Kolmogorov–Smirnov test. *Mon. Not. R. Astron. Soc.* **1987**, *225*, 155–170. [CrossRef]
27. Snoke, J.; Raab, G.M.; Nowok, B.; Dibben, C.; Slavkovic, A. General and specific utility measures for synthetic data. *J. R. Stat. Soc. Ser. A* **2018**, *181*, 663–688. [CrossRef]
28. Becker, B.; Kohavi, R. Adult. In *UCI Machine Learning Repository*; Department of Information and Computer Science, University of California: Irvine, CA, USA, 1996. [CrossRef]
29. Wang, W.Y. "Liar, Liar pants on fire": A new benchmark dataset for fake news detection. *arXiv* **2017**, arXiv:1705.00648.

30. Agrawal, R.; Srikant, R.; Thomas, D. Privacy preserving OLAP. In Proceedings of the Proceedings of the 2005 ACM SIGMOD International Conference on Management of Data, Baltimore, MD, USA, 14–16 June 2005; pp. 251–262.
31. Zemel, R.; Wu, Y.; Swersky, K.; Pitassi, T.; Dwork, C. Learning fair representations. In Proceedings of the International Conference on Machine Learning, PMLR, Atlanta, GA, USA, 17–19 June 2013; pp. 325–333.
32. Ding, F.; Hardt, M.; Miller, J.; Schmidt, L. Retiring adult: New datasets for fair machine learning. *Adv. Neural Inf. Process. Syst.* **2021**, *34*, 6478–6490.
33. Zhang, X.; Ghorbani, A.A. An overview of online fake news: Characterization, detection, and discussion. *Inf. Process. Manag.* **2020**, *57*, 102025. [CrossRef]
34. Kaliyar, R.K.; Goswami, A.; Narang, P. FakeBERT: Fake news detection in social media with a BERT-based deep learning approach. *Multimed. Tools Appl.* **2021**, *80*, 11765–11788. [CrossRef] [PubMed]
35. Nasir, J.A.; Khan, O.S.; Varlamis, I. Fake news detection: A hybrid CNN-RNN based deep learning approach. *Int. J. Inf. Manag. Data Insights* **2021**, *1*, 100007. [CrossRef]
36. Vaz, B.G. Using GANs to Create Synthetic Datasets for Fake News Detection Models. Master's Thesis, Universidade do Porto, Porto, Portugal, 2022.
37. Han, H.; Wang, W.Y.; Mao, B.H. Borderline-SMOTE: A new over-sampling method in imbalanced data sets learning. In Proceedings of the Advances in Intelligent Computing: International Conference on Intelligent Computing, ICIC 2005, Hefei, China, 23–26 August 2005; pp. 878–887.
38. Ping, H.; Stoyanovich, J.; Howe, B. Datasynthesizer: Privacy-preserving synthetic datasets. In Proceedings of the 29th International Conference on Scientific and Statistical Database Management, Chicago, IL, USA, 27–29 June 2017; pp. 1–5.
39. Xu, L.; Skoularidou, M.; Cuesta-Infante, A.; Veeramachaneni, K. Modeling tabular data using conditional gan. *Adv. Neural Inf. Process. Syst.* **2019**, *32*, 7335–7345.
40. Copula GAN Synthesizer. Available online: https://docs.sdv.dev/sdv/single-table-data/modeling/synthesizers/copulagansynthesizer (accessed on 17 March 2023).
41. Solatorio, A.V.; Dupriez, O. REaLTabFormer: Generating Realistic Relational and Tabular Data using Transformers. *arXiv* **2023**, arXiv:2302.02041.

Disclaimer/Publisher's Note: The statements, opinions and data contained in all publications are solely those of the individual author(s) and contributor(s) and not of MDPI and/or the editor(s). MDPI and/or the editor(s) disclaim responsibility for any injury to people or property resulting from any ideas, methods, instructions or products referred to in the content.

Article

Evaluation of Convergent, Discriminant, and Criterion Validity of the Cuestionario Burnout Granada-University Students

Elena Ortega-Campos [1], Gustavo R. Cañadas [2,*], Raimundo Aguayo-Estremera [3], Tania Ariza [4], Carolina S. Monsalve-Reyes [5], Nora Suleiman-Martos [6] and Emilia I. De la Fuente-Solana [7]

1. CEINSA-UAL, Universidad de Almería, Carretera de Sacramento s/n, La Cañada de San Urbano, 04120 Almería, Spain; elenaortega@ual.es
2. Department of Didactic of Mathematics, Faculty of Education Science, University of Granada, 18071 Granada, Spain
3. Departamento de Psicobiología y Metodología de las Ciencias del Comportamiento, Facultad de Psicología, Universidad Complutense de Madrid, Campus de Somosaguas, Ctra. De Húmera, s/n, Pozuelo de Alarcón, 28223 Madrid, Spain; r.aguayo@ucm.es
4. Department of Educational Psychology and Psychobiology, Faculty of Education, Universidad Internacional de la Rioja, Av. De la Paz, 137, 26006 Logroño, Spain; tania.ariza.castilla@unir.net
5. Departamento de Ciencias Sociales, Universidad Católica de La Santísima Concepción, Avenida Alonso de Ribera 2850, Concepción 4090541, Chile; carolinamonsalve@ucsc.cl
6. Faculty of Health Sciences, University of Granada, Avenida de la Ilustración, 60, 18016 Granada, Spain; norasm@ugr.es
7. Brain, Mind and Behaviour Research Center (CIMCYC), University of Granada, 18071 Granada, Spain; edfuente@ugr.es
* Correspondence: grcanadas@ugr.es

Citation: Ortega-Campos, E.; Cañadas, G.R.; Aguayo-Estremera, R.; Ariza, T.; Monsalve-Reyes, C.S.; Suleiman-Martos, N.; De la Fuente-Solana, E.I. Evaluation of Convergent, Discriminant, and Criterion Validity of the Cuestionario Burnout Granada-University Students. *Mathematics* 2023, *11*, 3315. https://doi.org/10.3390/math11153315

Academic Editor: Michael Voskoglou

Received: 9 June 2023
Revised: 13 July 2023
Accepted: 26 July 2023
Published: 28 July 2023

Copyright: © 2023 by the authors. Licensee MDPI, Basel, Switzerland. This article is an open access article distributed under the terms and conditions of the Creative Commons Attribution (CC BY) license (https://creativecommons.org/licenses/by/4.0/).

Abstract: Burnout is a health problem that affects professionals and students or professionals in training, especially those in health areas. For this reason, it is necessary that it is properly identified to prevent the impact it can have on the work and personal areas of the people who suffer from it. The aim of this work is to study the convergent, discriminant, and criterion validity of the Cuestionario Burnout Granada-University Students. The sample consisted of 463 undergraduate nursing students, selected by non-probabilistic convenience sampling, who participated voluntarily and anonymously in the study. The mean age of the participants was 21.9 (5.12) years, mostly female (74.1%), single (95.8%), and childless (95.6%). Information was collected face-to-face, and the instruments were completed on paper. Comparisons were made in the three dimensions of burnout of the CBG-USS between students with and without burnout, finding statistically significant differences in all three dimensions: Emotional Exhaustion ($p < 0.001$, d = 0.674), Cynicism ($p < 0.001$, d = 0.479), and Academic Efficacy ($p < 0.001$, d = −0.607). The Cuestionario Burnout Granada-University Students presents adequate reliability and validity indices, which demonstrates its usefulness in the identification of burnout. This syndrome has traditionally been measured in professionals, but students also present burnout, so it is necessary to have specific burnout instruments for students, since the pre-work situation and stressors of students are different from those of workers. In order to work on the prevention of university burnout, it is essential to have specific instruments for professionals in training that help in the detection of students with burnout.

Keywords: academic burnout syndrome; Cuestionario Burnout Granada-University Students; CBG-USS; MBI-SS; nursing students; reliability and validity

MSC: 62-11

1. Introduction

Burnout is a psychological disorder that develops from chronic exposure to stressors. Traditionally, this syndrome has been studied in professions characterized by a strong interaction between professionals and the beneficiaries of their work, especially among workers considered to be at high risk of suffering from the syndrome, such as doctors, nurses, teachers, and police officers. More recently, however, it has also been studied in other population groups, such as informal caregivers, housewives, and university students [1–7].

University students can be equated with the above workers in terms of their susceptibility to developing burnout syndrome [1,8]. In both cases, there is a relationship with an institution that offers products or services, the incentive for participation being monetary for working professionals and academic and social recognition for students.

The university environment produces many changes in students' lives, in areas such as relationships and physical and occupational contexts, and can be highly stressful. Students must develop a more independent life while preparing themselves for a professional future. Consequently, some may lack energy to continue, lose interest in their studies, and/or feel unable to meet their commitments and achieve their goals [1,9,10].

The prevalence of burnout in university students is high [11], professionals in training increasingly present symptoms compatible with burnout syndrome. Specifically, nursing students achieve higher average scores in burnout than the general population [12–15]. In light of this situation, it is crucial to conduct research on the occurrence of burnout among university students, particularly focusing on the variables that could serve as risk or protective factors for the syndrome. Additionally, it is important to develop and validate instruments that can effectively identify burnout in university students. Such endeavors are necessary to propose educational policies aimed at enhancing the conditions under which students pursue their university studies [16].

The most widely accepted conceptualization of burnout syndrome is based on the following three dimensions: Emotional Exhaustion, Depersonalization or Cynicism, and low Personal Accomplishment or Academic Efficacy [17–19]. In the university environment, students who perceive a loss of energy, who lack commitment to their studies, who disparage teachers and other students, and who are unable or unwilling to perform the tasks required of them, may be experiencing academic burnout. Indeed, numerous studies have reported high levels of academic burnout in university education [1,20]. This prevalence, together with the serious physical and psychological consequences that may arise from the disorder, underlines the need for an in-depth study of burnout regarding evaluation, prevention, and remedial intervention among university students [2,21,22].

The studies carried out with the aim of estimating the real relevance of burnout among university students indicate that between 9 and 21% of university students are at risk of developing burnout. These data are related to the COVID-19 pandemic, showing that students who have suffered psychological consequences derived from the pandemic present higher scores in Emotional Exhaustion and lower levels of Academic Efficacy than those who have not suffered them [20]. As a result of the COVID-19 pandemic, university students present high levels of burnout [23,24], uncertainty about their professional future [25], and higher levels in psychological variables, such as anxiety or depression [26,27].

In view of the reality of burnout syndrome in students, it is necessary to use instruments that are capable of identifying those students who present burnout in its different levels or intensities. The Maslach Burnout Inventory (MBI) [17,18] was created to measure burnout in workers. Subsequently, the MBI-SS [19] was created to measure burnout syndrome in students. The context and stressors of workers are far from being the same as those of students or workers in training, so the instruments for measuring burnout cannot be the same for both workers and students [1,28,29].

Although the MBI is the most widely used instrument for assessing burnout syndrome in both professional and non-professional samples [30], several studies examining its psychometric properties have identified some problems. For instance, it has been consistently found that the reliability of the Cynicism dimension falls below the recommended cutoff points [31]. Further, and more importantly, the MBI in its Spanish version was adapted back in 1997 [32]. However, the current survey is no longer accessible, making it legally infeasible to use. Additionally, there are no up-to-date scales available for evaluating the Spanish population. It is worth noting that the criteria used for assessment were developed in 1997, which raises concerns about their applicability in the present context. Due to the need to assess burnout syndrome in pre-professional groups such as students, the idea of adapting the CBG for university populations emerged. The theoretical framework of the MBI is well established and widely accepted in the field [31], so the CBG was developed under the same measurement structure.

In this work, we present the Cuestionario Burnout Granada-University Students (CBG-USS, for its Spanish abbreviation) in a sample of university students enrolled in undergraduate nursing. This sample has been chosen because, after the pandemic, students in health areas present high levels of burnout [20,23,24]. Nurses provide care to patients, often in contexts where they are at risk of experiencing physical and verbal aggression [33]. Furthermore, nurses encounter the challenging dichotomy of their daily work. While they are expected to exhibit a compassionate and caring demeanor, which is inherent to their vocation, they frequently find themselves compelled to be emotionally detached in order to make difficult decisions that require an objective approach devoid of emotional responses. For these reasons, nurses consistently face high levels of stress, which is known to contribute to the onset of burnout syndrome [34,35].

It is important to have instruments to measure burnout in workers, but without forgetting that it is also necessary to have specific instruments to measure burnout in students, since the students who present burnout today will be the workers with burnout in the future [36]. Given the need for instruments that adequately measure and discriminate academic burnout in university students, the aim of this study is to examine the reliability and some sources of validity of the Cuestionario Burnout Granada-University Students in a sample of nursing students. Specifically, reliability was analyzed based on internal consistency of the test items, and validity evidence was examined based on its relationship to other variables, namely, convergent, discriminant, and concurrent criterion-related validity.

1.1. Reliability Estimators

Reliability is closely related to measurement precision. In a broad sense, it refers to the degree to which test scores are free from errors of measurement [37]. The importance of measurement reliability is consistently significant, as the demand for precision is more crucial as the significance of decisions and interpretations amplifies.

A wide range of reliability estimators have been developed, depending on the measurement model of the test under study. Probably, the most used estimators are Cronbach's alpha and McDonald's total omega [38,39]. The alpha coefficient (α) was originally proposed for τ-equivalent models, and total omega coefficient (ω_t) for congeneric models. A discussion on these models can be consulted in another article of the current Special Issue [16]. Coefficient alpha is a lower bound to reliability, and its calculation is described by the following equation [40]:

$$\alpha = \frac{J}{J-1}\left[1 - \frac{\sum_{1 \leq i \neq k \leq J}^{J}(\sigma_{ik})}{\sigma_X^2}\right] \quad (1)$$

where J = number of items; σ_{ik} = covariance between item I and item k; σ_X^2 = test variance. The calculation of total omega coefficient is described by the following equation [41]:

$$\omega_t = \frac{\left(\sum_{i=1}^{J}\lambda_i\right)^2}{\left(\sum_{i=1}^{J}\lambda_i\right)^2 + \sum_{i=1}^{J}\sigma_{\varepsilon i}^2} \quad (2)$$

where λ_i = factor loading for item i; $\sigma_{\varepsilon i}^2$ = error variance for item i; $\lambda_i^{(g)}$ = factor loading of item i on the general factor g; $\lambda_i^{(s_1)}, \ldots, \lambda_i^{(s_h)}, \ldots \lambda_i^{(s_p)}$ = factor loadings of item i on the specific factors $s_1, \ldots, s_h, \ldots, s_p$, and the specific factors comprise $J_1, \ldots, J_h, \ldots, J_p$ items.

Alpha coefficient requires compliance with assumptions such as unidimensionality, τ-equivalence, and normality of the distribution of the items [42]. When the alpha coefficient assumptions are not met, it tends to underestimate the true reliability of the scale. For this reason, to deal with congeneric scales (which do not satisfy the assumption of τ-equivalence), the omega coefficient is usually recommended [43]. As for the analysis of reliability of the CBG-USS scores, both alpha and total omega coefficients were calculated and compared because they might give complementary information (a lower bond to reliability based on test scores, and a factor analysis-based estimation of reliability).

1.2. Evidence of Validity

There have been several authors who have developed the current ideas about validity [44,45], many of which have been embraced by the Standards for Educational and Psychological Testing (SEPT) [37]. According to the SEPT, validity refers to the extent to which theory and evidence support the interpretations of test scores for the intended uses of the test. Thus, any validation process must begin by clearly establishing the test's intended uses. In this line, the proposed uses for the CBG-USS focus on research and applied purposes. As for the research context, it is anticipated to be used in studying the psychometric properties of the test itself and in investigating burnout syndrome (theoretical predictions, prevalence, risk factors, etc.). In applied contexts, the instrument could be used to assess the extent to which university students exhibit burnout symptoms, aiming to detect potential academic difficulties without establishing a diagnosis.

The SEPT framework establishes five sources of validity that are based on the test content, the response processes of participants, the internal structure of the test, the relations of the test to other variables, and the consequences of testing. Being all sources of validity relevant to the validation process, there is no recommendation about prioritizing among them. In the present study, we aimed to analyze validity evidence based on the relationship between CBG-USS scores and other variables that are theoretically related to the three dimensions of the test.

Relation to other variables' validity provides evidence about the degree to which relationships between two measures are consistent with the common construct underlying the proposed test score interpretations. This type of evidence contributes to a better understanding of the construct and the nomological network in which it is theoretically embedded. Specifically, three strategies of this type of validity have been proposed: convergent, discriminant, and criterion.

To gather convergent validity evidence, a correlational analysis can be conducted, for example, including another test that assesses the same construct. Thus, we also measured burnout syndrome with the MBI that is conceptually equivalent to the CBG. Regarding discriminant validity evidence, the relationship with an external criterion is analyzed. One common strategy is creating (or selecting) groups and comparing the mean scores among these groups.

To gather criterion validity evidence of the CBG-USS, we used the area under the curve (AUC) analysis. AUC, commonly used to evaluate classification accuracies, is a preferred method that relies on predictive models [46]. It eliminates the need for subjective threshold decisions, making it a reliable choice.

AUC expands upon the receiver operating characteristics (ROC) curve, which summarizes the performance of label assignment [47]. It accomplishes this by integrating a confusion matrix (a 2 × 2 table that includes counts for true/false positive and true/false negative) at various threshold levels. These changes can affect classification accuracies. Figure 1 displays the confusion matrix and equations for various commonly calculated statistics derived from it. In the case of a binary classifier, a positive label is assigned when the predicted category of an instance is 1, while a negative label is assigned when the predicted category is 0. Correct predictions are labeled as true, while incorrect predictions are labeled as false. Consequently, each instance corresponds to a specific cell within the confusion matrix, which consolidates the instance counts for each of the four categories.

		\multicolumn{2}{c}{Ture class}
		1 (Positive)	0 (Negative)
Hypothesized class	1 (Positive)	True Positives	False Positives
	0 (Negative)	False Negatives	True Negatives

$$tp\ rate = \frac{TP}{TP + FN} = recall \qquad precision = \frac{TP}{TP + FP}$$

$$fp\ rate = \frac{FP}{FP + TN} \qquad F1 = \frac{2}{\frac{1}{precision} + \frac{1}{recall}}$$

$$specificity = \frac{TN}{FP + TN} = 1 - fp\ rate \qquad accuracy = \frac{TP + TN}{TP + TN + FP + FN}$$

Figure 1. Confusion matrix and usual performance statistics.

The ROC curve is plotted on a two-dimensional plane, where the horizontal axis represents the false positive rate and the vertical axis represents the true positive rate. A discrete classifier generates a single confusion matrix, corresponding to a specific point in the ROC space. In the case of probability or scoring classifiers, different thresholds can be used to obtain multiple confusion matrices, and each threshold value results in a distinct point on the ROC curve.

AUC transforms the ROC curve into a numerical measure of performance for a binary classifier. In essence, AUC combines the model's performance across all possible threshold values. AUC represents the area under the ROC curve and falls between 0 and 1: the highest value signifies a flawless classifier, while zero indicates that all predictions are incorrect. The AUC can be computed for a finite set of instances using the following steps: First, arrange the instances in descending order based on their predicted probabilities of being positive. Then, utilize these predicted probabilities as threshold values and calculate the corresponding true positive rate (TPR) and false positive rate (FPR). This process generates a series of points on the ROC plane, progressing upwards and to the right, thereby forming the ROC curve. Finally, the AUC is obtained by summing the areas of the trapezoids created between each instance point 'i' and the subsequent point 'i + 1' along with the horizontal axis (false positive rate). The calculation of AUC is described by the following equation [46]:

$$AUC = \frac{1}{2}\sum_{i=1}^{m-1}(FPR_{i+1} - FPR_i)(TPR_{i+1} + TPR_i) \qquad (3)$$

where m represents the number of instances.

The manuscript is organized as follows. First, the methodological aspects of the research are presented. Second, the results derived from the conducted analyses are described; specifically, descriptive statistics are presented first. Then, the prevalence of burnout syndrome among nursing students is estimated. Next, the results regarding the reliability of the CBG-USS are presented, followed by the results concerning convergent, discriminant, and concurrent criterion-related validity. In the Discussion section, the scope of the results in terms of psychometrics and the usefulness of the CBG-USS are presented.

2. Method

2.1. General Background

This paper presents an instrumental study to check the proper functioning of the Cuestionario Burnout Granada-University Students (CBG-USS) [48,49], whose objective is to identify burnout in university students. The Cuestionario Burnout Granada-University

Students (CBG-USS) has been developed following international guidelines for the construction of measurement instruments [37,50,51]. In creating this instrument, the definition and dimensions of burnout proposed by Maslach and Jackson [17] were taken as a reference. Thus, burnout is understood as a psychological problem arising from continual exposure to stressors related to the educational institution and to study activities. The syndrome is characterized by three dimensions: Emotional Exhaustion, Cynicism or Depersonalization, and low Academic Efficacy or personal accomplishment.

2.2. Participants

A total of 463 undergraduate nursing students formed the sample for this study. The participants were mostly women without a partner or children. To participate in this study, the students had to meet the following inclusion criteria: (a) be enrolled in the Bachelor's Degree in Nursing at the time of data collection; (b) agree to participate voluntarily in the research and complete the participation documentation, and (c) answer all the items of the CBG-USS and MBI-SS instruments.

2.3. Procedure

This research was conducted as an instrumental study [50,51]. The research project was made known to nursing students in information sessions. The announcement of the realization of the information sessions was made through the students' electronic learning platform. Those interested in participating were given information about the study, including the expected time for them to complete the questionnaire (15–20 min) and the privacy policy. Information was collected in person, individually, anonymously, and voluntarily. The students who decided to participate in the study completed a paper information collection notebook that included the sociodemographic variables and the burnout instruments used. The sampling technique used was non-probabilistic by convenience [51] reaching a response rate of 82.3%.

2.4. Ethical Considerations

The study was approved by the Ethics Committee of the University of Granada (393/CEIH2017). The participants were informed of the objective of the study. No personal data were collected from the students. Participation in the study was voluntary and anonymous. Written informed consent was requested before starting the study. The collection of information and its subsequent data processing was carried out completely anonymously.

2.5. Study Variables

The following sociodemographic variables were ascertained: sex, age, marital status, and number of children. With respect to the students' educational background, the following study data were obtained: degree course title, schedule (morning/evening classes), and previous university studies. Two workplace-related variables were included: extra-university work (yes/no) and, if so, the work schedule.

2.6. Instruments

The Cuestionario Burnout Granada-University Students (CBG-USS) is composed of 26 items grouped into three dimensions: Emotional Exhaustion (9 items), Cynicism (7 items), and Academic Efficacy (10 items). The items are presented using a Likert-type response format of five alternatives, where 1 = completely disagree and 5 = completely agree. In scoring the questionnaire items, the responses for 12 (items 1, 4, 10, 12, 13, 19, 20, 21, 23, 24, 25, and 26) were reverse coded. In order to calculate a total score for each dimension, it is necessary to sum all the item scores within that dimension. This provides the following interpretation: the higher the score, the greater the level of burnout experienced by the participant. The CBG-USS items can be consulted in the Supplementary Materials.

The Maslach Burnout Inventory-Student Survey (MBI-SS) [19] is the adaptation of the Maslach Burnout Inventory [17] for use with students. It contains 15 items grouped into three dimensions: Emotional Exhaustion (5 items), Cynicism (4 items), and Academic Efficacy (6 items). Each item is scored on a 7-point Likert scale as follows: Never = 0; A few times a year = 1; Once a month at most = 2; A few times a month = 3; Once a week = 4; A few times a week = 5; Every day = 6. In the present study, the Spanish version of the MBI-SS [52,53] was used.

2.7. Data Analysis

For the description of categorical variables, the relevant percentages and frequencies were calculated for each level of response. For the quantitative variables, descriptive statistics (mean, standard deviation, and minimum and maximum values) were obtained.

To estimate the prevalence of burnout, the Golembiewski, Munzerider, and Stevenson model was used; this classifies participants as high and low for each of the MBI-SS burnout dimensions [54,55].

Alpha and omega reliability coefficients together with their corresponding 95% confidence interval were calculated for each dimension of the CBG-USS and MBI-SS, taking as reference for the evaluation of the reliability coefficients the recommendations made in this regard by George and Mallery [56] and Aguayo et al. [57].

Regarding discriminant validity, differences between groups (students with and without burnout syndrome) on measures of burnout dimensions were detected. The procedure was as follows: First, we measured the dimensions of burnout syndrome using the MBI-SS (not the CBG-USS, which is the instrument we aim to validate). Second, based on the procedure proposed by Golembiewski and Munzenrider [54], we classified the participants into groups of students with high and low levels of burnout symptoms (i.e., students experiencing burnout and students not experiencing burnout). Subsequently, Student's t-tests for independent groups were conducted to determine if there were statistically significant differences in CBG-USS scores between the two groups. It is important to note that the instrument used to create the groups (MBI-SS) is different from the instrument used as the dependent variable in the hypothesis test (CBG-USS). If, as we expected, the CBG-USS is an instrument that discriminates between different levels of burnout, the results obtained with it would be similar to those obtained with the MBI-SS used as the dependent variable in the hypothesis tests. An effect size index (Cohen's d) and the corresponding 95% confidence intervals around the point estimation was also performed. The effect size index reports the degree of these differences obtained in the significance tests [58,59].

To gather evidence of convergent validity, a correlational analysis was performed including the dimensions of the MBI and the CBG. Thus, Pearson correlation coefficient and the corresponding 95% confidence interval between burnout dimensions were calculated. Adjusted correlations for reliability were also calculated following the formula by Gulliksen [60].

To study concurrent criterion validity of the CBG-USS, the area under the curve (AUC) and the corresponding 95% confidence interval were calculated for each dimension of burnout. The potential AUC score ranges from 0 (perfect negative prediction) to 1 (perfect positive prediction). An AUC of 0.50 reflects a prediction equal to chance; one of 0.56 to 0.64 represents a small effect; one of 0.64 to 0.71, a medium effect, and one higher than 0.71, a large effect [61]. All statistical analyses were performed using SPSS IBM (v.27).

3. Results

3.1. Sociodemographic Profile

The sample was composed of 463 nursing students, of whom 74.1% were women, aged from 18 to 59 years with a mean age of 21.9 (5.12) years; 95.8% were single and 95.6% had no children.

Regarding the participants' education variables, 30.9% were in the first year of the degree course, and 57.6% in the second year. Most had classes in the morning; 3.7% had previously obtained a university degree and 8.2% were currently working, as well as studying (of these, 32.4% were working full-time and 67.6% were working part-time) (Table 1).

Table 1. Categorical variables.

	% (n)		% (n)
Sex		**Degree year**	
Male	25.9 (119)	1st	30.9 (140)
Female	74.1 (341)	2nd	57.6 (261)
		3rd or 4th	11.5 (52)
Marital status		**Previous degree**	
Single	95.8 (435)	Yes	3.7 (17)
Married/Partnership	4.2 (19)	No	96.3 (439)
Children		**In employment**	
0	95.6 (345)	Yes	8.2 (37)
1 or more	4.4 (16)	No	91.8 (413)
Class schedule		**Work regime (if any)**	
Mornings	85.3 (221)	Full-time	32.4 (11)
Afternoons/Evenings	14.7 (38)	Part-time	67.6 (23)

3.2. Descriptive Data for the CBG-USS and the MBI-USS

Table 2 shows the descriptive statistics for burnout scores on the CBG-USS and the MBI-SS. Students obtained a mean score of 23.92 (SD = 5.48) and a range of 11 to 42 on the CBG-USS Emotional Exhaustion scale, 13.34 (SD = 3.86), with a range of 7 to 30, on CBG-USS Cynicism, and 40.33 (SD = 5.71) and a range of 15 to 50 on CBG-USS Academic Efficacy.

Table 2. Mean, SD, and minimum and maximum values for the CBG-USS and MBI-SS dimensions.

	Mean	SD	Minimum	Maximum	N
CBG-USS$_{\text{Emotional Exhaustion}}$	23.92	5.48	11	42	452
CBG-USS$_{\text{Cynicism}}$	13.34	3.86	7	30	456
CBG-USS$_{\text{Academic Efficacy}}$	40.33	5.71	15	50	453
MBI-SS$_{\text{Emotional Exhaustion}}$	13.23	5.91	0	30	452
MBI-SS$_{\text{Cynicism}}$	9.62	5.11	0	29	442
MBI-SS$_{\text{Academic Efficacy}}$	27.77	5.61	1	36	449

On the MBI-SS, the students obtained mean scores of 13.23 (SD = 5.91), ranging from 0 to 30 for Emotional Exhaustion, 9.62 (SD = 5.11), ranging from 0 to 29 for Cynicism, and 27.77 (SD = 5.61), and ranging from 1 to 36 for Academic Efficacy.

3.3. Prevalence of Burnout Syndrome

We used the phase model proposed by Golembiewski and Munzenrider [55] to estimate the prevalence of the burnout syndrome in nursing students, according to the scores obtained in the MBI-SS. These authors propose an 8-phase model in which students are placed in a phase according to their level (Low, High) in each of the Burnout dimensions (Emotional Exhaustion, Cynicism, and Academic Efficacy). Phases 1 and 2 reflect low levels of burnout; phases 3, 4, and 5 correspond to a moderate degree of burnout, and phases 6, 7, and 8 indicate high levels of burnout; 15.4% of the students presented mild levels of burnout (phases 1 and 2), 10.1% presented medium levels of burnout (phases 3 to 5), and 11.7% of the participants recorded high levels of burnout (phases 6 to 8) (Table 3).

Table 3. Prevalence of burnout according to the phases of the Golembiewski model.

Phase	1	2	3	4	5	6	7	8
Emotional Exhaustion	L	L	L	L	H	H	H	H
Cynicism	L	H	L	H	L	H	L	H
Academic Efficacy	L	L	H	H	L	L	H	H
N	61	10	19	9	19	19	13	22
%	13.2	2.2	4.1	1.9	4.1	4.1	2.8	4.8

Levels (high or low) in each of the burnout dimensions according to the model proposed by Golembiewski. H = High; L = Low.

3.4. Sample Reliability of the CBG-USS

We calculated alpha and omega coefficients to study sample reliability of the CBG-USS. The alpha coefficients for the CBG-USS dimensions showed the following values: Emotional Exhaustion = 0.769, 95%CI [0.736, 0.799], Cynicism = 0.809, 95%CI [0.781, 0.835], and Academic Efficacy = 0.820, 95%CI [0.794, 0.843]. The omega coefficients were: 0.771, 95%CI [0.740, 0.803] for Emotional Exhaustion, 0.753, 95%CI [0.719, 0.787] for Cynicism, and 0.823, 95%CI [0.799, 0.847] for Academic Efficacy.

In order to be able to make comparisons, we present here the reliability results concerning the MBI-SS. The alpha coefficients obtained for each of the scales of the MBI-SS for our study sample were: Emotional Exhaustion = 0.735, 95%CI [0.695, 0.771], Cynicism = 0.603, 95%CI [0.544, 0.657], and Academic Efficacy = 0.773, 95%CI [0.740, 0.802]. The omega coefficients were: 0.739, 95%CI [0.694, 0.781] for Emotional Exhaustion, 0.662, 95%CI [0.602, 0.714] for Cynicism, and 0.780, 95%CI [0.730, 0.817] for Academic Efficacy.

3.5. Convergent and Discriminant Validity of the CBG-USS

Evidence of convergent and discriminant validity of the CBG-USS was studied by calculating the correlation coefficients between the MBI-SS dimensions and those of the CBG-USS with the following results: Emotional Exhaustion r = 0.620, 95%CI [0.559, 0.674]; Cynicism, r = 0.281, 95%CI [0.192, 0.365], and Academic Efficacy, r = 0.487, 95%CI [0.413, 0.556] (see Table 4).

Table 4. Correlation coefficients for the CBG-USS and MBI-SS dimensions.

	MBI-SS Emotional Exhaustion	MBI-SS Cynicism	MBI-SS Academic Efficacy
CBG-USS$_{Emotional\ Exhaustion}$	0.620 ** (0.821)	0.314 ** (0.421)	−0.122 * (−0.156)
CBG-USS$_{Cynicism}$	0.166 ** (0.232)	0.281 ** (0.398)	−0.384 ** (−0.520)
CBG-USS$_{Academic\ Efficacy}$	−0.316 ** (−0.407)	−0.380 ** (−0.496)	0.487 ** (0.608)

Values in parentheses are adjusted correlations for reliability. ** $p < 0.01$; * $p < 0.05$.

We also studied evidence of convergent validity of the CBG-USS by comparing the results of the burnout dimensions in two groups: one that suffers from burnout syndrome, and another one that does not. The 54 students who were classified as corresponding to phases 6–8 of the Eight-Phase Model [54,55] were assumed to present burnout, and the remaining 398 students (corresponding to phases 1–5) were classed as not presenting burnout. With respect to the latter, the students who presented burnout recorded higher scores for Emotional Exhaustion and Cynicism, on both the MBI-SS and the CBG-USS instruments, as follows: [MBI-SS Emotional Exhaustion: 20.13 (3.62); CBG-USS Emotional Exhaustion: 26.98 (5.11)], [MBI-SS Cynicism: 14.90 (5.79); CBG-USS Cynicism: 14.92 (3.70)]. By contrast, the no-burnout group presented higher scores for Academic Efficacy [MBI-SS Academic Efficacy: 28.18 (5.40); CBG-USS Academic Efficacy: 40.77 (5.47)]. There were statistically significant differences between the with/without burnout groups for all the

MBI-SS and CBG-USS dimensions. The following effect sizes were measured for each scale: MBI-SS Emotional Exhaustion: 2.01, 95%CI [1.696, 2.323]; MBI-SS Cynicism: 1.062, 95%CI [0.77, 1.355] and MBI-SS Academic Efficacy: −0.557, 95%CI [−0.844, −0.271]; CBG-USS Emotional Exhaustion: 0.674, 95%CI [0.386, 0.962]; CBG-USS Cynicism: 0.479 95%CI [0.193, 0.765]; CBG-USS Academic Efficacy: −0.607 95%CI [−0.895, −0.32] (Table 5).

Table 5. Mean values, SD, and effect size for students with/without burnout for MBI-SS and CBG-USS.

	Burnout	M (SD)	p	Cohen's d	95%CI
CBG-USS$_{\text{Emotional Exhaustion}}$	No	23.510 (5.410)	<0.001	0.674	[0.386, 0.962]
	Yes	26.981 (5.119)			
CBG-USS$_{\text{Cynicism}}$	No	13.141 (3.837)	<0.001	0.479	[0.193, 0.765]
	Yes	14.923 (3.709)			
CBG-USS$_{\text{Academic Efficacy}}$	No	40.778 (5.470)	<0.001	−0.607	[−0.895, −0.32]
	Yes	36.942 (6.415)			
MBI-SS$_{\text{Emotional Exhaustion}}$	No	12.30 (5.539)	<0.001	2.01	[1.696, 2.323]
	Yes	20.13 (3629)			
MBI-SS$_{\text{Cynicism}}$	No	8.894 (4.554)	<0.001	1.062	[0.77, 1.355]
	Yes	14.907 (5.796)			
MBI-SS$_{\text{Academic Efficacy}}$	No	28.189 (5.405)	<0.001	−0.557	[−0.844, −0.271]
	Yes	24.759 (6.252)			

Absence of burnout: N = 398; presence of burnout: N = 54.

3.6. Concurrent Criterion Vailidity of the CBG-USS

Evidence of validity based on a concurrent criterion was analyzed by using area under the curve analysis. The area under the curve (AUC) was calculated for each dimension (MBI-SS and CBG-USS), obtaining the following results: Emotional Exhaustion, AUC = 0.893, 95%CI [0.862, 0.924] for MBI-SS and 0.687, 95%CI [0.610, 0.764] for CBG-USS; Cynicism, AUC = 0.807, 95%CI [0.733, 0.880] for MBI-SS and 0.654, 95%CI [0.581, 0.726] for CBG-USS, and Academic Efficacy, AUC = 0.338, 95%CI [0.250, 0.426] for MBI-SS and 0.318, 95%CI [0.243, 0.393] for CBG-USS (Table 6).

Table 6. AUC for each dimension of MBI-SS and CBG-USS.

	AUC	SE	95%CI
CBG-USS$_{\text{Emotional Exhaustion}}$	0.687 *	0.039	[0.610, 0.764]
CBG-USS$_{\text{Cynicism}}$	0.654 *	0.037	[0.581, 0.726]
CBG-USS$_{\text{Academic Efficacy}}$	0.318 *	0.038	[0.243, 0.393]
MBI-SS$_{\text{Emotional Exhaustion}}$	0.893 *	0.016	[0.862, 0.924]
MBI-SS$_{\text{Cynicism}}$	0.807 *	0.037	[0.733, 0.880]
MBI-SS$_{\text{Academic Efficacy}}$	0.338 *	0.045	[0.250, 0.426]

Total sample, N = 463; absence of burnout, N = 398; presence of burnout, N = 54. AUC = area under the curve; SE = standard error; CI = confidence interval; CBG-USS = Cuestionario Burnout Granada-University Students; MBI-SS = Maslach Burnout Inventory-Student Survey; * $p < 0.001$.

4. Discussion

The aim of this study was to obtain evidence of the validity of the Cuestionario Burnout Granada-University Students, in this case, in a sample of nursing students. The CBG-USS is introduced as a university student version of the Cuestionario Burnout Granada (CBG), which has been utilized and validated as a screening tool for burnout syndrome in Nurses and Police Officers [62–64]. Since the CBG-USS is a recently created instrument, a reference instrument has been used to demonstrate the correct functioning in the measurement and detection of burnout, in this case, the MBI-SS [19,53], the student version of the MBI [32]. The MBI-SS has been adapted and used in different countries [65–69] and has proven to be an adequate reference measure for the purpose of this work.

The results obtained corroborate the performance of the CBG-USS and provide evidence of its convergent, discriminant, and concurrent criterion-related validity. According to Cohen's [70] guidelines, high correlation coefficients were obtained between the Emotional Exhaustion and Academic Efficacy dimensions and moderate correlation coefficients for Cynicism between the dimensions of the CBG-USS and the MBI-SS. Given the problems concerning the low reliability values of the Cynicism dimension of the MBI-SS together with good reliability values of the CBG-USS, it was desirable to find a lower correlation coefficient compared with the other two dimensions. As expected by the Classical Test Theory, adjusted correlations for reliability were higher than non-adjusted correlations, leading to high correlations between all burnout dimensions. Both instruments measure burnout according to the three dimensions proposed by Maslach and Jackson [7,17–19].

Concerning reliability, the results of the alpha coefficient showed that in all dimensions of burnout, the CBG-USS obtained higher reliability values than the MBI-SS. This is especially relevant in the cynicism dimension, since in the MBI-SS, it was below the recommended cutoff point, a result that has been obtained on several occasions with various versions of the MBI [31]. Regarding the comparison between alpha and omega coefficients, the values were similar for all the dimensions of both measurement instruments. Specifically, while for the CBG-USS omega values were lower than alpha values, for the MBI-USS, the results showed the opposite (higher values for omega than for alpha). These results suggest that the alpha coefficient is not equivalent to the lower limit of reliability when assumptions, such as normality of item distribution or t-equivalence, are not met. Moreover, the alpha coefficient probably does not only underestimate the true reliability but also may overestimate it, as the results of the CBG-USS have shown. Only the Academic Efficacy dimension achieved the same values for alpha and omega coefficients. This was likely because this dimension has better psychometric properties than the other two dimensions, as the values of reliability indicated.

Statistically significant differences were observed between the with/without burnout groups. In general, the participants in the burnout group presented higher scores on the CBG-USS for Emotional Exhaustion and Cynicism, and lower ones for Academic Efficacy. Similar results were obtained with the MBI-SS. Moreover, this is in line with previous research findings concerning students following different degree courses [67]. These results corroborate the validity of the CBG-USS and confirm its ability to discriminate between students with/without burnout. The estimated AUC for the dimensions of the CBG-USS provide evidence of criterion validity of the instrument. The values found indicate adequate performance on all dimensions of the instrument. The results of this study support that the CBG-USS is an instrument that correctly measures and identifies burnout in university students.

Nursing students, professionals in training, have a high probability of developing burnout while pursuing their academic career. During their training, they must perform prolonged periods of internships in hospital centers, experiencing dissatisfaction with such internships or the organization of the same [71–73]; the gap between expectations and the reality of clinical practices or the perceived lack of support among peers and/or professors can facilitate the occurrence of burnout in nursing students [74,75].

Academic burnout is related to a higher probability of abandoning school/university and its impact may continue throughout one's working life. In addition, it can provoke psychological consequences and affect the professional performance of nursing professionals [76–78]. Among nursing students in particular, the COVID-19 pandemic aggravated the possibility of burnout, in terms of their learning capacity, their health and well-being, the quality of care provided, and the possible intention to abandon the profession after graduation [20,79].

It has been shown that nursing students are more vulnerable to burnout during the later years of their university education, and this has been especially so since the outbreak of the COVID-19 pandemic [79,80]. Moreover, the impact of burnout is often more intense towards the end of the degree course [81]. According to Hong [82], the level of burnout increases with the level of education and, for nursing students, the most severe conditions are experienced when clinical practice begins, although another study has reported that the level of burnout remains stable over time [83]. Final-year nursing students, moreover, may feel disappointed and believe themselves incapable of assuming the responsibilities of the profession [84].

The use of instruments that are able to measure and correctly identify the level or intensity of burnout syndrome in university students is important for both research and professional practice. Having calibrated instruments helps in the identification of students who should be the target of interventions aimed at reducing the symptomatology and consequences derived from the burnout syndrome they present. From a practical point of view, professionals need to be confident that the instruments they are going to use discriminate correctly between students with and without burnout; screening is the first step to identifying those future professionals who should be the object of interventions aimed at preventing and/or alleviating the effects caused by burnout in people who are affected by it.

Burnout syndrome prevention programs should be focused on the risk and protective factors presented by university students who suffer from it, so it would be advisable to conduct studies with follow-up measures to verify the stability of the identified needs. Research has described some of the factors to which attention should be paid in working with students with burnout; specifically, work should be conducted to increase resilience [80,85,86] and strengthen perceptions of professional usefulness and self-efficacy [24,87], among other aspects. The ultimate goal of intervention regarding students with burnout should be to improve the well-being of future nursing professionals.

Limitations and Future Directions

This research paper presents an approach to the functioning of the CBG-USS in a sample of nursing students. Among the limitations of this work, the following stand out: First, only the results related to the convergent, discriminant, and concurrent criterion validity of the CBG-USS are presented in this work. Second, the reliability of the CBG-USS has been estimated by taking a single measure; no test–retest was performed. Third, the sample used was selected by convenience sampling, and fourth, the sample used included only nursing students. The aspects related to the sample and sampling used derive from the design proposed for this research and the limitations inherent to carrying out a study in a university educational center in which students are not present all year round; there are vacation periods and students carry out exchanges between universities, which is why a test–retest measure was not considered or the sample was not extended to other degree programs. These aspects will be taken into account in future research; among the objectives of the research group is to extend the sample studied, that is, to apply the CBG-USS to university students from other degrees.

This present study addresses just one source of validity evidence, which is the one that is based on the relation to other variables. From the SEPT framework, there are four additional sources of validity. In future studies, we aim to carry out several analyses to gather new validity evidence on the functioning of the CBG-USS, such as analyzing its internal structure and measurement invariance in different samples.

5. Conclusions

The aim of this research was to examine and provide evidence on some psychometric properties of the Cuestionario Burnout Granada-University Students, specifically, the reliability and validity of this instrument. The study sample is formed by students enrolled in the Bachelor's Degree in Nursing, since health professionals present high rates of burnout. The results found provide evidence of the adequate functioning of the Cuestionario Burnout Granada-University Students, both in reliability and validity. Specifically, given the correlations between the dimensions of MBI and CBG, convergent validity evidence has been obtained, indicating that the burnout measure provided by CBG-USS appears to be appropriately aligned within the nomological network of the construct, given Maslach and Jackson's three-dimensional theory [17]. Discriminant validity evidence has also been obtained, as there were significant differences in CBG-USS scores between students with and without the syndrome. Thus, it appears that CBG-USS can be appropriately used to identify students who have burnout syndrome. Furthermore, criterion validity evidence was also obtained, considering the estimated values by the AUC analysis. This result, similar to the previous one, suggests that CBG-USS can be used to estimate the degree to which students experience burnout syndrome.

Among the findings of this work is the availability of an instrument that identifies burnout in university students, which can be used both for research purposes and for the identification of future professionals who present different levels of burnout, with the aim of intervening to reduce the levels of burnout and the minimization of the consequences derived from this syndrome.

In future research, we would like to improve some of the limitations of this work. Specifically, to enlarge the sample with students from other degrees, both from other health professions and non-health professions, and to include variables that modulate the intensity and severity of burnout that can help in the intervention carried out with those affected by this syndrome.

Supplementary Materials: The following supporting information can be downloaded at: https://www.mdpi.com/article/10.3390/math11153315/s1.

Author Contributions: Conceptualization and Investigation, E.O.-C. and E.I.D.l.F.-S.; Formal analysis, G.R.C.; Methodology, T.A.; Resources, E.O.-C.; Software, G.R.C. and T.A.; Supervision, R.A.-E. and E.I.D.l.F.-S.; Validation, G.R.C. and C.S.M.-R.; Visualization, G.R.C. and R.A.-E.; Writing—Original Draft, E.O.-C. and N.S.-M.; Writing—Review and Editing, R.A.-E. and E.O.-C. All authors have read and agreed to the published version of the manuscript.

Funding: This article has been funded by FEDER/Consejería de Universidad, Investigación e Inovación de la Junta de Andalucía, Project P20-00637.

Institutional Review Board Statement: The study was conducted in accordance with the Declaration of Helsinki and approved by the Institutional Review Board (or Ethics Committee) of the University of Granada (393/CEIH2017).

Data Availability Statement: Not applicable.

Conflicts of Interest: The authors declare no conflict of interest. The funders had no role in the design of the study; in the collection, analyses, or interpretation of data; in the writing of the manuscript, or in the decision to publish the results.

References

1. Aguayo, R.; Cañadas, G.R.; Assbaa, L.; Cañadas-De la Fuente, G.A.; Ramírez-Baena, L.; Ortega-Campos, E. A risk profile of sociodemographic factors in the onset of academic burnout syndrome in a sample of university students. *Int. J. Environ. Res. Public Health* **2019**, *16*, 707. [CrossRef] [PubMed]
2. Caballero, C.C.; Hederich, C.; Palacio, J.E. El burnout académico: Delimitación del síndrome y factores asociados con su aparición. *Rev. Latinoam. Psicol.* **2010**, *42*, 131–146.
3. Caballero, C.C.; Bresó, E.; González Gutiérrez, O. Burnout in university students. *Psicol. Caribe* **2015**, *32*, 424–441. [CrossRef]
4. Membrive-Jiménez, M.J. Prevalencia del Síndrome de Burnout e Identificación de Factores de Riesgo en el Personal de Enfermería Dedicado a la Administración y Gestión del Servicio Andaluz de Salud. Ph.D. Thesis, Universidad de Granada, Granada, Spain, 2022. Available online: https://digibug.ugr.es/handle/10481/76796 (accessed on 8 June 2023).
5. Ortega-Campos, E.; Vargas-Román, K.; Velando-Soriano, A.; Suleiman-Martos, N.; Cañadas-de la Fuente, G.; Albendín-García, L.; Gómez-Urquiza, J.L. Compassion fatigue, compassion satisfaction, and burnout in oncology nurses: A systematic review and meta-analysis. *Sustainability* **2020**, *12*, 72. [CrossRef]
6. Ramírez-Elvira, S.; Romero-Béjar, J.L.; Suleiman-Martos, N.; Gómez-Urquiza, J.L.; Monsalve-Reyes, C.; Albendín-García, L. Prevalence, risk-factors and burnout levels in intensive care unit nurses: A systematic review and meta-analysis. *Int. J. Environ. Res. Public Health* **2021**, *18*, 11432. [CrossRef]
7. Schaufeli, W.B.; Leiter, M.P.; Maslach, C. Burnout: 35 years of research and practice. *Career Dev. Int.* **2009**, *14*, 204–220. [CrossRef]
8. Lin, S.H.; Huang, Y.C. Life stress and academic burnout. *Active Learn. High. Educ.* **2014**, *15*, 77–90. [CrossRef]

9. Cho, S.; Lee, M.; Lee, S.M. Burned-Out Classroom Climate, Intrinsic Motivation, and Academic Engagement: Exploring Unresolved Issues in the Job Demand-Resource Model. *Psychol. Rep.* **2022**, *126*, 1954–1976. [CrossRef]
10. Heinen, I.; Bullinger, M.; Kocalevent, R.D. Perceived stress in first year medical students—Associations with personal resources and emotional distress. *BMC Med. Educ.* **2017**, *17*, 4. [CrossRef]
11. Ghods, A.A.; Ebadi, A.; Sharif Nia, H.; Allen, K.A.; Ali-Abadi, T. Academic burnout in nursing students: An explanatory sequential design. *Nurs. Open* **2023**, *10*, 535–543. [CrossRef]
12. Dos Santos Boni, R.A.; Paiva, C.E.; De Oliveira, M.A.; Lucchetti, G.; Fregnani, J.H.T.G.; Paiva, B.S.R. Burnout among medical students during the first years of undergraduate school: Prevalence and associated factors. *PLoS ONE* **2018**, *13*, e0191746.
13. Quina Galdino, M.J.; Preslis Brando Matos de Almeida, L.; Ferreira Rigonatti da Silva, L.; Cremer, E.; Rolim Scholze, A.; Trevisan Martins, J.; Haddad, F.L.; do Carmo, M. Burnout among nursing students: A mixed method study. *Investig. Educ. Enfermería* **2020**, *38*, e07. [CrossRef]
14. IsHak, W.; Nikravesh, R.; Lederer, S.; Perry, R.; Ogunyemi, D.; Bernstein, C. Burnout in medical students: A systematic review. *Clin. Teach.* **2013**, *10*, 242–245. [CrossRef]
15. Santen, S.A.; Holt, D.B.; Kemp, J.D.; Hemphill, R.R. Burnout in medical students: Examining the prevalence and associated factors. *South. Med. J.* **2010**, *103*, 758–763. [CrossRef]
16. Aguayo-Estremera, R.; Cañadas, G.R.; Ortega-Campos, E.; Pradas-Hernández, L.; Martos-Cabrera, B.; Velando-Soriano, A.; de la Fuente-Solana, E.I. Levels of Burnout and Engagement after COVID-19 among Psychology and Nursing Students in Spain: A Cohort Study. *Int. J. Environ. Res. Public Health* **2023**, *20*, 377. [CrossRef]
17. Maslach, C.; Jackson, S.E. The measurement of experienced burnout. *J. Organ. Behav.* **1981**, *2*, 99–113. [CrossRef]
18. Maslach, C.; Jackson, S.E.; Leiter, M.P. *Maslach Burnout Inventory Manual*, 4th ed.; Mind Garden, Inc.: Menlo Park, CA, USA, 2018.
19. Schaufeli, W.B.; Martínez, I.M.; Marques Pinto, A.; Salanova, M.; Bakker, A.B. Burnout and engagement in university students. *J. Cross-Cult. Psychol.* **2002**, *33*, 464–481. [CrossRef]
20. Aguayo-Estremera, R.; Cañadas, G.R.; Albendín-García, L.; Ortega-Campos, E.; Ariza, T.; Monsalve-Reyes, C.S.; De la Fuente-Solana, E.I. Prevalence of Burnout Syndrome and Fear of COVID-19 among Adolescent University Students. *Children* **2023**, *10*, 243. [CrossRef]
21. Barbosa, J.; Beresin, R. Burnout syndrome in nursing undergraduate students. *Einstein* **2007**, *5*, 225–230.
22. Bittar, C. Burnout y Estilos de Personalidad en Estudiantes Universitarios. Ph.D. Thesis, Universidad de las Islas Baleares, Palma, Spain, 2008. Available online: https://fci.uib.es/digitalAssets/177/177915_2.pdf (accessed on 29 June 2023).
23. Seperak-Viera, R.; Fernández-Arata, M.; Dominguez Lara, S. Prevalence and severity of academic burnout in college students during the COVID-19 pandemic. *Interacciones* **2021**, *7*, e199. [CrossRef]
24. Wang, J.; Bu, L.; Li, Y.; Song, J.; Lid, N. The mediating effect of academic engagement between psychological capital and academic burnout among nursing students during the COVID-19 pandemic: A cross-sectional study. *Nurse Educ. Today* **2021**, *102*, 104938. [CrossRef] [PubMed]
25. Capone, V.; Marino, L.; Park, M.S.A. Perceived Employability, Academic Commitment, and Competency of University Students during the COVID-19 Pandemic: An Exploratory Study of Student Well-Being. *Front. Psychol.* **2021**, *12*, 788387. [CrossRef] [PubMed]
26. Bai, W.; Xi, H.T.; Zhu, Q.; Ji, M.; Zhang, H.; Yang, B.X.; Cai, H.; Liu, R.; Zhao, Y.J.; Chen, L.; et al. Network analysis of anxiety and depressive symptoms among nursing students during the COVID-19 pandemic. *J. Affect. Disord.* **2021**, *294*, 753–760. [CrossRef] [PubMed]
27. García-González, J.; Ruqiong, W.; Alarcon-Rodriguez, R.; Requena-Mullor, M.; Ding, C.; Ventura-Miranda, M.I. Analysis of Anxiety Levels of Nursing Students Because of e-Learning during the COVID-19 Pandemic. *Healthcare* **2021**, *9*, 252. [CrossRef]
28. Obregon, M.; Luo, J.; Shelton, J.; Blevins, T.; MacDowell, M. Assessment of burnout in medical students using the Maslach Burnout Inventory-Student Survey: A cross-sectional data analysis. *BMC Med. Educ.* **2020**, *20*, 376. [CrossRef]
29. Popescu, B.; Maricuțoiu, L.P.; De Witte, H. The student version of the Burnout assessement tool (BAT): Psychometric properties and evidence regarding measurement validity on a romanian sample. *Curr. Psychol.* **2023**. [CrossRef]
30. De Beer, L.T.; Bianchi, R. Confirmatory factor analysis of the Maslach Burnout Inventory: A Bayesian structural equation modeling approach. *Eur. J. Psychol. Assess.* **2019**, *35*, 217–224. [CrossRef]
31. Aguayo, R.; Vargas, C.; de la Fuente, E.I.; Lozano, L.M. A meta-analytic reliability generalization study of the Maslach Burnout Inventory. *Int. J. Clin. Health Psychol.* **2011**, *11*, 343–361.
32. Seisdedos, N. *Manual del MBI, Inventario de Burnout de Maslach*; TEA: Madrid, Spain, 1997.
33. Gascon, S.; Leiter, M.P.; Andrés, E.; Santed, M.A.; Pereira, J.P.; Cunha, M.J.; Albesa, A.; Montero-Marín, J.; García-Campayo, J.; Martínez-Jarreta, E. The role of aggressions suffered by healthcare workers as predictors of burnout. *J. Clin. Nurs.* **2013**, *22*, 3120–3129. [CrossRef]
34. Campagne, D.M. When therapists run out of steam: Professional boredom or burnout? *Rev. Psicopatol. Psicol. Clin.* **2012**, *17*, 75–85.
35. Cañadas-de la Fuente, G.A.; San Luis, C.; Lozano, L.M.; Vargas, C.; García, I.; De la Fuente, E.I. Evidencia de validez factorial del Maslach Burnout Inventory y estudio de los niveles de burnout en profesionales sanitarios. *Rev. Latinoam. Psicol.* **2014**, *46*, 44–52. [CrossRef]

36. Dyrbye, L.N.; Thomas, M.R.; Huntington, J.L.; Lawson, K.L.; Novotny, P.J.; Sloan, J.A.; Shanafelt, T.D. Personal life events and medical student burnout: A multicenter study. *Acad. Med.* **2006**, *81*, 374–384. [CrossRef]
37. American Educational Research Association; American Psychological Association; National Council on Measurement in Education. *Standards for Educational and Psychological Testing*; AERA: Washington, DC, USA; APA: Washington, DC, USA; NCME: Mount Royal, NJ, USA, 2014.
38. Cho, E. The accuracy of reliability coefficients: A reanalysis of existing simulations. *Psychol. Methods* **2022**. [CrossRef]
39. Paniagua-Sánchez, D.; Alvarado, J.; Olivares, M.; Ruiz, I.; Romero-Suárez, M.; Aguayo-Estremera, R. Estudio de Seguimiento de las Recomendaciones sobre Análisis Factorial Exploratorio en RIDEP. *Rev. Iberoam. Diagn. Eval. Psicol.* **2022**, *66*, 127–140. [CrossRef]
40. Cronbach, L.J. Coefficient alpha and the internal structure of tests. *Psychometrika* **1951**, *16*, 297–334. [CrossRef]
41. McDonald, R.P. *Test Theory: A Unified Treatment*; L. Erlbaum Associates: Mahwah, NJ, USA, 1999.
42. Flora, D.B. Your coefficient alpha is probably wrong, but which coefficient omega is right? A tutorial on using R to obtain better reliability estimates. *Adv. Methods Pract. Psychol. Sci.* **2020**, *3*, 484–501. [CrossRef]
43. Holzinger, K.J.; Swineford, F. The bi-factor method. *Psychometrika* **1937**, *2*, 41–54. [CrossRef]
44. Kane, M.T. Validating the interpretations and uses of test scores. *J. Educ. Meas.* **2013**, *50*, 1–73. [CrossRef]
45. Messick, S. *Validity, in Educational Measurement*, 3rd ed.; Linn, R.L., Ed.; American Council on Education and McMillan: New York, NY, USA, 1989; pp. 13–103.
46. Han, Y.; Zhang, J.; Jiang, Z.; Shi, D. Is the Area Under Curve Appropriate for Evaluating the Fit of Psychometric Models? *Educ. Psychol. Meas.* **2023**, *83*, 586–608. [CrossRef]
47. Fawcett, T. An introduction to ROC analysis. *Pattern Recognit. Lett.* **2006**, *27*, 861–874. [CrossRef]
48. Montero, I.; León, O. A guide for naming research studies in Psychology. *Int. J. Clin. Health Psychol.* **2007**, *7*, 847–864.
49. Ato, M.; López-García, J.J.; Benavente, A. Un sistema de clasificación de los diseños de investigación en psicología. *An. Psicol.* **2013**, *29*, 1038–1059. [CrossRef]
50. Downing, S.M. Twelve steps for effective test development. In *Handbook of Test Development*; Downing, S.M., Haladyna, T.M., Eds.; Lawrence Erlbaum Associates Publishers: Mahwah, NJ, USA, 2006; pp. 3–25.
51. Muñiz, J.; Fonseca-Pedrero, E. Diez pasos para la construcción de un test. *Psicothema* **2019**, *31*, 7–16. [CrossRef] [PubMed]
52. Bresó, E.; Salanova, M.; Schaufeli, W.; Nogareda, C. NTP 732: Síndrome de Estar Quemado por el Trabajo "Burnout" (III): Instrumento de Medición. Ministerio de Trabajo y Asuntos Sociales (España) e Instituto Nacional de Seguridad e Higiene en el Trabajo. 2007. Available online: https://www.insst.es/documents/94886/326775/ntp_732.pdf (accessed on 29 June 2023).
53. Maslach, C.; Schaufeli, W.B.; Leiter, M.P. Job burnout. *Annu. Rev. Psychol.* **2001**, *52*, 397–422. [CrossRef] [PubMed]
54. Golembiewski, R.T.; Munzerider, R.F.; Stevenson, J.G. Stress in Organizations. In *Toward a Phase Model of Burnout*; Praeger Publishers: New York, NY, USA, 1986.
55. Golembiewski, R.T.; Munzenrider, R. *Phases of Burnout: Developments in Concepts and Applications*; Praeger: New York, NY, USA, 1988.
56. George, D.; Mallery, P. *SPSS for Windows Step by Step: A Simple Guide and Reference*, 4th ed.; 11.0 Update; Allyn & Bacon: Boston, MA, USA, 2003.
57. Aguayo-Estremera, R.; Cañadas, G.R.; Ortega-Campos, E.; Ariza, T.; De la Fuente-Solana, E.I. Validity Evidence for the Internal Structure of the Maslach Burnout Inventory-Student Survey: A Comparison between Classical CFA Model and the ESEM and the Bifactor Models. *Mathematics* **2023**, *11*, 1515. [CrossRef]
58. García, J.; Ortega, E.; De la Fuente Sánchez, L. The use of the effect size in JCR Spanish journals of psychology: From theory to fact. *Span. J. Psychol.* **2011**, *14*, 1050–1055. [CrossRef]
59. Vacha-Haase, T.; Nilsson, J.E.; Reetz, D.R.; Lance, T.S.; Thompson, B. Reporting practices and APA editorial policies regarding statistical significance and effect size. *Theory Psychol.* **2000**, *10*, 413–425. [CrossRef]
60. Gulliksen, H. *Theory of Mental Tests*; Lawrence Erlbaum: New York, NY, USA, 1987.
61. Rice, M.E.; Harris, G.T. Comparing effect sizes in follow-up studies: ROC area, Cohen's d, and r. *Law Hum. Behav* **2005**, *29*, 615–620. [CrossRef]
62. De la Fuente, E.I.; Lozano, L.M.; Carcía-Cueto, E.; San Luís, C.; Vargas, C.; Cañadas, G.R.; Cañadas-De la Fuente, G.A.; Hambleton, R.K. Development and validation of the Granada Burnout Questionnaire in Spanish police. *Int. J. Clin. Health Psychol.* **2013**, *13*, 216–225. [CrossRef]
63. De la Fuente, E.I.; García, J.; Cañadas, G.A.; San Luís, C.; Cañadas, G.R.; Aguayo, R.; de la Fuente, L.; Vargas, C. Psychometric properties and scales of the Granada Burnout Questionnaire applied to nurses. *Int. J. Clin. Health Psychol.* **2015**, *15*, 130–138. [CrossRef]
64. De La Fuente-Solana, E.I.; Ortega-Campos, E.; Vargas-Roman, K.; Cañadas-De la Fuente, G.R.; Ariza, T.; Aguayo-Extremera, R.; Albendín-García, L. Study of the Predictive Validity of the Burnout Granada Questionnaire in Police Officers. *Int. J. Environ. Res. Public. Health* **2020**, *17*, 6112. [CrossRef]
65. Bauernhofer, K.; Tanzer, N.; Paechter, M.; Papousek, I.; Fink, A.; Weiss, E.M. Frenetic, Underchallenged, and Worn-Out: Validation of the German "Burnout Clinical Subtypes Questionnaire"—Student Survey and Exploration of Three Burnout Risk Groups in University Students. *Front. Educ.* **2019**, *4*, 137. [CrossRef]

66. Morales-Rodríguez, F.M.; Pérez-Mármol, J.M.; Brown, T. Education Burnout and Engagement in Occupational Therapy Undergraduate Students and Its Associated Factors. *Front. Psychol.* **2019**, *10*, 2889. [CrossRef]
67. Portoghese, I.; Leiter, M.P.; Maslach, C.; Galletta, M.; Porru, F.; D'Aloja, E.; Finco, G.; Campagna, M. Measuring Burnout Among University Students: Factorial Validity, Invariance, and Latent Profiles of the Italian Version of the Maslach Burnout Inventory Student Survey (MBI-SS). *Front. Psychol.* **2018**, *9*, 2105. [CrossRef]
68. Souza, R.O.D.; Ricardo Guilherme, F.; Elias, R.G.M.; dos Reis, L.L.; Garbin de Souza, O.A.; Robert Ferrer, M.; Dos Santos, S.L.C.; Osiecki, R. Associated Determinants Between Evidence of Burnout, Physical Activity, and Health Behaviors of University Students. *Front. Sports Act. Living* **2021**, *3*, 733309. [CrossRef]
69. Wongtrakul, W.; Dangprapai, Y.; Saisavoey, N.; Sa-nguanpanich, N. Reliability and validity study of the Thai adaptation of the Maslach Burnout Inventory-Student Survey among preclinical medical students at a medical school in Thailand. *Front. Psychol.* **2023**, *14*, 1054017. [CrossRef]
70. Cohen, J. A power primer. *Psychol. Bull.* **1988**, *112*, 155–159. [CrossRef]
71. Gibbons, C. Stress, coping and burnout in nursing students. *Int. J. Nurs. Stud.* **2010**, *47*, 1299–1309. [CrossRef]
72. Arian, M.; Jamshidbeigi, A.; Kamali, A.; Dalir, A.; Ali-Abadi, T. The prevalence of burnout syndrome in nursing students: A systematic review and meta-analysis. *Teach. Learn. Nurs.* **2023**. [CrossRef]
73. Valero-Chillerón, M.J.; González-Chordá, V.M.; López-Peña, N.; Cervera-Gasch, Á.; Suárez-Alcázar, M.P.; Mena-Tudela, D. Burnout syndrome in nursing students: An observational study. *Nurse Educ. Today* **2019**, *76*, 38–43. [CrossRef] [PubMed]
74. Aghajari, Z.; Loghmani, L.; Ilkhani, M.; Talebi, A.; Ashktorab, T.; Ahmadi, M.; Borhani, F. The relationship between quality of learning experiences and academic burnout among nursing students of Shahid Beheshti University of Medical Sciences in 2015. *Electron. J. Gen. Med.* **2018**, *15*, 1–10. [CrossRef]
75. Cao, X.; Wang, L.; Wei, S.; Li, J.; Gong, S. Prevalence and predictors for compassion fatigue and compassion satisfaction in nursing students during clinical placement. *Nurse Educ. Pract.* **2021**, *51*, 102999. [CrossRef] [PubMed]
76. Chang, E.; Daly, J. *Transitions in Nursing Preparing for Professional Practice*; Elsevier: Chatswood, NSW, Australia, 2012.
77. Rudman, A.; Gustavsson, J.P. Burnout during nursing education predicts lower occupational preparedness and future clinical performance: A longitudinal study. *Int. J. Nurs. Stud.* **2012**, *49*, 988–1001. [CrossRef] [PubMed]
78. Salmela-Aro, K.; Upadyaya, K. School burnout and engagement in the context of demands-resources model. *Br. J. Educ. Psychol.* **2014**, *84 Pt 1*, 137–151. [CrossRef] [PubMed]
79. Sveinsdóttir, H.; Guðrún, B.; Hrönn, M.; Scheving, H.; Kort, G.; Bernharðsdóttir, J.; Svavarsdóttir, E.K. Predictors of university nursing students burnout at the time of the COVID-19 pandemic: A cross-sectional study. *Nurse Educ. Today* **2021**, *106*, 105070. [CrossRef]
80. Merino-Godoy, M.A.; Yot-Dominguez, C.; Conde-Jimenez, J.; Ramírez-Martin, P.; Lunar-Valle, P.M. The influence of emotional burnout and resilience on the psychological distress of nursing students during the COVID-19 pandemic. *Int. J. Ment. Health Nurs.* **2022**, *31*, 1457–1466. [CrossRef]
81. Al-Zayyat, A.S.; Al-Gamal, E. Perceived stress and coping strategies among Jordanian nursing students during clinical practice in psychiatric/mental health courses. *Int. J. Ment. Health Nurs.* **2014**, *23*, 326–335. [CrossRef]
82. Hong, C.M. The Relationship between burnout using MBI-SS and mental health in nursing students. *Asia-Pac. J. Multimed. Serv. Converg. Art Humanit. Sociol.* **2015**, *5*, 353–362. [CrossRef]
83. Deary, I.J.; Watson, R.; Hogston, R. A longitudinal cohort study of burnout and attrition in nursing students. *J. Adv. Nurs.* **2003**, *43*, 71–81. [CrossRef]
84. Pearcey, P.; Draper, P. Exploring clinical nursing experiences: Listening to student nurses. *Nurse Educ. Today* **2008**, *28*, 595–601. [CrossRef]
85. Horvath, C.; Grass, N. Pandemic, Economic Uncertainty, and Protests: What Will Happen to Student Registered Nurse Anesthetists-Resiliency or Burnout? *AANA J.* **2021**, *89*, 413–418.
86. Ríos-Risquez, M.I.; García-Izquierdo, M.; Sabuco-Tebar, E.L.Á.; Carrillo-Garcia, C.; Solano-Ruiz, C. Connections between academic burnout, resilience, and psychological well-being in nursing students: A longitudinal study. *J. Adv. Nurs.* **2018**, *74*, 2777–2784. [CrossRef]
87. Hu, Y.; Hu, J.; Li, L.; Zhao, B.; Liu, X.; Li, F. Development and preliminary validation of a brief nurses' perceived professional benefit questionnaire (NPPBQ). *BMC Med. Res. Methodol.* **2020**, *20*, 18. [CrossRef]

Disclaimer/Publisher's Note: The statements, opinions and data contained in all publications are solely those of the individual author(s) and contributor(s) and not of MDPI and/or the editor(s). MDPI and/or the editor(s) disclaim responsibility for any injury to people or property resulting from any ideas, methods, instructions or products referred to in the content.

Article

Dynamic Generation Method of Highway ETC Gantry Topology Based on LightGBM

Fumin Zou [1], Weihai Wang [1], Qiqin Cai [1,2,*], Feng Guo [1] and Rouyue Shi [1]

1. Fujian Key Laboratory for Automotive Electronics and Electric Drive, Fujian University of Technology, Fuzhou 350118, China; fmzou@fjut.edu.cn (F.Z.); 2211308003@smail.fjut.edu.cn (W.W.); n180310004@fzu.edu.cn (F.G.); 2211301008@smail.fjut.edu.cn (R.S.)
2. School of Mechanical Engineering and Automation, Huaqiao University, Xiamen 361021, China
* Correspondence: 20011080002@stu.hqu.edu.cn; Tel.: +86-181-4405-6202

Abstract: In Electronic Toll Collection (ETC) systems, accurate gantry topology data are crucial for fair and efficient toll collection. Currently, inaccuracies in the topology data can cause tolls to be based on the shortest route rather than the actual distance travelled, contradicting the ETC system's purpose. To address this, we adopt a novel Gradient Boosting Decision Tree (GBDT) algorithm, Light Gradient Boosting Machine (LightGBM), to dynamically update ETC gantry topology data on highways. We use ETC gantry and toll booth transaction data from a province in southeast China, where ETC usage is high at 72.8%. From this data, we generate a candidate topology set and extract five key characteristics. We then use Amap API and QGIS map analysis to annotate the candidate set, and, finally, apply LightGBM to train on these features, generating the dynamic topology. Our comparison of LightGBM with 14 other machine learning algorithms showed that LightGBM outperformed the others, achieving an impressive accuracy of 97.6%. This methodology can help transportation departments maintain accurate and up-to-date toll systems, reducing errors and improving efficiency.

Keywords: highway; ETC gantry; topology dynamic generation; LightGBM

MSC: 68T09

Citation: Zou, F.; Wang, W.; Cai, Q.; Guo, F.; Shi, R. Dynamic Generation Method of Highway ETC Gantry Topology Based on LightGBM. *Mathematics* **2023**, *11*, 3413. https://doi.org/10.3390/math11153413

Academic Editor: Alicia Cordero

Received: 11 July 2023
Revised: 30 July 2023
Accepted: 3 August 2023
Published: 4 August 2023

Copyright: © 2023 by the authors. Licensee MDPI, Basel, Switzerland. This article is an open access article distributed under the terms and conditions of the Creative Commons Attribution (CC BY) license (https://creativecommons.org/licenses/by/4.0/).

1. Introduction

Highways play a pivotal role in the modern transportation system, offering convenient transit services that fuel economic growth across various countries and regions. With societal advancement and technological evolution, toll collection methods on these highways are continually being modernized. Among these is the Electronic Toll Collection (ETC) system, which is renowned for its efficiency and which has been increasingly adopted across nations and regions. However, during practical implementation, the ETC system still faces numerous challenges, such as toll-related issues stemming from inaccurate topological data. This research aims to address this issue, proposing a method for the dynamic updating of the topological data of highway ETC gantries.

Serving as a pivotal piece of infrastructure within the realm of transportation, the highway gantry system gleans data via the On-Board Unit (OBU) and Roadside Unit (RSU) devices. The OBU, an electronic device mounted on vehicles, functions to communicate with the highway toll collection system. In contrast, the RSU, situated at highway toll booths or gantries, interacts with the OBU, facilitating automatic toll collection. The objective behind the installation of gantries is to compute the actual mileage traversed by vehicles. The gantry system operates by scanning the OBU mounted on vehicles, logging the vehicle's travel distance and time. By calculating the length of each gantry topological path that the vehicle traverses on the highway and the per-kilometer travel cost, the system determines the appropriate toll. According to the requirements of the "Implementation Plan for the

Full Promotion of Differentiated Toll Collection on Highways", China is actively promoting the reform of differentiated toll collection on highways to further improve the efficiency and service level of the highway network, reduce the cost of highway travel, and promote cost reduction and efficiency improvement in the logistics industry. To this end, China is implementing a series of strategies, including thoroughly summarizing the experience of differentiated toll collection pilot projects on highways, fully considering factors such as the structure and operating characteristics of local road networks, choosing suitable differentiated toll collection methods, innovating service models, and scientifically and accurately formulating differentiated toll collection schemes [1]. In addition, differentiated toll collection is being implemented by road section, mainly implementing flexible and diverse differentiated tolls on road sections where ordinary national and provincial trunk roads or urban roads are severely congested but parallel highways have small traffic flow, road sections where the traffic volume differs greatly between parallel highways, and road sections where the traffic volume is significantly lower than the design capacity. Although there have been some studies on the toll collection of highways, there are few studies on how to support these differentiated toll collection strategies better through efficient and accurate dynamic updates of topological data. In the domain of ETC gantry topology generation research, Cai and Yi et al. [2] introduced a pioneering approach known as the "Arch-Bridge topology." This method entails the generation of an initial topology candidate set through the analysis of discrete ETC data, followed by the identification and examination of abnormal topology features. Subsequently, the Dijkstra algorithm is applied to optimize the topology, resulting in a comprehensive ETC gantry topology.

A noteworthy limitation of their approach lies in their static nature, as they lack the capability for dynamic updates. Unlike our proposed methodology, which enables the real-time adjustment and refinement of gantry topology based on dynamically changing road network conditions, the Arch-Bridge topology is confined to performing solely static updates without considering real-time traffic variations. However, the topological information of highway ETC gantries is far from static. For instance, the addition or decommissioning of roads may lead to permanent changes in gantry topology, while road maintenance or unexpected events may cause temporary changes. This can result in inaccuracies in the topological data within some vehicle transaction data, thus making it impossible to calculate tolls based on the distance traveled. Instead, tolls are charged based on the shortest distance. This method of toll collection can lead to many problems. Firstly, it could result in financial losses, as drivers might be required to pay for distances that are longer or shorter than the actual distance traveled. Charging more is unfair to commercial vehicles that frequently use highways, and charging less can also lead to financial losses for ETC administrators. This method of toll collection deviates from the original intention of the ETC system, which is to provide drivers with more convenient, faster, and more accurate toll services. Secondly, the research on ETC gantry topology can provide richer application scenarios for intelligent transport systems. For example, based on the topological information of the ETC gantry, more refined applications, such as traffic condition prediction, congestion detection, and route planning, can be developed to provide drivers with more convenient travel services. At the same time, it can also provide optimization solutions for the fields of public transportation and logistics, reduce operational costs, and improve transportation efficiency.

To address the current issues in the dynamic update of highway ETC gantry topology, and to enhance the quality and efficiency of intelligent highway services, this paper proposes a method for dynamically generating highway ETC gantry topology based on LightGBM. Specifically, the method first generates a candidate set of gantry topology using the highway ETC transaction data of a province in China with a high ETC usage rate of 72.8% [3], extracts the candidate topology, and filters out incorrect topology. Then, five feature dimensions are extracted from each topology, including Topology Traffic Volume (*TTV*), Topological Passage Rate (*TPR*), Normalized Start Rate (*NSR*), Normalized End Rate (*NER*), and Topology Distance (*TD*). Next, by combining the Amap API [4] and QGIS [5] map analysis, the candidate topology set is marked, with real existing topology marked as 1 and non-existing topology marked as 0. In this case, the dynamic generation of gantry topology can be transformed into a typical binary classification problem in machine learning. This study uses supervised learning algorithms (such as SVM, Naive Bayes, Logistic Regression, etc.), tree model algorithms (such as LightGBM, RF, XGBoost, etc.), and ensemble learning algorithms (such as AdaBoost, Bagging, etc.) for the dynamic generation of gantry topology.

Our research makes the following key contributions:

- Introduction of an innovative methodology for dynamic updating of highway gantry topology based on ETC transaction data.
- Rectification of prevalent inaccuracies in topology data within vehicle transaction records, leading to more accurate fee computation for actual traversed distances.
- Utilization of the LightGBM model to facilitate dynamic updating of the gantry topology with an impressive accuracy rate of 97.6%.
- Universally applicable methodology and framework for dynamically updating highway ETC gantry topology, demonstrating extensive applicability and scalability.

The remainder of this paper is structured as follows. An exhaustive literature review graces Section 2. In Section 3, we expound upon the methodologies pertaining to data preprocessing and extraction of candidate topology sets, as well as dynamic generation techniques for gantry topology. Experiments along with a comprehensive analysis of the results form the substance of Section 4. Lastly, a cogent summary encapsulating the novelties and constraints of our research, along with a glance into future possibilities, concludes our discourse.

2. Literature Review

Within our research domain, which involves the dynamic generation of ETC gantry topology, there is currently a lack of directly related work due to the innovative and unique nature of this issue. As a result, we often draw upon methods and theories from other disciplines to inspire new perspectives and identify novel solutions. In this context, we have consulted related literature from fields such as "remote sensing image recognition methods" and "spatio-temporal trajectory data mining." Both of these areas have extensive research experience and significant achievements in dealing with large-scale, complex, and dynamic data. Their methodologies provide us with insightful perspectives. The inspiration drawn from their methods guides our direction and shapes our solutions, thereby enriching the results of our research.

2.1. Remote Sensing Image Recognition Method

Road network generation methods from remote sensing imagery can be broadly categorized as: (1) image-segmentation-based, (2) feature-extraction-based, and (3) methods reliant on machine learning and deep learning approaches.

Image segmentation involves dissecting remote sensing images into various regions, each comprising pixels with similar characteristics, such as color, intensity, or texture. After segmentation, the recognized road regions can be interconnected to formulate the road network. The Snake model, as proposed by Kass et al. [6], signifies a pivotal milestone. In their model, an energy function was defined on the image contour sketched by the user, then adjusted iteratively to converge the image contour at the minimum energy. Despite requiring human intervention, this method effectively addresses noise and large gaps in road images. This approach was further adopted in remote sensing imagery by Péteri and Ranchin [7] and Laptev et al. [8] with their multi-resolution Snake models for road extraction. Subsequent research, like the work of Gruen and Li [9], amalgamated a semi-automatic road extraction strategy utilizing wavelet decomposition for road sharpening alongside a model-driven linear feature extraction method based on dynamic programming. These methods could tackle more intricate road networks. While requiring human intervention, these methods successfully dealt with noise and large gaps in road images.

Regarding feature-extraction-based road recognition, these techniques identify road regions and generate the road network by extracting features like texture, edges, and color information from remote sensing images. For instance, Mokhtarzadeh and Zoej [10] employed artificial neural networks for the detection of roads within high-resolution satellite images, verifying the impact of disparate input parameters on the network's capability, thereby determining the superior network architecture. Moreover, Yager and Sowmya [11] were the trailblazers in utilizing support vector machines(SVM) to extract roads from remote sensing imagery based on road edges. While this technique boasts commendable integrity, its accuracy remains relatively inferior. The methods of feature extraction are incredibly sensitive to image quality and feature selection, with inappropriate features potentially resulting in imprecise identification. Although this methodology can utilize the abundant feature information within remote sensing imagery, it may encounter difficulties with low-resolution or subpar quality images.

In recent years, methods anchored in machine learning and deep learning have garnered attention. These methodologies train models to recognize and extract road regions, creating the road network using machine learning and deep learning algorithms. Deep learning has shown marked effectiveness in handling complex tasks, especially in remote sensing image processing. Mnih and Hinton [12] were the pioneers in applying deep learning techniques to extract road information from high-resolution aerial photographs. He et al. [13] blazed a new trail for training deep neural networks to tackle the vanishing gradient problem, proposing a deep residual learning framework and integrating identity mapping to streamline the training process. In addition, Saito et al. [14] successfully recognized roads and buildings in raw remote sensing imagery in an innovative manner, achieving commendable results on a road dataset from Massachusetts. The fruitful application of Fully Convolutional Networks (FCN) [15] in the semantic segmentation of high-resolution remote sensing imagery [16] is also noteworthy. Deep learning has also played an indispensable role in a myriad of complex remote sensing image tasks, such as automatic object detection [17], semantic labeling of satellite imagery [18], and image classification [19]. U-Net has also shown significant progress in road extraction studies. Inspired by U-Net's successful application in medical image processing, the U-Net model designed by Keramitsoglou et al. [20] has substantially enhanced the handling of complex road networks for road extraction tasks. However, these methodologies still demand a high quality and quantity of training data and computational resources, posing a challenge that has yet to be resolved.

2.2. Spatio-Temporal Trajectory Data Mining

In recent years, with the growing prevalence of vehicle GPS data, numerous researchers have begun to exploit these real-time and precise data to generate maps and road network models. The primary methodologies include: (1) Cluster-Based Approach, (2) Kernel Density Estimation (KDE), and (3) Intersection Linking.

The technique of cluster analysis, which includes clustering algorithms like K-means, is utilized in interpreting GPS sample points, consequently extracting basic points and edges of the road map. These elements are then assembled to create a road network. Initial explorations by Wagstaff and his colleagues [21] demonstrated the potential of this technique for identifying lanes from low-precision GPS data. They later designed a spatial clustering algorithm that independently inferred the map's connectivity structure sans initial inputs, thereby augmenting the process of lane division and merging [22]. In a mining environment devoid of clear road boundaries or lane markings, Worrall S and Nebot E [23] successfully adapted clustering algorithms, illustrating the method's wide-ranging utility and resilience. DBSCAN-centric point clustering techniques have also demonstrated their significance in map matching studies [24]. Edelkamp and Schrodl [25] took the pioneering step of employing K-means clustering on the position of trajectory samples. Each cluster's center was considered a road node, and the subsequent connection of these nodes resulted in a comprehensive road network. Chen C and his team [26] introduced "Traj-Meanshift", a noise reduction algorithm and a graph-oriented road segment clustering algorithm specifically for de-noising GPS data points to attain precise information processing. They further proposed a new graph-based road segment clustering algorithm that capitalized on prior knowledge of road smoothness to boost the precision and effectiveness of clustering. Huang J and his group [27] employed an ASCDT-based spatial clustering technique and assimilated spatial semantic data to construct roads and their topological relationships. Cluster analysis techniques are adept at handling vast quantities of GPS data without the need for intricate preprocessing. However, the outcome's quality is contingent upon the density and quality of the GPS data. Sparse or sub-standard data might compromise the accuracy of the generated road network.

Kernel Density Estimation (KDE) [28], a principal technique frequently enlisted for spatial data scrutiny and visualization, fundamentally aims at decoding and prognosticating potential event paradigms. In contemporary research, KDE has been synergistically harnessed in the realm of road network generation. Initially, KDE transmutes individual instances or trajectories into a discretized graphic representation, reflecting the density of samples or segments per pixel entity. It implements binary thresholds to actualize road binary illustrations within certain regions, thereby discovering central arterial lines of roads through an array of methodologies such as Voronoi segmentation. Researchers such as Fu Z and his colleagues [29] have successfully constructed an efficacious road network, circumventing the necessity for auxiliary parameter amendments, by employing kernel density analysis, Hidden Markov Models, and map matching techniques. Uduwaragoda E and his team [30] utilized non-parametric Kernel Density Estimation (KDE) to scrutinize the probability density distribution of trajectory nodes, subsequently generating geospatial representations containing lane centerlines. Similarly, Kuntzsch C and others [31] integrated heuristic methodologies with generative modeling, employing KDE to reconstruct an optimized rendition of road maps. Neuhold R and collaborators [32] utilized KDE to process low-precision GPS data, thereby accurately identifying lane centerlines for various road categorizations. KDE, by estimating the location and morphology of roads based on GPS data density, demonstrates an inherent adaptability to a broad spectrum of road types and structures. Furthermore, the KDE methodology is not contingent on a specific clustering algorithm, thus affording substantial flexibility during the processing of GPS data. Nonetheless, the KDE approach necessitates the selection of appropriate kernel functions and bandwidth parameters, which could require profound technical knowledge. Moreover, if the GPS data exhibit a heterogenous distribution, the precision of the KDE output may be compromised.

The Intersection Connection approach begins by identifying vertices at road intersections, subsequently establishing connections between these vertices and edges based on trajectory characteristics to detect road junctions. By interpolating the geometric configurations of trajectories, intersections are interconnected, thus creating and updating road networks. Huang Y and his team [33] deployed a priori knowledge regarding intersection typologies and turn restrictions for the detection of road segments. Deng and colleagues [34] proposed a clustering methodology predicated on hotspot analysis and Delaunay triangulation aimed at spatial coverage detection of road intersections. The accuracy of intersection detection was enhanced by the generation of structural models via K-segmentfit and common subsequence amalgamation. Wu J and others [35] proposed an intersection recognition mechanism founded on an augmented X-means algorithm, successfully implemented for the identification of road network intersections in Shenyang, Liaoning Province, China. Xie X and colleagues [36] introduced a novel intersection definition, delineating intersections as loci connecting three or more different directional road segments, and employed the Longest Common Subsequence (LCSS) for intersection detection under this definition. They managed to effectively identify intersections by discerning common sub-trajectories of multiple GPS trajectories. Fathi and Krumm [37] were among the pioneers who utilized Intersection Connection techniques for cartographic construction, revolutionizing the field with their groundbreaking research and attracting further scholarly attention and investigation. In subsequent research, they incorporated image processing techniques [35], performing skeletonization on binary trajectory matrices and conducting local "sub-path" detection. Finally, they employed Kernel Density Estimation (KDE) for the identification of road intersections, thereby offering a novel toolkit for intersection recognition and analysis. Karagiorgou and Pfoser [38] devised a heuristic algorithm capable of "bundling" trajectories in the vicinity of intersection nodes, thereby connecting these trajectories. Pu M and others [39] proposed a novel bi-stage road intersection detection framework dubbed RIDF composed of trajectory quality enhancement and intersection extraction. Zhao L and his team [40] pioneered calibration of road intersection impact area topology by introducing a tri-stage calibration framework, denoted as CITT. Qing R and colleagues [41] proposed a GPS trajectory-based road intersection detection methodology that leverages temporal–spatial feature extraction and their interactions to amplify the accuracy of intersection detection. Liu Y and his team [42] utilized the (xDeepFM) model to extract geometric and spatial features from GPS data, and integrated density-based spatial clustering of applications with noise (DBSCAN) and Delaunay triangulation for cluster and intersection radius computations, thereby enhancing the accuracy of road intersection recognition. The intersection connection method is adept at handling extensive GPS data and adapts well to dynamically changing road network environments. However, its reliance on intersection detection could be problematic if the GPS data quality is poor or if the data are sparse. Furthermore, for rural or mountainous roads with no distinct intersections, the intersection connection method might struggle to generate accurate road networks.

In a recent research endeavor, Cai and Yi et al. [2] presented a groundbreaking topology called the Arch-Bridge topology. This innovative construct offers a paradigm shift in the definition of highway network structures. It creates an initial topology candidate set via processing discrete ETC data, meticulously mines and analyzes anomalous topological features, and optimizes them utilizing Dijkstra's algorithm. The experimental results show superior performance in terms of recall, precision, F1 score, and the efficiency of topology generation. This seminal work has profoundly impacted our research, upon which we have further developed. Although this investigation has significantly contributed to topology generation, it exclusively accommodates updates in response to permanent topological changes and does not facilitate dynamic topology generation. Consequently, a comprehensive analysis and synthesis of various road network generation methodologies indicate that extant approaches either overemphasize static road network structure analysis, thus struggling to accurately track dynamic topological changes, or rely heavily on voluminous data and computational resources, limiting their practicality. Therefore, there is an exigent need for an innovative methodology that leverages a stable data source and efficiently captures real-time dynamic changes in the road network topology. To address these limitations, the paper at hand proposes a new method that employs real-time ETC data and the LightGBM algorithm to construct a dynamic generation model for ETC gantry topology, effectively mitigating the main limitations of the existing methodologies.

Our method generates and dynamically updates highway gantry topology information, providing accurate segment information to mitigate the economic losses of highway managers and ETC operators. This approach can offer the intelligent transportation domain a simplistic yet efficacious tool to compensate for the deficiencies of existing methods and enhance traffic management efficiency as well as the precision of segmented toll collection.

3. Methodology

3.1. Data Introduction and Relevant Explanation

The experimental dataset utilized in this study originates from the transaction records of an ETC system on a highway in a certain province of China from 1 to 5 June 2021. This includes transaction data from ETC gantries, entry toll booths, and exit toll booths, all of which are collectively referred to as ETC Transaction Data. Approximately 31 million transaction records were collected, involving around 1.33 million vehicles. These vehicles include four types of passenger vehicles, six types of freight vehicles, and six types of specialized operation vehicles. The classification of these vehicles is based on their purpose, structure, size, and passenger or freight capacity, following the Chinese transportation industry standard JT/T 489-2019, "Vehicle Classification of Toll for Highway" [43]. The four categories of passenger vehicles are classified according to their approved seating capacity: Category 1 (up to 9 seats), Category 2 (10–19 seats), Category 3 (20–39 seats), and Category 4 (40 seats and above). The six categories of freight vehicles are classified according to the total number of axles, length, and maximum permitted total mass, specifically including vehicles with 2, 3, 4, 5, 6, and more than 6 axles (for oversized transport vehicles). The six categories of specialized operation vehicles are similarly classified according to the total number of axles, length, and maximum permitted total mass. The first five categories follow the same classification method as freight vehicles, while the sixth category includes vehicles with no less than six axles. The ETC gantry transaction data (GData) encompasses anonymized vehicular identifiers, transactional timestamps, error codes related to transactions, and journey identifiers, supplemented by data pertaining to the gantry's identification number, name, type, and geographic coordinates (as shown in Table 1). The ETC entry toll booth data (EnData) includes anonymized vehicle identification, the identification number, names, types, and geographic coordinates of the entry toll booths, as well as the transactional timestamps, journey identifiers, and transaction error codes (as shown in Table 2). Similarly, the ETC exit toll booth data (ExData) includes anonymized vehicle identification, the identification number, names, types, and geographic coordinates of the exit toll booths, as well as the transactional timestamps, journey identifiers, and transaction error codes (as shown in Table 3). Through in-depth analysis and mining of these data, we can better understand the characteristics of the highway gantry topology and provide robust support for dynamically updating the gantry topology.

Table 1. Description of partial fields in ETC gantry transaction data.

Index	Field Name	Field Properties	Example
1	GantryID	gantry id number	340E11
2	GantryName	gantry name	Jinhai to Nanzhou
3	GantryType	gantry type	2
4	GantryCor	gantry geographic coordinates	(119.308, 25.888)
5	OBUid	vehicular identifiers	1452687261
6	PassID	journey identifiers	0142***561
7	TradeTime	transactional timestamps	2021/6/1 12:00:00
8	ErrorCode	transaction error codes	1

This table shows the descriptions of some ETC transaction data fields. PassID serves as a unique identifier used to precisely locate each journey; the ErrorCode field is used to record the status of each transaction, where 0 represents a normal transaction, and 1 signifies an abnormal transaction. In addition, OBUid is used to uniquely identify the OBU device number of a vehicle.

Table 2. Description of partial fields in ETC entrance toll booth data.

Index	Field Name	Field Properties	Example
1	EnBoothID	entrance booth id number	2100
2	EnBoothName	entrance booth name	Jinhai toll booth
3	EnBoothType	entrance booth type	0
4	EnBoothCor	entrance booth geographic coordinates	(118.318, 26.778)
5	OBUid	vehicular identifiers	1452687261
6	PassID	journey identifiers	0142***561
7	TradeTime	transactional timestamps	2021/6/1 12:00:00
8	ErrorCode	transaction error codes	1

Table 3. Description of partial fields in ETC exit toll booth data.

Index	Field Name	Field Properties	Example
1	ExBoothID	exit booth id number	2100
2	ExBoothName	exit booth name	Jinhai toll booth
3	ExBoothType	exit booth type	1
4	ExBoothCor	exit booth geographic coordinates	(118.316, 26.767)
5	OBUid	vehicular identifiers	1452687261
6	PassID	journey identifiers	0142***561
7	TradeTime	transactional timestamps	2021/6/1 12:00:00
8	ErrorCode	transaction error codes	1

In our analysis of ETC transaction data, we identified numerous fields with analogous attributes, such as identifiers, names, and types, across three datasets. Concurrently, these datasets share common fields, including transaction timestamps, geographic coordinates, PassID, and transaction error codes. These similarities aid in our comprehension and exploration of the relationships among these data, thereby facilitating the integration of data from ETC gantries and toll stations.

3.2. Preprocessing of ETC Transaction Data

3.2.1. Data Cleaning and Data Fusion

Data preprocessing is a vital step in data mining and analysis, capable of eliminating invalid data, reducing noise interference, and enhancing the accuracy and efficiency of data analysis. This section introduces how to preprocess ETC transaction data, including the filtering of error codes and transaction IDs, providing critical data support for subsequent updates to gantry topology.

Anomaly Removal: In the realm of ETC transactional data, anomalies may emerge that are attributable to equipment malfunctions or a plethora of diverse factors. Detrimental meteorological phenomena, impairment of hardware integrity, software dysfunctions, electromagnetic disruptions, network inconsistencies, and thermal overloads constitute potential confounding variables that could jeopardize the operational efficacy of RSUs and OBUs. Such circumstances may instigate instances within the ETC transactional dataset that significantly stray from the anticipated or normative patterns. For these divergent instances, the Errorcode field is designated as 1 during the data upload phase. To uphold the precision and authenticity of the ensuing data analysis, data entries where the Errorcode field is denoted as 1 are meticulously and preemptively excised from the dataset.

Related Data Filtering: The highway transaction data includes two types of transaction data, ETC and MTC (Manual Toll Collection). To concentrate on the transaction data of ETC gantries and toll booths, we need to filter out data unrelated to ETC. In this study, data starting with 01 in PassID were retained because a PassID beginning with 01 indicates that the data are ETC transaction data. Such filtering ensures that we focus only on data relevant to the research objectives, improving the specificity of the analysis.

After data cleaning, we used the SQL UNION ALL statement to merge the GData, EnData, and ExData into a single result set containing all data. Next, the merged data were sorted by PassID and TradeTime. Finally, the sorted result set was grouped by PassID. Through this integration, we obtained an ETC transaction data fusion data (EFusionData) containing ETC gantry data, ETC entry toll booth data, and ETC exit toll booth data.

In the ETC fused transaction data (EFusionData), we standardized the names of fields with similar attributes. Herein, we collectively referred to gantries and toll stations as ETC nodes (EtcNode). Based on this concept, the following fields were extracted: node ID, node name, node type, node geographic coordinates, transaction time, vehicle identification, and journey identifier. Importantly, since we removed data with a transaction error code of 1, only data with a transaction error code of 0 remained in the fusion table, so we discarded the transaction error code field in the data fusion table. Additionally, we need to explain the node type. There are four types of nodes, namely, entry toll booths, exit toll booths, intraprovincial gantries, entry provincial boundary gantries, and exit provincial boundary gantries, represented by numbers 0, 1, 2, 3, and 4, respectively. Table 4 shows the fields of the ETC transaction data fusion table, as well as their descriptions and examples. The detailed process of data cleaning and fusion is depicted in Algorithm 1.

Table 4. Description of fields in ETC transaction fusion data.

Index	Field Name	Field Properties	Example
1	NodeID	node id number	34E102
2	NodeName	node name	Jinhai toll booth
3	NodeType	node type	1
4	NodeCor	node geographic coordinates	(118.456, 26.657)
5	TradeTime	transactional timestamps	2021/6/1 12:00:00
6	OBUid	vehicular identifiers	1452687261
7	PassID	journey identifiers	0142***561

Algorithm 1 ETC Data Cleaning and Fusion

Input: GData, EnData, ExData
Output: EFusionData
1: **GData** = **GData**[GData['Errorcode'] != 1] # Anomaly Removal
2: **GData** = **GData**[GData['PassID'].startswith('01')] # Related Data Filtering
3: **EFusionData** = UNION_ALL(GData, EnData, ExData) # Data Fusion
4: **EFusionData** = **EFusionData**.sort_by('PassID', 'TradeTime').group_by('PassID') # Sorting and Grouping
5: # Standardization of field names
6: **For** each transaction in **EFusionData**:
7:     transaction.rename_fields(NodeID, NodeName, NodeType, NodeCor, TradeTime, OBUid, PassID)
8:     transaction.encode_node_types(0,1,2,3,4)
9: **End For**

3.2.2. Generation of ETC Vehicle Trajectory Set

Upon obtaining the ETC transaction fusion table, we can construct the driving trajectories of each vehicle on the highway according to the order of vehicle transaction records. This trajectory information includes the journey identifier (PassID), the transaction time at each passed node, the ID of each passed node, the name of each passed node, the type of each passed node, and the topology section passed. In Table 5, we present an example of a vehicle trajectory data table. The detailed procedure for generating the ETC Vehicle Trajectory Set is outlined in Algorithm 2.

Table 5. An example of a ETC Vehicle Trajectory Set.

PassID	Index	Transit Node Transaction Time	Transit Node ID	Transit Node Name	Transit Node Type	Transit Topological Segment
0142***561	1	2021-06-01 08:00:00	2100EN	Zhongnan Jinghai Booth	0	
	2	2021-06-01 09:00:00	350001	Jinghai to Xijin Hub	2	2100EN–350001
	3	2021-06-01 10:00:00	350003	Xijin Hub to Dongcheng	2	350001–350003
	4	2021-06-01 11:00:00	350005	Dongcheng to Xiyu Hub	2	350003–350005
	5	2021-06-01 12:00:00	350007	Xiyu Hub to Nancheng	2	350005–350007
	6	2021-06-01 13:00:00	2200EX	Zhongnan Nancheng Booth	1	350007–2200EX

Algorithm 2 Generation of ETC Vehicle Trajectory Set
Input: EFusionData
Output: ETC Vehicle Trajectory Set (EVTSet)
1: **EVTSet** = Initialize empty list   #
2: Unique_PassID_List = ExtractUniquePassID(EFusionData)
3: **For** each PassID in Unique_PassID_List:
4:    Transactions = ExtractTransactions(EFusionData, PassID) # Get transactions related to current PassID
5:    Sorted_Transactions = SortTransactions(Transactions) # Sort transactions by time
6:    Vehicle_Trajectory = GenerateVehicleTrajectory(Sorted_Transactions) # Generate trajectory from transactions
7:    **Append** Vehicle_Trajectory to **EVTSet**

3.3. Extraction of ETC Gantry and Toll Booth Node Set

After data preprocessing, we analyzed transaction data from ETC gantries, entry toll booths, and exit toll booths to extract ETC nodes, forming a node set of ETC gantries and toll booths, hereinafter referred to as the ETC node set. We selected gantries and in-province toll booths in normal operation to ensure that the extracted node set represents devices in actual operation, which facilitates further analysis. Through transaction data analysis, we identified and extracted valid gantry and toll booth nodes. We excluded the toll booths with abnormal ID codes, abandoned toll booths, and out-of-province toll booths. During the selection process, we found discrepancies between the entry toll booth set and the exit toll booth set. The booths in the discrepancies only had identifiers in the transaction data but no names, and, in fact, did not exist. This could be due to these toll booths being boundary toll booths, identifier changes, or names not displayed during maintenance. We eliminated these aberrant data. Ultimately, the ETC node set includes 1805 nodes, including 1051 gantry nodes and 754 toll booth nodes (378 entry toll booth nodes and 376 exit toll booth nodes), as shown in Figure 1. This ETC node set is the foundation for subsequent topology candidate set generation (Table 6).

Figure 1. The distribution map of ETC nodes in a certain province.

Table 6. Statistical count of ETC tollgate and toll booth nodes.

Category	Number of Nodes
Gantry Nodes	1051
Entry Toll Booth Nodes	378
Exit Toll Booth Nodes	376
Total	1805

3.4. Extraction and Selection of Candidate Topologies Set

Before proceeding to the extraction of the topological candidate set, it is paramount to familiarize ourselves with the basic topological structure that characterizes highways. This structure is a composite of ETC gantries, toll booths, and the highway lanes themselves. As depicted in Figure 2a, we utilize TB to represent toll booths, with TB^{ex} and TB^{ex} denoting the exit and entrance toll booths, respectively. G signifies ETC gantries, with G^{ul} and G^{dl} representing the gantries on the upward and downward lanes, respectively. A typical ETC topological structure is constituted by these nodes, along with the directional segments linking them. Following the procurement of the ETC vehicle trajectory and node datasets, we were equipped to generate a preliminary set of topological candidates. This assemblage was extracted from the sequential pairing of adjacent nodes within the vehicle trajectory dataset, which resulted in a comprehensive count of 31,379 potential topologies. Subsequently, the candidate set of topologies underwent data preprocessing to screen and eliminate a substantial number of erroneous topologies. Erroneous topologies include the following categories:

1. Topologies not included in the ETC node set: In these topologies, the starting or ending point, or both, are not part of the node set. Thus, these topologies can be directly eliminated.
2. Circular topologies: As depicted in Figure 2b, in these topologies, the start and end nodes are identical. This type of erroneous topology can be easily removed.
3. Bidirectional topologies (Figure 2c): These candidate topologies feature nodes from the upward (downward) lane that directly reach nodes of the downward (upward) lane.
4. Topologies terminating at toll booth entrances (Figure 2d): Similar to the case of topologies originating from exits, these topologies conclude at a toll booth entrance.
5. Topologies originating from toll booth exits (Figure 2e): These topologies commence from a toll booth exit. However, in actual trajectories, toll booth exits typically appear at the end, hence such topologies do not exist.
6. Topologies from an entrance toll booth to an exit toll booth (Figure 2f): These candidate topologies commence from an entrance toll booth and conclude at an exit toll booth. However, in actual trajectories, several gantries must be passed between the entrance and exit toll booths, rendering such topologies clearly erroneous and subject to direct elimination.
7. Topologies incorporating out-of-province toll booths (Figure 2g): These topologies feature a toll booth node outside the province as the starting point or endpoint. In the figure, the toll station outside the province is OTB. Due to the presence of inter-provincial ETC transaction data in the actual dataset, topologies may contain nodes of out-of-province toll booths. These nodes can be easily identified as their toll booth names differ from those of in-province nodes, despite having identical IDs.

Utilizing the aforementioned rules, we successfully identified and eliminated a substantial number of erroneous topologies, thus ensuring the accuracy of our analytical results. Following preprocessing, we retained a total of 13,598 topologies. Nevertheless, among these, 10,166 topologies had a vehicle traffic volume of less than 5 within a 7-day period and were confirmed as invalid topologies following inspection. Upon elimination of these invalid topologies, we retained 3432 valid topologies. These valid topologies form our candidate set of topologies; in Table 7, we present the fields of the candidate set of topologies, as well as their descriptions and examples.

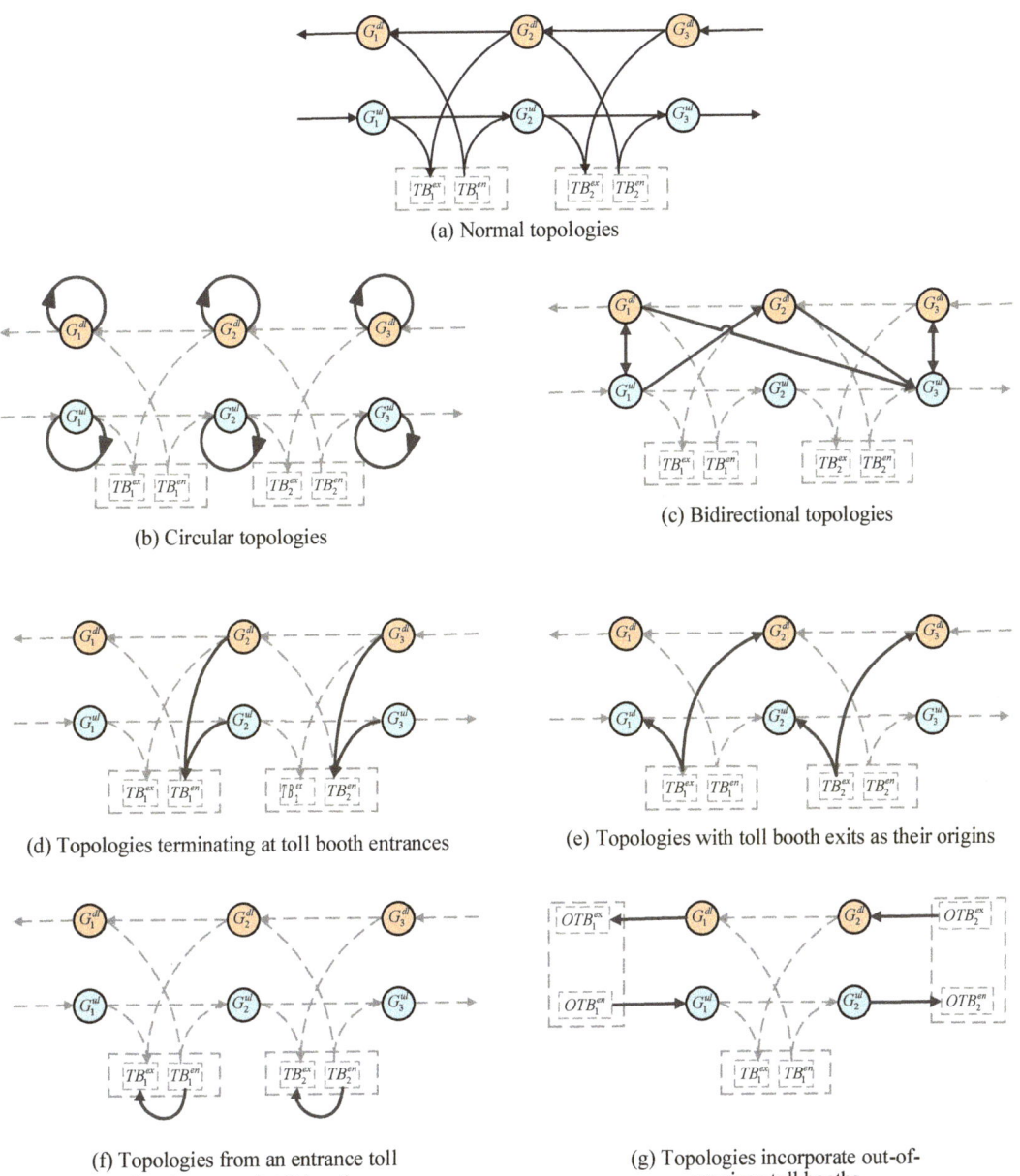

Figure 2. Illustration of normal and erroneous highway topologies.

Table 7. Description of fields in ETC topology data.

Index	Field Name	Field Properties	Example
1	Topology	topology array	['350E01', '350E03']
2	StartID	start node id	350E01
3	EndID	end node id	350E03
4	StartName	start node name	Jinhai to Nanzhou
5	EndName	end node name	Nanzhou to Xicheng
6	StartType	start node type	2
7	EneType	end node type	2
8	StartCor	start node geographic coordinate	(118.2434, 24.6884)
9	EndCor	end node geographic coordinate	(118.4107, 24.7195)
10	TrafficVolume	topology traffic volume	257,396
11	TopologyDistance	topology route distance	18,561 (m)

Herein, 'Traffic Volume' is derived based on the number of vehicles that traversed this topology within a 5-day period according to the vehicle trajectory dataset, and 'Topology Distance' is the planned route distance obtained from the Amap (Amap Map) API, using the geographic coordinates of the starting and ending nodes.

3.5. Feature Vector Modeling

In the analysis of the candidate topology set, we considered five feature dimensions: Topology Traffic Volume (*TTV*), Topological Passage Rate (*TPR*), Normalized Start Rate (*NSR*), Normalized End Rate (*NER*), and Topology Distance (*TD*). Taking the topology (a, b) as an example, we will illustrate the calculation methods for these features.

- Topology Traffic Volume (*TTV*): *TTV*, as a basic indicator of traffic flow, can reflect the importance of the topology in the traffic system. By analyzing this feature, we can understand the traffic differences in different topologies. *TTV* is calculated using Equation (1), where $T_i(a,b)$ represents the number of trajectories of topology (a, b) on the *i*-th day, and *N* represents the number of days considered. In this study, *N* = 5.

$$TTV_{(a,\ b)} = \sum_{i=1}^{N} T_i(a,\ b) \tag{1}$$

- Topological Passage Rate (*TPR*): To calculate this feature, first, we calculate the number of trajectories containing both nodes *a* and *b* within a 5-day span, regardless of whether these two nodes are directly connected. We refer to this as the Coexisting Nodes Trajectory Volume (*CNTV*). The calculation formula for *CNTV* is shown in Equation (2), where $C_i(a,b)$ represents the number of trajectories containing nodes A and B on the *i*-th day. The reason for choosing the *TPR* feature is that some topologies may have a large number of erroneous transactions or omissions, resulting in a high value of *CNTV*. By calculating the ratio of *CNTV* to *TTV*, we can more accurately assess the possibility of each topology in actual work. Then, we calculate the Topological Passage Rate, through the formula *TPR* = *TTV*/*CNTV*.

$$CNTV_{(a,b)} = \sum_{i=1}^{N} C_i(a,b) \tag{2}$$

$$TPR_{(a,b)} = \frac{TTV_{(a,b)}}{CNTV_{(a,b)}} \tag{3}$$

- Normalized Start Rate (*NSR*): In the actual high-speed ETC gantry topology, each starting node has 1 to 4 endpoints, and the rest of the topologies are likely to be generated by erroneous data. Therefore, by calculating the proportion of the traffic of candidate topology (*a*, *b*) in all the topologies starting from *a*, we can evaluate the possibility of this topology in actual work. Therefore, *NSR* is one of the crucial features for evaluating the existence probability of a topology. The computation formula for *NSR* is as shown in Equation (4), where *M* denotes the number of end nodes reached by the initial node *a* in the candidate topology.

$$NSR_{(a,b)} = \frac{TTV_{(a,b)}}{\sum_{j=1}^{M} TTV_{(a,j)}} \qquad (4)$$

- Normalized End Rate (*NER*): *NER* is the normalized rate of the end node, calculating the ratio of the traffic of topology (a, b) to the total traffic of all topologies ending at *b*. Since one end node may be connected to multiple start nodes, we need to consider all in-degree topologies of this end node. The calculation of *NER* is similar to that of *NSR*, but it is grouped by end node *b*. The computation formula for *NSR* is as shown in Equation (5).

$$NER_{(a,b)} = \frac{TTV_{(a,b)}}{\sum_{j=1}^{M} TTV_{(j,b)}} \qquad (5)$$

Topology Distance (*TD*): *TD* refers to the path distance between two gantries or toll booth nodes in the ETC highway system, which is used for traffic management and cost calculation. Its calculation is denoted as $TD_{(a,b)}$.

These five features allow us to construct the feature vector of candidate topology, as in Equation (6), and assess their likelihood of actual existence in the traffic system.

$$v(a,b) = \left\{ TTV_{(a,b)}, TPR_{(a,b)}, NSR_{(a,b)}, NER_{(a,b)}, TD_{(a,b)} \right\} \qquad (6)$$

3.6. Authenticity Verification and Accuracy Annotation of Candidate Topologies

Prior to deploying the LightGBM model for the dynamic updates of gantry topologies, it is necessary to substantiate the authenticity and annotate the accuracy of each candidate topology, to affirm their legitimate existence and correctness. The benchmarks for existence verification encapsulate: (1) No other gantries or toll booths should be present on the road the trajectory of the topology; (2) No actions indicative of highway exit should be observed within the topology trajectories. In order to attain these benchmarks, we initially utilize the AMap API to gather the trajectory data for each candidate topology. Following this, we project each topology trajectory along with the ETC node set onto the QGIS map, and, through visual inspection, we determine whether any candidate topology trajectories pass through other nodes or exhibit cases of highway exit. If a topology trajectory complies with the aforementioned criteria, it is validated as a legitimate topology. Otherwise, it is designated as a flawed topology.

3.7. Dynamic Generation Method of Highway ETC Gantry Topology Based on LightGBM

LightGBM, an efficient Gradient Boosting Decision Tree (GBDT) algorithm, was proposed by Ke et al., 2017 [44] and has since been extensively utilized in diverse data mining tasks such as classification, regression, and ranking. It exhibits superior performance in both efficiency and effectiveness, and it has a commendable capacity to handle non-linear data relationships. In the present study, we employ the LightGBM model to predict the authenticity of gantry topology, thereby enabling the dynamic update of gantry topology.

The primary steps of LightGBM are as follows:

1. Gradient Boosting: LightGBM is a model based on gradient boosting. Throughout the training process, it repetitively builds decision trees, striving to diminish the discrepancy between the predicted value, $f(x)$, and the true value, *y*, at every step, thereby continually enhancing the prediction accuracy of the model. Its loss function is defined as $L(y, f(x))$. The iterative model can be expressed as:

$$f_{t+1}(x) = f_t(x) + vh(x) \qquad (7)$$

where ν represents the learning rate.

2. Decision Tree Construction: Within LightGBM, the decision tree utilizes a depth-first approach for splitting, and it carries out efficient node splits according to the histogram of features. Additionally, LightGBM is capable of handling categorical features and employs a Gradient-based One-Side Sampling (GOSS) method [45] in feature selection, significantly enhancing training efficiency on high-dimensional data. GOSS is a sampling technique that preserves all large gradient samples and randomly selects a fraction of small gradient samples. This practice maintains the data's distribution while reducing computational cost. The update of tree nodes is represented in a form approximated by the least squares method, as shown in Equation (8):

$$c_j = -\frac{\sum_{x \in I_j} g_i}{\sum_{x \in I_j} h_i} \quad (8)$$

where g_i and h_i denote the first and second order gradients, respectively. To maximize the model's performance at each split, LightGBM seeks the optimal split point at every node division. The method of locating the best split point is achieved by maximizing the information gain. The specific calculation of information gain is as follows:

$$Gain = \left(\sum_{x \in I_L} [g_i + \nu h_i c^L]\right)^2 + \left(\sum_{x \in I_R} [g_i + \nu h_i c^R]\right)^2 - \left(\sum_{x \in I} [g_i + \nu h_i c]\right)^2 \quad (9)$$

where c^L, c^R, and c are the optimal output values for each leaf node.

3. Ensemble Prediction: After constructing N decision trees, LightGBM aggregates them for prediction. For a new input sample x, it is fed into each decision tree, and the obtained prediction result is the weighted average of all decision tree prediction results:

$$f(x) = \sum_{i=1}^{N} T_i(x) \quad (10)$$

4. Hyperparameter Optimization: A grid search is employed for hyperparameter optimization. The primary hyperparameters encompass the number of decision trees, the maximum depth of each tree, the learning rate, the number of features, etc. A parameter grid is defined, and the optimal hyperparameter combination is identified by iterating over potential parameter combinations.

In this study, we have employed the LightGBM model to predict the veracity of highway ETC gantry topology, thereby enabling the dynamic generation of gantry topology. LightGBM, an efficient Gradient Boosting Decision Tree (GBDT) algorithm, has made a significant contribution to our study.

Firstly, our study involves five feature dimensions, which may have complex interactive relationships. LightGBM is capable of effectively capturing these interactions, thereby enhancing the accuracy of the model's predictions. Within the context of highway ETC gantry topology data, the relationships between features may be complex and non-linear. For instance, the relationship between Topology Distance (TD) and Topology Traffic Volume (TTV) may not be linear. LightGBM is adept at handling these non-linear relationships, thereby further improving the predictive performance of the model.

Secondly, LightGBM is a robust classification algorithm that can effectively handle binary classification problems. In our study, we labelled the candidate gantry topology as either existing (marked as 1) or non-existing (marked as 0), and allowed LightGBM to train and predict on the candidate gantry topology. This is a typical binary classification problem. Through LightGBM, we are able to effectively solve this problem, thus actualizing the dynamic generation of gantry topology.

Moreover, LightGBM has the advantage of preventing model overfitting. It introduces regularization parameters (such as L1 and L2 regularization) and uses a Gradient-based

One-Side Sampling (GOSS) method, effectively preventing model overfitting and enhancing the model's generalization capability. Simultaneously, LightGBM provides a series of adjustable hyperparameters, such as the number of decision trees, the maximum depth of each tree, the learning rate, and the number of features, etc. In our study, we optimized these hyperparameters through a grid search method and identified the optimal hyperparameter combination, thereby further improving the performance of the model.

In summary, when dealing with large volumes of data and high-dimensional features, LightGBM often exhibits superior efficiency and performance. This is evident in our study, where we were able to utilize the well-trained LightGBM model to automatically generate predictive results for new gantry topology data, thereby actualizing the dynamic generation of gantry topology.

4. Experiment and Results Analysis

The experimental platform utilized an Intel (R) Core (TM) i9-10900K CPU with 10 cores and a base clock of 3.70 GHz, along with 64 GB RAM. The experiments were performed on the CentOS Linux release 7 September 2009 (Core) operating system and utilized Python 3.7.11 as the programming language. The experiment was implemented on Jupyter Notebook—an interactive programming IDE. For comprehensive information regarding Jupyter Notebook, please refer to its official website: https://jupyter.org/ (accessed on 30 July 2023).

4.1. Construction of Feature Vectors

In accordance with our feature vector model, we fabricated a training feature vector set for high-speed gantry topology generation, and several examples are demonstrated in Table 8. Each vector encompasses five-dimensional attributes and their respective sample classification labels. These five attributes include Topology Traffic Volume (TTV), Topology Passage Rate (TPR), Normalized Start Rate (NSR), Normalized End Rate (NER), and Topology Distance (TD). The sample classification label signifies the existence of the topology: 0 denotes the absence of the topology, whereas 1 indicates the presence of the topology in the actual traffic network.

Table 8. Sample of candidate gantry topological feature vector.

Candidate Topo	TTV	TPR	NSR	NER	TD	Label
['67**EN', '34**19']	6091	0.99918	0.2431	0.1157	1357	1
['67**EN', '35**03']	1212	0.999176	0.212	0.4994	13723	0
['34**07', '34**0B']	1199	0.999167	0.1612	0.3223	1357	1
['35**62', '35**5F']	314	0.996825	0.397	0.139	1802	1
['35**04', '35**11']	933	0.996795	0.1447	0.6839	1346	1
['34**15', '34**19']	2788	0.996782	0.0281	0.1053	1545	1
['79**EN', '35**23']	128	0.711111	0.0242	0.0006	2554	0
['35**13', '35**23']	211	0.710438	0.0179	0.0012	6955	0
['34**07', '35**5F']	571	0.710199	0.1008	0.0049	4523	0
['64**EN', '34**19']	236	0.571429	0.0977	0.278	3254	1
['47**EN', '34**19']	6091	0.99918	0.2431	0.1157	1357	1
['34**EN', '35**03']	1212	0.999176	0.212	0.4994	13723	0

Considering the sensitivity of the data, we have anonymized the toll gate numbers, in which ** represents two characters of the toll gate number, which could be either numbers or letters.

To establish a viable predictive model, an initial correlation analysis was conducted on the five features that define the topology generation problem, the results of which are depicted in Figure 3.

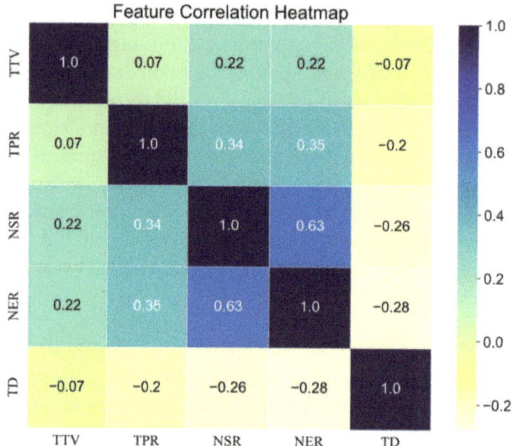

Figure 3. Correlation matrix of feature.

The correlation between *TTV* and *TPR* is 0.07, suggesting that the Topology Traffic Volume and the topology pass rate can independently reflect the characteristics of the topology without being tightly correlated. This finding can assist us in better understanding the uniqueness of each feature in depicting the properties of the topology. The correlation between *NSR* and *NER* is relatively high, reaching 0.63, which may indicate that these two features reflect the same or similar information to a certain extent. Therefore, when interpreting model results or understanding influencing factors, we can consider these two factors jointly. The correlation between feature *TD* and the other features is relatively low, implying that *TD* might contribute additional information necessary for the model. This lends a certain value to the *TD* feature when understanding its role in the model and interpreting prediction results.

4.2. Experimental Setup and Parameter Selection

In our experimental setup, parameter selection and optimization played a pivotal role in the performance outcomes of the LightGBM model. We focused primarily on four crucial categories of parameters: general parameters, core parameters, regularization parameters, and sampling parameters, employing a grid search methodology for fine-tuning. Firstly, general parameters encompass the 'number of estimators (n_estimators)' and 'learning rate'. These two parameters have a direct bearing on the model's learning capability and the pace at which it fits the data. An appropriately set 'number of estimators' ensures that the model possesses ample learning capacity to understand the data, while the 'learning rate' delineates the step size in the model's learning process. Secondly, core parameters dictate the basic structure and complexity of the model, which include 'maximum tree depth (max_depth)', 'number of leaves (num_leaves)', 'minimum child samples (min_child_samples)', 'minimum child weight (min_child_weight)', and 'minimum split gain (min_split_gain)'. In our experiments, the choice of 'maximum tree depth' was particularly salient, as it directly influences the model's complexity and fitting capacity. Further, regularization parameters, composed of 'L1 regularization term (reg_alpha)' and 'L2 regularization term (reg_lambda)', are utilized to inhibit overfitting phenomena in the model. Appropriate regularization helps prevent the model from overfitting the training data, thereby enhancing the model's generalization capacity. Lastly, sampling parameters, which include 'subsample ratio (subsample)', 'column sample by tree (colsample_bytree)', and 'subsample frequency (subsample_freq)', primarily control the sampling of data and features. These parameters aid in mitigating overfitting and augmenting the efficiency of the training process. To get the optimal parameter configuration, we used a grid search for hyperparameter tuning and combined it with 5-Fold Cross-Validation. This approach effectively prevents overfitting and improves the model's generalization on new data. Specifically, we divided the training set into five subsets, each time using four subsets as training data and the remaining one as validation data. In this way, we could obtain a robust parameter configuration for achieving the best model performance. This parameter optimization strategy has helped us find the best parameter combination in the search space, thereby greatly improving the model's performance. Specific search ranges, step sizes, and optimal values can be referred to in Table 9.

Table 9. Optimal combination of important parameters of LightGBM.

Parameter Categories	Parameter	Search Range	Step Size	Optimal Value
general parameters	n_estimators	[10, 500]	10	80
	learning_rate	[0.1, 0.01, 0.001]	-	0.1
core parameters	max_depth	[3, 10]	1	5
	num_leaves	[2, 50]	1	7
	min_child_samples	[5, 50]	5	5
	min_child_weight	[0.001, 0.01, 0.1]	-	0.001
	min_split_gain	[0, 0.1, 0.5]	-	0
regularization parameters	reg_alpha	[0, 0.1, 0.5]	-	0
	reg_lambda	[0, 0.1, 0.5]	-	0
sampling parameters	subsample	[0.5, 0.9]	0.2	0.5
	colsample_bytree	[0.5, 0.9]	0.2	0.5
	subsample_freq	[1, 5]	2	3

4.3. Empirical Outcomes and Integrated Appraisal

In our dataset, each sample is assumed to be independently and identically distributed (i.i.d), which implies that every sample originates from the same probability distribution and is independent of all other samples. Given this assumption, we chose classifiers that are known for effectively handling i.i.d multivariate feature data. To appraise the performance efficacy of various machine learning paradigms within the scope of the gantry topology generation task, we conducted a series of experiments. Alongside LightGBM, our evaluation paradigm incorporated a range of established machine learning methodologies, such as Logistic Regression (LR), Naive Bayes (NB), Linear Discriminant Analysis (LDA), K-Nearest Neighbors (KNN), Decision Tree (DT), Random Forest (RF), Support Vector Machine (SVM), Gradient Boosting (GB), AdaBoost (AB), Extreme Gradient Boosting (XGB), Quadratic Discriminant Analysis (QDA), Gaussian Process Classifier (GPC), Stochastic Gradient Descent (SGD), and Linear Support Vector Machine (Linear SVM). These methodologies were selected due to their demonstrated proficiency in handling datasets with complex, high-dimensional features, which bears a similarity to the nature of our ETC gantry topology data. These were juxtaposed with the LightGBM algorithm to establish comparative performance parameters. For maintaining the replicability of the empirical outcomes, we resorted to the use of default parameters during the algorithmic training phase and designated the random seed as 1. The specifications and configurations of each algorithm have been placed in Appendix A. For LightGBM, we elected the parameter combination that was subject to rigorous optimization to accomplish superior performance efficacy. The algorithmic iterations deployed in our study comprised scikit-learn version 1.0.2, XGBoost version 1.5.1, and LightGBM version 3.3.5.

To evaluate the performance of each model, the evaluation metrics selected in this study include *Accuracy*, *Precision*, *Recall*, and *F1-Score*. These are defined as follows:

$$Accuracy = \frac{TP + TN}{TP + FP + FN + TN} \quad (11)$$

$$Precision = \frac{TP}{TP + FP} \quad (12)$$

$$Recall = \frac{TP}{TP + FN} \qquad (13)$$

$$F1 - score = \frac{2 \times Recall \times Precision}{Recall + Precision} \qquad (14)$$

where TP (True Positives) are the count of topologies that our model correctly identifies as existing. FP (False Positives) represent the count of topologies incorrectly recognized by our model as existing when in reality, they do not exist. TN (True Negatives), on the other hand, refer to the count of topologies that our model correctly identifies as non-existing. Lastly, FN (False Negatives) represent the instances where our model incorrectly classifies existing topologies as non-existent.

In the ensuing phase, we executed experiments and evaluations on a homogenous dataset, drawing performance comparisons of each algorithm with respect to evaluation matrices such as accuracy, precision, recall, and F1-score, as delineated in Table 10.

Table 10. Evaluation metrics comparison of different algorithms.

Model	Accuracy	Precision	Recall	F1-Score
LR	0.8239	0.8918	0.8835	0.8877
NB	0.8705	0.8717	0.9797	0.9225
LDA	0.9403	0.9310	0.9982	0.9634
KNN	0.8574	0.8625	0.9741	0.9149
DT	0.9549	0.9688	0.9741	0.9714
RF	0.9651	0.9624	0.9945	0.9782
SVM	0.8617	0.8517	0.9982	0.9191
GB	0.9636	0.9624	0.9926	0.9773
AB	0.9578	0.9571	0.9908	0.9737
XGB	0.9651	0.9691	0.9871	0.9780
LightGBM	0.9709	0.9668	0.9982	0.9822
QDA	0.9607	0.9606	0.9908	0.9754
GPC	0.4032	0.9456	0.2569	0.4041
SGD	0.7365	0.9186	0.7301	0.8136
Linear SVM	0.7875	0.7875	1.0000	0.8811

In pursuit of offering an enriched visual illustration comparing the performance of LightGBM vis-a-vis other models, we have meticulously crafted a graphical representation (refer to Figure 3). This representation systematically encapsulates the relative efficacies of various algorithms gauged across a range of evaluation metrics. Within the purview of our empirical investigation, LightGBM emerged as a notably superior algorithm for the task of gantry topology generation, outperforming all contenders in terms of accuracy and F1 scores, with the exception of recall, where it trailed marginally behind Linear SVM. An accuracy metric of 0.9709 exemplifies LightGBM's adept capability in forecasting the topology of gantries with a high degree of precision. While LightGBM's recall metric, registered at 0.9982, is slightly outperformed by Linear SVM's optimal recall of 1.0000, it is crucial to note that Linear SVM lags significantly behind in terms of accuracy (0.7875), precision (0.7875), and the F1 score (0.8811). Consequently, in this holistic context, LightGBM's recall metric still demonstrates remarkable excellence. This metric underscores the model's proficiency in identifying true positive gantry topologies, a critical factor within the task of gantry topology generation. Simultaneously, LightGBM's F1 score of 0.9822 epitomizes an exemplary equilibrium between minimizing false positives (indicative of high precision) and maximizing true positives (indicative of high recall). Despite a precision score of 0.9668, which surpasses several models, it only ranks third across the evaluated models, indicating a relative underperformance of LightGBM in identifying negative instances. Nonetheless, it is important to highlight that, despite a marginally lower precision and being second to Linear SVM in recall, these factors do not compromise the overarching prowess of LightGBM. Particularly in the context of gantry topology generation, recall takes precedence, as the goal is to identify the maximum count of true gantries, even if it involves potential false positives. Therefore, despite certain minor shortcomings, LightGBM's remarkable performance in terms of accuracy, recall, and the F1 score, coupled with the precise requirements of gantry topology generation, unequivocally endorse LightGBM as the recommended algorithm for this task.

In the endeavor to augment the efficacy of our model, we adopted the stratagem of 5-Fold Cross-Validation. This rigorous validation technique not only provided a multifaceted assessment of our model's performance, but also furnished insights into its generalizability after each training iteration. As delineated in Figure 4, we observed a consistent decrement in the log-loss values across all five partitions, underscoring the robust learning capability of our model and the successful mitigation of overfitting. Furthermore, the attainment of lower log-loss values across all folds during the terminal rounds signified the absence of underfitting.

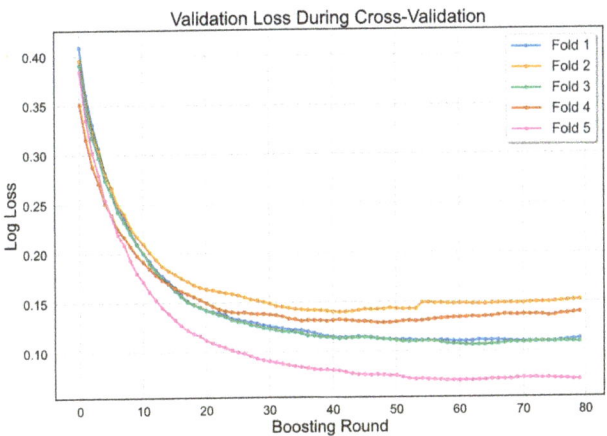

Figure 4. Validation log-loss during training.

Throughout the 5-Fold Cross-Validation process, we meticulously recorded the key performance metrics of accuracy, precision, recall, and F1 score for each fold (refer Table 11). The fifth fold, in particular, demonstrated superior accuracy and precision metrics when juxtaposed with the model devoid of cross-validation. To be precise, the 5-Fold Cross-Validation reported an accuracy of 0.976321 and a precision of 0.980306, both surpassing their counterparts from the non-cross-validated model, which registered an accuracy of 0.9709 and precision of 0.9668. While the recall metric from 5-Fold Cross-Validation (0.991150) marginally trailed behind the non-cross-validated model (0.9982), the F1 score (0.985699) outperformed the non-cross-validated counterpart (0.9822). These outcomes not only demonstrate the quintessential role of 5-Fold Cross-Validation in model appraisal and selection but also testify to its superiority in most scenarios, notwithstanding the volatility it introduces.

Table 11. Evaluation metric performance during 5-Fold Cross-Validation.

Fold	Accuracy	Precision	Recall	F1-Score
1	0.9672	0.9631	0.9978	0.9801
2	0.9599	0.9614	0.9911	0.9760
3	0.9599	0.9577	0.9956	0.9763
4	0.9563	0.9667	0.9831	0.9748
5	**0.9763**	**0.9803**	**0.9912**	**0.9857**

The bolded sections represent the best-performing round across all evaluation metrics.

In order to assess the time efficiency of our model when dealing with datasets of varying sizes, we designed a series of experiments to approximate its time complexity. We first established a range of test sets, starting from 10% of the total data volume, incrementing by 1% each time, until the entire dataset was encompassed. Subsequently, we trained and made predictions with the LightGBM model on each test set, precisely recording the execution time of each operation, as shown in Figure 5. As depicted in the graph, the time taken for prediction remains relatively constant as the percentage of data used increases, oscillating slightly around the average time. This observation suggests that the time complexity of our algorithm tends to be a constant, i.e., O(1). This result indicates that our model maintains a high level of execution efficiency even when faced with expanding data scales.

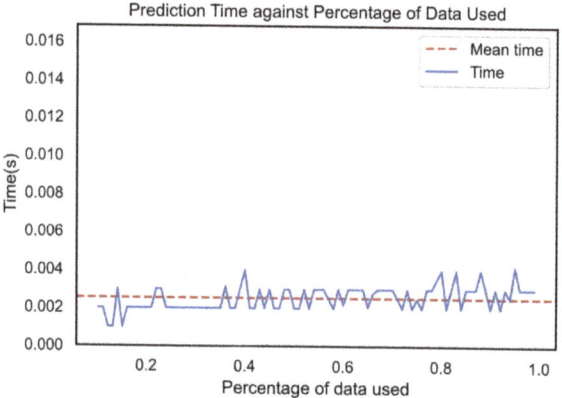

Figure 5. Prediction time against percentage of data used.

In comparison to a closely related study by Cai and Yi et al. [2], our method demonstrates evident superiority in terms of time complexity, evaluation metrics, and operational efficiency, as shown in Table 12.

Table 12. Comparison of our method with Cai and Yi et al. [2].

Metric	Our Method	Cai and Yi et al. [2]
Accuracy	0.9763	-
Precision	0.9803	0.966
Recall	0.9912	0.982
F1-Score	0.9857	0.974
Time Complexity	O(1)	O(n)
Average Time to Generate a Topology (ms)	0.00142	2
Total Time to Generate all Topologies (s)	0.004	5.76

Specifically, our method achieved scores of 0.9763, 0.9803, 0.9912, and 0.9857 on accuracy, precision, recall, and F1-score, respectively. Although the approach of Cai and Yi et al. [2] also reported high scores of 0.966, 0.982, and 0.974 for precision, recall, and F1-score, respectively, they did not present an accuracy score. Thus, our method surpasses theirs across all reported metrics.

From the standpoint of time complexity, our method exhibits an almost constant runtime with an increasing amount of data, implying a time complexity nearing O(1). Conversely, the runtime of the method proposed by Cai and Yi et al. [2] escalates linearly with the data volume, indicating a time complexity of O(n). Hence, our method offers superior time efficiency.

Additionally, our method excels in operational efficiency. Specifically, our approach generates 2819 topologies within a mere 0.004 s, averaging less than 0.00142 milliseconds per generated topology. In contrast, Cai and Yi et al. [2]'s method takes 5.76 s to generate 2950 topologies, averaging less than 2 milliseconds per topology. Therefore, our method significantly outperforms that of Cai and Yi et al. [2] in terms of the efficiency of topology generation.

On this basis, we endeavored to understand the decision-making process of the model and ascertain the most salient features influencing gantry topology generation through a feature importance analysis. This examination encompassed all the features utilized during the training of the LightGBM model, quantifying the contribution of each feature to the predictive performance of the model. The detailed results are displayed in Figure 6. As observed from Figure 6, the feature 'PN' boasts the highest importance ratio, highlighting its substantial influence on gantry topology generation. The feature 'TD' trails 'PN', but still maintains a relatively high importance ratio. On the other hand, features 'PR', 'NSR', and 'NER' have lower importance ratios. These findings suggest that the features 'TTV' and 'TD' are indispensable to our model's prediction, given their substantial contributions to the predictive capacity for the target variable. Conversely, the importance of the 'TPR', 'NER', and 'NSR' features is less pronounced, indicating a potentially smaller contribution to the model's predictive capabilities.

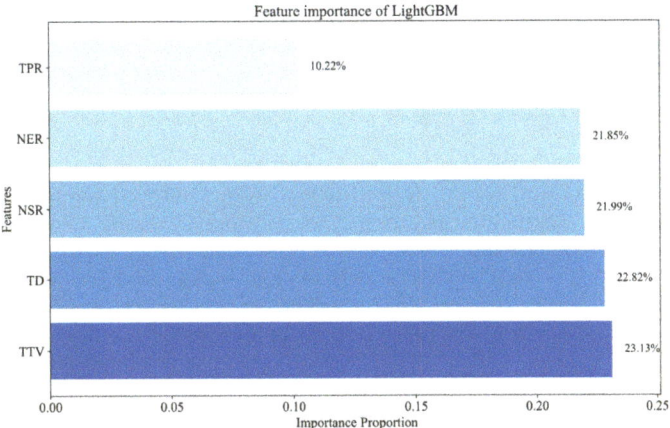

Figure 6. Feature importance of LightGBM.

4.4. Results of Gantry Topology Generation

We apply the trained LightGBM model to classify and predict the candidate topology, and use the predicted topology as the updated gantry topology set.

In this result, we apply the trained LightGBM model to classify and predict the candidate topology, and we use the predicted topology as the updated gantry topology set. As illustrated in Figure 7, we initially generated a preliminary set of candidate topologies and, subsequently, discerned the final topological framework through algorithmic filtration. Among the initially generated topologies in Figure 7a, despite the presence of a large number of erroneous topologies such as long-distance topologies, we still assumed that all 3432 preliminary topologies were correct. This assumption served as our baseline model, which preset all labels as 1. Under these circumstances, the model's accuracy and precision were both 0.825, the recall rate was 1, and the F1 score was 0.904. However, this all-positive-prediction model could not accurately distinguish between the positive and negative categories in the data, thereby exhibiting severe imbalance.

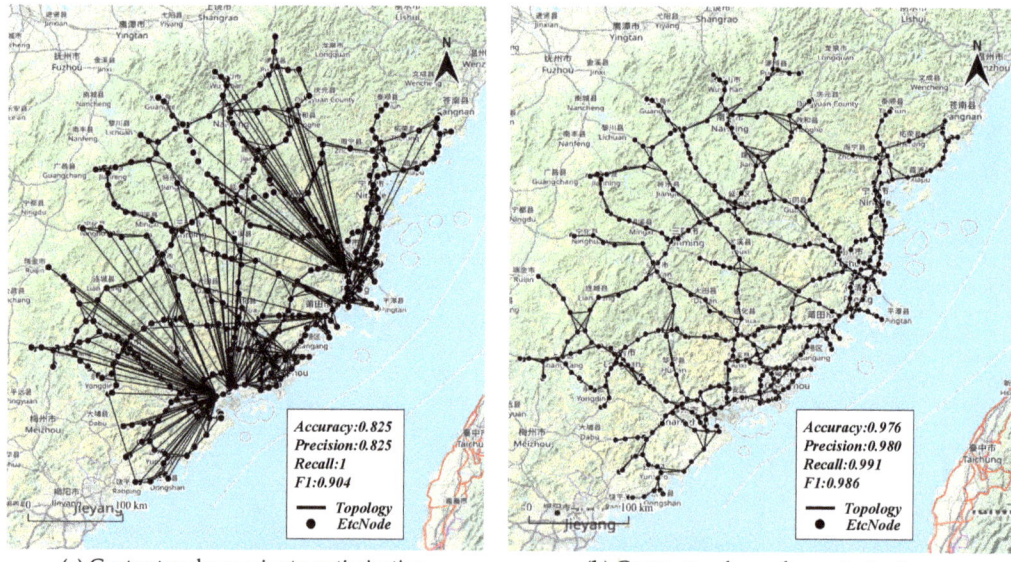

(**a**) Gantry topology prior to optimization (**b**) Gantry topology after optimization.

Figure 7. Comparison diagram of gantry topology pre- and post-LightGBM model generation.

After applying the LightGBM model to generate the gantry topology, we successfully identified 2819 legitimate topologies from these candidates. As shown in Figure 7b, the distribution of gantry topology more accurately reflected reality and was closer to the distribution of highways. In this case, the model's accuracy increased to 0.976, precision increased to 0.980, and the F1 score increased to 0.986. Although the recall rate decreased to 0.991, this actually reflected the model's improved ability to distinguish between positive and negative categories, rather than simply predicting all samples as positive. It also demonstrated that our model could maintain high precision while still achieving a high recall rate. This significant performance improvement fully demonstrated the effectiveness of the proposed method.

5. Conclusions

In the domain of Electronic Toll Collection (ETC) gantry topology dynamic updating on highways, this research has instituted the following pivotal contributions and advancements:

This study introduces an innovative methodology for dynamic updating of highway gantry topology predicated on ETC transaction data. This effectively ameliorates prevalent inaccuracies in topology data within vehicle transaction records, thus rectifying the existing predicament of solely charging based on minimum distance due to the inability to compute fees according to actual traversed distances. By extrapolating potential topologies and systematically eliminating erroneous variants, we have furnished a robust and reliable data source for the genesis and iterative refinement of gantry topology. In tandem with the integration of the Amap API and QGIS mapping analytics, we have substantiated the veracity of our candidate topologies, thereby safeguarding the precision of the resultant gantry topology. We have employed the LightGBM model to facilitate the dynamic updating of the gantry topology. Empirical evidence suggests that this approach yields commendable outcomes, registering an accuracy rate of 97.6%, thereby satisfactorily meeting the requisites for dynamic updating of ETC gantry topology on highways. This research promulgates a universally applicable methodology and framework for dynamically updating highway ETC gantry topology, underscoring its extensive applicability and scalability. In practical implementation, the methodologies propounded by this research can be fine-tuned and optimized to accommodate actual needs, thereby catering to the varying requirements for updating ETC gantry topology in diverse scenarios.

In summary, this research has heralded groundbreaking methods and techniques for the dynamic generation of highway ETC gantry topology, providing significant reinforcement for highway management and road network analytics. Concurrently, it lays a formidable foundation for the evolution of intelligent transportation systems and enhancement of the overall quality of transportation services.

Author Contributions: Conceptualization, F.Z. and W.W.; methodology, W.W.; software, W.W., Q.C. and R.S.; validation, F.Z., W.W. and Q.C.; formal analysis, W.W. and F.Z.; investigation, W.W.; resources, F.Z.; data curation, W.W. and F.Z.; writing—original draft preparation, W.W. and Q.C.; writing—review and editing, F.Z., W.W. and F.G.; visualization, W.W. and R.S.; supervision, F.Z. and F.G.; project administration, F.Z. and W.W.; funding acquisition, F.Z. All authors have read and agreed to the published version of the manuscript.

Funding: This work was funded by the National Natural Science Foundation of China (Funding number: 41971340), the 2020 Fujian Province "Belt and Road" Technology Innovation Platform (Funding number: 2020D002), the Provincial Candidates for the Hundred, Thousand and Ten Thousand Talent of Fujian (Funding number: GY-Z19113), the Municipal Science and Technology project (Funding number: GY-Z22006, GY-Z220230), the Open Fund project (Funding number: KF-X1902, KF-19-22001), the Patent Grant project (Funding number: GY-Z20074), and the Crosswise project (Funding number: GY-H-21021, GY-H-20077).

Data Availability Statement: The ETC transaction data utilized in this study were obtained from Fujian Expressway Information Technology Co., Ltd. Restrictions apply to the availability of these data, which were used under license for this study and are not publicly available. Data are available from the authors with the permission of Fujian Expressway Information Technology Co., Ltd. All data processing and analyses were conducted in compliance with relevant data protection and privacy laws. No individual or personal data were used in this study.

Conflicts of Interest: The authors declare no conflict of interest.

Glossary

Acronym	Full Form
AB	Adaptive Boosting
API	Application Programming Interface
DBSCAN	Density-Based Spatial Clustering of Applications with Noise
DT	Decision Tree
ETC	Electronic Toll Collection
GBDT	Gradient Boosting Decision Tree
GB	Gradient Boosting
GPC	Gaussian Process Classifier
GNSS	Global Navigation Satellite System
GPS	Global Positioning System
KDE	Kernel Density Estimation
KNN	k-Nearest Neighbors
LCSS	Longest Common Subsequence
LDA	Linear Discriminant Analysis
LightGBM	Light Gradient Boosting Machine
LR	Logistic Regression
MTC	Manual Toll Collection
NB	Naive Bayes Classifier
NSR	Normalized Start Rate
OBU	On-Board Unit
QDA	Quadratic Discriminant Analysis
RF	Random Forest
RSU	Road Side Unit
SGD	Stochastic Gradient Descent

SVM	Support Vector Machine
TD	Topology Distance
TTV	Topology Traffic Volume
XGB	XGBoost (Extreme Gradient Boosting)
5	0.9763

Appendix A

In this study, we used the default parameter settings of various models for comparison. The following is a detailed list of the individual model parameters:

Table A1. LR (Logistic Regression) parameter settings.

Parameter	Description	Value
C	Inverse regularization	1
class_weight	Class weights	None
dual	Dual formulation	False
fit_intercept	Add constant to function	True
intercept_scaling	For solver 'liblinear'	1
l1_ratio	Elastic-Net mixing	None
max_iter	Max iterations	100
multi_class	Multiclass option	Auto
n_jobs	CPU cores for parallel	None
penalty	Norm in penalization	L2
random_state	Random number seed	None
solver	Optimization algorithm	Lbfgs
tol	Stopping criteria	0.0001
verbose	Verbose for liblinear	0
warm_start	Reuse previous solution	False

Table A2. NB (Naive Bayes) parameter settings.

Parameter	Description	Value
priors	Prior probabilities	None
var_smoothing	Portion of the largest variance	1×10^{-9}
priors	Prior probabilities	None

Table A3. LDA (Linear Discriminant Analysis) parameter settings.

Parameter	Description	Value
covariance_estimator	Covariance estimator	None
n_components	Number of components	None
priors	Prior probabilities	None
shrinkage	Shrinkage parameter	None
solver	Solver for computation	Svd
store_covariance	If True, compute covariance	False
tol	Tolerance for stopping criteria	0.0001

Table A4. KNN (Decision Tree) parameter settings.

Parameter	Description	Value
algorithm	Algorithm used	Auto
leaf_size	Leaf size	30
metric	Distance metric	Minkowski
metric_params	Metric params	None
n_jobs	Num of jobs	None
n_neighbors	Num of neighbors	5
p	Power parameter	2
weights	Weight function	Uniform

Table A5. DT (Decision Tree) parameter settings.

Parameter	Description	Value
ccp_alpha	Cost complexity pruning	0
class_weight	Class weights	None
criterion	Criterion to split	Gini
max_depth	Max depth of tree	None
max_features	Max features for split	None
max_leaf_nodes	Max leaf nodes	None
min_impurity_decrease	Node impurity decrease	0
min_samples_leaf	Min samples at leaf	1
min_samples_split	Min samples to split	2
min_weight_fraction_leaf	Min weight fraction	0
random_state	Random seed	None
splitter	Split strategy	Best

Table A6. RF (Random Forest) parameter settings.

Parameter	Description	Value
bootstrap	Bootstrap samples	True
ccp_alpha	Cost complexity pruning	0
class_weight	Class weights	None
criterion	Split criterion	None
max_depth	Max tree depth	None
max_features	Max features	None
max_leaf_nodes	Max leaf nodes	None
max_samples	Max samples	None
min_impurity_decrease	Min impurity decrease	0
min_samples_leaf	Min samples at leaf	1
min_samples_split	Min samples to split	2
min_weight_fraction_leaf	Min weight fraction	0
n_estimators	Num of trees	100
n_jobs	Num of jobs	None
oob_score	OOB score	False
random_state	Random seed	None
verbose	Logging level	0
warm_start	Reuse previous solution	False

Table A7. SVM (Support Vector Machines) parameter settings.

Parameter	Description	Value
C	Penalty parameter	1
break_ties	Break ties	False
cache_size	Cache size	200
class_weight	Class weights	None
coef0	Kernel coef	0
decision_function_shape	Decision function	Ovr
degree	Kernel degree	3
gamma	Kernel coef	Scale
kernel	Kernel type	Rbf
max_iter	Max iterations	−1
probability	Estimate prob	True
random_state	Random seed	None
shrinking	Use shrinking	True
tol	Tolerance	0.001
verbose	Verbose	False

Table A8. GB (Gradient Boosting) parameter settings.

Parameter	Description	Value
ccp_alpha	Pruning parameter	0
criterion	Split criterion	Friedman_mse
init	Initial estimator	None
learning_rate	Learning rate	0.1
loss	Loss function	Deviance
max_depth	Max depth	3
max_features	Max features	None
max_leaf_nodes	Max leaf nodes	None
min_impurity_decrease	Min impurity decrease	0
min_samples_leaf	Min samples at leaf	1
min_samples_split	Min samples to split	2
min_weight_fraction_leaf	Min weight fraction	0
n_estimators	Num of estimators	100
n_iter_no_change	Iterations no change	None
random_state	Random seed	None
subsample	Subsample fraction	1
tol	Tolerance	0.0001
validation_fraction	Validation fraction	0.1
verbose	Verbose	0
warm_start	Reuse previous solution	False

Table A9. AB (AdaBoost) parameter settings.

Parameter	Description	Value
algorithm	Algorithm type	Samme.r
base_estimator	Base estimator	None
learning_rate	Learning rate	1
n_estimators	Num of estimators	50
random_state	Random seed	None
verbose	Verbose for liblinear	0
warm_start	Reuse previous solution	False

Table A10. XGB (XGBoost) parameter settings.

Parameter	Description	Value
use_label_encoder	Use label encoder	False
enable_categorical	Categorical data	False
eval_metric	Evaluation metric	Logloss
objective	Objective function	Binary:logistic
n_estimators	Num of estimators	100

In order to not cause redundancy, we did not put the value of empty parameters into the table, The parameters set to null values are: booster, colsample_bylevel, colsample_bynode, colsample_bytree, gamma, gpu_id, importance_type, interaction_constraints, learning_rate, max_delta_step, max_depth, min_child_weight, missing, monotone_constraints, n_jobs, num_parallel_tree, predictor, random_state, reg_alpha, reg_lambda, scale_pos_weight, subsample, tree_method, validate_parameters, verbosity.

Table A11. QDA (Quadratic Discriminant Analysis) parameter settings.

Parameter	Description	Value
priors	Class priors	None
reg_param	Regularization	0
store_covariance	Store covariance	False
tol	Tolerance	0.0001

Table A12. GPC (Gaussian Process Classifier) parameter settings.

Parameter	Description	Value
copy_X_train	Copy training data	True
kernel	Kernel function	None
max_iter_predict	Max iterations	100
multi_class	Multi-class strategy	One_vs_rest
n_jobs	Num of jobs	None
n_restarts_optimizer	Num of restarts	0
optimizer	Optimizer	Fmin_l_bfgs_b
random_state	Random seed	None
warm_start	Reuse previous solution	False

Table A13. SGD (Stochastic Gradient Descent) parameter settings.

Parameter	Description	Value
alpha	Regularization param	0.0001
average	Average coef	False
class_weight	Class weights	
early_stopping	Early stopping	False
epsilon	Epsilon	0.1
eta0	Initial learning rate	0
fit_intercept	Fit intercept	True
l1_ratio	L1 ratio	0.15
learning_rate	Learning rate	Optimal
loss	Loss function	Hinge
max_iter	Max iterations	1000
n_iter_no_change	Iterations no change	5
n_jobs	Num of jobs	None
penalty	Penalty	L2
power_t	Power t	0.5
random_state	Random seed	None
shuffle	Shuffle	True
tol	Tolerance	0.001
validation_fraction	Validation fraction	0.1
verbose	Verbose	0
warm_start	Reuse previous solution	False

Table A14. Linear SVM parameter settings.

Parameter	Description	Value
C	Regularization param	1
class_weight	Class weights	None
dual	Dual formulation	True
fit_intercept	Fit intercept	True
intercept_scaling	Intercept scaling	1
loss	Loss function	Squared_hinge
max_iter	Max iterations	1000
multi_class	Multi-class strategy	None
penalty	Penalty	L2
random_state	Random seed	None
tol	Tolerance	0.0001

References

1. Ministry of Transport, National Development and Reform Commission, Ministry of Finance. Notice on Issuing the Implementation Plan for the Full Promotion of Differentiated Toll Collection on Highways. Jiaogongluhan No. 228. 2021. Available online: https://www.gov.cn/zhengce/zhengceku/2021-06/15/content_5617919.htm (accessed on 30 July 2023).
2. Cai, Q.; Yi, D.; Zou, F.; Wang, W.; Luo, G.; Cai, X. An Arch-Bridge Topology-Based Expressway Network Structure and Automatic Generation. *Appl. Sci.* **2023**, *13*, 5031. [CrossRef]
3. Fujian Provincial State-Owned Assets Supervision and Administration Commission. Fujian: ETC Usage Rate Ranks First in the Country. 12 December 2019. Available online: http://www.sasac.gov.cn/n2588025/n2588129/c13072896/content.html (accessed on 30 July 2023).
4. Amap. Web Service API. Amap API. Available online: https://lbs.amap.com/api/webservice/summary/ (accessed on 27 July 2023).
5. QGIS Development Team. QGIS Desktop User Guide/Manual (QGIS 3.28). Available online: https://docs.qgis.org/3.28/en/docs/user_manual/ (accessed on 27 July 2023).
6. Kass, M.; Witkin, A.; Terzopoulos, D. Snakes: Active contour models. *Int. J. Comput. Vis.* **1988**, *1*, 321–331. [CrossRef]
7. Péteri, R.; Ranchin, T. Extraction of network-like structures using a multiscale representation. *IEEE Geosci. Remote Sens. Lett.* **2005**, *2*, 402–406.
8. Laptev, I.; Caputo, B.; Schuldt, C.; Lindeberg, T. Local velocity-adapted motion events for spatio-temporal recognition. *Comput. Vis. Image Underst.* **2004**, *108*, 207–229. [CrossRef]
9. Gruen, A.; Li, H. Linear feature extraction with dynamic programming and Globally Least Squares. *ISPRS J. Photogramm. Remote Sens.* **1995**, *50*, 23–30.
10. Mokhtarzade, M.; Zoej, M.V. Road detection from high-resolution satellite images using artificial neural networks. *Int. J. Appl. Earth Obs. Geoinf.* **2007**, *9*, 32–40. [CrossRef]
11. Yager, K.; Sowmya, A. Road detection from aerial images using SVMs. *Mach. Vis. Appl.* **2008**, *19*, 261–274.
12. Mnih, V.; Hinton, G.E. *Learning to Detect Roads in High-Resolution Aerial Images*; Springer: Berlin/Heidelberg, Germany, 2010; pp. 210–223.
13. He, K.; Zhang, X.; Ren, S.; Sun, J. Deep residual learning for image recognition. In Proceedings of the IEEE Conference on Computer Vision and Pattern Recognition, Las Vegas, NV, USA, 27–30 June 2016; pp. 770–778.
14. Saito, S.; Yamashita, T.; Aoki, Y. Multiple object extraction from aerial imagery with convolutional neural networks. *Electron. Imaging* **2016**, *2016*, 1–9. [CrossRef]
15. Long, J.; Shelhamer, E.; Darrell, T. Fully convolutional networks for semantic segmentation. In Proceedings of the IEEE Conference on Computer Vision and Pattern Recognition, Boston, MA, USA, 7–12 June 2015; pp. 3431–3440.
16. Maggiori, E.; Tarabalka, Y.; Charpiat, G.; Alliez, P. Fully convolutional neural networks for remote sensing image classification. In Proceedings of the 2016 IEEE International Geoscience and Remote Sensing Symposium (IGARSS), Beijing, China, 10–15 June 2016; pp. 5071–5074.
17. Ševo, I.; Avramović, A. Convolutional Neural Network Based Automatic Object Detection on Aerial Images. *IEEE Geosci. Remote Sens. Lett.* **2016**, *13*, 740–744. [CrossRef]
18. Volpi, M.; Tuia, D. Dense Semantic Labeling of Subdecimeter Resolution Images with Convolutional Neural Networks. *IEEE Trans. Geosci. Remote Sens.* **2017**, *55*, 881–893. [CrossRef]
19. Maggiori, E.; Tarabalka, Y.; Charpiat, G.; Alliez, P. Convolutional neural networks for large-scale remote-sensing image classification. *IEEE Trans. Geosci. Remote Sens.* **2017**, *55*, 645–657. [CrossRef]
20. Keramitsoglou, I.; Kontoes, C.; Sifakis, N.; Konstantinidis, P.; Fitoka, E. Deep learning for operational land cover mapping using Sentinel-2 data. *J. Appl. Remote Sens.* **2020**, *14*, 014503.
21. Wagstaff, K.; Cardie, C.; Rogers, S.; Schroedl, S. Constrained k-means clustering with background knowledge. In Proceedings of the Eighteenth International Conference on Machine Learning (ICML), San Francisco, CA, USA, 28 June–1 July 2001; Volume 1, pp. 577–584.

22. Schroedl, S.; Wagstaff, K.; Rogers, S.; Langley, P.; Wilson, C. Mining GPS traces for map refinement. *Data Min. Knowl. Discov.* **2004**, *9*, 59–87. [CrossRef]
23. Worrall, S.; Nebot, E. Automated process for generating digitised maps through GPS data compression. In Proceedings of the Australasian Conference on Robotics and Automation (ACRA), Brisbane, Australia, 10–12 December 2007; Volume 6.
24. Sasaki, Y.; Yu, J.; Ishikawa, Y. Road segment interpolation for incomplete road data. In Proceedings of the IEEE International Conference on Big Data and Smart Computing, Kyoto, Japan, 27 February–2 March 2019; pp. 1–8.
25. Edelkamp, S.; Schrödl, S. Route planning and map inference with global positioning traces. In *Computer Science in Perspective*; Springer: Berlin/Heidelberg, Germany, 2003; pp. 128–151.
26. Chen, C.; Lu, C.; Huang, Q.; Yang, Q.; Gunopulos, D.; Guibas, L. City-scale map creation and updating using GPS collections. In Proceedings of the 22nd ACM SIGKDD International Conference on Knowledge Discovery and Data Mining, San Francisco, CA, USA, 13–17 August 2016; pp. 1465–1474.
27. Huang, J.; Zhang, Y.; Deng, M.; He, Z. Mining crowdsourced trajectory and geo-tagged data for spatial-semantic road map construction. *Trans. GIS* **2022**, *26*, 735–754. [CrossRef]
28. Silverman, B.W. *Density Estimation for Statistics and Data Analysis*; Routledge: Abingdon, UK, 2018.
29. Fu, Z.; Fan, L.; Sun, Y.; Tian, Z. Density adaptive approach for generating road network from GPS trajectories. *IEEE Access* **2020**, *8*, 51388–51399. [CrossRef]
30. Uduwaragoda, E.; Perera, A.S.; Dias, S.A.D. Generating lane level road data from vehicle trajectories using kernel density estimation. In Proceedings of the 16th International IEEE Conference on Intelligent Transportation Systems (ITSC 2013), The Hague, The Netherlands, 6–9 October 2013; pp. 384–391.
31. Kuntzsch, C.; Sester, M.; Brenner, C. Generative models for road network reconstruction. *Int. J. Geogr. Inf. Sci.* **2016**, *30*, 1012–1039. [CrossRef]
32. Neuhold, R.; Haberl, M.; Fellendorf, M.; Pucher, G.; Dolancic, M.; Rudigier, M.; Pfister, J. Generating a lane-specific transportation network based on floating-car data. In *Advances in Human Aspects of Transportation*; Springer: Cham, Switzerland, 2017; pp. 1025–1037.
33. Huang, Y.; Xiao, Z.; Yu, X.; Wang, D. Road network construction with complex intersections based on sparsely sampled private car trajectory data. *ACM Trans. Knowl. Discov. Data (TKDD)* **2019**, *13*, 1–28. [CrossRef]
34. Deng, M.; Huang, J.; Zhang, Y.; Liu, H.; Tang, L.; Tang, J.; Yang, X. Generating urban road intersection models from low-frequency GPS trajectory data. *Int. J. Geogr. Inf. Sci.* **2018**, *32*, 2337–2361. [CrossRef]
35. Wu, J.; Zhu, Y.; Ku, T.; Wang, L. Detecting Road Intersections from Coarse-gained GPS Traces Based on Clustering. *J. Comput.* **2013**, *8*, 2959–2965. [CrossRef]
36. Xie, X.; Philips, W. Road intersection detection through finding common sub-tracks between pairwise GNSS traces. *ISPRS Int. J. Geo-Inf.* **2017**, *6*, 311. [CrossRef]
37. Fathi, A.; Krumm, J. Detecting road intersections from GPS traces. In Proceedings of the International Conference on Geographic Information Science, Zurich, Switzerland, 14–17 September 2010; Springer: Berlin/Heidelberg, Germany, 2010; pp. 56–69.
38. Karagiorgou, S.; Pfoser, D. On vehicle tracking data-based road network generation. In Proceedings of the 20th International Conference on Advances in Geographic Information Systems, Redondo Beach, CA, USA, 6–9 November 2012; pp. 89–98.
39. Pu, M.; Mao, J.; Du, Y.; Shen, Y.; Jin, C. Road intersection detection based on direction ratio statistics analysis. In Proceedings of the 2019 20th IEEE International Conference on Mobile Data Management (MDM), Hong Kong, China, 10–13 June 2019; pp. 288–297.
40. Zhao, L.; Mao, J.; Pu, M.; Liu, G.; Jin, C.; Qian, W.; Zhou, A.; Wen, X.; Hu, R.; Chai, H. Automatic calibration of road intersection topology using trajectories. In Proceedings of the 2020 IEEE 36th International Conference on Data Engineering (ICDE), Kuala Lumpur, Malaysia, 20–24 April 2020; pp. 1633–1644.
41. Qing, R.; Liu, Y.; Zhao, Y.; Liao, Z.; Liu, Y. Using feature interaction among GPS Data for road intersection detection. In Proceedings of the 2nd International Workshop on Human-Centric Multimedia Analysis, Cheng Du, China, 20–24 October 2021; pp. 31–37.
42. Liu, Y.; Qing, R.; Zhao, Y.; Liao, Z. Road Intersection Recognition via Combining Classification Model and Clustering Algorithm Based on GPS Data. *ISPRS Int. J. Geo-Inf.* **2022**, *11*, 487. [CrossRef]
43. Ministry of Transport of the People's Republic of China. Vehicle Classification of the Toll for Highway (JT/T 489-2019). 2019. Available online: https://jtst.mot.gov.cn/kfs/file/read/0de9ee528422ee3a99ff87b1c1295e8e (accessed on 30 July 2023).
44. Ke, G.; Meng, Q.; Finley, T.; Wang, T.; Chen, W.; Ma, W.; Ye, Q.; Liu, T.Y. Lightgbm: A highly efficient gradient boosting decision tree. *Adv. Neural Inf. Process. Syst.* **2017**, *30*, 3149–3157. [CrossRef]
45. Meng, Q.; Ke, G.; Wang, T.; Chen, W.; Ye, Q.; Ma, Z.M.; Liu, T.Y. A communication-efficient parallel algorithm for decision tree. *Adv. Neural Inf. Process. Syst.* **2016**, *29*, 1279–1287. [CrossRef]

Disclaimer/Publisher's Note: The statements, opinions and data contained in all publications are solely those of the individual author(s) and contributor(s) and not of MDPI and/or the editor(s). MDPI and/or the editor(s) disclaim responsibility for any injury to people or property resulting from any ideas, methods, instructions or products referred to in the content.

Article

BLogic: A Bayesian Model Combination Approach in Logic Regression

Yu-Chung Wei

Graduate Institute of Statistics and Information Science, National Changhua University of Education, No. 1, Jin-De Road, Changhua City 500207, Taiwan; weiyuchung@cc.ncue.edu.tw; Tel.: +886-4-7232105 (ext. 3236)

Abstract: With the increasing complexity and dimensionality of datasets in statistical research, traditional methods of identifying interactions are often more challenging to apply due to the limitations of model assumptions. Logic regression has emerged as an effective tool, leveraging Boolean combinations of binary explanatory variables. However, the prevalent simulated annealing approach in logic regression sometimes faces stability issues. This study introduces the BLogic algorithm, a novel approach that amalgamates multiple runs of simulated annealing on a dataset and synthesizes the results via the Bayesian model combination technique. This algorithm not only facilitates predicting response variables using binary explanatory ones but also offers a score computation for prime implicants, elucidating key variables and their interactions within the data. In simulations with identical parameters, conventional logic regression, when executed with a single instance of simulated annealing, exhibits reduced predictive and interpretative capabilities as soon as the ratio of explanatory variables to sample size surpasses 10. In contrast, the BLogic algorithm maintains its effectiveness until this ratio approaches 50. This underscores its heightened resilience against challenges in high-dimensional settings, especially the large p, small n problem. Moreover, employing real-world data from the UK10K Project, we also showcase the practical performance of the BLogic algorithm.

Keywords: Bayesian model combination; ensemble learning; logic tree; logic regression; machine learning; simulated annealing; UK10K project; variable interactions

MSC: 62C10; 62G08; 62M99; 62R07; 68T09

1. Introduction

The development of mathematical and computational models is fundamental in dissecting the intricate nature of relationships within sets of data. Constructing models that describe the relationship between explanatory variables (denoted as X, also called an independent variable, predictor variable, feature, or input) and response variables (denoted as Y, also called a dependent variable, label, or output) has been a continuously evolving topic in the field of mathematics and data science. Whether delving into traditional statistical models, which have been the bedrock of quantitative analysis for centuries, or navigating the waters of the rapidly growing domain of modern machine learning algorithms, researchers and practitioners constantly seek robust methodologies. Many studies, spanning decades and even centuries, aim to establish the relationship between explanatory and response variables based on various theoretical concepts. These models, underpinned by a rich tapestry of mathematical theories, are widely used in many practical fields, from economics to biology, and from physics to social sciences.

The value of these models lies in their versatility. They can be tailored to answer specific questions pertinent to the field of study. For instance, an economist might use such models to gauge the impact of fiscal policy changes on GDP, while a biologist could employ similar methodologies to explore the relationship between genetic markers and susceptibility to certain diseases [1].

Citation: Wei, Y.-C. BLogic: A Bayesian Model Combination Approach in Logic Regression. *Mathematics* **2023**, *11*, 4353. https://doi.org/10.3390/math11204353

Academic Editors: Jose Antonio Sáez Muñoz and José Luis Romero Béjar

Received: 28 August 2023
Revised: 10 October 2023
Accepted: 16 October 2023
Published: 19 October 2023

Copyright: © 2023 by the author. Licensee MDPI, Basel, Switzerland. This article is an open access article distributed under the terms and conditions of the Creative Commons Attribution (CC BY) license (https://creativecommons.org/licenses/by/4.0/).

These models, which diligently work to describe the relationship between X and Y, pivot around two core facets: predictability and interpretability. Predictability signifies the model's ability to accurately forecast the response variable given explanatory variables. This capacity to anticipate is not just a theoretical endeavor but is rooted in practical needs. For instance, forecasting stock prices or predicting weather patterns can have tangible economic impacts. Making accurate predictions can lead to saving resources, both monetary and human, and in some cases, such as medical diagnoses, can even save lives.

On the flip side, interpretability delves deeper, seeking to unearth the underlying dynamics, mechanisms, and intricacies that bind the explanatory variable to the response variable. This is not merely about drawing a line of best fit but rather understanding the forces and factors that sculpt this relationship. Understanding the 'why' and 'how' is pivotal. For instance, in a clinical setting, knowing a drug works is essential, but understanding how it works can pave the way for refining its efficacy or reducing side effects [2].

Many modern machine learning models, with their intricate architectures and algorithms, prioritize predictability. They voraciously consume data, sifting through it and teasing out patterns that might elude the human eye. By adopting data-driven approaches, these models can often achieve breathtaking accuracy in their predictions. However, this comes at a cost. The sheer complexity of some of these models, often labeled as "black boxes", can shroud their inner workings, making decisions opaque. This opacity can be a significant impediment, especially in fields such as biomedical research, where understanding the correlation between disease risk factors and the incidence of disease is paramount. Not just predicting, but understanding these correlations can lead to better preventative strategies.

In contrast, traditional statistical models offer a more transparent lens into these relationships. By allowing users to specify relationships and then rigorously testing these assumptions, these models lend themselves to greater scrutiny. The interplay of estimation and hypothesis testing serves as a robust mechanism to assess the significance of each explanatory variable. Some models, particularly tree-based and rule-based ones, are specifically architected to emulate human decision-making processes [3,4]. They set discernible rules for explanatory variables to predict the response variable. While occasionally their predictive accuracy might be eclipsed by machine learning models, their transparency and elucidative prowess often render them more suitable for specific research undertakings.

Recognizing interactions among explanatory variables is pivotal. In the digital age, the ubiquity of big data has transformed the landscape of research. Data sets have ballooned in size, often housing a plethora of explanatory variables. These variables, far from existing in isolation, often entwine in a complex choreography of interactions. Recognizing and understanding these interactions is no longer a luxury but a necessity. Furthermore, fields such as genomic epidemiology stand as a testament to this complexity [1]. Research highlights that certain genetic variations exert substantial influence on disease individually. Conversely, while some variations might not present significant main effects when considered in isolation, their interactive synergy can significantly alter disease outcomes.

In practical decision-making frameworks, there is a prevalent tendency to translate explanatory variables into a binary schema. This approach augments the clarity of discerning how these variables and their synergistic interactions influence the response variable. Illustrative transformations encompass binary explanatory variables (such as smoking status), categorical explanatory variables (for instance, single-nucleotide polymorphism genotypes coded as either dominant or recessive), and continuous explanatory variables (such as determining whether blood pressure exceeds a designated threshold).

In the pursuit of modeling these interaction effects, statistical approaches, with their precision and rigor, offer critical insights. However, these methods frequently demand predefined models, a requirement that becomes daunting when navigating the complex landscape of high-dimensional data. Logic regression (LR) emerges as an invaluable alternative in such scenarios. Through its use of Boolean combinations of binary explanatory variables, known as a logic tree, LR circumvents the need for presetting interaction types.

This nimbleness enhances its interpretability, making it a valuable tool in a researcher's arsenal. Consequently, LR, along with methods derived from its foundational model, has been widely applied to areas emphasizing interpretability, such as medical and genomic topics [5–7], public health and social sciences [8], network systems [9], and robot grasping systems [10]. Beyond its commendable interpretability, it has also been proven to possess exemplary predictive capabilities [11].

However, its reliance on simulated annealing (SA) as an optimization strategy has raised concerns, primarily owing to perceived stability issues. This instability is not just a theoretical concern; it has practical ramifications. Some studies attempt to avoid the instability of SA by using alternative complex solution approaches, while others have pivoted towards ensemble learning methods. Yet, the core issue, the shaky foundation of SA, often remains unaddressed.

Given the aforementioned context, this study is primarily motivated by the necessity of addressing the instability found in simulated annealing within the realm of logic regression. Such instability presents noteworthy challenges, especially considering the crucial role of logic regression in identifying significant interactions among explanatory variables and providing valuable predictive insights. With this understanding, our study pursues two main objectives. Firstly, we aim to illuminate the factors contributing to the instability of simulated annealing within logic regression by leveraging simulation studies. Secondly, we introduce the BLogic algorithm, which incorporates the principles of Bayesian model combination (BMC) to aggregate results from multiple iterations of SA-based logic regression models. This integration aims to mitigate the concerns associated with SA's instability, aspiring to enhance the predictability of a single SA run in logic regression while maintaining the model's prized interpretability.

Subsequent sections of this manuscript have been methodically organized to shepherd readers through our investigation. Section 2 explains the fundamentals of logic regression, describes the logic tree structure, and discusses the use of simulated annealing for optimization. Additionally, we discuss the Bayesian model combination, an ensemble method for integrating multiple models. Section 3 introduces the BLogic algorithm conceived in this study, elucidating its theoretical foundation, detailing its forecasting methods, and showcasing the important scores of interactions. Section 4, supported by simulation studies and experimental data analysis, examines our research objectives, comparing results from individual SA analyses with the combined results of multiple SA iterations merged using the BLogic algorithm. Finally, Section 5 summarizes our findings and suggests possible directions for future research.

2. Preliminaries

In this study, we aim to delve into the potential instability of logic regression when simulated annealing (SA) is used to find the optimal solution. Furthermore, we aspire to strategically employ the concept of Bayesian model combination (BMC) to consolidate the outcomes from numerous logic regression models generated through repeated SA executions, seeking a more steadfast model. Accordingly, the Preliminaries section will expound upon the two central models anchoring our research: logic regression and Bayesian model combination.

2.1. Logic Regression

Logic regression (LR) is a statistical method tailored for analyzing situations where a response variable (Y) is modulated by specific Boolean combinations of binary explanatory variables, denoted as $\{X_1, X_2, \ldots X_p\}$, each taking values of either 0 or 1.

The mathematical formulation characterizing the relationship between the response variable and its predictors is given by $g(E(Y)) = \beta_0 + \sum_{k=1}^{K} \beta_k T_k$. Within this framework, the primary objective of LR is to identify a specific Boolean function, also referred to as a 'logic tree', denoted as T_k for $k = 1, \ldots, K$. These functions encapsulate logical conjunctions of the predictors using operations such as AND, OR, and NOT. For example, a logic tree might be interpreted as (the conjugate of X_5 OR X_3) AND X_1, which denoted as $(X_5^c \vee X_3) \wedge X_1$. However, these logic trees can be represented in different forms. For uniform representation, Boolean expressions are typically articulated in the Disjunctive Normal Form (DNF) [12]. DNF is fundamentally a series of prime implicants (PIs) connected by OR operations. PIs are either a single explanatory variable or multiple explanatory variables and their conjugates linked through AND operations. The transformation mentioned above is the DNF of a given tree, where both subsets are PIs. For example, $(X_5^c \wedge X_1) \vee (X_3 \wedge X_1)$ is the DNF for the tree $(X_5^c \vee X_3) \wedge X_1$, and the subset of the DNF $(X_5^c \wedge X_1)$ and $(X_3 \wedge X_1)$ are PIs. This transformation into PIs offers insight into the complex interactions among specific sets of explanatory variables.

LR is versatile, accommodating a wide range of response variable types, such as continuous, categorical, and even survival outcomes. The bridge between the systematic component and the response is forged through an aptly chosen link function, $g(.)$. Notably, when Y is binary, logic regression can be simplified to a single logic tree T, denoted as $E(Y) = T$. In this study, we initially adopt this more streamlined model, ensuring both clarity and depth in our analyses. We believe that the findings and conclusions drawn from this research can be extrapolated to a broader range of logic regression models.

The process of estimating the regression structure leans heavily on optimization techniques. Specifically, simulated annealing often serves to explore the solution space [13,14], optimizing the configurations of logic trees in LR. SA, derived from the Metropolis–Hastings algorithm, employs a Monte Carlo method. When transitioning between a current solution and a neighboring solution, the decision hinges on their objective value differences and a parameter reminiscent of temperature. If the neighboring solution is superior to the current one, the algorithm shifts to the neighboring solution. Otherwise, a transition probability is set, allowing a potential shift. Initially, a high temperature value is adopted, allowing acceptance of subpar solutions. Over time, as the temperature decreases, the algorithm becomes more stringent, leading to convergence. In LR's landscape, SA assists in navigating the vast potential logic tree configurations. Starting with an initial logic tree, SA refines the combinations, highlighting pivotal predictors and their optimal logical relationships. The gamut of moves employed in this context is well documented in relevant LR literature [15,16].

However, using SA in logic regression presents certain challenges. Studies have shown that SA can sometimes stagnate at local optima within the vast space of logical combinations [15,16]. This convergence can yield suboptimal logic models, potentially misrepresenting underlying data patterns. The intricacies of SA's occasional erratic behavior, particularly when interpreting complex predictor variable interactions such as in single nucleotide polymorphism (SNP) datasets, are further underscored [17]. Additionally, when logic regression models, disrupted by SA's occasional inconsistencies, are integrated into ensemble frameworks such as Logic Forest [18] and LogicFS [19], the overall classifier's performance may decline. While some research, such as MCLR [20], FBLR [17], and GMJMCMC [21], has discussed the instability of SA, suggesting the adoption of Markov chain Monte Carlo methods in lieu of the traditionally used SA, there are challenges. These MCMC-based models are quite intricate, necessitating the setting of parameters to simplify the model upfront. This in turn limits the dimensionality of interactions. Moreover, some models only display results from each iteration during the Monte Carlo process without synthesizing all findings, posing difficulties for practical uses in both prediction and interpretation. Consequently, SA remains the preferred solution for logic regression and its related models.

These insights emphasize the inherent uncertainties in employing SA in logic regression, necessitating thorough exploration and careful management. This understanding is vital as it significantly impacts the predictive and interpretative capacities of models associated with or built upon logic regression.

2.2. Bayesian Model Combination

The paradigm of ensemble learning has significantly shaped the way we approach machine learning problems [22]. Ensemble methods, by definition, aim to consolidate predictions from multiple models to produce a final output that is often more robust and accurate than a prediction from any individual model. Various strategies have been developed to achieve this confluence, broadly categorized under different ensemble learning techniques such as bagging, boosting, and the Bayesian perspective.

Bagging, or bootstrap aggregating, involves generating multiple versions of a predictor by training on subsets of the data [23]. It is based on the principle of leveraging the variance among these different models to produce an aggregated result. Boosting, on the other hand, is an iterative technique that adjusts the weight of an observation based on the last classification [24]. It aims to convert weak individual learners into strong combined learners. Both methods, though distinct, come with their own strengths and challenges.

Contrastingly, Bayesian approaches to model combination, such as Bayesian model averaging (BMA) and Bayesian model combination (BMC), utilize the complete dataset instead of subsets. The Bayesian framework offers a probabilistic mechanism that facilitates the merging of prior knowledge with observed data. In BMA, despite its name suggesting an averaging technique, it behaves more like model selection, emphasizing the identification of the 'best' model [25–27]. BMC, however, truly embodies the essence of model averaging, where each model is weighted based on different strategic considerations [25]. Our research pinpoints a peculiar behavior when simulated annealing is used in logic regression. Due to its inherent stochastic nature, the results of simulated annealing in logic regression manifest instability. Such instability resonates with the notion that relying solely on BMA to discern a single 'optimal' model might not be judicious. Instead, a combination of models through BMC presents a more robust approach.

Delving further into Bayesian model combination, the methodology is anchored on the understanding that it is often more advantageous to amalgamate several models rather than pinpointing the singular best one. Suppose B models have been constructed, denoted as $H = \{h_1, \ldots, h_B\}$. Let $E = \{e_1, \ldots, e_J\}$ represent a spectrum of potential model combinations, wherein an element e_j is perceived as a weight vector for the B models, that is, $e_j = (w_{j1}, \ldots, w_{jB})$ for $j = 1, \ldots, J$. The combination can then be articulated as:

$$p(y_i|\mathbf{x}_i, D, H, E) = \sum_{j=1}^{J} p(y_i|\mathbf{x}_i, H, e_j) p(e_j|D) \propto \sum_{j=1}^{J} p(y_i|\mathbf{x}_i, H, e_j) p(e_j) p(D|e_j) \quad (1)$$

Here, $p(y_i|\mathbf{x}_i, D, H, E)$ represents the probability of predicting Y for the ith individual, conditioned on its corresponding explanatory variable $\mathbf{x}_i = (x_{i1}, \ldots, x_{ip})$, the entire training dataset D, the suite of formulated models H, and the diverse combination strategies encapsulated in E. In cases where the response variable is binary, the category that maximizes this probability is designated.

The predictive probability comprises an ensemble across J combination strategies. For each distinct combination strategy e_j, the predictive probability is essentially the weighted average of the results from the B models, steered by the weight vector $(w_{j1}, \ldots, w_{jB})$ corresponding to e_j. Crucially, the posterior probability of e_j is in direct proportion to the multiplication of its prior $p(e_j)$ and its likelihood $p(D|e_j)$. The formulation of this likelihood function can be predicated upon the predictive accuracy under the guidance of the combination strategy e_j.

The advantages of Bayesian model combinations are numerous. Primarily, it alleviates the instability inherent in individual models. Given that each model might have its own set of strengths and weaknesses, a combination approach ensures that the collective strength is leveraged while minimizing individual model vulnerabilities. Notably, literature substantiates that even rudimentary Bayesian model combination strategies surpass conventional bagging and boosting methodologies and also outperform Bayesian model averaging [25]. Additionally, it furnishes a structured resolution to the conundrum of model selection, an aspect critically pivotal in instances demanding the utmost model stability and trustworthiness.

Synthesizing the above exposition, the Bayesian model combination paradigm emerges as a robust structure, particularly apt for circumstances marked by model volatility, such as when implementing simulated annealing in logic regression. Thus, subsequent sections of this research not only delve into the latent instabilities associated with the use of simulated annealing in logic regression but also construct a methodology BLogic, inspired by the BMC ethos, to amalgamate multiple logic regression models discovered through multiple runs of simulated annealing, addressing the challenges posed by its inherent instability.

3. Methods

From the aforementioned introduction, it is evident that numerous studies have observed that when simulated annealing (SA) is employed to ascertain optimal solutions for logic regression, instability issues arise. This stands in stark contrast to many methods derived from logic regression, which simply acknowledge this instability without delving into its root causes or attempting to rectify them. To address this gap, our research directly confronts the inherent instability encountered when using SA for logic regression. In response to this challenge, our study adopts the Bayesian model combination (BMC) approach, merging multiple logic regression models that arise from repeated SA methods. The algorithm we have developed, termed BLogic, is anticipated to effectively mitigate the detrimental effects of SA instability on both predictability and interpretability. Furthermore, by amalgamating the recognized significant interactions from each SA-driven logic regression model through BMC, we enhance our ability to pinpoint the most salient explanatory variable interactions in the comprehensive model.

3.1. The BLogic Model Structure

For the sake of simplicity and clarity in this exposition, we primarily use the most elementary architecture of logic regression as an exemplar. This encompasses a binary response variable Y with only a single logic tree T encapsulated within the model, signified by $E(Y) = T$. We posit that the techniques introduced herein can be extended to standard logic regression formats and other response variable types.

Given a dataset of sample size n, each observation contains p binary explanatory variables $\mathbf{x}_i = (x_{i1}, \ldots, x_{ip})$ and one unique response variable y_i for $i = 1, 2, \ldots, n$. Each observation can be represented as $(\mathbf{x}_i, y_i)$. The entire training dataset can be defined as the set $D = \{(\mathbf{x}_i, y_i): i = 1, 2, \ldots, n\}$. When constructing logic regression using SA, let us assume we repetitively formulate B models, $H = \{h_1, \ldots, h_B\}$. As each model in this context only includes one logic tree, H can be written as $H = \{h_1 = T_1, \ldots, h_B = T_B\}$.

Expanding upon the foundation of these B logic trees (or equivalently, the B models), the BLogic model is constructed by leveraging the concepts of BMC. Analyzing Equation (1), it becomes evident that it is directly tied to the summation across the J combination strategies, namely, $p(y_i | \mathbf{x}_i, D, H, E)$. Given the assumption that each combination strategy e_j has a uniform prior, our focus narrows down to determining two primary components within the equation: $p(y_i | \mathbf{x}_i, H, e_j)$ and $p(D | e_j)$. Moving forward, we interpret the jth combination strategy e_j as a weight vector designated for the B models, denoted as $e_j = (w_{j1}, \ldots, w_{jB})$. Subsequent sections will elucidate the systematic configuration of the combination strategy.

The term $p(y_i | \mathbf{x}_i, H, e_j)$ represents the predicted probability of the ith data point y_i, given the collection H of B logic trees generated from training data and the jth weight combination $e_j = (w_{j1}, \ldots, w_{jB})$. As the model used here solely contains one logic tree, each model predicts Y based on the Boolean expressions of its explanatory variables, giving an outcome of either 0 or 1. Hence, $p(y_i | \mathbf{x}_i, H, e_j)$ is defined as $w_{j1}\hat{y}_{i1} + w_{j2}\hat{y}_{i2} + \ldots + w_{jB}\hat{y}_{iB}$, where $\hat{y}_{ib}$ is the predicted outcome for model b. In more generic logic regression scenarios, $p(y_i | \mathbf{x}_i, H, e_j)$ can be the weighted average of the predicted probabilities from each model.

Additionally, $p(D|e_j)$ represents the likelihood function of the training data given the jth combination strategy D. Here, we adopt the commonly assumed "uniform class noise model" in Bayesian model combination strategies [26]. This implies that each instance of training data is independent, and under the combination of e_j, the predictive error rate remains constant at ε_j. Consequently, $p(D|e_j) = \prod_{i=1}^{n} p(\mathbf{x}_i, y_i|e_j) = \varepsilon_j^{n-r_j}(1-\varepsilon_j)^{r_j}$, where r_j signifies the number of correctly predicted samples within the training data under the specific combination strategy e_j.

In addition to the two components previously discussed, the systematic configuration of combination strategies $E = \{e_1, \ldots, e_J\}$ must be considered. The method outlined in the original BMC literature [25] was adopted as the default method to set the combination strategies for the BLogic algorithm. Moreover, the present study offers a clearer explanation of the weight-sampling technique than what is provided in the original literature, incorporating minor adjustments to the sampling method for improved clarity and understanding. It is important to note, however, that the combination strategy methodologies within the algorithm still maintain an open framework. This structure allows users the flexibility to define strategies at their discretion.

The weights for the first q combinations, such as $e_1 = (w_{11}, \ldots, w_{1B})$, $e_2 = (w_{21}, \ldots, w_{2B})$, and so forth up to $e_q = (w_{q1}, \ldots, w_{qB})$, are generated randomly from the Dirichlet distribution with parameters $(\alpha_1, \ldots, \alpha_B)$. We have meticulously set each α_b based on the accuracy of the training data for the bth model, added by one. This ensures that the data's fit to the model is encapsulated in the weight extraction, providing a more rational approach to weight configuration. Subsequently, strategies derived from these q weight combinations are considered.

Following the consideration of strategies derived from the first q weight combinations, we compute the posterior probability of each combination strategy given the data. However, since we assume equal priors for all combination strategies, the posterior probability will be exclusively influenced by the likelihood function discussed in the preceding section. Consequently, we compute the likelihoods $p(D|e_1)$, $p(D|e_2)$, ..., $p(D|e_q)$ and identify the combination strategy that maximizes the likelihood function, denoted as $e_{j^*} = (w_{j^*1}, \ldots, w_{j^*B})$.

The weight configurations for the subsequent q combination strategies, i.e., $(e_{q+1}, e_{q+2}, \ldots, e_{2q})$, are then randomly drawn from a Dirichlet distribution characterized by parameters $(w_{j^*1}+1, \ldots, w_{j^*B}+1)$. The generation of new combination strategies ceases either when the likelihood functions calculated across Q consecutive iterations are identical or when a predefined maximum number of iterations is reached. This results in a total of J combination strategies, collectively denoted as $E = \{e_1, \ldots, e_J\}$. After generating these J combination strategies and their respective posterior probabilities, it is imperative to normalize the posterior probabilities. This ensures that the sum of posterior probabilities across the J combination strategies equates to 1.

After establishing all the essential components, Equation (1) can be employed to compute the probability $p(y_i|\mathbf{x}_i, D, H, E)$ for an individual data point. In scenarios where the response variable is binary, if this probability equates to or exceeds 0.5, the predicted value of the data point y is assigned as 1, otherwise, it is set to 0. Figure 1 displays a diagram of the algorithm's structure.

3.2. Determining the PI Importance Score within BLogic

Beyond the intricate construction of the BLogic model for response variable prediction, the essence of logic regression is retained, describing the importance of explanatory variables and their interactions. Thus, we introduce the computation of the prime implicant (PI) importance score, enhancing the interpretability of our ensemble model. Here, PI refers to Boolean expressions of the logic tree transmuted into the disjunctive normal form. PIs connote the interactions between features, symbolized by AND operations.

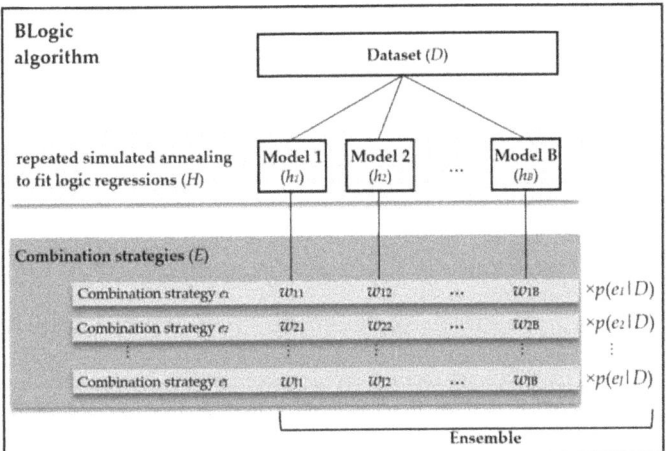

Figure 1. The workflow for the BLogic algorithm.

Suppose a specific PI_l is incorporated into the bth model, also known as the logic tree. In this tree, the importance score of PI_l is denoted as $VIMP_b(PI_l)$. This score is calculated using a permutation-based importance measure. Essentially, all explanatory variables within this PI_l are randomly permuted in the dataset. The difference in accuracy before and after this random permutation signifies the importance score of PI_l. A larger discrepancy in accuracy between pre-permutation and post-permutation implies a higher importance of the PI_l.

Integrating results from B trees and J combination strategies, the importance score for a specific PI_l in BLogic is formulated using the predetermined J combination strategies, alongside their normalized posterior probabilities. This is mathematically represented as:

$$VIMP.BLogic(PI_l) = \sum_{j=1}^{J} \sum_{b:PI_l \in b} VIMP_b(PI_l) \times w_{jb} \times p(e_j|D) \qquad (2)$$

After determining the importance scores for all PIs included in the BLogic model, a comprehensive bar chart can be generated. This visual representation effectively highlights the critical interactions.

4. Results

4.1. Simulation

In this analysis, we employ simulated data as a foundation to illuminate the potential instabilities of logic regression when using the simulated annealing (SA) method. Furthermore, we highlight the efficacy of our BLogic algorithm. This approach consolidates multiple logic regression models generated via SA, underscoring its prowess in both predictive and explanatory capacities. Consequently, our benchmarking primarily juxtaposes foundational logic regression with our novel BLogic.

Although numerous established methods have evolved from logic regression, we specifically chose not to compare them in our study, focusing instead only on the fundamental and original logic regression. One primary reason is that MCMC-based approaches do not employ SA, and the instability of SA is one of the main issues we aim to address. Additionally, due to the complexity of these methods, there is a need to preset parameters to simplify the model, creating a different comparison baseline. Most importantly, some of these methods only provide the outcomes of each MCMC iteration and lack a comprehensive strategy to amalgamate these results for practical prediction and vital interaction interpretation. Consequently, comparing them with our method in terms of predictability and interpretability becomes problematic.

Furthermore, certain methodologies, such as LogicFS and Logic Forest, blend ensemble learning with logic regression. These techniques leverage the bootstrap aggregation (bagging) method to amalgamate several logic regression models. However, their modeling is grounded in bootstrap samples, not the complete dataset. Moreover, they recognize the instability inherent in SA without addressing or amending it. Their primary goals do not align with ours. Given that models born from the bagging process usually employ in-bag and out-of-bag validation methods, and considering our study does not harness the bagging approach, our evaluation criteria differ. Hence, these methods were set aside in our benchmarking.

The subsequent subsections are structured as follows: Section 4.1.1 provides details on the parameter configurations for the simulated data. Section 4.1.2 explores potential factors leading to instability in the SA technique when applied to logic regression. Section 4.1.3 delves into the BLogic algorithm's method of amalgamating multiple logic regression models derived from SA, emphasizing its proficiency in both predictability and interpretability.

4.1.1. Parameter Settings for Simulated Data

To investigate the impact of data composition characteristics on model performance, design parameters for simulations were set. Two total sample sizes, n, of 200 and 1000 were considered. For both sizes, samples included individuals designated as $y = 1$ (cases) and $y = 0$ (controls). A case-to-control ratio in two configurations for each sample size was established: 1:1 and 1:2. Recognizing the vital interplay between the number of predictors and the sample size, the number of predictors, p, was set at eleven relative levels relative to n: $0.1n, 0.25n, 0.5n, 1n, 2.5n, 5n, 10n, 25n, 50n, 100n$, and $250n$.

In alignment with the literature [18], explanatory variables and their corresponding response variables were generated for every scenario. It is important to mention that each dataset was designed to contain a single true prime implicant (PI) that represented the interaction affecting the response variable. These PIs ranged from two-way to eight-way interactions among the explanatory variables, with the response variable being determined through a Boolean operation on the PI. Each explanatory variable was distributed independently and identically, adhering to a Bernoulli distribution. The parameters for this distribution were determined based on the given n, p, and PI setups. For a thorough subsequent analysis, we generated 100 training datasets and an independent testing dataset for each parameter combination.

For illustration, let us consider a scenario where there is a true two-way interaction serving as the PI. In this case, X_1 and X_2, which constitute the true PI, are independently generated through a Bernoulli distribution with identical parameter values. The process of generation produces a number of samples that surpasses the initially set sample size. From this extended pool, samples for case and control groups are selected based on the predetermined size and case-to-control ratio requirements. Samples in which (X_1 AND X_2) equal 1 are randomly selected until the count meets the number previously established for the case group ($y = 1$). In a similar fashion, samples where (X_1 AND X_2) equal 0 are randomly chosen to reach the predetermined count for the control group ($y = 0$). After this careful selection, values of X_1 and X_2 for the necessary samples are ascertained. Subsequently, additional explanatory variables, such as $X_3, X_4, \ldots, X_p$, which are not part of the PI, are generated for each sample. These additional variables are independently drawn from a Bernoulli distribution with a parameter of 0.5, ensuring alignment with the parameters for n, p, and PI previously specified for the ensuing analysis.

4.1.2. Instabilities in Logic Regression via Simulated Annealing

This subsection focuses on exploring the intrinsic data attributes that might lead to simulated annealing instabilities in logic regression. We bypass discussions on SA hyperparameters, including the initial temperature (Temp$_{start}$), the final temperature (Temp$_{end}$), and the iteration count (iter$_{SA}$). It should be noted, however, that we posit that the choice of these hyperparameters in SA could, to some degree, impact the stability of logic regression. Still, compared to the properties of the dataset, this is likely a minor effect. Moreover, general users often adhere to the default settings provided by the package, making repeated adjustments to hyperparameters and re-executing SA impractical. For our analysis, the SA hyperparameters were Temp$_{start}$ = 100, Temp$_{end}$ = 0.1, and iter$_{SA}$ = 50,000 [28]. Moreover, we set an upper limit of leaves = 8 for the number of leaves in the logic tree, in line with the original recommendations for logic regression [15].

In examining SA's potential instabilities, each of the 100 training datasets underwent 100 iterations of SA-based logic regression, with the mean performance across iterations subsequently evaluated. Though initial simulations considered both n = 200 and 1000, analogous trends appeared for both. As would be expected, the results for larger samples (n = 1000) were predictably more stable, making it challenging to clearly investigate the instability trends associated with simulated annealing. Therefore, we limit our presentation to the results for n = 200. Additionally, both 1:1 and 1:2 case-to-control ratios were explored. The observations indicate analogous trends, though the 1:2 ratio slightly underperformed. We thus center our discussion on the equal-proportion scenario. Initially, both two-ways to eight-ways true interactions were considered. As anticipated, the performance for lower-ways surpassed that of higher-ways. Therefore, we present the two-ways results in the manuscript.

Regarding predictive performance, Figure 2a,b depict training and testing dataset accuracy for logic regression, represented by gray dots and lines, respectively. These figures illustrate that both training and testing performances decline and become notably more varied as the number of explanatory variables increases, particularly when surpassing the sample size n. Given that logic regression is a subset of statistical regression techniques, regression methods might lack unique solutions or struggle to find optimal ones when the number of explanatory variables (or parameters to estimate) exceeds the sample size. This challenge could contribute to inconsistent model outcomes and reduced predictive capabilities with SA. Notably, when the number of explanatory variables substantially exceeds the sample size, the accuracy for the training dataset can outperform the testing dataset by 10–20%. This potential overfitting aligns with documented challenges when using logic regression for prediction in certain contexts [16]. The F1-scores for the training and testing datasets follow trends akin to the accuracy measures and as such have been omitted from the figures to maintain focus and conciseness in the presentation of results.

In the context of model interpretability, Figure 3a scrutinizes the capability of the model to correctly identify true PIs through 100 repeated iterations of simulated annealing in logic regression. Our analysis reveals that when the count of explanatory variables is either equal to or less than the sample size, a substantial majority—exceeding 90%—of the models generated via these 100 SA iterations are successful in pinpointing the true PIs. However, this rate of successful identification undergoes a steep decline as the number of explanatory variables starts to outnumber the sample size. For instance, when the number of explanatory variables is tenfold of the sample size, the mean detection rate drops precipitously to 48.13%. Moreover, when the ratio of explanatory variables to sample size scales between 100 and 250, discerning genuine interactions becomes exceptionally difficult.

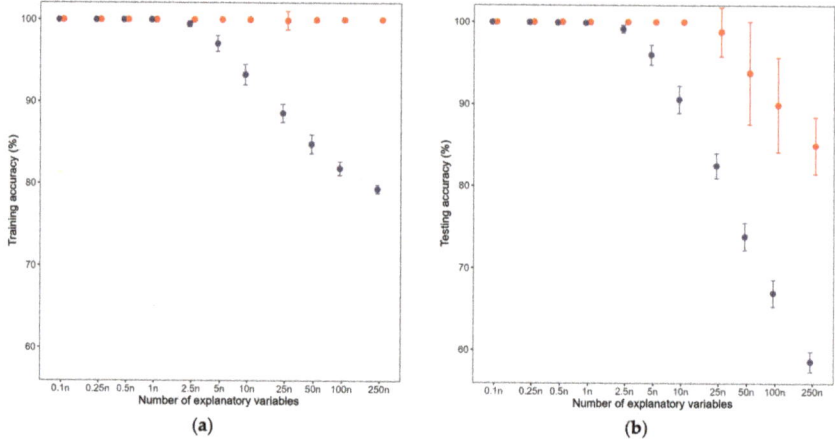

Figure 2. Predictability assessment of logic regression (represented by gray dots and lines) and BLogic algorithm (represented by red dots and lines). The X-axis of each panel displays the number of explanatory variables on a logarithmic scale. Each bar displays the mean ± 1 standard deviation. (**a**) Training accuracy; (**b**) Testing accuracy.

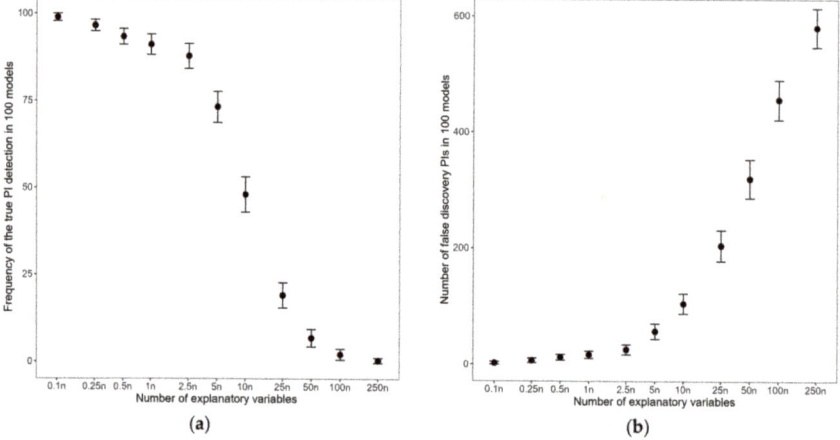

Figure 3. Analysis of instability in logic regression via repeated simulated annealing. The X-axis of each panel displays the number of explanatory variables on a logarithmic scale. Each bar displays the mean ± 1 standard deviation. (**a**) Frequency of the true PI detection; (**b**) Number of false discovery PIs.

These observations highlight the inherent limitations of individual SA runs in uncovering true PIs, especially in cases where the predictor variables vastly outnumber the samples. They also shed light on the potential shortcomings of more complex ensemble techniques, such as Logic Forest. Despite its aggregation of multiple models generated from bootstrap samples, Logic Forest might still face challenges in identifying true interactions due to the constraints of individual SA implementations within each model.

Furthermore, Figure 3b denotes the number of detected PIs during the 100 SA iterations, excluding the true PI. These additional PIs can be regarded as false discoveries. The graph reveals that as the number of explanatory variables significantly surpasses the sample size, each SA iteration might detect varying PIs. This variability reaffirms the instability of SA executions.

From a computational standpoint, our analyses were conducted on the Taiwania 1 supercomputer, hosted by the National Center for High-performance Computing within the National Applied Research Laboratories in Taiwan. The machine is equipped with dual Intel Xeon Gold 6148 2.40 GHz CPUs and offers configurations of either 192 GB or 384 GB of memory. We noted a discernible increase in computational time as the number of explanatory variables expanded. When the number of explanatory variables reached the size of the sample, the average time required was approximately 0.19 s. As the number of explanatory variables ranged from 2.5 to 25 times the sample size, computation times varied between 0.22 and 0.5 s. When the number of explanatory variables reached 50 and 100 times the sample size, the average time rose to around one second. This climbed sharply to an average of approximately 5.08 s when the number of explanatory variables was 250 times the sample size.

In conclusion, the ratio of explanatory variables to samples is a decisive factor impacting the efficacy of SA in identifying optimal logic regression solutions. The performance degrades rapidly when the number of explanatory variables surpasses the sample size, influencing predictability, interpretability, and computation times. While other factors might subtly impact performance, such as the complexity of the explanatory variable structure influencing response variables, the overall results are consistently poorer in more intricate scenarios.

4.1.3. Performance of BLogic Algorithm

From the previous subsection, it was established that the performance of logic regression constructed by simulated annealing can be unstable under certain data characteristics, leading to unsatisfactory predictive and interpretative outcomes. In this section, we delve into simulated data with the same settings to examine how the BLogic algorithm, by merging multiple logic regression models obtained through repeated simulated annealing via a Bayesian model combination (BMC) approach, can enhance the performance of a single logic regression constructed by SA both in prediction and interpretation.

In the BLogic algorithm, each data point undergoes repeated simulated annealing to obtain 100 logic tree models (i.e., $B = 100$). Throughout the iterations where a Dirichlet distribution sampling determines the weights of combination strategies, ten combination strategies are sampled in each iteration (i.e., $q = 10$). If the maximum prediction accuracy remains consistent over three consecutive iterations or when the iteration count reaches its threshold ($Q = 10{,}000$), the algorithm halts its generation of further combination strategies. The final ensemble consists of 100 models pinpointed by SA, integrated with the sampled combination strategies by the BLogic algorithm for predicting and pinpointing vital PI.

Regarding predictive performance, the red dots and lines in Figure 2a demonstrate that BLogic delivers remarkable results in terms of the accuracy of the training dataset, regardless of the ratio of samples to explanatory variables. This performance might be attributed to the fact that BLogic's design integrates the training dataset's performance over a range of combination strategies, thus shaping a likelihood function and, furthermore, the posterior probability. For the testing dataset, BLogic sustains commendable performance even when the number of explanatory variables exceeds the sample size. This distinction is particularly evident when comparing the red dots and lines representing BLogic in Figure 2b to the gray dots and lines representing logic regression. With the explanatory variables being ten times the number of samples, BLogic's testing accuracy almost invariably nears a remarkable 100%. Such prowess markedly surpasses a solitary instance of SA logic regression, which averages around 90%. Furthermore, as the tally of explanatory variables skyrockets to a staggering 250 times the sample size, BLogic achieves a testing accuracy of 84.97%, dramatically outperforming the 58.60% mean accuracy garnered from a standalone SA-based logic regression. The F1-scores for the training and testing sets still show similar trends to accuracy metrics, so we have left them out of the figures for clarity and brevity.

For interpretability, our examination utilizes two metrics. The first metric is the "Ranking of the true PI detected by BLogic". Considering that each dataset contains only a single true PI, an average rank close to 1 indicates the true PI's correct detection. Insights from Figure 4a delineate that the true PI's consistent detection and its crowning rank as the most paramount are evident when the number of explanatory variables is equal to or less than 25 times the sample size. However, as this ratio escalates to 50, 100, or even 250, the true PI's importance ranking may experience fluctuations, occasionally missing the top berth but on average landing within the top two or three positions. Hence, when the number of explanatory variables significantly outnumbers the sample size, it is recommended to observe multiple top-ranked PIs based on their importance scores.

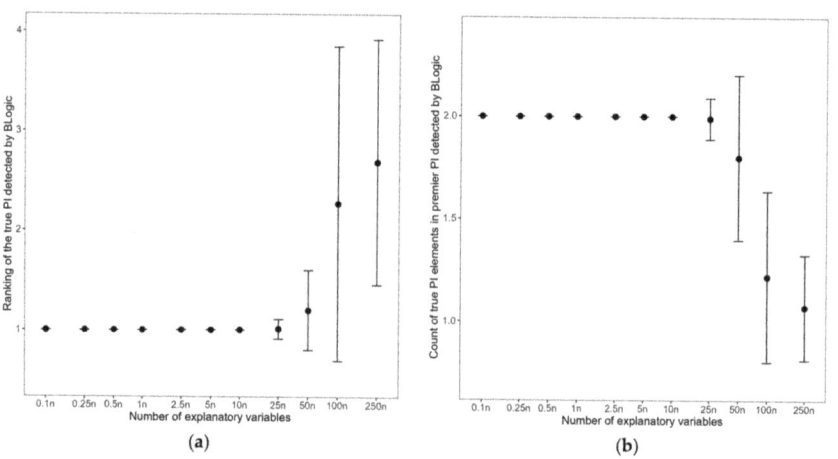

Figure 4. Interpretability assessment of the BLogic algorithm. The X-axis of each panel represents the number of explanatory variables on a logarithmic scale. Each bar displays the mean ± 1 standard deviation. (**a**) Ranking of the true PI; (**b**) Count of true PI elements in premier PI.

On the other hand, since we have only set a single true PI, if the model boasts great interpretability, the premier PI detected by BLogic should ideally be the true PI. Yet, in scenarios where there is an abundance of explanatory variables or when the true PI encompasses complex higher-way interactions, it may become challenging for the premier PI to fully capture all the variables within the true PI. Nonetheless, it remains imperative that the detected premier PI at least embodies elements of the true PI instead of being entirely disparate. To quantify this nuance, we introduced a metric termed the "count of true PI elements in premier PI detected by BLogic." For illustration, if the true PI was a two-way interaction $(X_1 \wedge X_3)$, and BLogic's premier PI embraces both X_1 & X_3, the metric equals 2. If it contains only one of them, the value is 1, and if neither, the value is 0. Observations from Figure 4b indicate that when the number of explanatory variables is less than or up to roughly 25 times the sample size, BLogic's premier PI always encompasses elements of the true PI. However, as this ratio increases, the detected premier PI may begin to incorporate other non-critical explanatory variables. When the count of explanatory variables hits 250 times the sample size, the premier PI might only contain one element of the true PI. The recommendations based on Figure 4 indicate that when the number of explanatory variables significantly outweighs the sample size, the premier PI may comprise some elements of the true PI but may not entirely represent it. Observing multiple top-ranked PIs based on their importance scores is suggested.

4.2. Experimental Data

To assess the capabilities of our developed BLogic algorithm, we drew upon next-generation sequencing data from the UK10K Project [29], housed within the European Genome-Phenome Archive. These data encompass both patients with specific diseases and healthy control samples, allowing us to employ a case-control study approach. From the provided sequencing data, we underwent a series of preprocessing steps to extract SNPs, which then served as explanatory variables for our model. The following subsections delve into the specifics of the data and preprocessing techniques and showcase the comparative analytical outcomes between the BLogic algorithm and logic regression.

4.2.1. Data Overview and Preprocessing

To evaluate the model, data from the UK10K Project was curated and organized into a case-control study design, with cases and controls delineated as binary response variables. Ninety-seven individuals diagnosed with severe insulin resistance (SIR), a rare condition, were chosen as the case group. To balance data between the case and control cohorts, 100 control samples were drawn from the TwinsUK subset within the UK10K Project, considering factors such as dizygotic twinning, sequencing quality, and depth.

All selected samples underwent preprocessing to identify uniform genomic variations as explanatory variables. Since the UK10K Project utilized whole-genome sequencing for controls and exome sequencing for cases, the analysis focused on target regions defined by exome sequencing. Following preprocessing guidelines set by tools such as Samtools [30] and the Genome Analysis Toolkit (GATK) [31], SNPs were extracted for consideration. Due to model constraints limiting the number of explanatory variables, only SNPs from a single chromosome were used for subsequent analysis. Specifically, 28,144 SNPs from chromosome 19 were chosen, as this chromosome is recognized for containing genes associated with the genomic mechanisms of SIR, as evidenced by scientific research [32–40].

Furthermore, based on typical data selection criteria in genome-wide association studies [41], we chose the final samples and SNPs for the model. These criteria included a minor allele frequency of ≥ 0.01, passing the Hardy–Weinberg Equilibrium test with a p-value $> 5.7 \times 10^{-7}$, and an identical-by-state value of less than 86%. Population stratification issues were overlooked as all samples were from the UK population. After exclusions, the dataset comprised 86 SIR patients and 100 control samples. Each SNP was converted into two binary dummy variables, signifying dominant and recessive effects. After removing non-informative features, 21,387 features were left for further analysis.

4.2.2. Analysis Results

To thoroughly assess the predictive performance of our proposed BLogic algorithm in comparison to logic regression with single simulated annealing, we employed a repeated 10-fold cross-validation, conducted 50 times. Common hyperparameters for both BLogic and logic regression were uniformly set, encompassing SA parameters (Temp$_{start}$ = 100, Temp$_{end}$ = 0.1, and iter$_{SA}$ = 50,000) and a maximum of 8 leaves for the logic tree. For BLogic-specific hyperparameters, each data batch was configured to undergo SA 100 times ($B = 100$), sampling ten combination strategies ($q = 10$) during each cycle. The algorithm ceases operation either when maximum prediction accuracy remains stable across three successive iterations or upon reaching its predetermined threshold ($Q = 10,000$).

After obtaining the accuracy and F1-score for each of the 10-fold cross-validations, the results from the fifty repetitions were averaged to compute the mean and standard deviation (sd) to gauge the predictive performance of the BLogic algorithm against the single SA in logic regression. These outcomes are detailed in Table 1. The BLogic algorithm substantially surpasses logic regression in both mean accuracy and F1-score. Additionally, the standard deviations underscore the enhanced stability of BLogic, further distinguishing it from logic regression in terms of consistency in performance. With regards to the training set, although the mean accuracy and F1-score of BLogic are only marginally superior to those of logic regression, this subtle edge might be due to the effective fitting of the logic regression model to the training data. However, a closer examination of the standard deviation for both metrics unequivocally demonstrates that BLogic consistently maintains a higher level of stability compared to its counterpart. These predictive results align with the findings from the simulation

Table 1. Prediction performance on experimental data.

		Training Set		Testing Set	
	(%)	Accuracy	F1-Score	Accuracy	F1-Score
BLogic	mean	100	100	99.88	99.85
	(sd)	(0)	(0)	(0.22)	(0.32)
logic regression	mean	99.84	99.82	95.17	94.36
	(sd)	(0.10)	(0.11)	(0.60)	(0.89)

On the interpretive front, to illustrate the crucial SNPs and their interactions identified by BLogic within this dataset, the unpartitioned data was processed again using the BLogic algorithm, keeping the hyperparameter settings consistent with those previously mentioned. This process culminated in the generation of 30 combination strategies, with iterations ceasing when no further enhancements in predictive accuracy were observed. Additionally, we calculated the PI importance scores, as mentioned earlier, and highlighted the most significant PIs in descending order based on these scores. It is worth noting that the results of the PIs detected during individual runs of SA in logic regression are not provided here. This decision was made due to the inconsistent PI outcomes from each individual SA run in logic regression, deeming them unsuitable for presentation.

Figure 5 displays the top 10 PIs obtained from the BLogic algorithm. In the PI notation, an '!' prefix to the SNP rs number signifies a complement set, while the suffixes '_1' and '_2' indicate dominant and recessive coding, respectively. The top-ranking PIs often involve interactions between two or three SNPs. A search using the Genome Data Viewer at the National Center for Biotechnology Information revealed that all SNPs included in the top 10 PIs are located within the genomic bands of 19p13 and 19q13. Numerous studies have pinpointed gene mutations within the 19p13 region that influence the insulin receptor, subsequently leading to insulin resistance [32–36]. Additionally, some research confirms that genetic variants on 19q13 can cause severe insulin resistance [37,38], as well as Type 2 diabetes mellitus associated with insulin abnormalities [39,40]. The PI with the highest importance score identified by BLogic, which has a score notably higher than other PIs, involves an interaction between the SNPs rs162124 (located at 19q13.41) and rs11085209 (located at 19p13.2). Experts are thus recommended to not only examine the regions of 19p13 and 19q13 independently but also to further probe into the potential impacts on the SIR mechanism pathway resulting from mutations within these regions. Moreover, within the top 10 PIs, the SNPs rs11085209 (located at 19p13.2) and rs10415889 (located at 19p13.11) frequently interact with other SNPs. Hence, this outcome suggests that experts might consider a more in-depth investigation of the genetic variants within the 19p13 sub-bands, particularly 19p13.2 and 19p13.11, and their impact on the SIR mechanism.

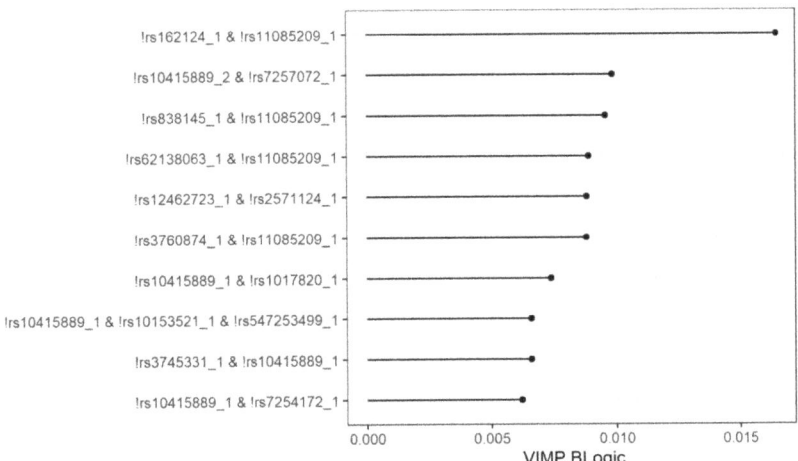

Figure 5. Top 10 prime implicants by BLogic from the UK10K project on severe insulin resistance.

5. Conclusions and Discussions

Logic regression presents a unified model that utilizes Boolean combinations of binary explanatory variables to predict response variables. This structure inherently identifies crucial interactions among the explanatory variables. It is in stark contrast to conventional statistical methods, which necessitate predefined interaction categories. Thus, it is particularly advantageous when seeking to uncover significant interactions among many explanatory variables.

Simulated annealing (SA) is commonly adopted to find the optimal solution in logic regression. Numerous studies have acknowledged the instability of SA in this context, but the underlying causes of this instability remain largely unexplored. In our research, we specifically employed simulated data to probe the characteristics underlying the instability of SA in logic regression. In our simulation studies, when only one two-way interaction PI is set in the data, we found that the challenges arising from SA's instability begin to manifest when the number of explanatory variables is more than ten times the sample size. Furthermore, when the quantity of explanatory variables greatly exceeds the sample size, the instability in SA becomes profoundly evident. This results in a significant drop in predictive precision and a failure to pinpoint vital interactions. We believe that when the data contain more complex interactions, the performance of the model might be adversely affected by SA's instability, even when the number of explanatory variables does not greatly exceed the sample size. Such instability undermines the unique advantage of logic regression: its inherent capability to autonomously detect interactions.

To address this issue, our study proposed the BLogic algorithm. This method relies on repeatedly employing SA to construct various logic regression models on the same dataset. The Bayesian model combination (BMC) approach, suitable for assimilating multiple unstable models, is then employed to combine each logic regression using different combination strategies. This methodology has demonstrated superiority in prediction accuracy compared to relying solely on a single unstable result from SA in logic regression. Furthermore, through this theoretical method, we can systematically consolidate all interactions identified by multiple SAs and evaluate the relative importance of each interaction. Moreover, the influence of the ratio between explanatory variables and sample size is attenuated in this approach. In the setting of our simulation study, it is only when the number of explanatory variables exceeds 50 times the sample size that prediction and interpretability begin to show some minor effects.

Based on our research findings, there is substantial scope for further investigation. One noteworthy area of interest is determining the optimal number of logic regression models to be constructed within BLogic, specifically deciding on the appropriate setting for the hyperparameter B. While we highlighted findings with $B = 100$ in our simulations, additional tests with $B = 200, 300, 400$, and 500 ($n = 200$, case-to-control ratios 1:1, and two-way true interactions) have also been conducted. The results are presented in Tables 2 and 3, focusing on predictability and interpretability, respectively. Even though the patterns in predictability and interpretability remained relatively stable across varying B values, larger B values seemed to slightly better mitigate the issues arising from a high ratio of explanatory variables to sample size. While there was no significant difference in performance across our chosen B values ranging from 100 to 500, the selection of B still might influence the analysis results. Opting for a smaller B might not fully capture the breadth of potential SA outcomes, potentially making the merging of unstable SA instances ineffective. Conversely, a larger B, while encompassing varied SA results, would demand more computational resources due to repeated SA runs and the necessity to determine weights for each model within the BLogic algorithm. Additionally, in scenarios where the sample size significantly outnumbers the explanatory variables, as demonstrated in our simulations, SA often generates nearly identical logic tree outputs. As a result, executing SA numerous times might be inefficient. For the time being, we have set a default value, drawing inspiration from the commonly used default value of 100 in random forests [42]. Future work might consider adjusting the value of B based on factors such as the total number of explanatory variables in the data, the number of leaves in each logic tree, the number of potential key explanatory variables, and the order of interactions. Subsequent investigations could seek to methodically understand the interconnectedness of these considerations. As an example, with a predetermined number of explanatory variables, pinpointing the lowest B value essential to identifying a specific sequence of PI might be of interest. Employing simulation studies could also be valuable in analyzing the patterns and relations of these parameters.

Table 2. Impact of hyperparameter B on the predictability of the BLogic algorithm.

	B		\multicolumn{11}{c}{Number of Explanatory Variables}
			$0.1n$	$0.25n$	$0.5n$	$1n$	$2.5n$	$5n$	$10n$	$25n$	$50n$	$100n$	$250n$
Training accuracy (%)	100	Mean	100	100	100	100	100	100	100	99.89	99.98	99.99	100
		(sd)	(0)	(0)	(0)	(0)	(0)	(0)	(0)	(1.15)	(0.21)	(0.05)	(0)
	200	Mean	100	100	100	100	100	100	100	99.87	99.99	100	100
		(sd)	(0)	(0)	(0)	(0)	(0)	(0)	(0)	(1.20)	(0.07)	(0)	(0)
	300	Mean	100	100	100	100	100	100	100	99.70	99.96	100	100
		(sd)	(0)	(0)	(0)	(0)	(0)	(0)	(0)	(1.79)	(0.31)	(0)	(0)
	400	Mean	100	100	100	100	100	100	100	99.65	99.995	100	100
		(sd)	(0)	(0)	(0)	(0)	(0)	(0)	(0)	(1.88)	(0.05)	(0)	(0)
	500	Mean	100	100	100	100	100	100	100	99.38	100	100	100
		(sd)	(0)	(0)	(0)	(0)	(0)	(0)	(0)	(2.61)	(0.05)	(0)	(0)
Testing accuracy (%)	100	Mean	100	100	100	100	100	100	99.995	98.81	93.80	89.91	84.97
		(sd)	(0)	(0)	(0)	(0)	(0)	(0)	(0.05)	(3.01)	(6.26)	(5.78)	(3.46)
	200	Mean	100	100	100	100	100	100	100	96.64	90.54	89.61	86.14
		(sd)	(0)	(0)	(0)	(0)	(0)	(0)	(0)	(5.24)	(8.32)	(7.19)	(3.33)
	300	Mean	100	100	100	100	100	100	100	95.20	90.69	89.55	86.96
		(sd)	(0)	(0)	(0)	(0)	(0)	(0)	(0)	(6.73)	(9.45)	(7.64)	(3.34)
	400	Mean	100	100	100	100	100	100	100	93.88	89.71	89.15	86.99
		(sd)	(0)	(0)	(0)	(0)	(0)	(0)	(0)	(8.07)	(9.71)	(7.95)	(3.49)
	500	Mean	100	100	100	100	100	100	100	93.57	89.74	89.14	87.12
		(sd)	(0)	(0)	(0)	(0)	(0)	(0)	(0)	(8.48)	(9.82)	(7.62)	(3.65)

Table 3. Impact of hyperparameter B on the interpretability of the BLogic algorithm.

	B		\multicolumn{11}{c}{Number of Explanatory Variables}
			$0.1n$	$0.25n$	$0.5n$	$1n$	$2.5n$	$5n$	$10n$	$25n$	$50n$	$100n$	$250n$
Ranking of the true PI detected by BLogic	100	Mean	1	1	1	1	1	1	1	1.01	1.20	2.27	2.69
		(sd)	(0)	(0)	(0)	(0)	(0)	(0)	(0)	(0.10)	(0.40)	(1.58)	(1.23)
	200	Mean	1	1	1	1	1	1	1	1.01	1.28	1.87	3.00
		(sd)	(0)	(0)	(0)	(0)	(0)	(0)	(0)	(0.10)	(0.45)	(0.56)	(1.76)
	300	Mean	1	1	1	1	1	1	1	1	1.29	1.96	2.99
		(sd)	(0)	(0)	(0)	(0)	(0)	(0)	(0)	(0)	(0.46)	(0.55)	(1.93)
	400	Mean	1	1	1	1	1	1	1	1.01	1.22	1.98	2.85
		(sd)	(0)	(0)	(0)	(0)	(0)	(0)	(0)	(0.10)	(0.42)	(0.51)	(1.22)
	500	Mean	1	1	1	1	1	1	1	1.02	1.25	1.99	3.34
		(sd)	(0)	(0)	(0)	(0)	(0)	(0)	(0)	(0.14)	(0.43)	(0.48)	(2.68)
Count of true PI elements in premier PI detected by BLogic	100	Mean	2	2	2	2	2	2	2	1.99	1.80	1.22	1.07
		(sd)	(0)	(0)	(0)	(0)	(0)	(0)	(0)	(0.10)	(0.40)	(0.42)	(0.26)
	200	Mean	2	2	2	2	2	2	2	1.99	1.72	1.23	1.03
		(sd)	(0)	(0)	(0)	(0)	(0)	(0)	(0)	(0.10)	(0.45)	(0.42)	(0.17)
	300	Mean	2	2	2	2	2	2	2	2	1.71	1.17	1.01
		(sd)	(0)	(0)	(0)	(0)	(0)	(0)	(0)	(0)	(0.46)	(0.38)	(0.10)
	400	Mean	2	2	2	2	2	2	2	1.99	1.78	1.14	1.01
		(sd)	(0)	(0)	(0)	(0)	(0)	(0)	(0)	(0.10)	(0.42)	(0.35)	(0.10)
	500	Mean	2	2	2	2	2	2	2	1.98	1.75	1.12	1.01
		(sd)	(0)	(0)	(0)	(0)	(0)	(0)	(0)	(0.14)	(0.44)	(0.33)	(0.10)

Secondly, overfitting remains a concern. Previous literature, along with the gray lines and dots in Figure 2, suggests that a single logic regression model can overfit under certain data conditions. The red lines and dots in the figure indicate that while BLogic has marginally better training accuracy than testing, it does not amplify the overfitting problem. This discrepancy in BLogic's training and testing accuracy may arise from the inherent instability of a logic regression run with a single SA. A potential solution for future research might involve the application of cross-validation, segmenting the full dataset into training and validation sets. The training set could be used to develop the logic regression model with SA, and the validation set could help determine the likelihood function and subsequent posterior probabilities for each combination strategy. This approach might mitigate overfitting by decreasing the contribution of the training set. However, such a modification would necessitate a thorough re-evaluation of the theoretical framework of the BMC model due to the dataset division.

Thirdly, the BLogic algorithm adopts the method outlined in the original BMC literature as the default approach for configuring combination strategies within BMC. This method uses the performance of individual combination strategies as a basis to update the parameters within the Dirichlet distribution, continually iterating to generate a series of combination strategies. While theoretically plausible, there are reservations cautiously acknowledged regarding its consistent ability to yield optimal combination strategies. The effectiveness of this method might vary due to differences in data or potential correlations with other hyperparameters within the model. The optimal strategies of model combination for BLogic remain an open area of research and require further study.

A straightforward comparison is presented with a naive approach, which assigns equal weights to each logic regression model obtained from a single SA run and the combination strategies deployed by the BLogic algorithm. This comparison, generated with true two-way PIs, a sample size of 200, and a 1:1 case-to-control ratio, deliberately selects a scenario with a reduced number of SA runs, namely 50, to spotlight the effective performance of the BMC technique. To keep the article's focus sharp, Figure 6a represents predictability solely through testing accuracy, while Figure 6b illustrates the ranking of true PI, shedding light on their interpretability. In most scenarios, the combination method of BLogic's model (denoted by red dots and lines) not only consistently secures higher accuracy but also assigns the highest importance scores to the true PI, thereby correctly identifying it as a priority, compared to the equal weight approach (signified by gray dots and lines). This illustration emphasizes the effectiveness of employing the BMC technique for combinations. However, it is crucial to acknowledge that alternative combination strategies necessitate further and more detailed exploration.

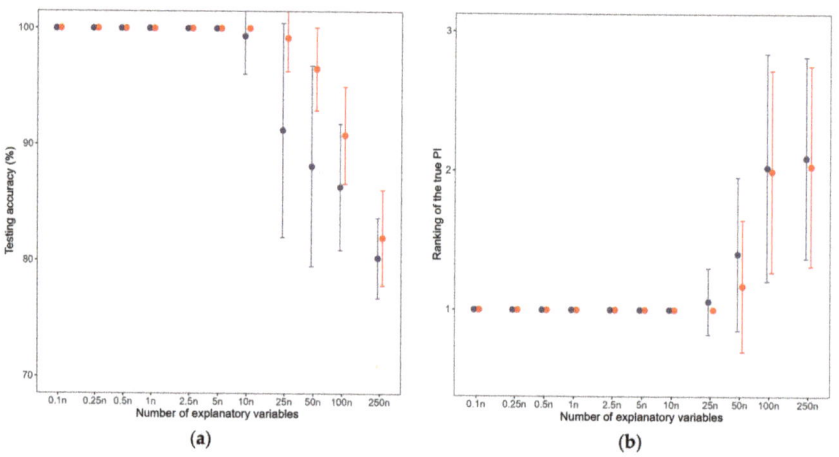

Figure 6. Performance assessment of equal weight combination (represented by gray dots and lines) and BLogic via BMC default combination strategy (represented by red dots and lines). The X-axis of each panel displays the number of explanatory variables on a logarithmic scale. Each bar displays the mean ± 1 standard deviation. (**a**) Testing accuracy; (**b**) Ranking of the true PI.

Lastly, a significant contribution of our research lies in the revelation through simulated studies that SA outcomes become notably unstable when the number of explanatory variables greatly exceeds the sample size. This "large p small n" dilemma, common in datasets generated through bioinformatics techniques such as microarrays or next-generation sequencing, poses challenges [43]. For instance, while genomic variations can run into tens of thousands, sample sizes remain relatively smaller. Though our study demonstrates that using BLogic, by repeatedly applying SA and then combining the results using BMC, can address this, the sheer volume of explanatory variables still presents a hurdle. When the genuine key explanatory variables are substantially fewer than the total variables, directly applying SA might fail to find the true solution. Therefore, when confronted with datasets abundant in explanatory variables, various methods of integrating or transforming variable information become worth considering. Techniques such as feature selection [44,45], feature extraction [46,47], weighting variables [48,49], regularization [50], and split-and-merge [51] approaches can be integrated. Incorporating these into the our proposed BLogic algorithm may set the stage for a more streamlined inclusion of genuinely pivotal explanatory variables into the logic regression model, concluding our quest for enhanced predictability and interpretability in this field.

Funding: This work was partially supported by grants from the National Science and Technology Council of Taiwan (NSTC 112-2118-M-018-004).

Acknowledgments: The author thanks the National Center for High-performance Computing (NCHC) of the National Applied Research Laboratories (NARLabs) of Taiwan for providing computational and storage resources. Moreover, this study makes use of data generated by the UK10K Consortium, derived from samples from the Cambridge Severe Insulin Resistance Study Cohort and the Twins UK Cohort. A full list of the investigators who contributed to the generation of the data is available at www.UK10K.org. Funding for UK10K was provided by the Wellcome Trust under award WT091310. Special acknowledgment goes to Yu-Xiang Liu and Chin-Yi Chao for their invaluable assistance in downloading, cleaning, and validating the data. Furthermore, the author extends sincere thanks to the diligent reviewers for their rigorous scrutiny and insightful feedback through multiple rounds of review. Their invaluable advice was crucial for refining the manuscript, resulting in a more coherent, robust, and valuable final paper.

Conflicts of Interest: The author declares no conflict of interest.

References

1. Cordell, H.J. Detecting gene–gene interactions that underlie human diseases. *Nat. Rev. Genet.* **2009**, *10*, 392–404. [CrossRef] [PubMed]
2. Tekin, E.; Savage, V.M.; Yeh, P.J. Measuring higher-order drug interactions: A review of recent approaches. *Curr. Opin. Syst. Biol.* **2017**, *4*, 16–23. [CrossRef]
3. Kuhn, M.; Johnson, K.; Kuhn, M.; Johnson, K. Classification trees and rule-based models. In *Applied Predictive Modeling*; Springer: Berlin/Heidelberg, Germany, 2013; pp. 369–413.
4. Apté, C.; Weiss, S. Data mining with decision trees and decision rules. *Future Gener. Comput. Syst.* **1997**, *13*, 197–210. [CrossRef]
5. Kocbek, S.; Kocbek, P.; Gosak, L.; Fijačko, N.; Štiglic, G. Extracting new temporal features to improve the interpretability of undiagnosed type 2 diabetes mellitus prediction models. *J. Pers. Med.* **2022**, *12*, 368. [CrossRef]
6. Bellavia, A.; Rotem, R.S.; Dickerson, A.S.; Hansen, J.; Gredal, O.; Weisskopf, M.G. The use of logic regression in epidemiologic studies to investigate multiple binary exposures: An example of occupation history and amyotrophic lateral sclerosis. *Epidemiol. Methods* **2020**, *9*, 20190032. [CrossRef] [PubMed]
7. Meijsen, J.J.; Rammos, A.; Campbell, A.; Hayward, C.; Porteous, D.J.; Deary, I.J.; Marioni, R.E.; Nicodemus, K.K. Using tree-based methods for detection of gene–gene interactions in the presence of a polygenic signal: Simulation study with application to educational attainment in the Generation Scotland Cohort Study. *Bioinformatics* **2019**, *35*, 181–188. [CrossRef]
8. Yoneoka, D.; Eguchi, A.; Nomura, S.; Kawashima, T.; Tanoue, Y.; Murakami, M.; Sakamoto, H.; Maruyama-Sakurai, K.; Gilmour, S.; Shi, S. Identification of optimum combinations of media channels for approaching COVID-19 vaccine unsure and unwilling groups in Japan. *Lancet Reg. Health–West. Pac.* **2022**, *18*, 100330. [CrossRef]
9. Rocco, C.M.; Hernandez-Perdomo, E.; Mun, J. Application of logic regression to assess the importance of interactions between components in a network. *Reliab. Eng. Syst. Saf.* **2021**, *205*, 107235. [CrossRef]
10. Li, T.; Sun, X.; Shu, X.; Wang, C.; Wang, Y.; Chen, G.; Xue, N. Robot grasping system and grasp stability prediction based on flexible tactile sensor array. *Machines* **2021**, *9*, 119. [CrossRef]
11. Lau, M.; Wigmann, C.; Kress, S.; Schikowski, T.; Schwender, H. Evaluation of tree-based statistical learning methods for constructing genetic risk scores. *BMC Bioinform.* **2022**, *23*, 1–30. [CrossRef]
12. Ruczinski, I. Logic Regression and Statistical Issues Related to the Protein Folding Problem. Ph.D. Thesis, University of Washington, Washington, DC, USA, 2001.
13. Otten, R.H.; van Ginneken, L.P. *The Annealing Algorithm*; Springer Science & Business Media: Berlin/Heidelberg, Germany, 2012; Volume 72.
14. Aarts, E.H. *Simulated Annealing: Theory and Applications*; Reidel: Dordrecht, The Netherlands, 1987.
15. Kooperberg, C.; Ruczinski, I.; LeBlanc, M.L.; Hsu, L. Sequence analysis using logic regression. *Genet. Epidemiol.* **2001**, *21*, S626–S631. [CrossRef]
16. Ruczinski, I.; Kooperberg, C.; LeBlanc, M. Logic regression. *J. Comput. Graph. Stat.* **2003**, *12*, 475–511. [CrossRef]
17. Fritsch, A.; Ickstadt, K. Comparing logic regression based methods for identifying SNP interactions. In Proceedings of the International Conference on Bioinformatics Research and Development, Berlin, Germany, 12–14 March 2007; pp. 90–103.
18. Wolf, B.J.; Hill, E.G.; Slate, E.H. Logic forest: An ensemble classifier for discovering logical combinations of binary markers. *Bioinformatics* **2010**, *26*, 2183–2189. [CrossRef] [PubMed]
19. Schwender, H.; Ickstadt, K. Identification of SNP interactions using logic regression. *Biostatistics* **2008**, *9*, 187–198. [CrossRef]
20. Kooperberg, C.; Ruczinski, I. Identifying interacting SNPs using Monte Carlo logic regression. *Genet. Epidemiol.* **2005**, *28*, 157–170. [CrossRef]
21. Hubin, A.; Storvik, G.; Frommlet, F. A novel algorithmic approach to Bayesian logic regression (with discussion). *Bayesian Anal.* **2020**, *15*, 263–333. [CrossRef]
22. Sagi, O.; Rokach, L. Ensemble learning: A survey. *Wiley Interdiscip. Rev. Data Min. Knowl. Discov.* **2018**, *8*, e1249. [CrossRef]
23. Breiman, L. Bagging predictors. *Mach. Learn.* **1996**, *24*, 123–140. [CrossRef]
24. Freund, Y.; Schapire, R.E. Experiments with a new boosting algorithm. In Proceedings of the Thirteenth International Conference on International Conference on Machine Learning, Bari, Italy, 3–6 July 1996; pp. 148–156.
25. Monteith, K.; Carroll, J.L.; Seppi, K.; Martinez, T. Turning Bayesian model averaging into Bayesian model combination. In Proceedings of the 2011 International Joint Conference on Neural networks, San Jose, CA, USA, 31 July–5 August 2011; pp. 2657–2663.

26. Domingos, P. Bayesian averaging of classifiers and the overfitting problem. In Proceedings of the International Conference on International Conference on Machine Learning, Stanford University, Stanford, CA, USA, 29 June–2 July 2000; pp. 223–230.
27. Minka, T.P. Bayesian Model Averaging Is Not Model Combination. 2002. Available online: https://tminka.github.io/papers/minka-bma-isnt-mc.pdf (accessed on 21 February 2021).
28. Kooperberg, C.; Ruczinski, I.; Kooperberg, M.C. Package 'LogicReg'. Comprehensive R Archive Network. 2015. Available online: http://cran.fhcrc.org/web/packages/LogicReg/LogicReg.pdf (accessed on 1 March 2021).
29. The UK10K Consortium. The UK10K project identifies rare variants in health and disease. *Nature* **2015**, *526*, 82–90. [CrossRef]
30. Li, H.; Handsaker, B.; Wysoker, A.; Fennell, T.; Ruan, J.; Homer, N.; Marth, G.; Abecasis, G.; Durbin, R. Genome Project Data Processing Subgroup. The sequence alignment/map format and SAMtools. *Bioinformatics* **2009**, *25*, 2078–2079. [CrossRef]
31. McKenna, A.; Hanna, M.; Banks, E.; Sivachenko, A.; Cibulskis, K.; Kernytsky, A.; Garimella, K.; Altshuler, D.; Gabriel, S.; Daly, M. The Genome Analysis Toolkit: A MapReduce framework for analyzing next-generation DNA sequencing data. *Genome Res.* **2010**, *20*, 1297–1303. [CrossRef] [PubMed]
32. Joshi, S.R.; Pendyala, G.S.; Shah, P.; Pustake, B.; Mopagar, V.; Padmawar, N. Severe insulin resistance syndrome–A rare case report and review of literature. *Natl. J. Maxillofac. Surg.* **2021**, *12*, 100. [CrossRef] [PubMed]
33. Longo, N.; Wang, Y.; Smith, S.A.; Langley, S.D.; DiMeglio, L.A.; Giannella-Neto, D. Genotype–phenotype correlation in inherited severe insulin resistance. *Hum. Mol. Genet.* **2002**, *11*, 1465–1475. [CrossRef] [PubMed]
34. Sinnarajah, K.; Dayasiri, M.; Dissanayake, N.; Kudagammana, S.; Jayaweera, A. Rabson Mendenhall Syndrome caused by a novel missense mutation. *Int. J. Pediatr. Endocrinol.* **2016**, *2016*, 21. [CrossRef] [PubMed]
35. Kosztolanyi, G. Leprechaunism/Donohue syndrome/insulin receptor gene mutations: A syndrome delineation story from clinicopathological description to molecular understanding. *Eur. J. Pediatr.* **1997**, *156*, 253. [CrossRef]
36. Al-Beltagi, M.; Bediwy, A.S.; Saeed, N.K. Insulin-resistance in paediatric age: Its magnitude and implications. *World J. Diabetes* **2022**, *13*, 282. [CrossRef]
37. Tan, K.; Kimber, W.A.; Luan, J.a.; Soos, M.A.; Semple, R.K.; Wareham, N.J.; O'Rahilly, S.; Barroso, I. Analysis of genetic variation in Akt2/PKB-β in severe insulin resistance, lipodystrophy, type 2 diabetes, and related metabolic phenotypes. *Diabetes* **2007**, *56*, 714–719. [CrossRef]
38. An, P.; Freedman, B.I.; Hanis, C.L.; Chen, Y.-D.I.; Weder, A.B.; Schork, N.J.; Boerwinkle, E.; Province, M.A.; Hsiung, C.A.; Wu, X. Genome-wide linkage scans for fasting glucose, insulin, and insulin resistance in the National Heart, Lung, and Blood Institute Family Blood Pressure Program: Evidence of linkages to chromosome 7q36 and 19q13 from meta-analysis. *Diabetes* **2005**, *54*, 909–914. [CrossRef]
39. Van Tilburg, J.; Sandkuijl, L.; Strengman, E.; Van Someren, H.; Rigters-Aris, C.; Pearson, P.; Van Haeften, T.; Wijmenga, C. A genome-wide scan in type 2 diabetes mellitus provides independent replication of a susceptibility locus on 18p11 and suggests the existence of novel Loci on 2q12 and 19q13. *J. Clin. Endocrinol. Metab.* **2003**, *88*, 2223–2230. [CrossRef]
40. Dorajoo, R.; Liu, J.; Boehm, B.O. Genetics of type 2 diabetes and clinical utility. *Genes* **2015**, *6*, 372–384. [CrossRef] [PubMed]
41. Uffelmann, E.; Huang, Q.Q.; Munung, N.S.; De Vries, J.; Okada, Y.; Martin, A.R.; Martin, H.C.; Lappalainen, T.; Posthuma, D. Genome-wide association studies. *Nat. Rev. Methods Primers* **2021**, *1*, 59. [CrossRef]
42. Breiman, L. Random forests. *Mach. Learn.* **2001**, *45*, 5–32. [CrossRef]
43. Schena, M.; Shalon, D.; Davis, R.W.; Brown, P.O. Quantitative monitoring of gene expression patterns with a complementary DNA microarray. *Science* **1995**, *270*, 467–470. [CrossRef] [PubMed]
44. Chen, J.; Aseltine, R.H.; Wang, F.; Chen, K. Tree-guided rare feature selection and logic aggregation with electronic health records data. *arXiv* **2022**, arXiv:2206.09107.
45. Li, J.; Cheng, K.; Wang, S.; Morstatter, F.; Trevino, R.P.; Tang, J.; Liu, H. Feature selection: A data perspective. *ACM Comput. Surv.* **2017**, *50*, 1–45. [CrossRef]
46. Khalid, S.; Khalil, T.; Nasreen, S. A survey of feature selection and feature extraction techniques in machine learning. In Proceedings of the 2014 Science and Information conference, London, UK, 27–29 August 2014; pp. 372–378.
47. Guyon, I.; Gunn, S.; Nikravesh, M.; Zadeh, L.A. *Feature Extraction: Foundations and Applications*; Springer: Berlin/Heidelberg, Germany, 2008; Volume 207.
48. Maudes, J.; Rodríguez, J.J.; García-Osorio, C.; García-Pedrajas, N. Random feature weights for decision tree ensemble construction. *Inf. Fusion* **2012**, *13*, 20–30. [CrossRef]
49. Chen, Y.-C. *An Ensemble Logic Regression Approach for Detecting Important Genes and Interactions*; National Changhua University of Education: Changhua, Taiwan, 2023.
50. Lim, M.; Hastie, T. Learning interactions via hierarchical group-lasso regularization. *J. Comput. Graph. Stat.* **2015**, *24*, 627–654. [CrossRef]
51. Huang, W.-H.; Wei, Y.-C. A split-and-merge deep learning approach for phenotype prediction. *Front. Biosci. Landmark* **2022**, *27*, 78. [CrossRef] [PubMed]

Disclaimer/Publisher's Note: The statements, opinions and data contained in all publications are solely those of the individual author(s) and contributor(s) and not of MDPI and/or the editor(s). MDPI and/or the editor(s) disclaim responsibility for any injury to people or property resulting from any ideas, methods, instructions or products referred to in the content.

Article

Research on Emotional Infection of Passengers during the SRtP of a Cruise Ship by Combining an SIR Model and Machine Learning

Gaohan Xiong [1], Wei Cai [2], Min Hu [2,*] and Zhiyan Yu [1]

1. School of Naval Architecture, Ocean and Energy Power Engineering, Wuhan University of Technology, Wuhan 430063, China; xionggaohan@whut.edu.cn (G.X.); yuzhiyan@whut.edu.cn (Z.Y.)
2. Green and Smart River-Sea-Going Ship, Cruise Ship and Yacht Research Center, Wuhan University of Technology, Wuhan 430063, China; wcai@whut.edu.cn
* Correspondence: hu_min@whut.edu.cn

Abstract: The Safe Return to Port issue regarding cruise ships has been extensively researched, covering aspects such as performance, operations, and electrical systems. However, an often overlooked aspect is the potential eruption of negative emotions among passengers during SRtP. This study aims to investigate the prediction of collective emotions to facilitate timely safety planning and enhance the safety of the Safe Return to Port process. To achieve this objective, an improved susceptible-infectious-recovered model with bidirectional infection is proposed to describe the emotional contagion process during the Safe Return to Port process. This model classifies the population into five emotional (extremely anxious–anxious–normal–calm–very calm) states and introduces two sources of infection. Moreover, it allows for emotions to transition both positively and negatively, making it a more realistic representation of scenarios resembling long-term refuge scenarios. In this study, questionnaire data, collected and statistically analyzed, serve as the primary dataset. A machine learning technique (the weighted random forest algorithm) is integrated with the model to make predictions. The accuracy, precision, recall, and the F-measure of prediction results demonstrate good performance. Additionally, through simulation, this study illustrates the fluctuating nature of emotional changes during the Safe Return to Port process of the cruise ship and analyzes the effects of varying parameters. The findings suggest that the improved susceptible-infectious-recovered model proposed in this paper can provide valuable insights for cruise ship emergency planning and positively contribute to maintaining passenger emotional stability during the Safe Return to Port process.

Keywords: emotional contagion; improved susceptible-infectious-recovered model; two sources of infection; machine learning; long-term refuge scenarios

MSC: 91D25

Citation: Xiong, G.; Cai, W.; Hu, M.; Yu, Z. Research on Emotional Infection of Passengers during the SRtP of a Cruise Ship by Combining an SIR Model and Machine Learning. *Mathematics* 2023, 11, 4461. https://doi.org/10.3390/math11214461

Academic Editors: Shih-Wei Lin and Daniel-Ioan Curiac

Received: 5 September 2023
Revised: 6 October 2023
Accepted: 23 October 2023
Published: 27 October 2023

Copyright: © 2023 by the authors. Licensee MDPI, Basel, Switzerland. This article is an open access article distributed under the terms and conditions of the Creative Commons Attribution (CC BY) license (https://creativecommons.org/licenses/by/4.0/).

1. Introduction

1.1. Background

The introduction of Safe Return to Port (SRtP) into the International Convention for Safety of Life at Sea (SOLAS) convention is based on the principle that "the ship itself is its best lifeboat" [1]. SRtP refers to the fact that a passenger ship can rely on its own power to return to the nearest port within the accident limit of an event, such as a fire or water ingress, and that onboard safety meets the basic needs of its passengers and crew. Increasing the number of people that passenger ships can hold has brought enormous challenges to emergency evacuation and rescue work after marine accidents.

The existing regulations lack a specific timeframe for determining the duration of the SRtP. This duration typically depends on factors such as the ship's condition and the specific route. In the case of ocean-going cruise ships, the average SRtP duration is approximately

10 days. The SRtP designs usually adopt appropriate means (such as separation, double sets, redundancy, protection, or a combination of these) to achieve the objectives of the given specification. Therefore, due to various cost and construction considerations, a part of the passenger ship's main vertical areas is usually selected as its safety areas, while targeted redundancy and separation protection designs are also carried out. Safety areas need to safely accommodate all personnel on board, protect them from life threats, and provide them with basic services. The specification requires that the per capita area be no less than 2 m^2 [1]. Therefore, for cruise ships, there might be scenarios during the SRtP process where crowds are gathered in limited spaces for an extended period.

1.2. Related Work

In order to ensure SRtP and prevent passenger ship accidents, human factors in emergencies are important factors that should be considered. The key problem is how to understand the role of group emotions in decision-making processes. In general, group emotions directly affect group behavior in decision-making processes. Currently, research into crowd emotions is divided into two main directions. The first involves monitoring, analyzing, and predicting abnormal emotions in groups based on image recognition. These studies focus on detecting abnormal behaviors and emotions by integrating relevant theories on emotions, 2D [2], 3D convolutional neural networks [3], and deep learning methods [4]. Predictive models are constructed using classifiers, and multiple deep learning frameworks can be integrated with psychological fuzzy computing [5] for crowd behavior detection and prediction. The other direction involves semantic recognition, which starts with specific events and combines emotional analysis theory and semantic analysis to process, discover, and infer the spatiotemporal emotions involved in the semantic expressions of the event [6]. This approach provides situational awareness of the event. However, these studies often ignore the impact of emotional contagion, and their prediction accuracy depends on the size of their sample datasets and computational power. When the sample dataset is small, the prediction accuracy may be decreased.

In emergency situations, it is crucial for decision makers to comprehend the impact of crowd emotions. One method for researching the spread of emotions within a group involves constructing an infection model that tracks the evolution of emotional states over time. This model simulates how emotions are transmitted within a specific population. As machine learning technologies have advanced, many scholars are increasingly turning to machine learning to mathematically model these temporal state transitions in various practical scientific and engineering problems. Researchers employ a variety of methods and techniques in diverse fields, including physical systems [7], climate and environmental data analysis [8], and structural health monitoring [9,10], to address complex real-world challenges. These approaches provide powerful tools for modeling and predicting changes in state over time. They enable the extraction of essential hidden variables and the representation of state transition rules in a comprehensible manner, ultimately contributing to applications and research in the realms of science and engineering.

The infectious characteristics and influencing factors of emotions can vary across different scenarios. This highlights the importance of conducting research tailored to specific scenarios, including those that involve adversarial situations [11], queuing scenarios [12], and emergency evacuations [13]. Additionally, different boundary conditions can have an impact on the transmission process [14]. The research purpose of the emotional transmission process is to explore the impact of various influencing factors on the transmission process. The other aim is to develop reasonable interventions within an appropriate timeframe to prevent the spread of negative emotions that could lead to serious consequences [15].

The emotional transmission in crowds is often difficult to predict and varies depending on the context, making it important to establish valid anxiety and emotional transmission models. Computational models of emotional contagion typically include three aspects: understanding, prediction, and control. Over the past decade, researchers have developed several models based on the hypothesis that emotional contagion is similar to infectious diseases, with many relying on the susceptible-infectious-recovered (SIR) model. The SIR model divides the human population into three categories: the susceptible population (S), those who are infected and can transmit the disease (I), and the removed population (R), which is typically assumed to have recovered [16–20]. However, for some diseases, individuals who have recovered can still transmit the disease to others, leading to the development of the extended SIR model, the (susceptible—infected—removed—susceptible) SIRS model. For diseases with latent periods, the (suspected–exposed–affected–removed) SEIR model has been proposed.

With the advancement of research in the field, the limitations of the traditional SIR model and the SEIR model have become apparent. To increase their adaptability to complex scenarios, many scholars have proposed improvements to the traditional SIR model from various perspectives. For instance, certain researchers have added an alert state [21] or two healing states [22] to the model, while others have classified the original state in greater detail [23]. Furthermore, additional factors have been considered, such as nodes [24], positive emotions [25], and other psychological effects [26]. Additionally, longitudinal expansions have been pursued by integrating the SIR model into other theoretical frameworks [27].

While certain scholars have researched the infection parameters and transmission routes of the model to improve its single infection rate, in actual emotional transmission, other factors can also influence the rate. These include trust differences between groups [28], deviations in emotional information transmission [29], and the external environment [30]. Additionally, the incidence of infection sources [31] and the recurrence of infection terminals [32] are also crucial factors affecting emotional transmission. Thus, it is essential to consider all of these factors when studying emotional transmission.

Based on the summary of relevant literature, it can be observed that there are two main limitations in the current research on the SIR model. Firstly, there are comparatively fewer studies on the impact of positive and negative emotions. As related psychology research advances, an increasing number of studies report the significant influence of positive emotions on emotional contagion. In various situations, including appropriate positive emotional guidance can prevent the swift dissemination of negative emotions. Secondly, there are insufficient studies on long-term emotional transmission. The existing body of research on emotional contagion has primarily concentrated on short-term, sudden events occurring within a timeframe of a few hours. However, there has been a notable dearth of studies examining prolonged situations, such as long-term refuge behavior lasting for several days.

With the implementation of SRtP regulations, researchers have gradually carried out various studies on issues related to the SRtP. Currently, research on the safe return of ships mainly focuses on the impact of ship design, system reliability, electromechanical equipment, engine design, fire alarm devices, and other aspects. Further, some studies have introduced various systems concepts. Due to the need to concentrate the entire passengers in a safe area during the SRtP, the possibility of passengers experiencing depression and anxiety gradually increases as the time they spend in the area increases, which increases the possibility of serious incidents and threatens the overall safety of the ship.

In the SRtP of cruise ships, the severity of the accident is not always immediately apparent, resulting in different emotional responses among all persons. Some may not initially feel anxious, but as time passes, they may gradually become more anxious or adapt to the situation. Both negative emotions and positive emotions, such as calmness and optimism, can spread, leading individuals to transition to a positive emotional state. Traditional models, like the SIR, susceptible-infectious-susceptible (SIS), and susceptible-exposed-infectious-recovered (SEIR) models, may no longer be applicable in these cases. Moreover, the SRtP may last for several days, and passengers may have negative emotions during these days.

To address these challenges, this study first constructs an improved SIR model. This model defines two sources of infection, allowing for bidirectional emotional transitions.

For instance, in situations where accidents are not exceptionally severe, most individuals' emotions tend to fluctuate, and extremely anxious or very calm states can influence the emotions of those around them. Subsequently, this model is integrated with the random forest algorithm, utilizing questionnaire survey data as the primary dataset, to predict the emotional states of the population. Finally, this study employs simulation techniques to visualize the dynamics of the model and analyze the effects of relevant parameters.

The contributions of this paper can be summarized as follows:

(1) This paper introduces a novel model to investigate changes in collective emotions, especially in long-term refuge scenarios.
(2) By combining the model with machine learning, the problem of predicting emotional states is transformed into a classification task, facilitating the prediction of the emotional states of the population.
(3) The proposed model is validated through simulation software, enabling the visualization of emotional transitions.

The rest of this paper is structured as follows. Section 2 constructs an emotional contagion model that conforms to the SRtP scenario based on the SIR model theory. Section 3 analyzes the questionnaire data. Section 4 constructs a prediction algorithm and analyzes the results. The last section presents the conclusions and future work.

2. Problem Description

According to the relevant regulations on SRtP in reference [1], it can be seen that accidents are not serious when cruise ships execute the SRtP program. Moreover, for ocean-going cruise ships, the duration of SRtP scenarios is typically longer. Consequently, in the context of SRtP, it is highly probable that passenger emotions will undergo reciprocating changes. This stands in stark contrast to the scenario where nearly everyone experiences panic during the execution of an abandon ship procedure.

Uncertainty in emotion state scoring is a common challenge in emotion research due to the inherent complexity of this psychological phenomenon. Several factors contribute to this uncertainty and should be carefully considered:

1. Subjectivity: Emotions are highly subjective experiences, meaning that individuals can have varying emotional responses to the same situation. Consequently, when assigning emotion state scores, researchers may be influenced by their own subjective biases, resulting in inconsistent ratings.
2. Variability: Emotions are dynamic and can change over time. Different stimuli can lead to diverse changes in emotional states, further complicating the scoring process.
3. Difficulty in Quantification: Quantifying emotions poses a significant challenge, especially when categorizing them into discrete levels. Generally, employing a finer-grained classification system with more levels can help reduce uncertainty.
4. External Factors: Emotional states are susceptible to external influences, including cultural, individual, and societal factors. These external factors can lead to variations in emotional responses among individuals facing the same situation, thus augmenting scoring uncertainty.

Taking into account the above issues, this section introduces a novel emotional contagion model. This model aims to provide a more realistic simulation of the emotional contagion process within a population. It accomplishes this by incorporating dual contagion sources and taking into account bidirectional emotional transitions. An approach involving self-scoring through questionnaires is adopted to minimize the impact of subjective biases across participants. Additionally, the traditional SIR model, which typically encompasses three levels, is expanded to include five levels, providing a more nuanced and comprehensive understanding of emotional states. This expansion aims to enhance the overall clarity and accuracy of emotion state assessment in the research.

2.1. Model Building

When a cruise ship executes the SRtP, passengers begin to realize the emergency condition of the ship and move to the safety area. Owing to variations in individuals' psychological thresholds, emotional states typically fall into three broad categories: positive emotional state, neutral emotional state, and negative emotional state [33]. Nonetheless, the contagiousness of these emotional states at different levels also varies, with a proportional increase in infectious potential corresponding to the extremity of the emotional state. To refine the analysis, all positive and negative emotional states have been further subdivided, resulting in the identification of five distinct emotional states, namely extreme anxiety (E), general anxiety (A), normal (N), calm (C), and very calm (V). To facilitate statistical and computational analysis, assign a score of 1 to 5 to these five emotional states, ranging from negative to positive, as shown in Table 1. These five emotional states can transform into each other, as shown in Figure 1. The transformation process can be described as follows:

- During the initial stage of the SRtP in response to an accident on a cruise ship, passengers may experience five different emotional states with varying probabilities after they have gathered in a safety area. These states represent the initial distribution of emotions among the crowd.
- Over time, the emotional states of passengers undergo changes, influenced by varying probabilities of transitioning between different emotional states. The transition probability between two states is denoted by α_{ij}, where i and j represent the scores of each state, respectively. Thus, the state transition matrix is given by Equation (1), where P_{ij} is the transition probability from state i to j.

Table 1. Emotional states score.

Emotional State	Score
Extremely anxious	1
Anxious	2
Normal	3
Calm	4
Very calm	5

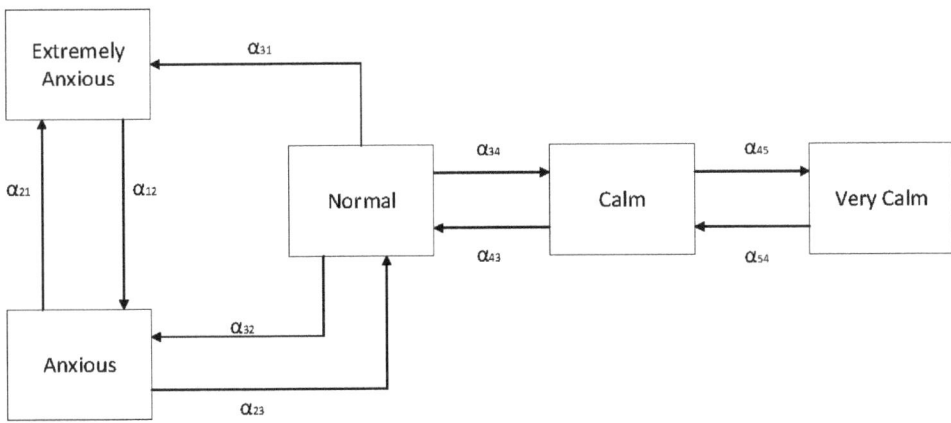

Figure 1. Relationship between emotional state transitions (α_{ij} is the probability of transition from state i to state j).

To provide a clearer description of the passenger emotion contagion process, the following definitions are established:

$$P_{ij} = \begin{bmatrix} 1-\alpha_{12} & \alpha_{12} & 0 & 0 & 0 \\ \alpha_{21} & 1-\alpha_{21}-\alpha_{23} & \alpha_{23} & 0 & 0 \\ \alpha_{31} & \alpha_{32} & 1-\alpha_{31}-\alpha_{32}-\alpha_{34} & \alpha_{34} & 0 \\ 0 & 0 & \alpha_{43} & 1-\alpha_{43}-\alpha_{45} & \alpha_{45} \\ 0 & 0 & 0 & \alpha_{54} & 1-\alpha_{54} \end{bmatrix}, \quad (1)$$

Definition 1. *Definition of the crowd's emotional state. The emotional states of passengers can be divided into five categories at time t: extremely anxious, anxious, normal, calm and very calm, where t represents the time point at which a specific phase concludes. The calculation of total number of people is shown in Equation (2), where T is a constant total number of people. $E(t)$, $A(t)$, $N(t)$, $C(t)$ and $V(t)$ represent the number of people in extremely anxious, anxious, normal, calm and very calm states at time t.*

$$T = E(t) + A(t) + N(t) + C(t) + V(t), \quad (2)$$

Definition 2. *State transition events. A state transition involves a passenger transitioning from one state to another state. Let $T_{(i)}(t)$ and $T_{(j)}(t)$ represent the number of people in state i and j at time t; $T_{(i,j)}(\Delta t)$ represents the number of people whose status has changed from the state i to j in a time interval Δt. Thus, the number of people at time $t + \Delta t$ is given by Equations (3) and (4), where $T_{(i)}(t + \Delta t)$ and $T_{(j)}(t + \Delta t)$ are the number of people in state i and j at time $t + \Delta t$. Moreover, each person in a state can only transition to the nearest positive or negative state. For example, people in an anxious state can only transition to an extremely anxious state or a normal state, and cannot directly transition to a calm state or a very calm state. Only people in a normal state can directly transition to an extremely anxious state.*

$$T_{(i)}(t + \Delta t) = T_{(i)}(t) - T_{(i,j)}(\Delta t), \quad (3)$$

$$T_{(j)}(t + \Delta t) = T_{(j)}(t) + T_{(i,j)}(\Delta t), \quad (4)$$

Definition 3. *The source of infection is set as extremely anxious people and very calm people. The other states are not contagious.*

Definition 4. *Due to the impact of environment and time, people in any given state have the potential to transition to an adjacent state. For instance, even if people in a state of calm are not directly affected by someone experiencing extreme anxiety or people who are extremely calm, there exists a certain probability that they may transition to a very calm state or return to normal during the SRtP.*

Definition 5. *$R(E)$ denotes the coefficient of the rate of change, indicating how passengers in states other than the negative state are influenced by passengers experiencing extreme anxiety. For example, if $R(E) = 2$, it implies that, under this influence, the probability of passengers transitioning to a negative state will become twice as high as the original probability. The function $R(E)$ can be characterized as a temporal and spatial function, represented by Equation (5), in which the variable d signifies the linear distance between the passenger in a state of extreme anxiety and the target passenger.*

$$R(E) = R_1(d,t), \quad (5)$$

$R(V)$ is the coefficient that represents the influence of individuals in a very calm state on the probability of transitioning the other person from a different state to a positive state. $R(V)$ can be characterized as a temporal and spatial function, which can be expressed as Equation (6).

$$R(V) = R_2(d,t), \qquad (6)$$

While the proposed model is not a traditional SIR model, it incorporates some similar concepts to describe the spread and evolution of emotions.

1. Emotional State Classification: Unlike the infected state in the SIR model, this model categorizes emotional states into five distinct categories, ranging from extreme anxiety to very calm, and assigns scores from 1 to 5 to represent negative to positive emotional states. These different states can be thought of as different "infection" states within the crowd.
2. State Transitions: Similar to the infection rate in the SIR model, the state transition probabilities α_{ij} in this model represent the likelihood of transitioning from one emotional state to another. These transition probabilities constitute a state transition matrix used to describe the spread and change of emotions between different emotional states, resembling the transmission process in the SIR model.
3. Temporal Evolution: Like the SIR model, this model also considers the evolution over time. At the initial moment, passengers are in different emotional states, representing the initial distribution of emotions. Over time, passengers' emotional states change influenced by the transition probabilities between different emotional states, akin to the infection spread process in the SIR model.

While this model is used to describe the spread and evolution of emotions, rather than the transmission of infectious diseases, it employs similar concepts of probabilistic transitions and state changes to describe how emotions propagate within a crowd in response to an emergency situation. The following advanced models also use the SIR model concept: the cyber-physical society-oriented recurrent emotional contagion (CPS–REC) model, stochastic event-based emotional contagion (SEEC) model, and emotional contagion-aware deep reinforcement learning model for antagonistic crowd simulation (ACSED) model (Appendix C).

The CPS–REC model takes into consideration the influence of emotional recurrence on the emotional contagion process, aiming to provide a more comprehensive understanding of crowd behavior. The formula of degree-based Mean-Field Equations is presented. These equations describe the dynamic evolution of the number of individuals within crowds while accounting for their heterogeneity.

The SEEC model introduces the occurrence intensity of infection/recovery events and constructs a state transition matrix to calculate the crowd state evolution. There are two categories within the crowd: susceptible individuals (i.e., individuals without negative emotions) and infected individuals (i.e., individuals with negative emotions) in this model.

The ACSED is a method designed to investigate the intricate interactions between emotions and decision-making within adversarial environments. Its emotional contagion module is constructed using the enhanced SIS model. What sets it apart from previous studies on emotional contagion is its integration of deep q network (DQN). ACSED leverages DQN to estimate individuals' inclinations towards engaging in adversarial behavior, and then analyze the rationality underlying behavioral predictions.

2.2. Calculation of Crowd States

According to the relevant theories of the SIR Model [34], the system dynamics differential equations of the emotion model studied in this paper can be constructed. After time interval Δt, the population in each state can be calculated by the number of existing passengers, passengers transferred into the state, and passengers transferred out of the state. The following is the specific calculation formula:

$$\frac{dE}{dt} = \frac{E(t + \Delta t) - E(t)}{\Delta t} = N(t) \cdot \alpha_{31} + A(t) \cdot \alpha_{21} - E(t) \cdot \alpha_{12}, \tag{7}$$

$$\frac{dA}{dt} = \frac{A(t + \Delta t) - A(t)}{\Delta t} = E(t) \cdot \alpha_{12} + N(t) \cdot \alpha_{32} - A(t) \cdot (\alpha_{21} + \alpha_{23}), \tag{8}$$

$$\frac{dN}{dt} = \frac{N(t + \Delta t) - N(t)}{\Delta t} = A(t) \cdot \alpha_{23} + C(t) \cdot \alpha_{43} - N(t) \cdot (\alpha_{31} + \alpha_{32} + \alpha_{34}), \tag{9}$$

$$\frac{dC}{dt} = \frac{C(t + \Delta t) - C(t)}{\Delta t} = N(t) \cdot \alpha_{34} + V(t) \cdot \alpha_{54} - C(t) \cdot (\alpha_{43} + \alpha_{45}), \tag{10}$$

$$\frac{dV}{dt} = \frac{V(t + \Delta t) - V(t)}{\Delta t} = C(t) \cdot \alpha_{45} - V(t) \cdot \alpha_{54}, \tag{11}$$

$$\frac{dE}{dt} + \frac{dA}{dt} + \frac{dN}{dt} + \frac{dC}{dt} + \frac{dV}{dt} = 0, \tag{12}$$

$\frac{dE}{dt}$, $\frac{dA}{dt}$, $\frac{dN}{dt}$, $\frac{dC}{dt}$ and $\frac{dV}{dt}$ represent the change rate of the number of people in each emotional state, respectively.

Equation (7) describes the rate of change of the number of people in state E over time. It considers the change in the number of people in this state as a result of passengers transitioning into this state from state N and state A (with probabilities α_{31} and α_{21}, respectively) and those transitioning out of this state to the normal state A (with a probability of α_{12}).

Equation (8) represents the rate of change in the number of people in state A over time. It accounts for people moving into this state from state E and state N (with probabilities α_{12} and α_{32}, respectively) and those transitioning out to either state E or state N (with probabilities α_{21} and α_{23}).

Equation (9) represents the rate of change in the number of people in state N over time. It considers individuals transitioning into this state from state A and state C (with probabilities α_{23} and α_{43}, respectively) and those transitioning out to state E, state A, and state C (with probabilities α_{31}, α_{32}, and α_{34}).

Equation (10) describes the rate of change for the number of people in state C over time. It accounts for individuals transitioning into this state from state N and state V (with probabilities α_{34} and α_{54}, respectively) and transitioning out to state N and state V (with probabilities α_{43} and α_{45}, respectively).

Equation (11) represents the rate of change in the number of people in state V over time. It considers individuals transitioning into this state from state C (with probabilities α_{45}, respectively) and those transitioning out to the state C (with probabilities α_{54}, respectively).

Equation (12) reflects the conservation of the total population within the emotional states. In other words, the sum of the rate of change of people in all emotional states equals zero, indicating that the total number of passengers remains constant over time.

These equations collectively model how the population in each emotional state changes over time based on transition probabilities between different emotional states.

As inferred from the preceding context, α_{ij} denotes the probability of transitioning from state i to state j. In cases where a direct transition between two states is not feasible, α_{ij} is assigned a value of 0. When multiple passengers are in extreme emotional states, their influence on passengers in other states becomes cumulative. Let $\alpha_{ij}(t)$ represent the transition probability from state i to state j at time t, and let $\alpha'_{ij}(t + \Delta t)$ signify the transition probability at time $t + \Delta t$. The relationship between $\alpha_{ij}(t)$ and $\alpha'_{ij}(t + \Delta t)$ can be described by Equation (13).

$$\alpha'_{ij}(t + \Delta t) = \alpha_{ij}(t) \cdot \left[\sum_0^{T_E} R(E), i > j + \sum_0^{T_V} R(V), i < j + \delta_{ij} \right], \tag{13}$$

When $i > j$, this indicates the impact of passengers experiencing heightened anxiety levels, and the probability of transitioning towards negative emotional states is increased. At time $t + \Delta t$, the probability of state transition can be calculated as the sum of the rate of change coefficients for passengers in a state of extreme anxiety, multiplied by $\alpha_{ij}(t)$. Similarly, when $i < j$, the probability of state transition can be computed as the sum of rate of change coefficients for passengers in a notably calm state, again multiplied by $\alpha_{ij}(t)$. If $i = j$, it indicates the maintenance of the existing state. T_E represents the number of passengers with extreme anxiety in the enclosed space, and T_V represents the number of passengers who are very calm in the enclosed space. δ_{ij} represents 1 when $i = j$, and otherwise 0.

Normalization can be achieved by using Equation (14).

$$\alpha_{ij}(t + \Delta t) = \frac{\alpha'_{ij}(t + \Delta t)}{\sum_{j=1}^{5} \alpha'_{ij}(t + \Delta t)}, \tag{14}$$

where $i \in [1, 5]$, and $\sum_{j=1}^{5} \alpha'_{ij}(t + \Delta t)$ represents the sum of the probabilities of the transition from the i state to the other states. $\alpha_{ij}(t + \Delta t)$ is the result after normalization.

3. Data and Analysis

3.1. Questionnaire Experimental Data

A significant portion of the existing research in this field has primarily concentrated on examining emotional contagion within relatively few hours, rendering it inadequate for application to the SRtP of cruise ships, which can span tens of days. This study is based on the emotional contagion model established in Section 2.1, and a questionnaire is developed using the control variable method (Appendix A). In the SRtP of a cruise ship, there is a scenario that all passengers are concentrated in safe areas, and the specification requires that the per capita area be no less than 2 m² [1]. In large cruise ships, the large public areas are about 1000 m², which can accommodate more than 500 passengers. In the event of a mass incident involving 500 passengers, the safety of the ship and passengers could be seriously compromised. By separating these areas into smaller zones, the number of passengers and crowd density can be effectively regulated. In order to study the influence of the total population and population density on emotional contagion in a small area, the control variates are used to establish different scenario control groups. In each scenario, the moment when all passengers complete the assembly in the safety area is recorded as moment $t = 0$. The emotional state collected at 0:00 on the second day of the assembly is noted as the emotional state for Day 1. Similarly, the emotional states for Day 3, Day 5, Day 7, and Day 10 correspond to the emotional states collected at 0:00 on the day after these days. Scenario 1 to Scenario 3 set the density to 2 m²/person and form a comparison group by gradually increasing the number of passengers. In scenarios 1, 4, and 5, maintain the total number of passengers unchanged, and create another comparison group by altering the crowd density. The scores are set for various emotional states from extreme anxiety to very calm states, ranging from 1 to 5 points, as shown in Table 1. In addition, to filter out questionnaires that have not been filled out diligently, an attention mechanism screening question will be included in the questionnaire. In total, 527 valid questionnaires were collected through the China Questionnaires Star Corporation. The results of the questionnaire are shown in Figures 2–7.

Based on the data presented in Figure 2, it can be observed that, when guest rooms on a cruise ship are unavailable, most passengers (59.7%) prefer to seek refuge in restaurants, even if these restaurants are unable to offer regular catering services. Meanwhile, 18.6% of passengers tend to choose commercial areas as a place of refuge. Additionally, 21.7% of passengers tend to opt for gangway and stairway landings as a refuge.

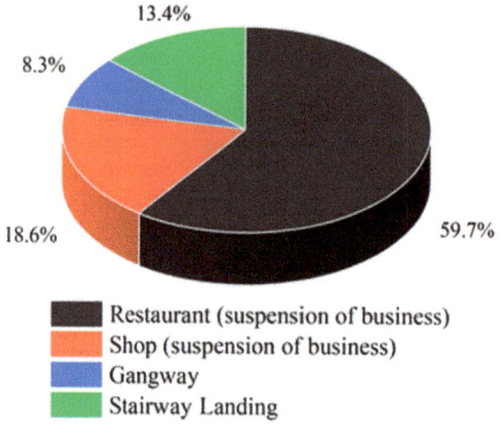

Figure 2. Distribution of public space selection.

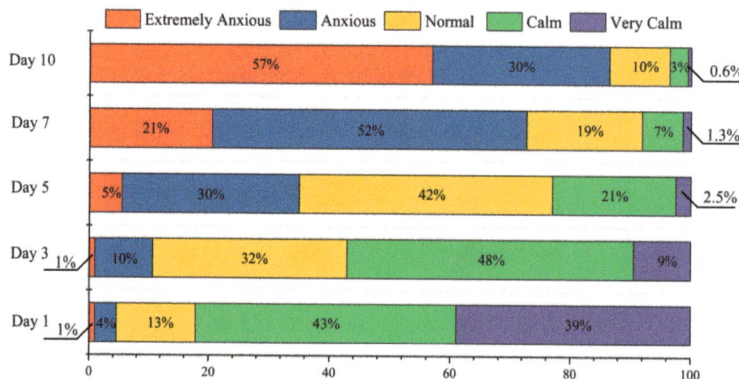

Figure 3. Emotional state distribution in Scenario 1 ($\rho = 2 \text{ m}^2/\text{person}, n = 30$).

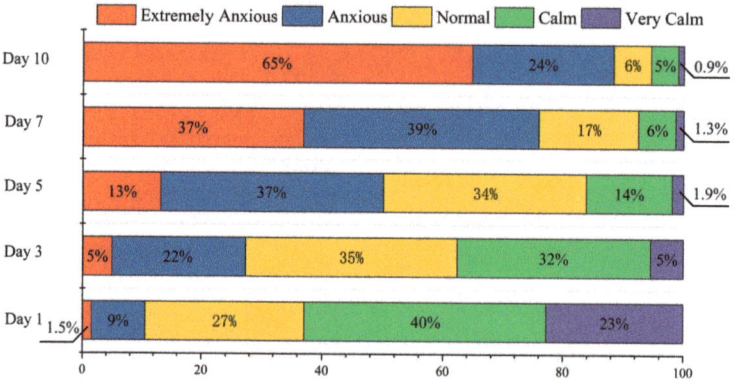

Figure 4. Emotional state distribution in Scenario 2 ($\rho = 2 \text{ m}^2/\text{person}, n = 50$).

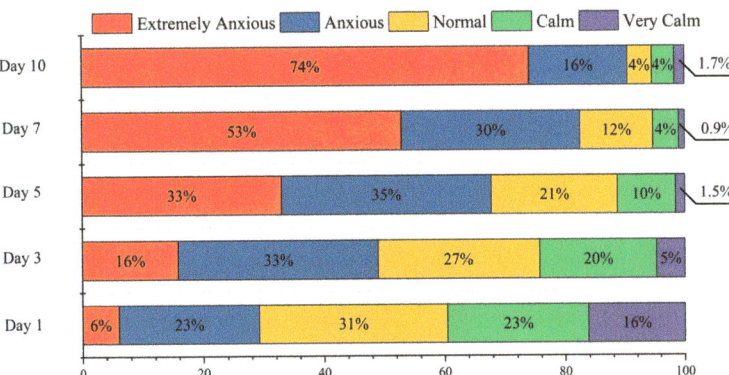

Figure 5. Emotional state distribution in Scenario 3 ($\rho = 2$ m^2/person, $n = 100$).

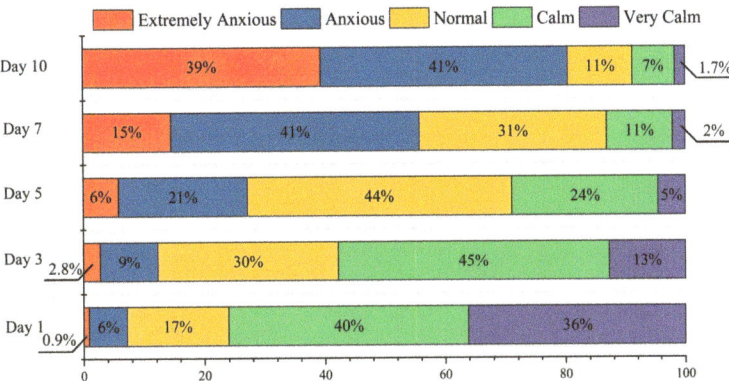

Figure 6. Emotional state distribution in Scenario 4 ($\rho = 3$ m^2/person, $n = 30$).

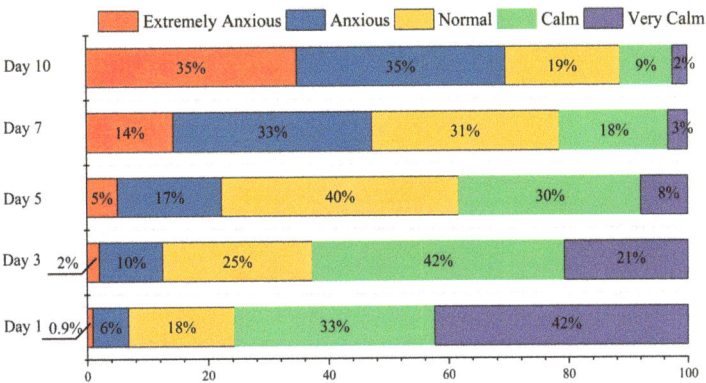

Figure 7. Emotional State Distribution in SCENARIO 5 ($\rho = 4$ m^2/person, $n = 30$).

Figures 3–5 illustrate the emotional state distributions of three scenarios with equal density but varying total number of passengers. Figure 3 illustrates a notable surge in the population experiencing the extremely anxious state from Day 7 to Day 10, while the number of passengers in the very calm state decreases significantly from Day 1 to Day 3. Upon comparing Figures 3–5, a discernible correlation emerges: changes in the number of passengers exert a significant influence upon their emotional states. It is observed that, with a constant density, an increase in the total number of passengers results in an upward trend and a higher peak value for the number of passengers in an extremely anxious state. However, the proportion of passengers in the very calm state exhibits a downward trend. The proportion of passengers in the anxious state fluctuates, with the peak time gradually advancing. Moreover, the fluctuation of the normal and calm states becomes gradually smooth over time.

Upon a comparison of Figures 3, 6 and 7, evident correlations surface between fluctuations in passenger density and resultant changes in their emotional states. Based on the data presented in these three figures, an obvious trend emerges wherein the proportion of individuals in an extremely anxious state experiences a pronounced surge from Day 7 through to Day 10. In contrast, during the initial Day 1 to Day 3 period, there is a discernible reduction in the proportion of individuals manifesting a very calm state. Despite the total number of passengers remaining constant, the peak number of those in the extremely anxious state decreases as the per capita area increases. Conversely, the peak number of the very calm state shows an upward trend. According to the data depicted in these three figures, two distinct trends become evident: as the per capita area increases, there is a recognizable postponement in the onset of the proportion of passengers in an anxious state; simultaneously, there is a corresponding decrease in the magnitude of the peak. The normal state exhibits relatively smooth fluctuation patterns. The state of calm also demonstrates relatively smooth fluctuations but with an increasing peak. Available data indicate a multifaceted correlation between population density and emotional states, with different emotional states exhibiting varying degrees of response to changes in population density.

3.2. Reliability and Validity Analysis of the Questionnaire

To ensure the quality of questionnaires, researchers often use reliability and validity measures [35]. When a questionnaire demonstrates good reliability and validity, it suggests that the data obtained from the questionnaire are internally consistent and accurate, making it suitable for further analysis. It is necessary to conduct a comprehensive evaluation of the reliability and validity of the questionnaire, as it consists of multiple scale questions.

Table A3 presents the reliability calculation table, which shows that the reliability coefficient value of the Cronbach α is 0.916. This value is greater than 0.9, indicating that the reliability quality of the research data is high [36]. Additionally, the value of the Corrected Item–Total Correlation (CITC) is also analyzed to indicate the degree of association between the items. It is found that the CITC values corresponding to questions 1, 3, and 4 are all less than 0.2. This suggests that the relationships between these three questions and the rest of the analysis items are weak. This is mainly due to the fact that these questions involve pre-test analysis. Overall, the reliability of the research data enhances the credibility of the study's findings.

Validity research is used to analyze whether a research item is reasonable and meaningful. The validity level of data can be analyzed through indicators such as the Kaiser–Meyer–Olkin (KMO), commonality, variance interpretation rate, and factor load coefficient values. The KMO value is used to determine the suitability of information extraction, the commonality value is used to exclude unreasonable research items, the variance interpretation rate value is used to explain the level of information extraction, and the factor load coefficient is used to measure the corresponding relationship between factors (dimensions) and items [37]. The results of the validity analysis are presented in Table A4. It can be seen from the table that most research items have commonality values over 0.4, except for question 4. This indicates effective extraction of research item information. Question 4 is less than 0.4, indicating that the research item information is not able to be effectively expressed, mainly because it involves the collection of intentions. The variance interpretation rates of the five factors are 21.3%, 20.1%, 19.4%, 8.9%, and 3.8%, respectively. The cumulative variance interpretation rates after rotation are 73.346% > 50%. This means that the amount of information in the research item can be effectively extracted. If the

p-value of Bartlett's Test of Sphericity is less than 0.05, it indicates that it has passed the Bartlett sphericity test and has validity [37]. The p-value in this study is less than 0.001, indicating that the questionnaire has successfully undergone validity analysis, confirming its adequate validity.

3.3. Correlation Analysis

Correlation analysis refers to the analysis of two or more correlated variable elements, which is used to measure the degree of correlation between two variable factors. Based on a designated control group, the study examines the correlations between population density and emotional states at different times, as well as the correlations between total population and emotional states at different times.

Table 2 shows the results of a Non-parametric test comparing emotional states across different population densities. Q_1 and Q_3 denote the lower quartile and upper quartile, respectively. The differences in density can be seen for five emotional states at five different time points. It is evident that the density values can be divided into two groups, with values of 2.0 and 3.0, respectively. The second and third columns present the lower quartile, median, and upper quartile values for daily emotional states under two different densities: $2\ m^2$/person and $3\ m^2$/person. To illustrate, see the data in the first row of the second column, where, under a density of $2\ m^2$/person, the median for emotional score in Day 1 registers as 2, the lower quartile as 3, and the upper quartile as 5. Notably, within the second and third columns, instances arise where the median is equal to either the lower or upper quartile. This occurrence signifies that, for a specific emotional state, the count of individuals with that emotional score encompasses at least 25% of the total population. For example, in the first row of the third column, both the median and lower quartile are reported as 4. This observation implies that, under the density of $3\ m^2$/person, when individuals are arranged in ascending order based on their emotional scores, those scoring 4 represent a range spanning at least 25% to 50% of the total population.

Table 2. Non-parametric test analysis for different densities.

Items	$p_{Median}(Q_1, Q_3)$		Mann–Whitney U	Mann–Whitney z	p
	2.0	3.0			
Day 1 Emotional States	4.000 (3.0, 5.0)	4.000 (4.0, 5.0)	666,584.500	−9.116	<0.01
Day 3 Emotional States	3.000 (2.0, 4.0)	4.000 (3.0, 4.0)	601,989.000	−12.634	<0.01
Day 5 Emotional States	2.000 (2.0, 3.0)	3.000 (3.0, 4.0)	564,124.000	−14.654	<0.01
Day 7 Emotional States	2.000 (1.0, 2.0)	2.000 (2.0, 3.0)	549,802.500	−15.549	<0.01
Day 10 Emotional States	1.000 (1.0, 2.0)	2.000 (1.0, 3.0)	586,807.000	−12.559	<0.01

To analyze these groups, a Mann–Whitney test is needed. However, if there are more than two groups, a Kruskal–Wallis test is necessary [38]. The results indicate that the emotional states at all five time points vary significantly across different population densities ($p < 0.05$). This suggests that samples with different densities display significant differences in emotional states at all five time points. Further analysis reveals the following:

(1) Based on the study results, it appears that population density has a significant effect on emotional state on Day 1, with a p-value less than 0.01 indicating a significant difference. Additionally, the comparison of median differences suggests that the source of the differences is due to different data distributions. Figure 8 shows a block diagram of emotional state data at different densities, revealing that the mean emotional state on the first day is around 4.1 when the density is $2\ m^2$/person, while it is around 4 when the density is $3\ m^2$/person. These findings suggest that higher population densities may lead to a decrease in emotional state on Day 1.

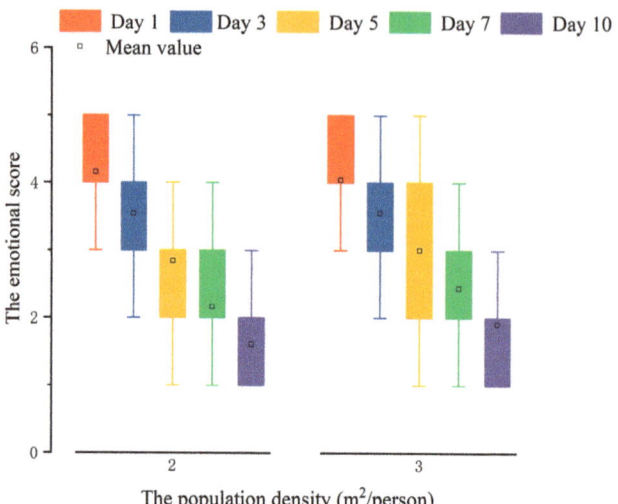

Figure 8. Box chart of daily emotional states at different population densities.

(2) The population density shows a significance level for the emotional states on Day 5, and the mean value of 2 m^2/person is significantly lower than that of 3 m^2/person.

(3) The analysis shows that population density has a significant impact on emotional states on Day 7. Based on Figure 8, it can be observed that, when the population density is 2 m^2/person, the mean emotional state score on Day 7 is approximately 2.1. However, the mean emotional state score on Day 7 is around 2.5 when the density is 3 m^2/person. From these two mean values, it can be inferred that, when the population density is 2 m^2/person, the emotional state for Day 7 is more likely to be distributed with a score of 2.

(4) Passenger density exhibits a notable correlation with emotional states on Day 10. The median differences further demonstrate that the average density of 2 m^2/person is significantly lower than that of 3 m^2/person. Specifically, in the scenario where the density is 3 m^2/person, there appears to be a higher count of individuals exhibiting comparatively lower emotional scores. This observation is particularly evident when contrasting it with the crowd characterized by a density of 2 m^2/person.

Through the Mann–Whitney test, it can be found that the samples with different densities showed significant differences in emotional states at different days.

Table 3 presents the results of a non-parametric test for the emotional states of different population samples, indicating the differences in the total number of passengers in five emotional states at different days. The Kruskal–Wallis test is used to analyze the data since the total number of samples exceeds two groups. The results show that there are no significant differences in emotional states among different population samples on Day 1, Day 3, Day 5, and Day 7 ($p > 0.05$), indicating consistent emotional patterns across these time periods. Further analysis is required from the box plots. Figure 9 shows the box plots of emotional states for different population samples. However, significant differences ($p < 0.01$) are observed in one emotional state on Day 10, suggesting that the emotional state differs among different total sample sizes on this day. Based on Figure 9, the following results can be obtained.

Table 3. Non-parametric test analysis for different total population samples.

Items	n_{Median} (Q_1, Q_3)			Kruskal-Wallis H	p
	30.0	50.0	100.0 (n = 527)		
Day 1 Emotional States	4.000 (4.0, 5.0)	4.000 (3.0, 5.0)	3.000 (2.0, 4.0)	−501.523	1.000
Day 3 Emotional States	4.000 (3.0, 4.0)	3.500 (3.0, 4.0)	3.000 (2.0, 3.0)	−587.672	1.000
Day 5 Emotional States	3.000 (2.0, 4.0)	3.000 (2.0, 4.0)	2.000 (1.0, 3.0)	−747.935	1.000
Day 7 Emotional States	2.000 (2.0, 3.0)	2.000 (1.0, 3.0)	1.000 (1.0, 2.0)	−305.579	1.000
Day 10 Emotional States	2.000 (1.0, 2.0)	2.000 (1.0, 2.0)	1.000 (1.0, 2.0)	41.962	< 0.01

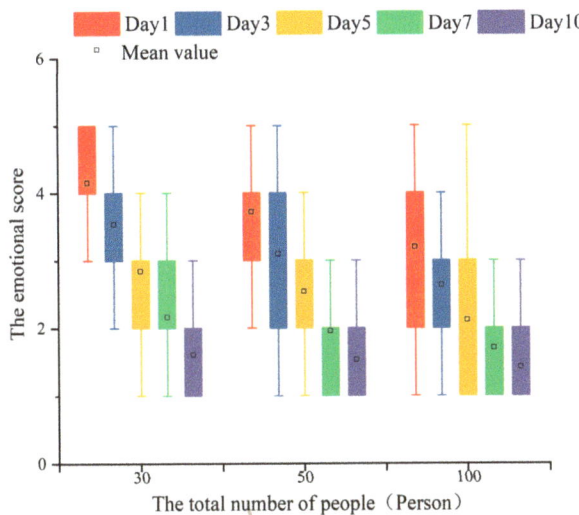

Figure 9. Box chart of emotional states of different total passengers.

(1) Regarding the emotional state scores on Day 1 shown in Figure 9, as the total number of passengers changes from 30 to 50, the lowest score decreases from 3 to 2. Simultaneously, the interquartile range, representing the concentration interval, narrows from (4, 5) to (3, 4), and the mean score also decreases. When the number of passengers changes from 50 to 100, the concentration interval expands from (3, 4) to (2, 4), and the mean score decreases further. A decrease in the mean implies a rise in the proportion of passengers with lower scores. Therefore, overall, an increase in the total number of passengers has a negative impact on the emotional state on Day 1.

(2) For the emotional state scores on Day 3, as the total number of passengers changes from 30 to 50, the concentration interval expands from (3, 4) to (2, 4), and the mean score decreases significantly. When the number of passengers changes from 50 to 100, the lowest score decreases from 2 to 1, the concentration interval narrows from (2, 4) to (2, 3), and the mean score also decreases. Thus, an increase in the number of passengers has a negative effect on the emotional state on Day 3.

(3) Regarding the emotional state scores on Day 5, as the number of passengers changes from 30 to 50, the extreme value and concentration interval remain the same, but the mean score decreases, indicating that the group is shifting towards lower emotional state scores. When the number of passengers changes from 50 to 100, the concentration interval expands from (2, 3) to (1, 3), and the mean score decreases further. This suggests that an increase in the total number of passengers has a negative impact on the emotional state on Day 5.

(4) For the emotional state scores on Day 7, as the number of passengers change from 30 to 50, the concentration interval decreases from (2, 3) to (1, 2), and the mean score

decreases. When the number of passengers changes from 50 to 100, the extreme value and concentration interval remain the same, but the mean score decreases, indicating that the group is shifting towards lower emotional state scores. Therefore, an increase in the number of people has a negative effect on the emotional state on Day 7.

(5) Regarding the emotional state scores on Day 10, as the number of passengers change from 30 to 50, the extreme value and concentration interval remain the same, but the mean score slightly decreases. When the number of passengers change from 50 to 100, the highest score decreases from 3 to 2, the concentration interval remains the same, and the mean score decreases. Thus, an increase in the number of passengers has a negative impact on the emotional state on Day 10.

In summary, an increase in the number of passengers will lead to a decrease in the emotional score, reflecting a negative impact on emotions. Moreover, different numbers of passengers may result in different ranges of emotional fluctuations.

Tables 4 and 5 are summary tables of model regression coefficients, where SE represents standard error, $z(CR)$ represents critical ratio, and p represents significance. As can be seen in Table 4, a significant and positive impact relationship is revealed between emotional states at adjacent time points. The standardized path coefficients, approximately 0.7, emphasize this relationship.

Table 4. Summary table of model regression coefficients for adjacent time.

Items 1	Items 2	Unstandardized Path Coefficient	SE	z (CR)	p	Standardized Path Coefficient
Day 7 Emotional States	Day 10 Emotional States	0.741	0.012	63.026	< 0.001	0.775
Day 5 Emotional States	Day 7 Emotional States	0.748	0.012	61.752	< 0.001	0.769
Day 3 Emotional States	Day 5 Emotional States	0.758	0.012	61.104	< 0.001	0.766
Day 1 Emotional States	Day 3 Emotional States	0.784	0.012	64.018	< 0.001	0.780

Table 5. Summary table of model regression coefficients for non-adjacent time.

Items 1	Items 2	Unstandardized Path Coefficient	SE	z (CR)	p	Standardized Path Coefficient
Day 1 Emotional States	Day 10 Emotional States	0.200	0.018	11.381	< 0.001	0.216
Day 1 Emotional States	Day 7 Emotional States	0.433	0.017	25.671	< 0.001	0.447
Day 1 Emotional States	Day 5 Emotional States	0.630	0.015	42.016	< 0.001	0.633
Day 1 Emotional States	Day 3 Emotional States	0.784	0.012	64.019	< 0.001	0.780

Furthermore, an assessment of emotional states on Day 1 and subsequent days was executed using standardized path analysis, as depicted in Table 5. The outcomes of this analysis show a gradual reduction in the strength of the relationship between emotional states on the initial day and those on all subsequent days.

In conclusion, these results suggest two key points. Firstly, the emotional states from the prior time period can serve as valuable indicators for predicting emotional states in immediate successive time periods. Notably, the substantial standardized path coefficients of around 0.7 reinforce this predictive relationship. Secondly, while the emotional states on Day 1 provide predictive utility for subsequent days, this predictability diminishes as the temporal lag increases. Hence, direct inference of emotional states on Days 5, 7, and 10 from the emotional state of Day 1 is not feasible.

To address this limitation, an iterative forecasting approach emerges as a viable strategy. Although direct prediction of emotional states on later days from the emotional state of Day 1 is constrained, a step-by-step iterative approach can be employed. This iterative forecasting method would involve predicting emotional states on Day 2 based on Day 1, then using the predicted Day 2 emotional state to predict Day 3, and so forth. This approach accommodates the temporal dynamics of emotional state progression and gains more accurate predictions.

4. Simulation and Results Analysis

The questionnaire data collected are employed as machine learning samples to construct a random forest algorithm for prediction purposes. From the model constructed in Section 2.1, it can be seen that the prediction essentially consisted of a classification problem. Random forest is a commonly used method in machine learning that employs decision trees for data analysis. The random forest (RF) algorithm can avoid overfitting to a certain extent in classification problems and is suitable for parallel operations [39]. Finally, testing and visualization are conducted through simulation.

4.1. Model Parametric Construction

4.1.1. Initial State of Each Scenario

1. Grid division of the scenarios

The emotional contagion within a population in physical space is closely related to the distribution of individuals. To avoid excessively long queues in the setting, the grid of the scenarios determines the number of columns by taking the square root of the total number of passengers and rounding it down, while the number of rows is determined by rounding it up.

2. Distance between passengers

Passenger spatial distribution is categorized according to different densities, as illustrated in Figure 10.

For a density of 2 m^2/person, the spatial allocation entails a longitudinal gap of 1 m in the front–back direction and a lateral gap of 0.5 m in the left–right direction.

In the case of a density of 3 m^2/person, the spatial distribution involves a longitudinal gap of 1 m in the front–back direction, coupled with a lateral gap of 0.75 m in the left–right direction.

When considering a density of 4 m^2/person, the spatial configuration encompasses a longitudinal gap of 1 m in the front–back direction and a lateral gap of 1 m in the left–right direction. The personnel arrangement shape in each scene should be as close to a square as possible.

3. Initial number of passengers and initial transition probabilities

The initial passenger count is determined by rounding the proportions of individuals in various emotional states on Day 1, as per the statistical data collected from the surveys. Priority is given to preserving the proportion of individuals in extreme emotional states, with subsequent adjustments made for individuals in other states.

The initial values for the emotion transition rates were set based on data gathered through a questionnaire. However, it should be noted that the respondents may not have fully understood the impact of emotional contagion among passengers in different scenarios, so the data may be more biased toward passenger emotional transitions. The rate of conversion for each emotion from Day 1 to Day 3 was calculated to determine the initial value for each emotion conversion rate. For example, in the questionnaire for Scenario 1, there were 67 passengers recorded in a normal state on Day 1. On Day 3, 1 passenger transitioned to the extreme anxiety state, 25 passengers transitioned to the anxious state, 22 passengers remained in the normal state, and 14 passengers transitioned to the calm state. If the transition probabilities at Day 1 are not affected by extreme emotions, the values of $(\alpha_{31}, \alpha_{32}, \alpha_{33}, \alpha_{34}, \alpha_{35})$ are $(0.1, 0.4, 0.3, 0.2, 0)$.

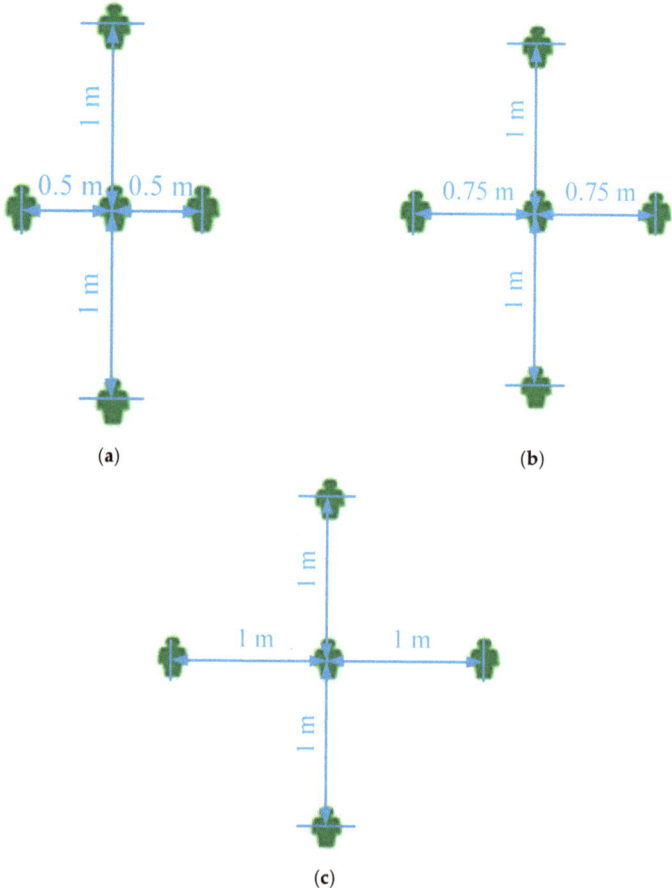

Figure 10. The population space; (**a**) 2 m^2/person; (**b**) 3 m^2/person; (**c**) 4 m^2/person.

The specific initial passenger numbers and transition probabilities for each scenario are as follows:

$$T_1 = \begin{bmatrix} 0 & 2 & 3 & 13 & 12 \end{bmatrix}, \tag{15}$$

$$T_2 = \begin{bmatrix} 1 & 5 & 14 & 20 & 10 \end{bmatrix}, \tag{16}$$

$$T_3 = \begin{bmatrix} 6 & 26 & 30 & 22 & 16 \end{bmatrix}, \tag{17}$$

$$T_4 = \begin{bmatrix} 0 & 2 & 5 & 12 & 11 \end{bmatrix}, \tag{18}$$

$$T_5 = \begin{bmatrix} 0 & 2 & 5 & 10 & 13 \end{bmatrix}, \tag{19}$$

$$P_1 = \begin{bmatrix} 0 & 1 & 0 & 0 & 0 \\ 0.32 & 0.63 & 0.05 & 0 & 0 \\ 0.01 & 0.39 & 0.37 & 0.23 & 0 \\ 0 & 0 & 0.59 & 0.39 & 0.02 \\ 0 & 0 & 0 & 0.8 & 0.2 \end{bmatrix}, \quad (20)$$

$$P_2 = \begin{bmatrix} 0.63 & 0.37 & 0 & 0 & 0 \\ 0.35 & 0.6 & 0.05 & 0 & 0 \\ 0.01 & 0.5 & 0.38 & 0.11 & 0 \\ 0 & 0 & 0.65 & 0.33 & 0.02 \\ 0 & 0 & 0 & 0.81 & 0.19 \end{bmatrix}, \quad (21)$$

$$P_3 = \begin{bmatrix} 0.88 & 0.12 & 0 & 0 & 0 \\ 0.4 & 0.56 & 0.04 & 0 & 0 \\ 0.04 & 0.58 & 0.32 & 0.06 & 0 \\ 0 & 0 & 0.67 & 0.28 & 0.05 \\ 0 & 0 & 0 & 0.79 & 0.21 \end{bmatrix}, \quad (22)$$

$$P_4 = \begin{bmatrix} 0.4 & 0.6 & 0 & 0 & 0 \\ 0.44 & 0.53 & 0.03 & 0 & 0 \\ 0 & 0.29 & 0.56 & 0.15 & 0 \\ 0 & 0 & 0.5 & 0.48 & 0.02 \\ 0 & 0 & 0 & 0.67 & 0.33 \end{bmatrix}, \quad (23)$$

$$P_5 = \begin{bmatrix} 0.6 & 0.4 & 0 & 0 & 0 \\ 0.27 & 0.63 & 0.1 & 0 & 0 \\ 0.02 & 0.25 & 0.6 & 0.13 & 0 \\ 0 & 0 & 0.42 & 0.56 & 0.02 \\ 0 & 0 & 0 & 0.54 & 0.46 \end{bmatrix}, \quad (24)$$

where $(T_1, T_2, T_3, T_4, T_5)$ represent the initial number of passengers from scenario 1 to scenario 5 and $(P_1, P_2, P_3, P_4, P_5)$ represent the transition probabilities from scenario 1 to scenario 5.

4.1.2. Other Model Parameters

According to the definition in the previous part, $R(E)$ and $R(V)$ are functions of distance and time. Emotional transmission is a complex phenomenon that is influenced by a combination of factors, including non-verbal communication, social influences, and group effects. There are fewer studies on the expression of specific functions between emotional contagion and distance and time. However, some research suggests that emotional contagion may be more likely to occur when people are in close proximity to each other [33]. Emotional contagion can also spread quickly in a relatively short period of time and spread further over time [40]. To enhance the analytical and computational processes while ensuring a tractable problem formulation, we introduce postulates for the functions $R(E)$ and $R(V)$, as delineated below:

$$R(E) = \frac{a_1}{d_1} + b_1 \times t^2, \quad (25)$$

$$R(V) = \frac{a_2}{d_2} + b_2 \times t^2, \quad (26)$$

where a_1, a_2, b_1, b_2 are coefficients greater than 0, of which all initial values are 1. d_1 and d_2 are the linear distances from passengers in the extremely anxious state to passengers in the very calm state. d_1 denotes the Euclidean distance between a passenger and one of the passengers exhibiting extreme anxiety. Conversely, d_2 represents the Euclidean distance between a passenger and one of the passengers displaying significant calmness.

4.2. Algorithm Flow

Based on an improved weighted random forest algorithm [41–45], the prediction problem of emotional contagion in the SRtP process of cruise ships could be constructed as a classification problem. Figure 11 is the algorithm flowchart and the detailed process is as follows:

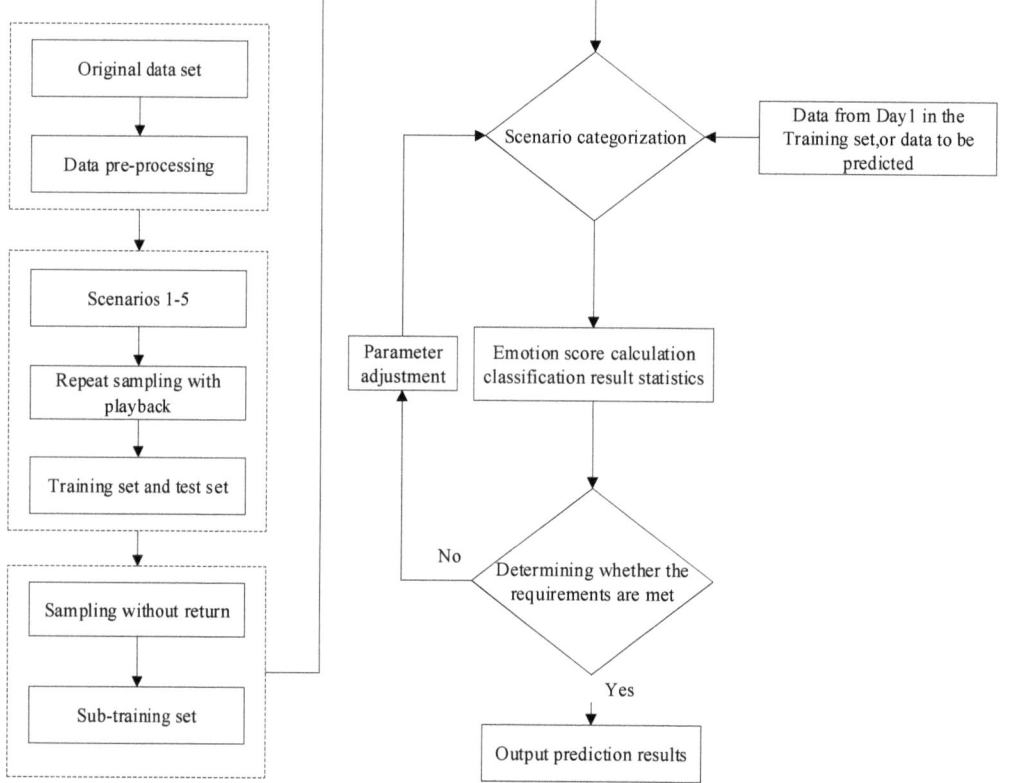

Figure 11. The algorithm flowchart.

Step 1: Data pre-processing of the original dataset, mainly including data cleaning and feature encoding.

Step 2: Repeat sampling with playback is used to extract $K+1$ datasets, wherein K constitutes a training set and the other constitutes a test set, and the capacity of each dataset is equal to the total sample capacity, and where the value of K is 100.

Step 3: By simulating a realistic area separation, each scene is sampled without replacement from the sample set based on the predetermined number of passengers until the collection is complete. Each sample is taken as a sub-training set.

Step 4: Scenario categorization. Each decision tree is divided into five scenarios set by the questionnaire. F_m represents the degree of similarity between the scenario m and the scenario i. The smaller the F_m, the more similar it is to the training scenario i, which is introduced into this scenario. The specific calculation of F_m is as follows:

$$F_m = \varepsilon_1 \cdot \frac{|\rho_m - \rho_i|}{\rho_i} + \varepsilon_2 \cdot \frac{|n_m - n_i|}{n_i}, \qquad (27)$$

It is assumed that the parameters for the passenger m of Day 1 to be classified are (ρ_m, n_m, e_m), where ρ_m represents the population density of the group in which the passenger to be classified belongs, n_m is the total number of passengers in the group, e_m indicates the emotional state score, ρ_i represents the population density of the scenario i, and n_i is the total number of passengers in scenario i. ε_1 and ε_2 are weight coefficients with an initial value of 0.5, which are adjusted through computational calculations. If multiple scenarios have equal F_m values, they participate in the subsequent voting session together.

Step 5: Emotion score calculation and classification result statistics. According to each possible transition direction and the corresponding infection rate equations given in the previous infectious disease model, the next stage score can be calculated. The results of the calculations are then weighted and disaggregated into statistics.

G_k is the weighting coefficient for the k-th training set which can be calculated by Equation (28):

$$G_k = \sigma_1 \cdot \frac{n_k - n_{k1}}{n_k} + \sigma_2 \cdot \frac{n_k - n_{k5}}{n_k}, \tag{28}$$

where σ_1 and σ_2 are weight coefficients with an initial value of 0.5, which are adjusted through computational calculations. n_k is the total number of passengers in the k-th training set scenario. n_{k1} is the number of passengers with a score of 1 in the k-th training set, and n_{k5} is the number of passengers with a score of 5 in the k-th training set. By using comprehensive weighting coefficients to vote on all sub-training sets, then classification results could be obtained.

Step 6: Determining whether the requirements are met. Compare the final proportions of each emotional state with the test set. If the results do not meet the specified requirements, make parameter adjustments and return to step 4. If the results meet the requirements, output the final result.

Table 6 shows the values of the parameters (e.g., number of trees, etc.) of the applied random forest algorithm.

Table 6. Values of the parameters.

Parameters	Values
bagSizePercent	100
batchSize	100
numIterations	100
n_estimators	500
max_depth	5

4.3. Results and Discussion

4.3.1. Evaluation of Models

According to the previous study, the situation on Day 1 is used as the initial value for prediction learning over the following days. The confusion matrix of the model is shown in Figures 12–15, with the horizontal axis representing the predicted emotional score and the vertical axis representing the actual emotional score. The confusion matrix diagram can clearly reflect the accuracy of predicting various emotional states and the distribution of misjudgments.

From Figures 12–15, it is evident that the prediction accuracy for the extreme anxiety and very calm states is high. The prediction accuracy for the very calm state gradually improves over time, especially on Day 10, where it reaches 100%. There are several reasons for this result:

(1) The high prediction accuracy for the very calm state is due to the fact that it has only two transition directions: maintaining the current status or transitioning into the calm state. This makes it relatively easy to predict. Additionally, the changing trend of the very calm state is relatively fixed, with an overall transition towards a calm state over time. Furthermore, the probability of transitioning from the calm state to the very calm state is low, which also contributes to the high prediction accuracy.

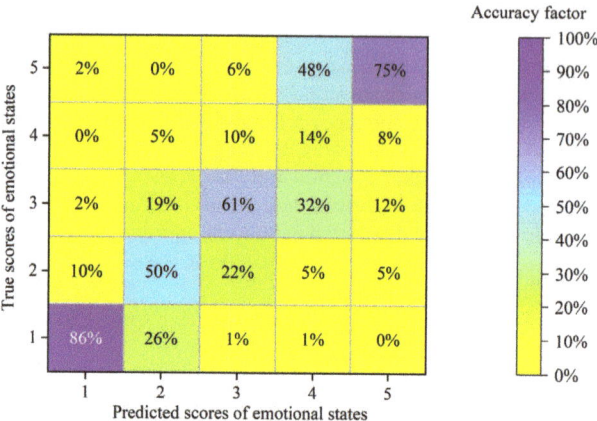

Figure 12. Confusion matrix of forecast results for Day 3.

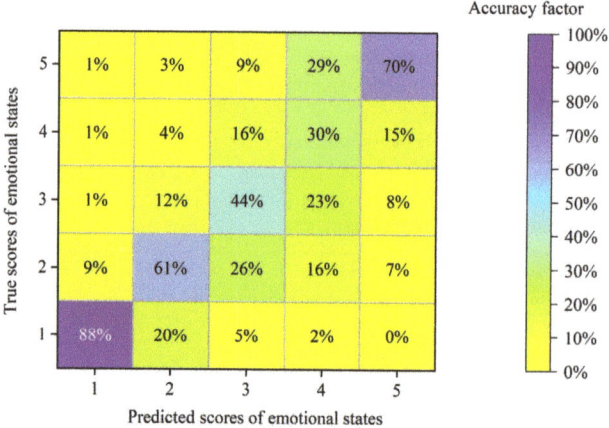

Figure 13. Confusion matrix of forecast results for Day 5.

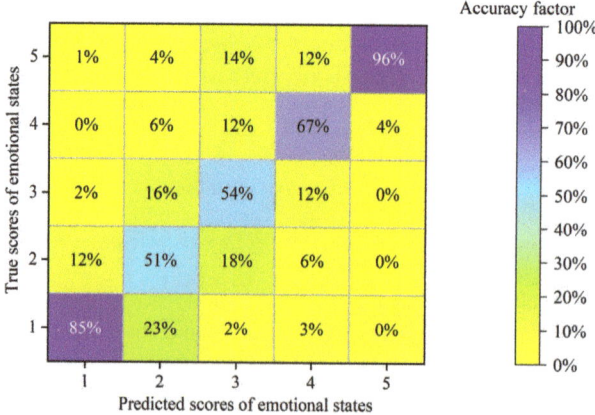

Figure 14. Confusion matrix of forecast results for Day 7.

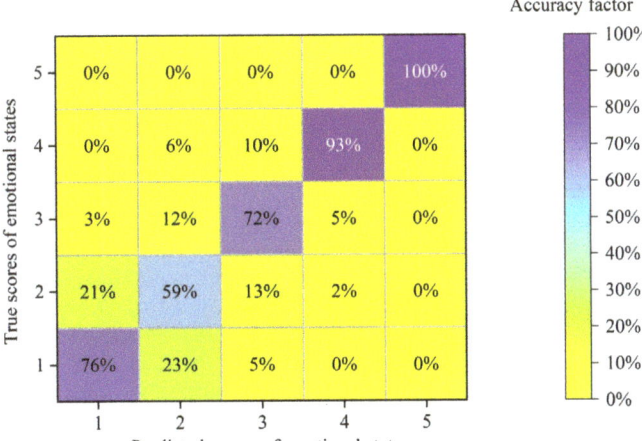

Figure 15. Confusion matrix of forecast results for Day 10.

(2) In the absence of intervention measures, the number of passengers in the very calm state decreases rapidly until it reaches zero. As a result, the model's prediction accuracy for the very calm state reaches 100% on Day 10.

(3) The prediction accuracy of the extreme anxiety state remained at around 80%, with some fluctuations. This is because there are three possible transition directions for this state, making it not easy to predict accurately. In another, in the absence of interventions, the extreme anxiety state will gradually dominate. There is a reciprocal change between the extreme anxiety state and the anxiety state, leading to fluctuations in prediction accuracy.

(4) According to the findings presented in Figures 12 and 13, it is evident that the initial few days exhibit a significantly low prediction accuracy for the calm state, with accuracy levels not surpassing 30%. A significant portion of these misclassifications involves predicting a very calm state when the actual state is a calm state. One possible explanation for this observation is that the RF algorithm possesses an inherent tendency to predict extreme emotional states.

Furthermore, Figures 14 and 15 exhibit a notable improvement in the prediction accuracy for the calm state in the later days, surpassing 90%. Moreover, there is a distinct enhancement in the prediction accuracy for the calm state observed in Figures 14 and 15, particularly in the later days, where it consistently exceeds 90%. This significant advancement can be predominantly attributed to the precipitous reduction in the count of passengers experiencing the calm state as time progresses, eventually converging towards zero.

(5) The predictive precision of both states, anxious and normal, demonstrates a heightened trend of fluctuation. This phenomenon may be attributed to an enhanced tendency of the underlying transitional dynamics between these two states, resulting in frequent and oscillating transitions.

Table 6 presents a statistical table of algorithm indicators, including accuracy, precision, recall, and the F-measure. From Table 7, it can be seen that the previously introduced model can effectively capture the characteristics of passenger emotion transmission during the SRtP process.

Table 7. Algorithm performance.

	Accuracy	Precision	Recall	F1-Score
Day3	0.57	0.55	0.57	0.54
Day5	0.56	0.54	0.56	0.54
Day7	0.68	0.67	0.69	0.67
Day10	0.81	0.80	0.81	0.81

4.3.2. Visualization and Analysis of Emotional Infections in Questionnaire Scenarios

Utilizing the previously introduced model, we simulate and visualize the five scenarios presented in the questionnaire using the AnyLogic software [46]. In order to visualize the process of emotion transmission and examine the variations in its parameters, there are a total of 39 sets of simulation experiments. In Scenario 5, where passengers are equally spaced in front, behind, left, and right, there is no requirement to account for the conversion of spatial grid distribution into rows and columns, unlike the other scenarios. Tables 8–12 show the parameter settings for each simulation experiment for Scenarios 1 to 5. The value of (a_1, b_1, a_2, b_2) can be adjusted through machine learning techniques, where the value of (a_1, b_1, a_2, b_2) for $(Sim1-1, Sim2-1, Sim3-1, Sim4-1, Sim5-1)$ are obtained through manual tuning. Grid space is the hyperparameter of this model. We examine how hyperparameters affect the model by configuring various grid spaces such as $(Sim1-1, Sim1-5\ to\ 9)$.

Table 8. Parameter settings of Scenario 1.

Simulation No.	(a_1, b_1, a_2, b_2)	Grid Space ($Row \times Column$)
$Sim1-1$	$a_1 = 0.8,\ b_1 = 0.8,$ $a_2 = 0.1,\ b_2 = 0.1,$	6×5
$Sim1-2$	$a_1 = 1,\ b_1 = 0,$ $a_2 = 1,\ b_2 = 0,$	6×5
$Sim1-3$	$a_1 = 0,\ b_1 = 1,$ $a_2 = 0,\ b_2 = 1,$	6×5
$Sim1-4$	$a_1 = 1,\ b_1 = 1,$ $a_2 = 1,\ b_2 = 1,$	6×5
$Sim1-5$	$a_1 = 0.8,\ b_1 = 0.8,$ $a_2 = 0.1,\ b_2 = 0.1$	5×6
$Sim1-6$	$a_1 = 0.8,\ b_1 = 0.8,$ $a_2 = 0.1,\ b_2 = 0.1$	3×10
$Sim1-7$	$a_1 = 0.8,\ b_1 = 0.8,$ $a_2 = 0.1,\ b_2 = 0.1$	10×3
$Sim1-8$	$a_1 = 0.8,\ b_1 = 0.8,$ $a_2 = 0.1,\ b_2 = 0.1$	2×15
$Sim1-9$	$a_1 = 0.8,\ b_1 = 0.8,$ $a_2 = 0.1,\ b_2 = 0.1$	15×2

Table 9. Parameter settings of Scenario 2.

Simulation No.	(a_1, b_1, a_2, b_2)	Grid Space ($Row \times Column$)
$Sim2-1$	$a_1 = 0.5,\ b_1 = 0.6,$ $a_2 = 0.2, b_2 = 0.2,$	8×7 (The space is not fully occupied)
$Sim2-2$	$a_1 = 1,\ b_1 = 0,$ $a_2 = 1,\ b_2 = 0,$	8×7 (The space is not fully occupied)
$Sim2-3$	$a_1 = 0,\ b_1 = 1,$ $a_2 = 0,\ b_2 = 1,$	8×7 (The space is not fully occupied)
$Sim2-4$	$a_1 = 1,\ b_1 = 1,$ $a_2 = 1,\ b_2 = 1,$	8×7 (The space is not fully occupied)

Table 9. Cont.

Simulation No.	(a_1, b_1, a_2, b_2)	Grid Space (Row×Column)
$Sim2-5$	$a_1=0.5, b_1=0.6,$ $a_2=0.2, b_2=0.2,$	8×7 (The space is not fully occupied)
$Sim2-6$	$a_1=0.5, b_1=0.6,$ $a_2=0.2, b_2=0.2,$	5×10
$Sim2-7$	$a_1=0.5, b_1=0.6,$ $a_2=0.2, b_2=0.2,$	10×5
$Sim2-8$	$a_1=0.5, b_1=0.6,$ $a_2=0.2, b_2=0.2,$	2×25

Table 10. Parameter settings of Scenario 3.

Simulation No.	(a_1, b_1, a_2, b_2)	Grid Space (Row×Column)
$Sim3-1$	$a_1=0.2, b_1=0.2,$ $a_2=0.1, b_2=0.1,$	10×10
$Sim3-2$	$a_1=1, b_1=0, a_2=1, b_2=0,$	10×10
$Sim3-3$	$a_1=0, b_1=1, a_2=0, b_2=1,$	10×10
$Sim3-4$	$a_1=1, b_1=1, a_2=1, b_2=1,$	10×10
$Sim3-5$	$a_1=0.2, b_1=0.2, a_2=0.1, b_2=0.1,$	5×20
$Sim3-6$	$a_1=0.2, b_1=0.2, a_2=0.1, b_2=0.1,$	20×5
$Sim3-7$	$a_1=0.2, b_1=0.2, a_2=0.1, b_2=0.1,$	4×25
$Sim3-8$	$a_1=0.2, b_1=0.2, a_2=0.1, b_2=0.1,$	25×4

Table 11. Parameter settings of Scenario 4.

Simulation No.	(a_1, b_1, a_2, b_2)	Grid Space (Row×Column)
$Sim4-1$	$a_1=0.3, b_1=0.3, a_2=0.1, b_2=0.1,$	6×5
$Sim4-2$	$a_1=1, b_1=0, a_2=1, b_2=0,$	6×5
$Sim4-3$	$a_1=0, b_1=1, a_2=0, b_2=1,$	6×5
$Sim4-4$	$a_1=1, b_1=1, a_2=1, b_2=1,$	6×5
$Sim4-5$	$a_1=0.3, b_1=0.3, a_2=0.1, b_2=0.1$	5×6
$Sim4-6$	$a_1=0.3, b_1=0.3, a_2=0.1, b_2=0.1$	3×10
$Sim4-7$	$a_1=0.3, b_1=0.3, a_2=0.1, b_2=0.1$	10×3
$Sim4-8$	$a_1=0.3, b_1=0.3, a_2=0.1, b_2=0.1$	2×15
$Sim4-9$	$a_1=0.3, b_1=0.3, a_2=0.1, b_2=0.1$	15×2

Table 12. Parameter settings of Scenario 5.

Simulation No.	(a_1, b_1, a_2, b_2)	Grid Space (Row×Column)
$Sim5-1$	$a_1=0.3, b_1=0.3, a_2=0.1, b_2=0.1,$	6×5
$Sim5-2$	$a_1=1, b_1=0, a_2=1, b_2=0,$	6×5
$Sim5-3$	$a_1=0, b_1=1, a_2=0, b_2=1,$	6×5
$Sim5-4$	$a_1=1, b_1=1, a_2=1, b_2=1,$	6×5
$Sim5-5$	$a_1=0.3, b_1=0.3, a_2=0.1, b_2=0.1$	3×10
$Sim5-6$	$a_1=0.3, b_1=0.3, a_2=0.1, b_2=0.1$	2×15

Figures 16–20 provide a depiction of how the trend in passenger emotions evolves with varying values of the parameters (a_1, b_1, a_2, b_2). The x-axis in these figures represents the number of days, while the y-axis illustrates the proportions of passengers in different emotional states. These figures make it readily apparent how different emotional states change over time and highlight the disparities in emotional state proportions within each day. Furthermore, these figures offer a clear and intuitive representation of the fluctuations in passenger emotional states.

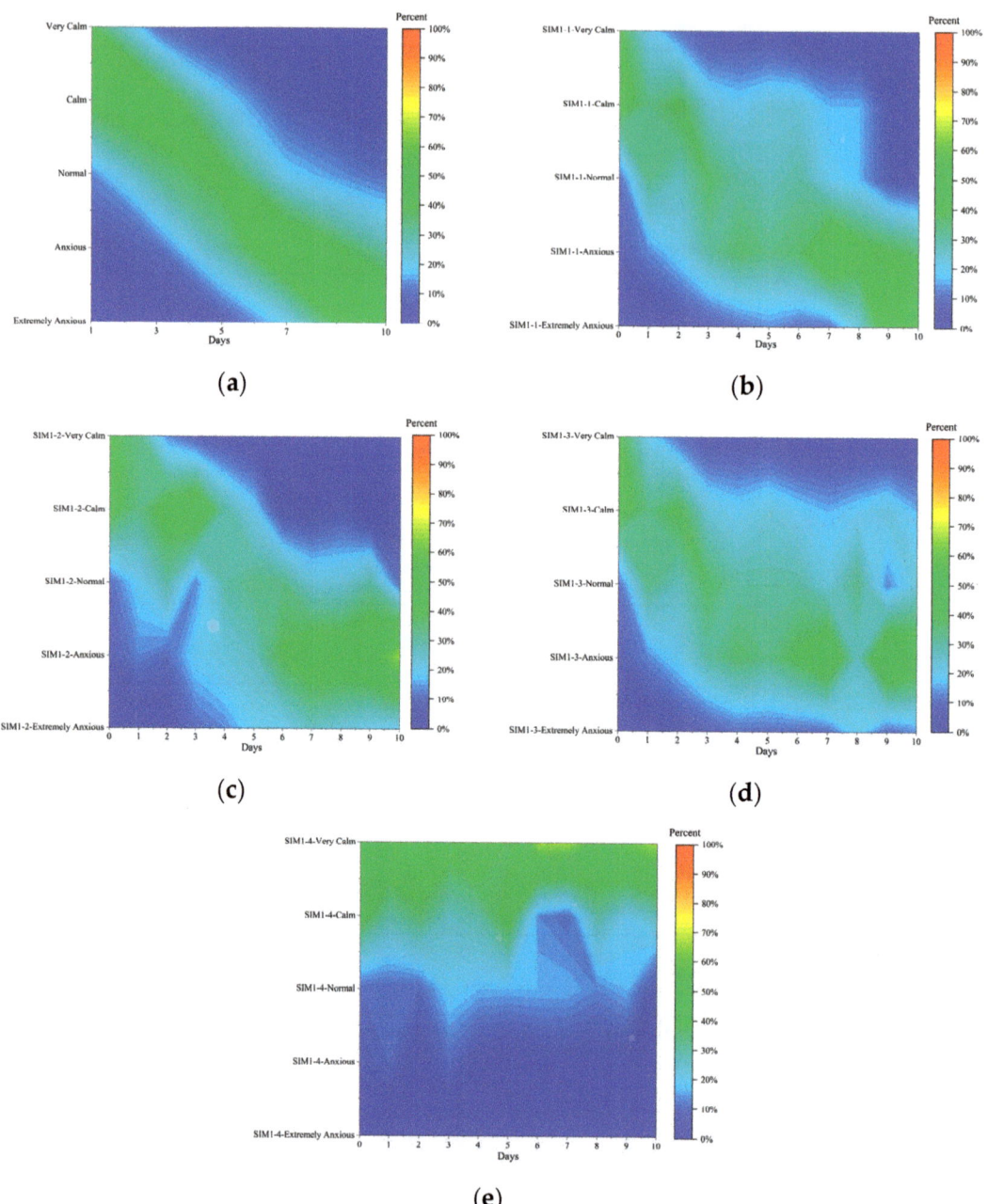

Figure 16. Simulation result of Scenario 1; (**a**) Initial data; (**b**) $Sim1-1$; (**c**) $Sim1-2$; (**d**) $Sim1-3$; (**e**) $Sim1-4$.

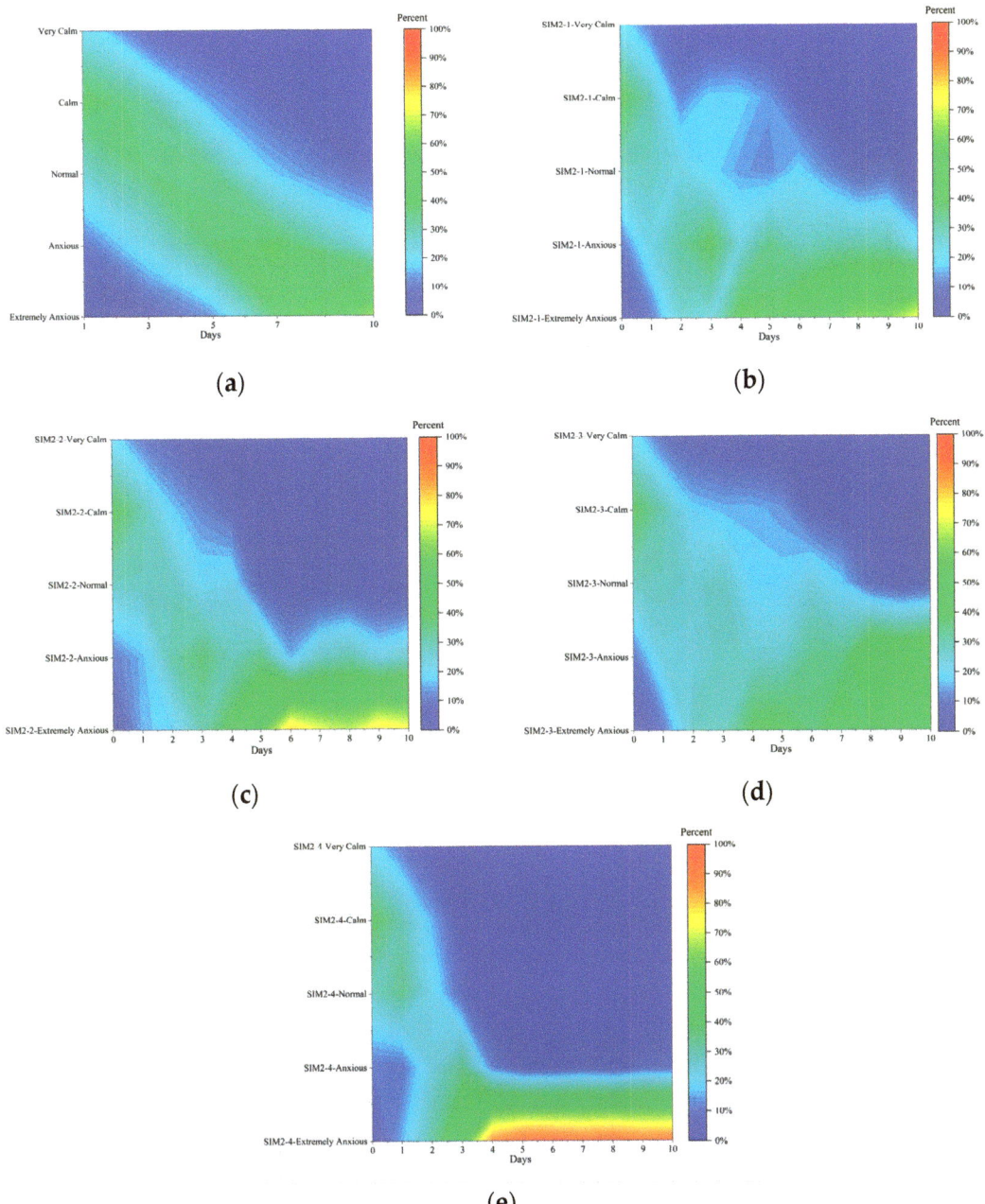

Figure 17. Simulation result of Scenario 2; (**a**) Initial data; (**b**) $Sim2-1$; (**c**) $Sim2-2$; (**d**) $Sim2-3$; (**e**) $Sim2-4$.

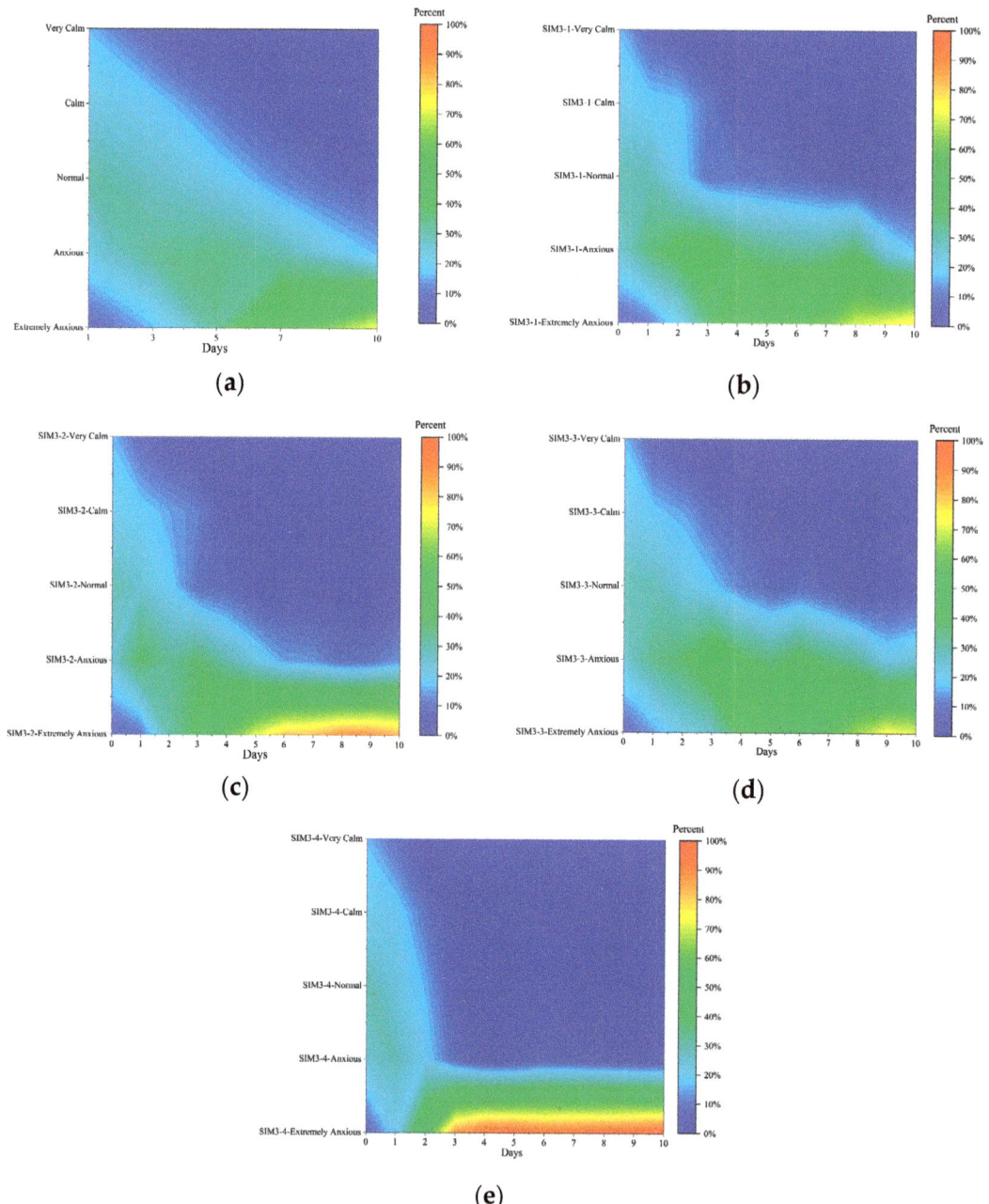

Figure 18. Simulation result of Scenario 3; (**a**) Initial data; (**b**) $Sim3-1$; (**c**) $Sim3-2$; (**d**) $Sim3-3$; (**e**) $Sim3-4$.

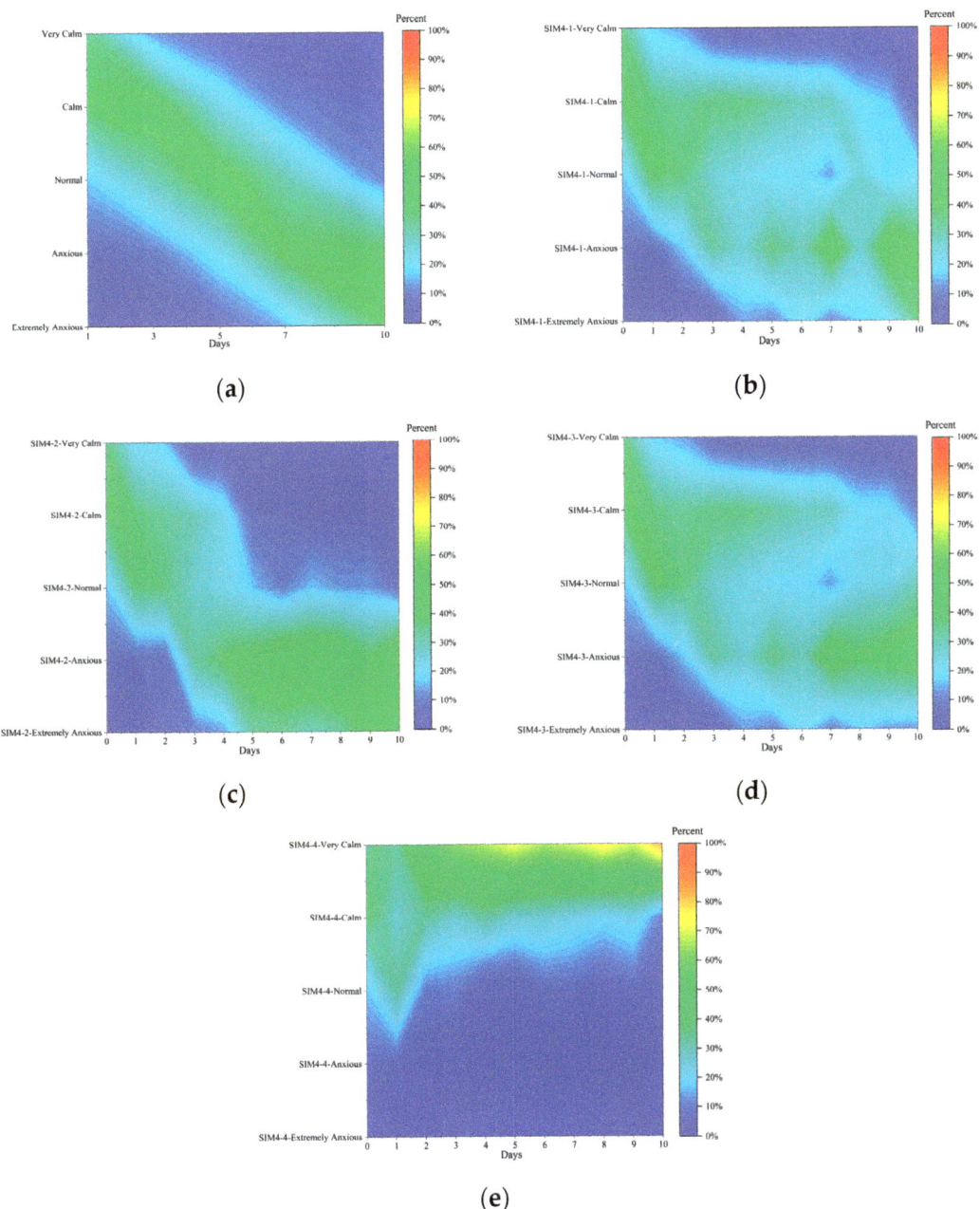

Figure 19. Simulation result of Scenario 4; (**a**) Initial data; (**b**) $Sim4-1$; (**c**) $Sim4-2$; (**d**) $Sim4-3$; (**e**) $Sim4-4$.

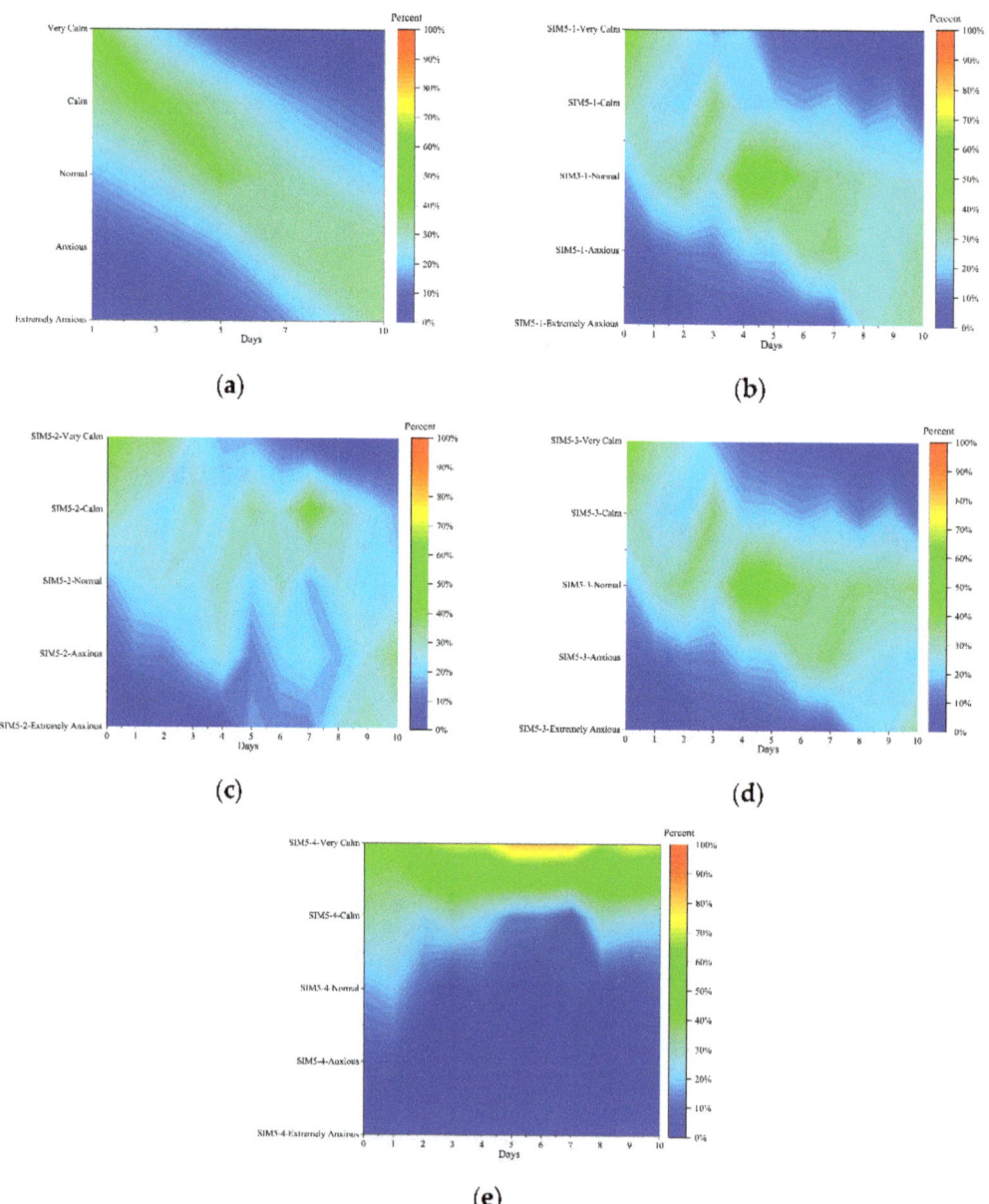

Figure 20. Simulation result of Scenario 5; (**a**) Initial data; (**b**) $Sim5 - 1$; (**c**) $Sim5 - 2$; (**d**) $Sim5 - 3$; (**e**) $Sim5 - 4$.

Figure 16 presents the distribution of different emotional states among passengers in Scenario 1 for different values of the parameter (a_1, b_1, a_2, b_2). When comparing Figure 16c–e, we observe that, despite the identical coefficients for $R(E)$ and $R(V)$ in these three graphs, there are notable distinctions in the distribution of passenger proportions. In Figure 16c, although the overall distribution closely aligns with the original data, there is a noticeable increase in fluctuation trends. Figure 16d exhibits even more pronounced fluctuation trends, while the distribution in Figure 16e differs substantially from the original data. By contrasting the values of parameter (a_1, b_1, a_2, b_2) in these graphs, we can deduce that in Scenario 1, time exerts a predominant influence in comparison to the inter-individual distances within the population.

Figure 17 illustrates the distribution of passengers in various emotional states in Scenario 2, under different values of the parameter (a_1, b_1, a_2, b_2). Similarly, when comparing Figure 17c–e, it is evident that they closely resemble the trends observed in the original data. Figure 17c,e seem to project potential future states of the original data, with e exhibiting a notably accelerated rate of development. In contrast, Figure 17d demonstrates more significant fluctuations compared to the original dataset. By examining the values of parameter (a_1, b_1, a_2, b_2) in these graphs, it becomes evident that in Scenario 2, the impact of time is considerably more pronounced.

With the continual expansion of the passenger population, the impact of time becomes notably more significant, as clearly evident in Figure 18c,e. Interestingly, when comparing Figure 18c–e, it becomes apparent that the fluctuations are gradually decreasing. One possible explanation for this trend is that as the total number of passengers increases, there is a directed influence causing passenger emotions to shift towards anxiety.

Comparing Figures 16 and 19 both horizontally and vertically reveals a notable trend: as the distance between passengers increases, the influence of distance decreases significantly. Additionally, there is a consistent shift toward calmer emotional states among passengers overall. When we contrast Scenario 1 with Scenario 4, a few key observations emerge. In Scenario 4, the proportion of passengers experiencing extreme anxiety further decreases, while the proportion of those in very calm and calm emotional states rises. This highlights a significant shift towards emotional calmness in Scenario 4 as compared to Scenario 1.

Upon a comprehensive comparison of Figures 16, 19 and 20, a significant trend becomes evident: as the distance between passengers increases, it intensifies the fluctuations among several positive emotional states. Additionally, this widening distance leads to a further reduction in the proportion of passengers experiencing extreme anxiety.

Based on the patterns analyzed earlier, an attempt is made to manually fine-tune the parameters, as shown in the (b) subfigures in Figures 16–20. Further statistical analysis was conducted to assess the tuning results. Figures 21–25 illustrate the difference-values (D-value) in emotional state distributions corresponding to different parameters over time, while Tables 13–17 provide a concise summary of the statistical analysis. Figures 26–30 show the simulation results of manual parameter adjustment.

Table 13. Descriptive Statistics of Extremely Anxious.

	Mean	Standard Deviation	Mean SE	Sum	Harmonic Mean	Minimum	Maximum
Extremely Anxious	0.16948	0.23723	0.10609	0.8474	0.02103	0.0094	0.5687
SIM1-1-Extremely Anxious	0.17333	0.22534	0.10077	0.86667	0	0	0.56667
SIM1-2-Extremely Anxious	0.15333	0.10435	0.04667	0.76667	0	0	0.26667
SIM1-3-Extremely Anxious	0.08	0.05055	0.02261	0.4	0	0	0.13333
SIM1-4-Extremely Anxious	0.02	0.01826	0.00816	0.1	0	0	0.03333

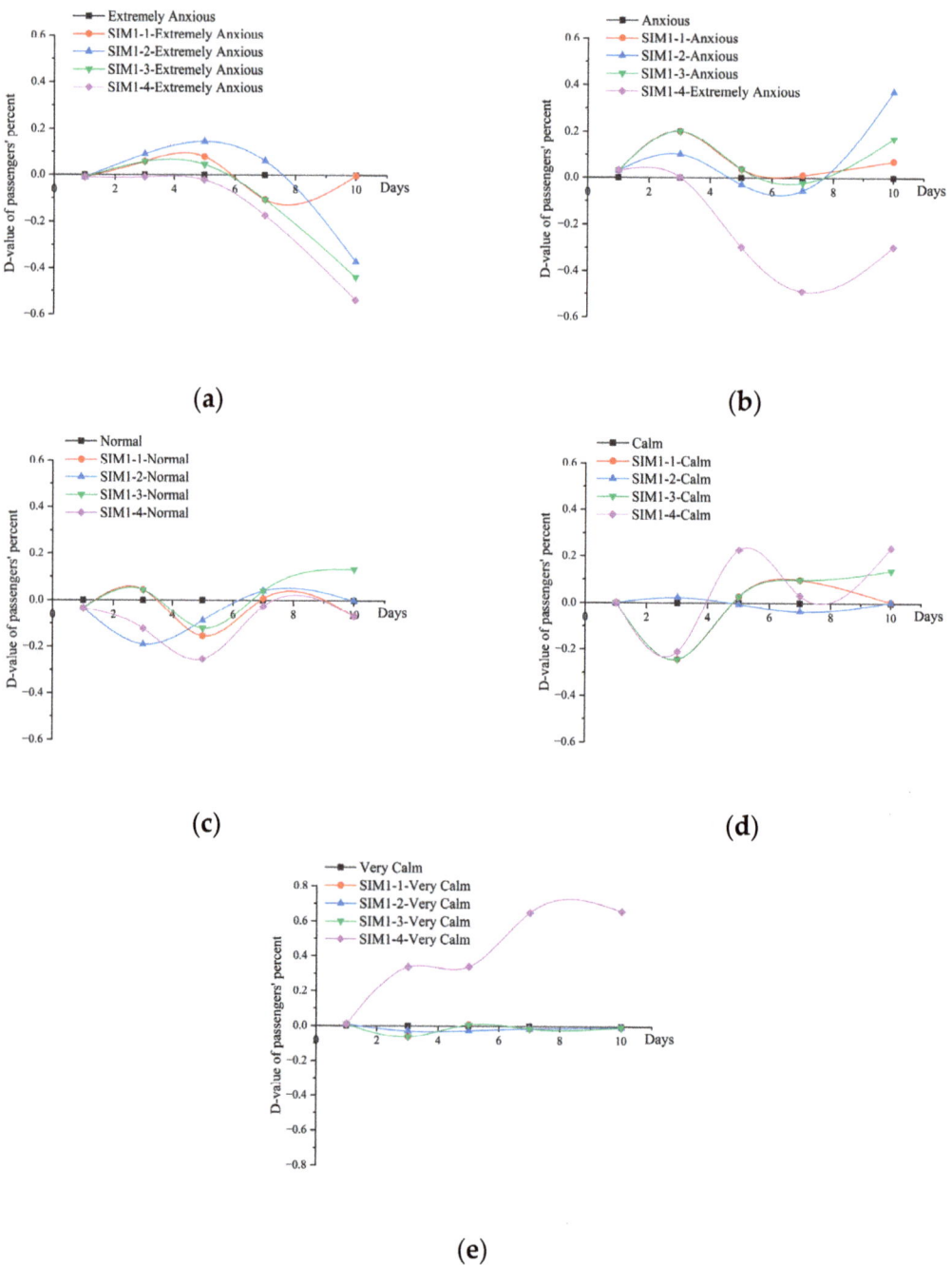

Figure 21. D-value of different emotional states in Scenario 1; (**a**) Extremely Anxious; (**b**) Anxious; (**c**) Normal; (**d**) Calm; (**e**) Very Calm.

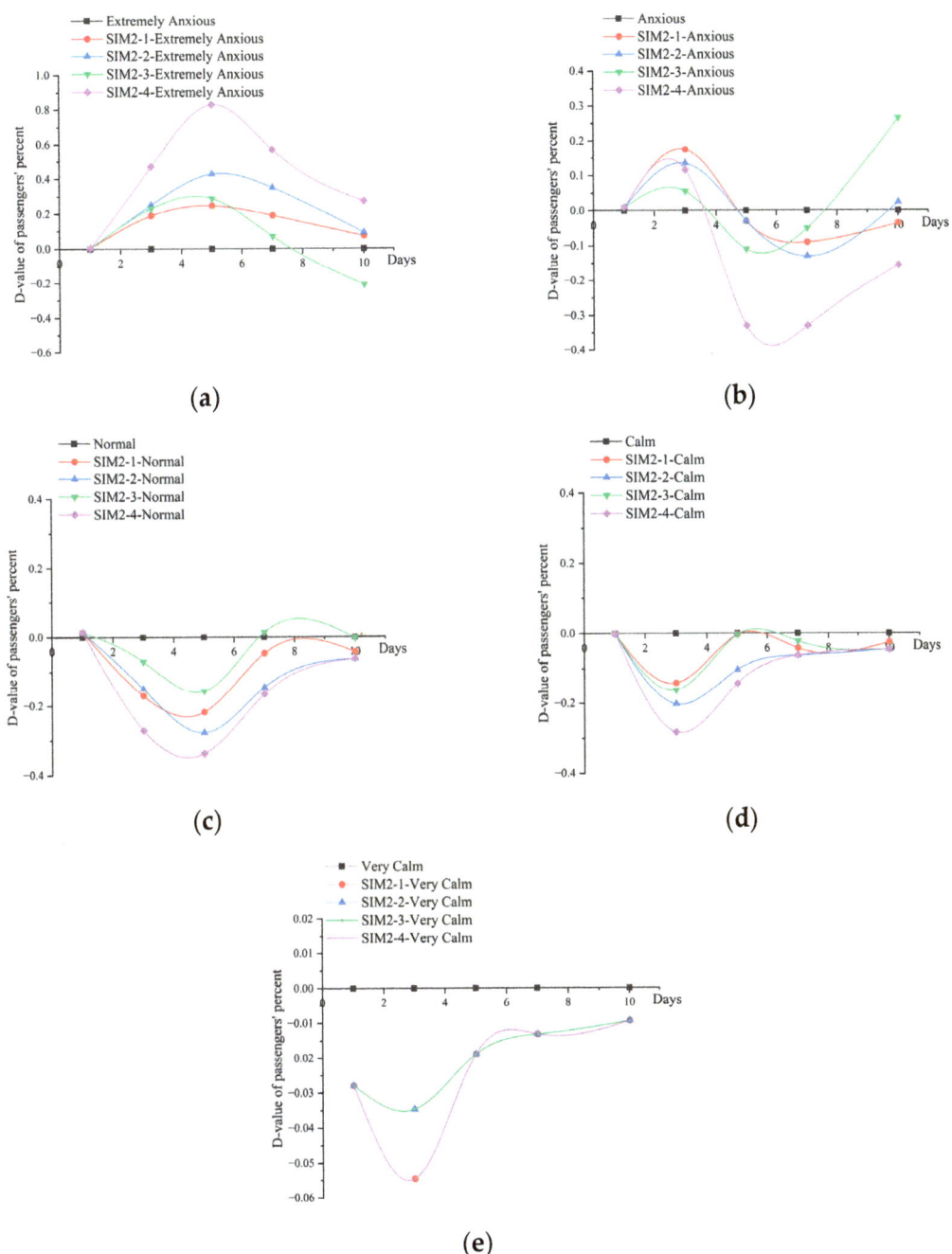

Figure 22. D-value of different emotional states in Scenario 2; (**a**) Extremely Anxious; (**b**) Anxious; (**c**) Normal; (**d**) Calm; (**e**) Very Calm.

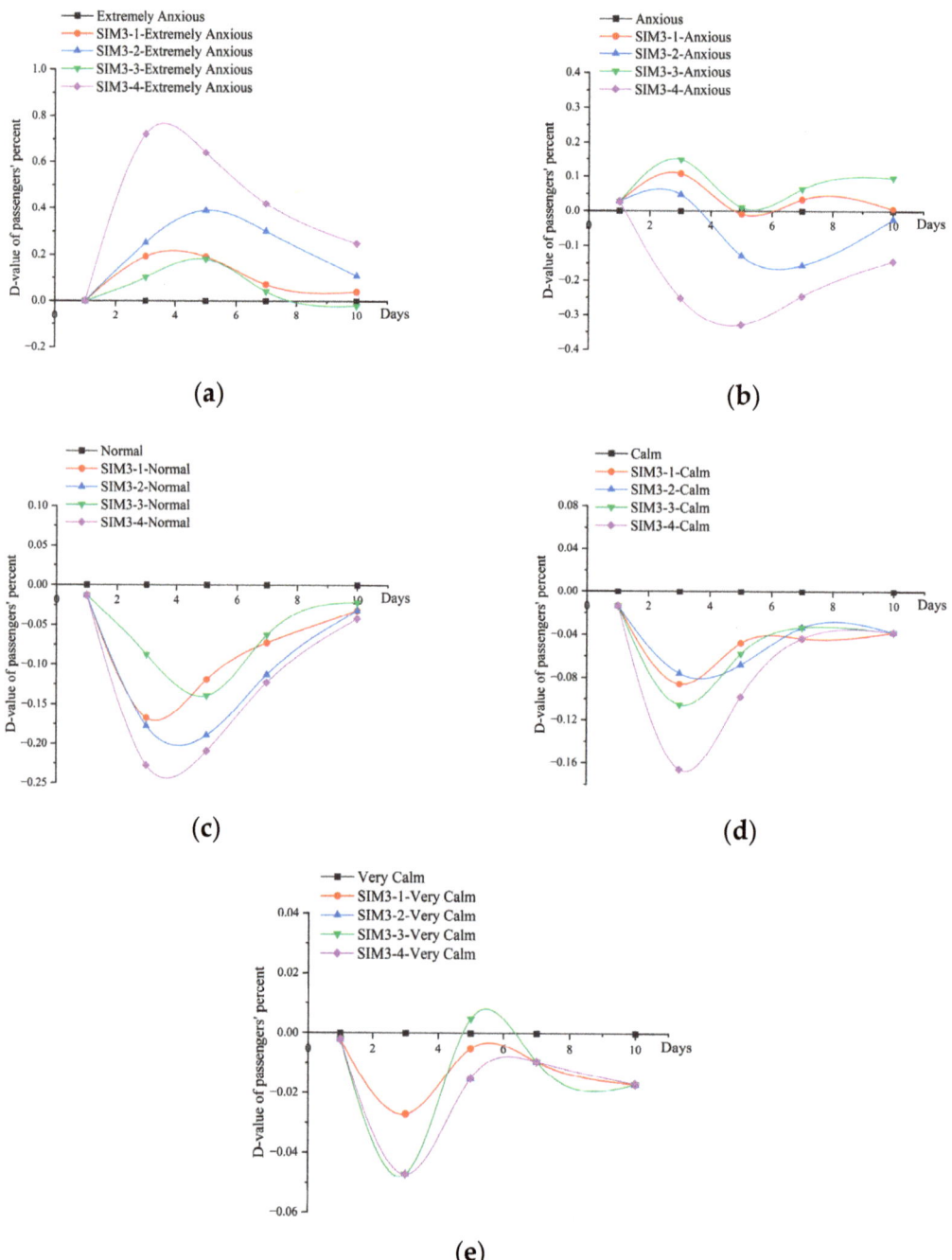

Figure 23. D-value of different emotional states in Scenario 3; (**a**) Extremely Anxious; (**b**) Anxious; (**c**) Normal; (**d**) Calm; (**e**) Very Calm.

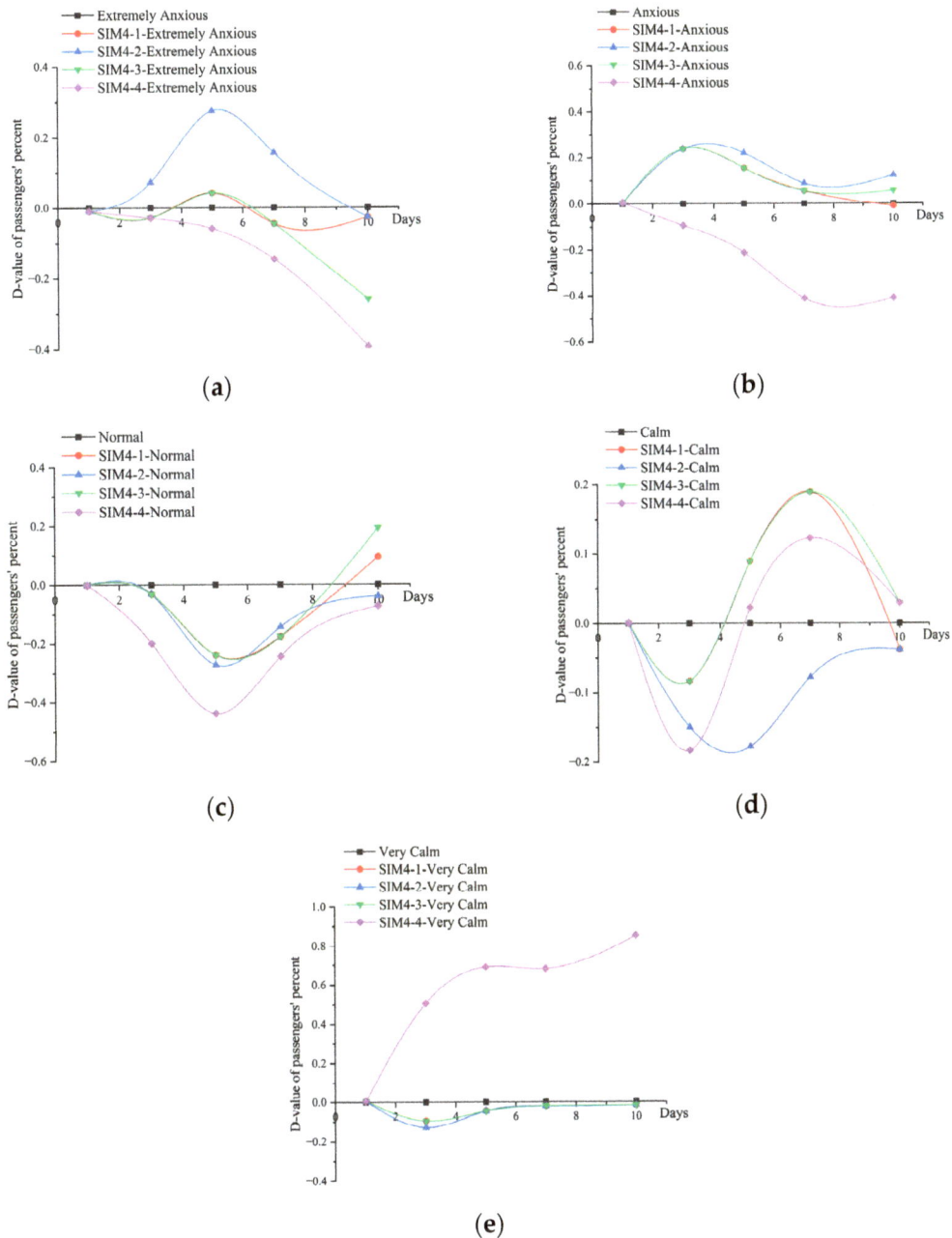

Figure 24. D-value of different emotional states in Scenario 4; (**a**) Extremely Anxious; (**b**) Anxious; (**c**) Normal; (**d**) Calm; (**e**) Very Calm.

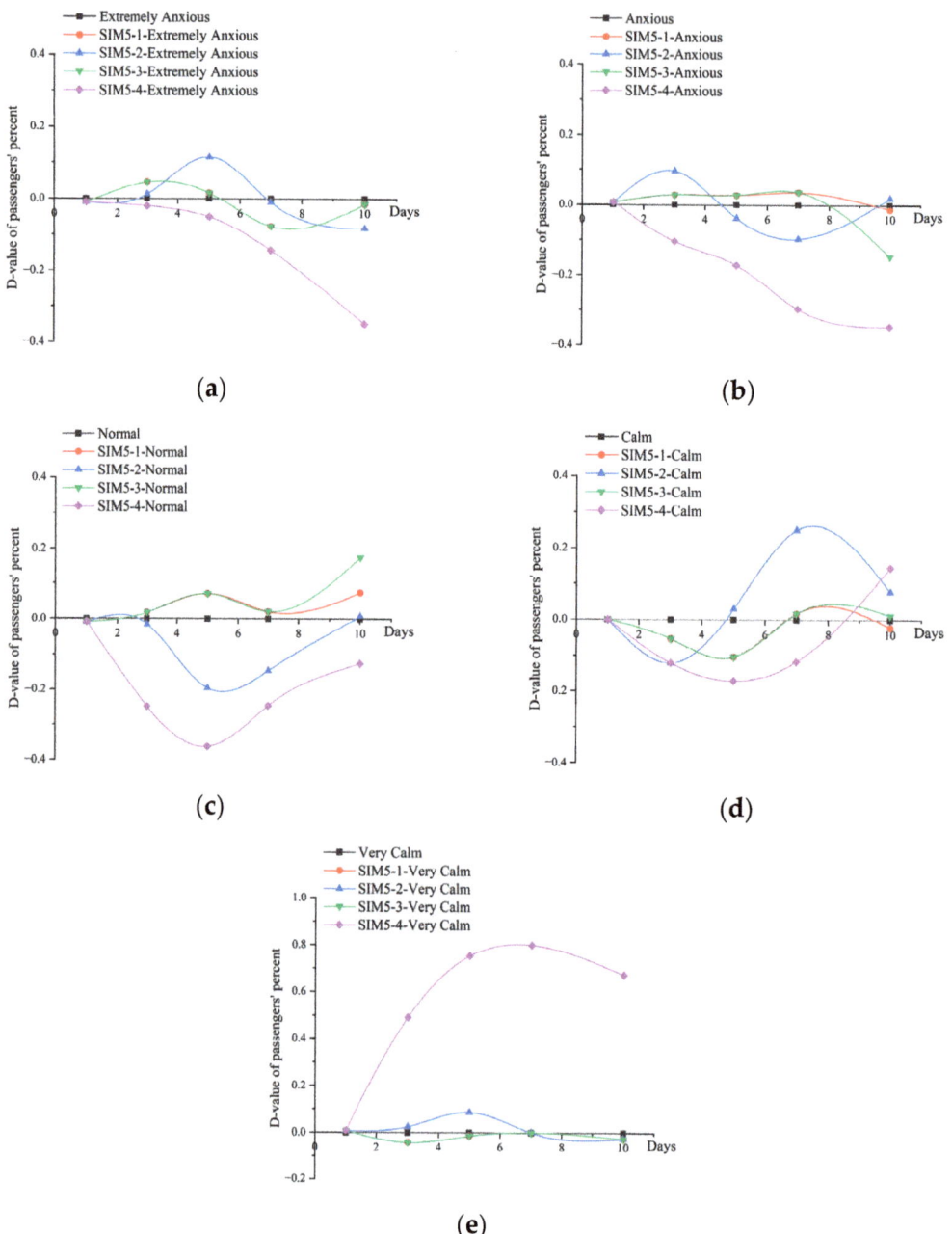

Figure 25. D-value of different emotional states in Scenario 5; (**a**) Extremely Anxious; (**b**) Anxious; (**c**) Normal; (**d**) Calm; (**e**) Very Calm.

Table 14. Descriptive Statistics of Anxious.

	Mean	Standard Deviation	Mean SE	Sum	Harmonic Mean	Minimum	Maximum
Anxious	0.24936	0.19171	0.08574	1.2468	0.10677	0.0358	0.5217
SIM1-1-Anxious	0.33333	0.14337	0.06412	1.66667	0.27121	0.13333	0.53333
SIM1-2-Anxious	0.34667	0.21807	0.09752	1.73333	0.25135	0.13333	0.66667
SIM1-3-Anxious	0.34667	0.14644	0.06549	1.73333	0.27815	0.13333	0.5
SIM1-4-Anxious	0.04667	0.05055	0.02261	0.23333	0	0	0.1

Table 15. Descriptive Statistics of Normal.

	Mean	Standard Deviation	Mean SE	Sum	Harmonic Mean	Minimum	Maximum
Normal	0.23352	0.13432	0.06007	1.1676	0.17736	0.0998	0.42
SIM1-1-Normal	0.24	0.13208	0.05907	1.2	0.11242	0.03333	0.36667
SIM1-2-Normal	0.2	0.09129	0.04082	1	0.16787	0.1	0.33333
SIM1-3-Normal	0.29333	0.05963	0.02667	1.46667	0.28357	0.23333	0.36667
SIM1-4-Normal	0.13333	0.06667	0.02981	0.66667	0.08772	0.03333	0.2

Table 16. Descriptive Statistics of Calm.

	Mean	Standard Deviation	Mean SE	Sum	Harmonic Mean	Minimum	Maximum
Calm	0.2422	0.20458	0.09149	1.211	0.08732	0.0301	0.4765
SIM1-1-Calm	0.2	0.11055	0.04944	1	0.10511	0.03333	0.33333
SIM1-2-Calm	0.22	0.20083	0.08981	1.1	0.07143	0.03333	0.5
SIM1-3-Calm	0.22667	0.06831	0.03055	1.13333	0.21212	0.16667	0.33333
SIM1-4-Calm	0.27333	0.11879	0.05312	1.36667	0.21607	0.1	0.43333

Table 17. Descriptive Statistics of Very Calm.

	Mean	Standard Deviation	Mean SE	Sum	Harmonic Mean	Minimum	Maximum
Very Calm	0.10546	0.1628	0.07281	0.5273	0.01622	0.0056	0.3898
SIM1-1-Very Calm	0.05333	0.08367	0.03742	0.26667	0	0	0.2
SIM1-2-Very Calm	0.08	0.14453	0.06464	0.4	0	0	0.33333
SIM1-3-Very Calm	0.05333	0.08367	0.03742	0.26667	0	0	0.2
SIM1-4-Very Calm	0.52667	0.13622	0.06092	2.63333	0.49826	0.36667	0.66667

It is evident from the presented charts and tables that the manual parameter tuning yielded highly favorable outcomes. Moreover, during the manual tuning process, it is noted that parameter (a_1, b_1, a_2, b_2) demonstrate closer alignment with the initial data when all its values are less than 1. This observation could be attributed to the fact that the incidents triggering the cruise ship's execution of the SRtP procedure are not overly severe, resulting in relatively minor effects on emotional transitions. However, as the incident duration increases, the impact accelerates.

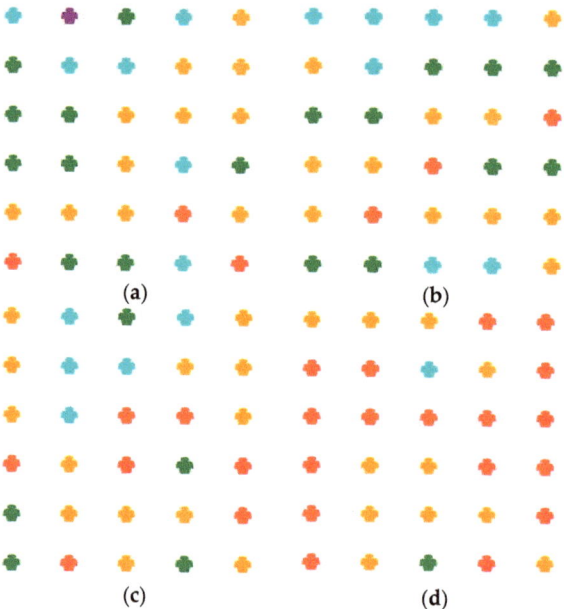

Figure 26. Simulation diagram of $Sim1-1$; (**a**) Day 3; (**b**) Day 5; (**c**) Day 7; (**d**) Day 10.

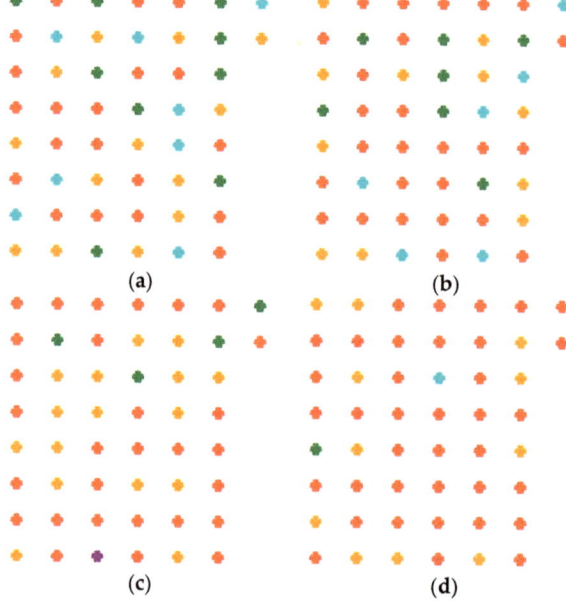

Figure 27. Simulation diagram of $Sim2-1$; (**a**) Day 3; (**b**) Day 5; (**c**) Day 7; (**d**) Day 10.

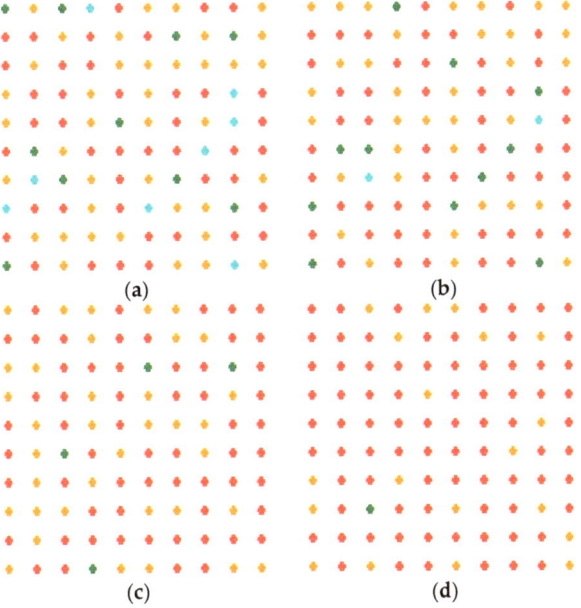

Figure 28. Simulation diagram of $Sim3-1$; (**a**) Day 3; (**b**) Day 5; (**c**) Day 7; (**d**) Day 10.

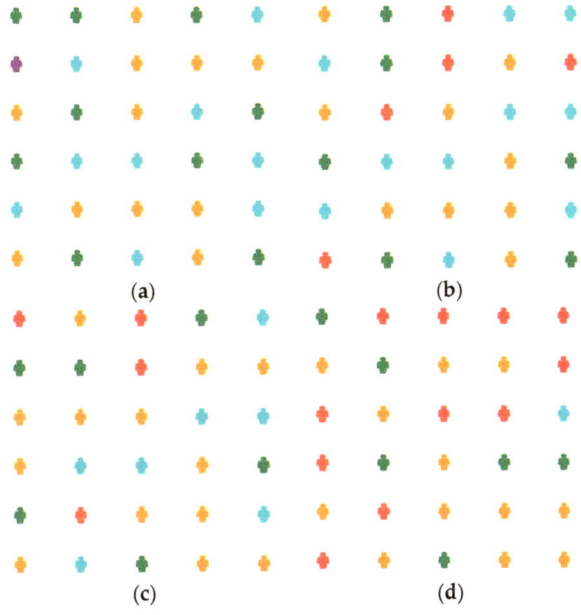

Figure 29. Simulation diagram of $Sim4-1$; (**a**) Day 3; (**b**) Day 5; (**c**) Day 7; (**d**) Day 10.

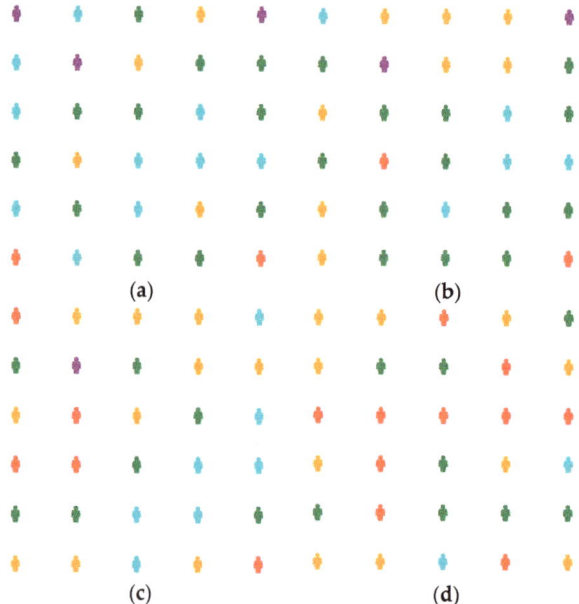

Figure 30. Simulation diagram of $Sim5 - 1$; (**a**) Day 3; (**b**) Day 5; (**c**) Day 7; (**d**) Day 10.

The scenario parameters and simulation results are shown in Figures 16–25. In the scene simulation diagrams, red indicates extremely anxious passengers, orange indicates generally anxious passengers, green indicates normal passengers, light blue indicates generally calm passengers and purple indicates very calm passengers.

Figures 31–35 present an analysis of the impact of spatial grid distribution on emotional transitions based on the results of manual parameter tuning, specifically exploring the influence of model hyperparameters.

In Figure 31, when comparing subfigures horizontally, it becomes evident that the conversion of rows and columns in two different spatial grid configurations affects emotional state transitions. Vertical comparisons, on the other hand, reveal that a complete alteration of the spatial grid distribution may or may not influence emotional state transitions.

A similar trend is observed in Figure 32. In horizontal comparisons, there exists a set of spatial grid row–column conversions that do not produce discernible impacts. In vertical comparisons, instances of a radical shift in grid distribution do not result in corresponding changes in emotional state distributions.

Figure 33 highlights this phenomenon more prominently, with emotional state distributions across all comparison groups undergoing only minimal changes.

Both Figures 34 and 35, whether involving the conversion of rows and columns within the spatial grid or an entire overhaul of the grid, show no notable changes in the distribution of emotional states.

Several factors may contribute to these observations:

1. In the model developed in this study, the distance component is relatively minor compared to the temporal component, and its influence diminishes as time progresses.
2. The original data and simulations are based on a daily time scale, with no exploration of scenarios where t falls between 0 and 1. According to the proposed model, distance effects become dominant only when t is small.

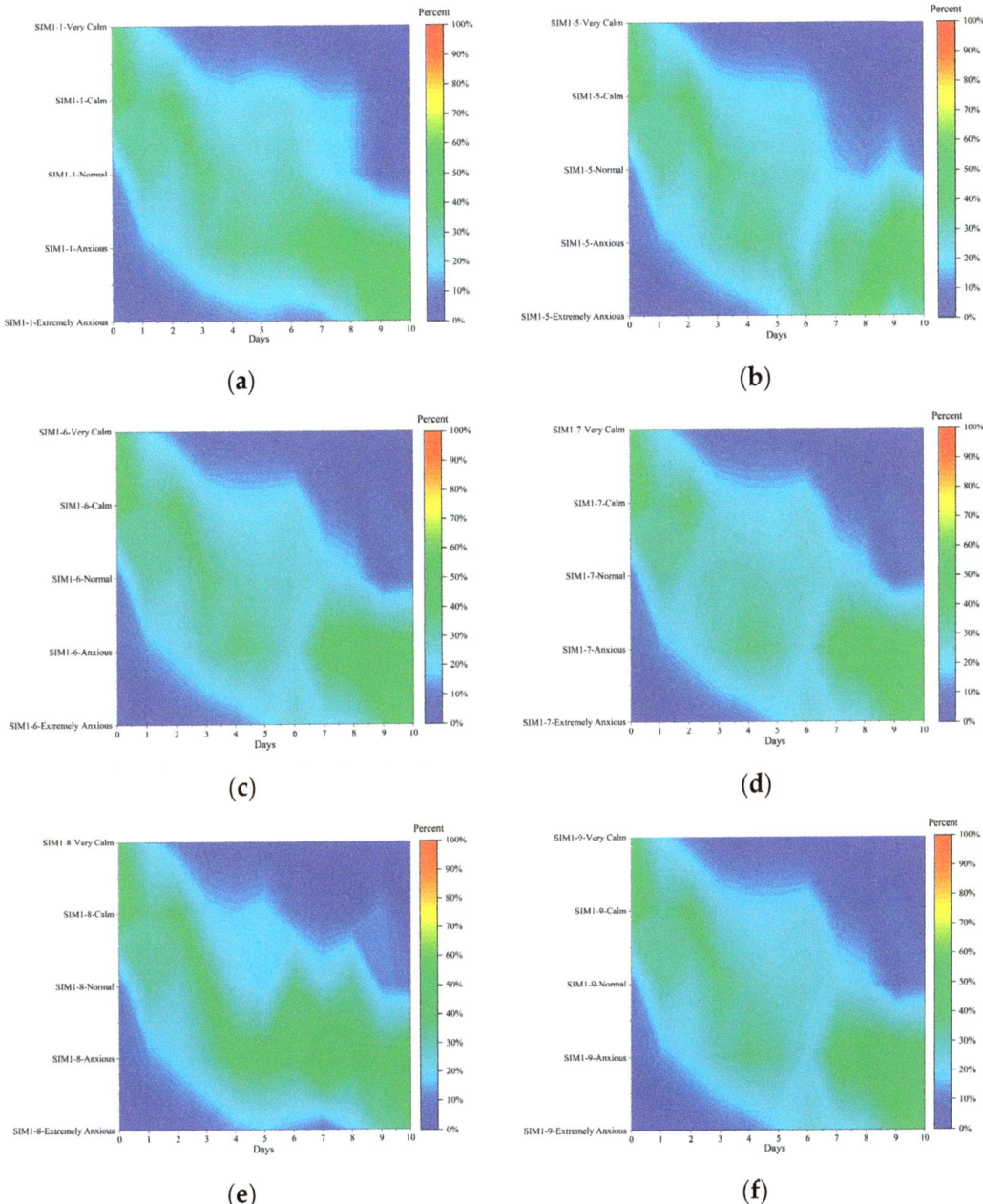

Figure 31. Simulation result of Scenario 1; (**a**) $Sim1-1$; (**b**) $Sim1-5$; (**c**) $Sim1-6$; (**d**) $Sim1-7$; (**e**) $Sim1-8$; (**f**) $Sim1-9$.

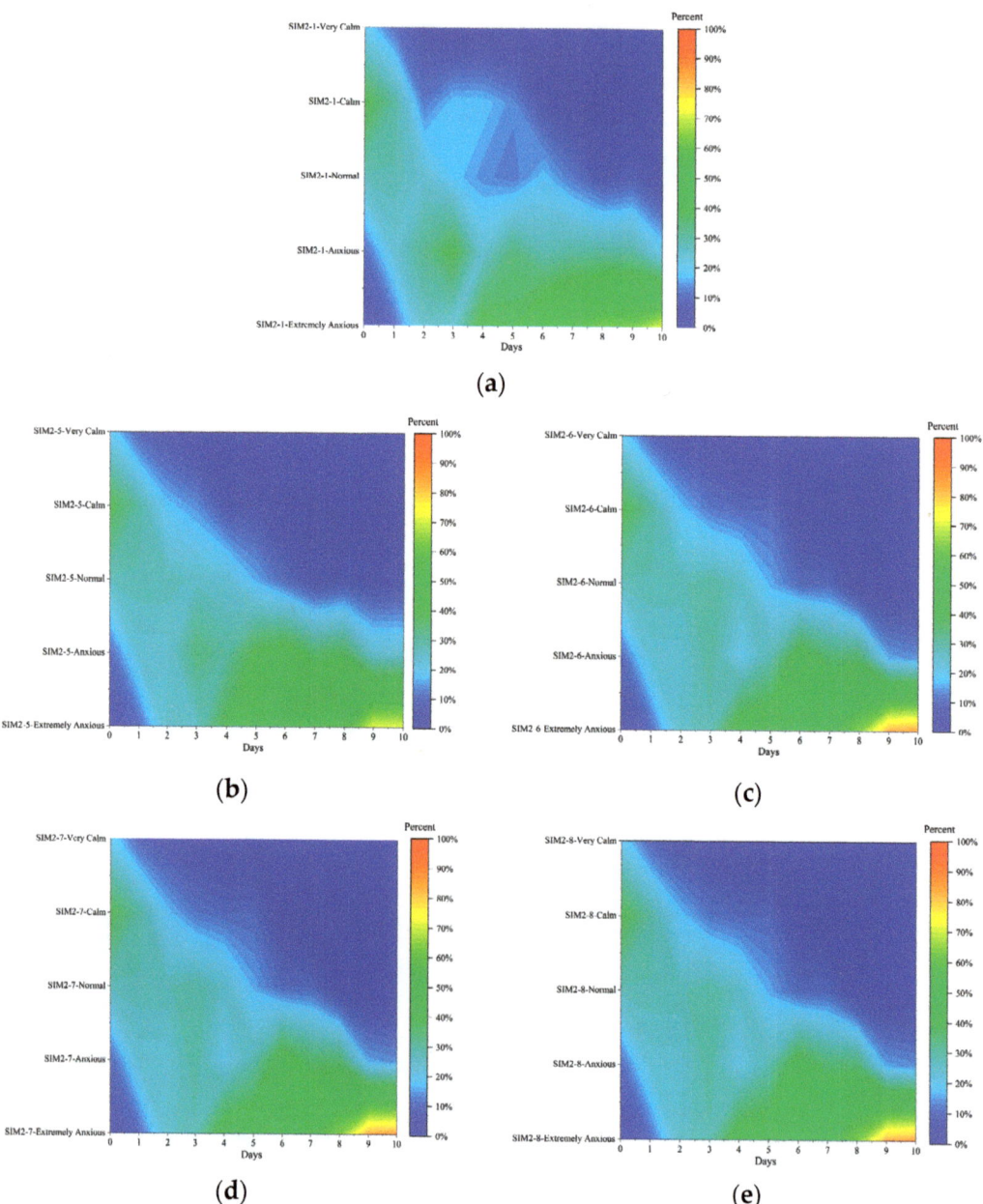

Figure 32. Simulation result of Scenario 2; (**a**) $Sim2-1$; (**b**) $Sim2-5$; (**c**) $Sim2-6$; (**d**) $Sim2-7$; (**e**) $Sim2-8$.

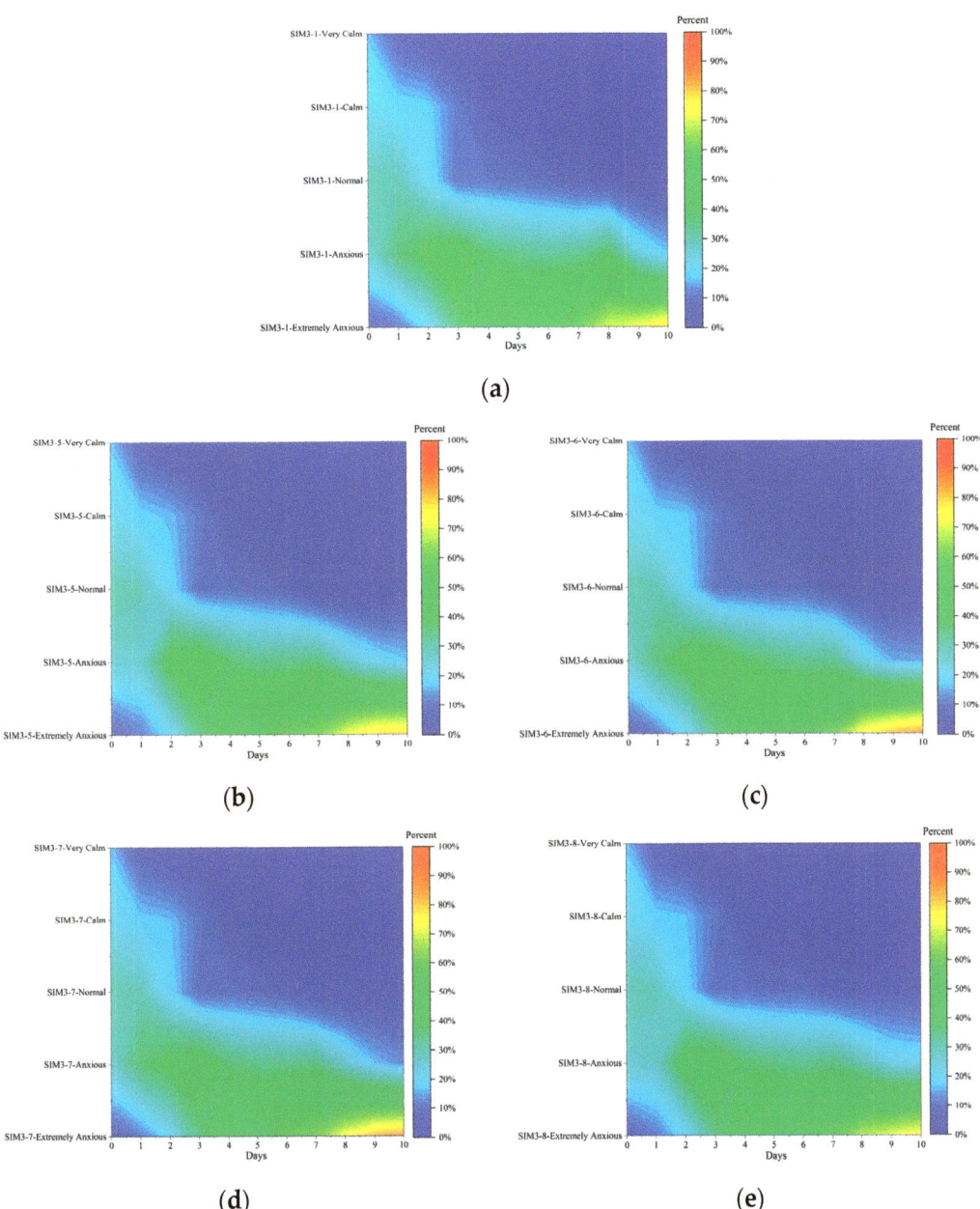

Figure 33. Simulation result of Scenario 3; (**a**) $Sim3-1$; (**b**) $Sim3-5$; (**c**) $Sim3-6$; (**d**) $Sim3-7$; (**e**) $Sim3-8$.

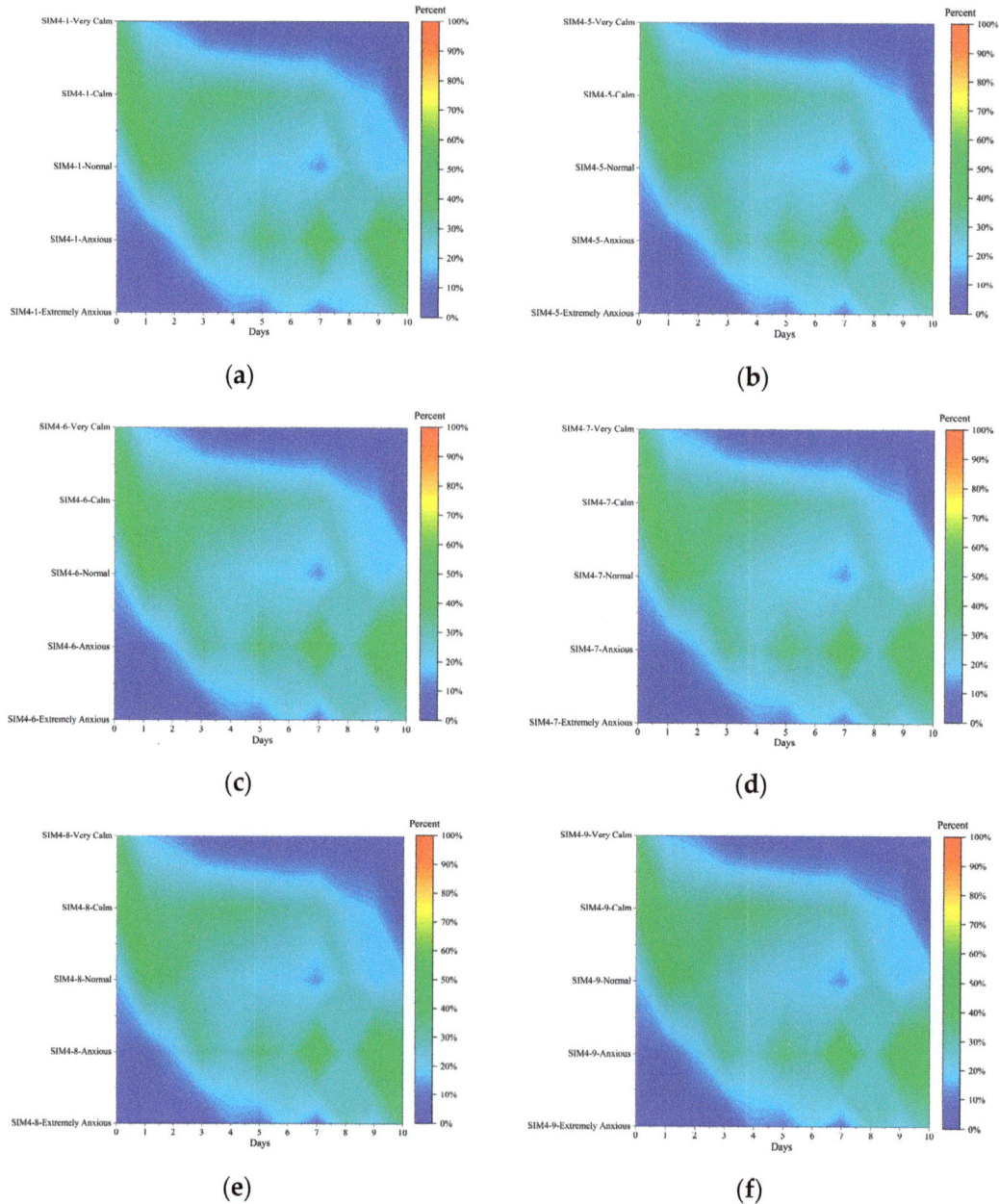

Figure 34. Simulation result of Scenario 4; (**a**) $Sim4-1$; (**b**) $Sim4-5$; (**c**) $Sim4-6$; (**d**) $Sim4-7$; (**e**) $Sim4-8$; (**f**) $Sim4-9$.

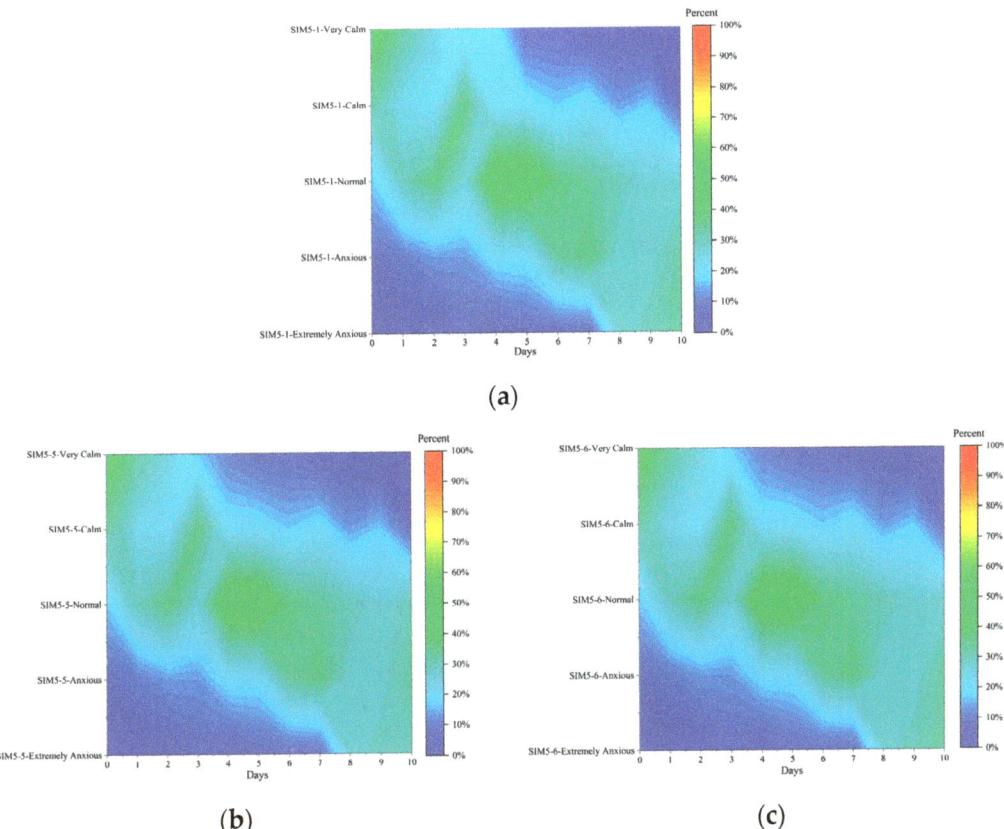

Figure 35. Simulation result of Scenario 5; (**a**) $Sim5-1$; (**b**) $Sim5-5$; (**c**) $Sim5-6$.

3. Emotional transmission emanates from the infection source and radiates outward, potentially affecting eight, five, or three individuals in the first level of transmission. Calculations using Euclidean distance indicate that, when conducting row–column conversions within spatial grids, the overall dynamics of the first-level transmission remain unaltered. However, the impact gradually becomes noticeable in the second-level transmission. Notably, the magnitude of this effect increases with greater disparities in row and column distances within the spatial grid.

The aforementioned observations indicate that, in prolonged shelter-in-place scenarios akin to SRtP, the spatial distribution of individuals exerts a minimal influence on the overall emotional state transitions. This implies that safety planning for such scenarios can incorporate more flexible designs for the shape of spaces.

Upon comparing Figures 31–33, it becomes apparent that, as the total number of passengers increases, altering the spatial grid distribution does not significantly reduce the likelihood of emotional transitions towards anxiety. However, strategically implementing spatial separation to reduce the total number of individuals in each space can indeed diminish this probability.

Further comparisons involving Figures 31, 34 and 35 reveal that appropriately increasing the distance between individuals does contribute to a lower overall likelihood of emotional transitions towards anxiety.

5. Conclusions and Future Work

This study is centered on the emotional contagion process during the cruise ship SRtP (Ship-to-Rescue-Platform) procedure and presents an improved SIR (Susceptible-Infection-Removal) model. In comparison to other emotional contagion models, the one proposed in this paper expands its scope across emotional state divisions, contagion source configurations, and transition directions. By combining it with the weighted random forest algorithm for emotional state distribution prediction, the results demonstrate that this model adeptly captures the fluctuating characteristics inherent in the emotional transition process. Through simulation experiments, we visualize these fluctuating characteristics during emotional transition, employing multiple control groups with varying parameters to analyze the effects of parameter variations.

Although this research focuses on a specific passenger group, the proposed model exhibits applicability to similar extended-duration emergency scenarios. In summary, the research findings explore and substantiate the efficacy of the enhanced SIR model in modeling the emotional contagion process among passengers during the cruise ship SRtP procedure.

In future research endeavors, we contemplate the integration of individual traits and demographic factors for further model refinement.

Author Contributions: Conceptualization, G.X. and Z.Y.; methodology, G.X.; writing—original draft preparation, G.X. and M.H.; writing—review and editing, W.C. and M.H. All authors have read and agreed to the published version of the manuscript.

Funding: This research was funded by the Design Technology of High-Tech Ocean Passenger Ships for the Safe Return to Port System in the Ministry of Industry and Information Technology of China, grant number [2019] No.331.

Institutional Review Board Statement: According to Article 9 of the *Ethical Review Measures for Life Sciences and Medical Research Involving Human Beings* issued by the Science and Technology Education Department of the National Health Commission of the People's Republic of China (http://www.nhc.gov.cn/qjjys/s3582/202302/23de06e70e8b4c9e86695f6877f3c248.shtml (accessed on 18 April 2023)): Considering that most basic research activities do not directly involve human trials, and some studies do not directly involve clinical diagnosis and treatment information of research participants, drawing on international practices, in order to improve review efficiency and reduce unnecessary burden on researchers, the "Measures" stipulate that "under the premise of using human information data or biological samples, not causing harm to the human body, and not involving sensitive personal information or commercial interests", some cases involving human life sciences and medical research can be exempted from ethical review, mainly including: (1) Using legally obtained public data or conducting research through observation without interfering with public behavior; (2) Conducting research using anonymous information data; (3) Using existing human biological samples to carry out research, the source of biological samples used complies with relevant laws and ethical principles, the relevant content and purpose of research are within the scope of standardized informed consent, and do not involve the use of human germ cell, embryos and reproductive cloning, chimerism, heritable gene manipulation and other activities; (4) Conducting research using human cell lines or cell lines derived from biological sample banks, with relevant content and objectives within the authorized scope of the provider, and without involving activities such as human embryonic and reproductive cloning, chimerism, and heritable gene manipulation. The research conducted in this paper does not cause harm to the human body, nor does it involve sensitive personal information or commercial interests. Its questionnaire survey is conducted through completely anonymous centroids. So ethical review can be exempted.

Informed Consent Statement: Patient consent is waived due to REASON. (The research conducted in this paper does not cause harm to the human body, nor does it involve sensitive personal information or commercial interests. Its questionnaire survey is conducted through completely anonymous centroids.).

Data Availability Statement: Not applicable.

Conflicts of Interest: The authors declare no conflict of interest.

Appendix A. Questionnaire of the Survey

Questionnaire on Adverse Emotions in the SRtP of Cruise Ship

The setting scenario is: An accident occurs during a cruise ship's voyage, and the accident causes a certain degree of damage to the cruise ship, requiring a nearby port call. At this point, the accident has been handled, and the cruise ship has the ability to navigate independently and dock at the port. It is necessary to centralize personnel in a fixed area to facilitate the provision of daily meals and management. All restaurants, shops, and entertainment activities have ceased normal operation, and unified food distribution has been implemented instead.

Table A1. The basic information questionnaire of passengers.

Category	Content	Supplementary Description	The Specific Opinions		
Basic information	Gender		☐ Male	☐ Female	
	Age				
	Which deck do you prefer to stay on?	Deck 4 is the evacuation deck, and the accident occurred on the Deck 6			
	Which area would you prefer to be assigned to except guest room?		☐ Restaurant (food and beverage supply suspended)	☐ Store (suspended sales of goods)	☐ Gangway

Table A2. The emotion questionnaire of passengers.

Category	Scenario No.	Scenario Setting	Extremely Anxious	Anxious	Normal	Calm	Very Calm
Emotional states score	Day 1	Scenario 1 $\rho = 2~m^2/person$, $n = 30$	1	2	3	4	5
	Day 3		1	2	3	4	5
	Day 5		1	2	3	4	5
	Day 7		1	2	3	4	5
	Day 10		1	2	3	4	5
	Day 1	Scenario 2 $\rho = 2~m^2/person$, $n = 50$	1	2	3	4	5
	Day 3		1	2	3	4	5
	Day 5		1	2	3	4	5
	Day 7		1	2	3	4	5
	Day 10		1	2	3	4	5
	Day 1	Scenario 3 $\rho = 2~m^2/person$, $n = 100$	1	2	3	4	5
	Day 3		1	2	3	4	5
	Day 5		1	2	3	4	5
	Day 7		1	2	3	4	5
	Day 10		1	2	3	4	5
	Day 1	Scenario 4 $\rho = 3~m^2/person$, $n = 30$	1	2	3	4	5
	Day 3		1	2	3	4	5
	Day 5		1	2	3	4	5
	Day 7		1	2	3	4	5
	Day 10		1	2	3	4	5
	Day 1	Scenario 5 $\rho = 4~m^2/person$, $n = 30$	1	2	3	4	5
	Day 3		1	2	3	4	5
	Day 5		1	2	3	4	5
	Day 7		1	2	3	4	5
	Day 10		1	2	3	4	5

Appendix B. The Results of Reliability and Validity Analysis of the Questionnaire

Table A3. Reliability statistics.

Items		Corrected Item–Total Correlation (CITC)	Cronbach α
Gender		−0.026	
Question3		0.060	
Question4		0.063	
Scenario 1 $\rho = 2\ m^2/\text{person},\ n = 30$	Day1	0.505	
	Day3	0.636	
	Day5	0.726	
	Day7	0.691	
	Day10	0.563	
Scenario 2 $\rho = 2\ m^2/\text{person},\ n = 50$	Day1	0.580	
	Day3	0.671	
	Day5	0.737	
	Day7	0.695	
	Day10	0.567	
Scenario 3 $\rho = 2\ m^2/\text{person},\ n = 100$	Day1	0.602	0.916
	Day3	0.675	
	Day5	0.673	
	Day7	0.611	
	Day10	0.484	
Scenario 4 $\rho = 3\ m^2/\text{person},\ n = 30$	Day1	0.536	
	Day3	0.596	
	Day5	0.670	
	Day7	0.691	
	Day10	0.566	
Scenario 5 $\rho = 4\ m^2/\text{person},\ n = 30$	Day1	0.505	
	Day3	0.525	
	Day5	0.631	
	Day7	0.635	
	Day10	0.579	

Cronbach α (Standardized): 0.933.

Table A4. Validity analysis.

Items		Factor Loadings					Communalities
		Factor 1	Factor 2	Factor 3	Factor 4	Factor 5	
Gender		0.025	−0.083	−0.015	0.022	0.647	0.427
Question3		0.044	−0.060	0.057	0.126	−0.631	0.423
Question4		0.032	−0.036	0.095	0.055	0.440	0.208
Scenario 1 $\rho = 2\ m^2/\text{person}$, n = 30	Day1	−0.135	0.501	0.266	0.622	−0.054	0.731
	Day3	0.203	0.372	0.236	0.723	−0.022	0.758
	Day5	0.460	0.219	0.326	0.645	−0.000	0.782
	Day7	0.663	0.056	0.322	0.489	0.017	0.785
	Day10	0.774	−0.091	0.247	0.272	−0.001	0.742
Scenario 2 $\rho = 2\ m^2/\text{person}$, n = 50	Day1	−0.147	0.409	0.610	0.464	−0.000	0.776
	Day3	0.048	0.300	0.682	0.436	−0.006	0.748
	Day5	0.377	0.163	0.666	0.356	0.066	0.743
	Day7	0.574	−0.008	0.605	0.250	0.037	0.759
	Day10	0.721	−0.124	0.465	0.052	−0.034	0.756
Scenario 3 $\rho = 2\ m^2/\text{person}$, n = 100	Day1	−0.080	0.298	0.819	0.235	0.049	0.823
	Day3	0.139	0.196	0.876	0.136	0.058	0.847
	Day5	0.304	0.087	0.850	0.057	0.006	0.826
	Day7	0.480	−0.056	0.766	−0.037	−0.027	0.822
	Day10	0.648	−0.193	0.520	−0.099	−0.083	0.743

Table A4. Cont.

Items		Factor Loadings					Communalities
		Factor 1	Factor 2	Factor 3	Factor 4	Factor 5	
Scenario 4 $\rho = 3$ m^2/person, n = 30	Day1	−0.073	0.815	0.240	0.195	−0.052	0.768
	Day3	0.171	0.851	0.131	0.105	−0.049	0.784
	Day5	0.465	0.716	0.120	0.100	0.048	0.755
	Day7	0.719	0.505	0.123	0.020	0.018	0.787
	Day10	0.859	0.255	0.051	−0.039	0.030	0.807
Scenario 5 $\rho = 4$ m^2/person, n = 30	Day1	−0.092	0.824	0.162	0.244	−0.018	0.773
	Day3	0.110	0.858	0.046	0.120	−0.053	0.767
	Day5	0.438	0.762	0.027	0.082	0.002	0.780
	Day7	0.669	0.586	−0.007	0.052	0.026	0.794
	Day10	0.834	0.350	−0.020	0.007	0.048	0.821
Eigenvalues (Rotated)		5.958	5.615	5.426	2.488	1.051	-
Variance (Rotated)		21.277%	20.054%	19.379%	8.885%	3.752%	-
Cum. Variance (Rotated)		21.277%	41.331%	60.710%	69.594%	73.346%	-
KMO				0.928			-
p-value of Bartlett's Test of Sphericity				< 0.001			

Appendix C. Theoretical Description and Analysis of Alternative Models

Appendix C.1. CPS-REC Model

The CPS-REC model closely aligns with our theoretical framework. It effectively addresses the aspects of heterogeneity and recurrent infections. However, there is a key distinction between the CPS-REC model and our own approach. The CPS-REC model necessitates a comprehensive consideration of emotional contagion in both the network and physical space, which may lead to a greater emphasis on individual emotional changes. In contrast, our model primarily focuses on the physical space of the scenario and excels in examining the emotional shifts within the group as a whole.

Furthermore, our model introduces the concept of dual infection sources and bidirectional infection, which closely mirrors the actual dynamics observed in the study of group emotions within refuge scenarios. This distinction underscores our model's relevance and applicability in capturing the complexities of emotional contagion in emergency situations of refuge scenarios.

Let a six-tuple $S(i) = (State(i;t), \beta, \mu;, \delta; P_0)$ denote the attributes of individuals i in space, where $State(i;t)$ represents the state of individuals i at any time t, β represents the contagion rate of infected individuals, μ represents the cure rate of infected individuals, δ represents the recurrence rate of temporarily immune individuals, and P_0 represents individual spontaneous infection.

$$State(i,t) = (WS, WI, WR, XS, XI, XR), \tag{A1}$$

where WS represents the susceptible individuals in physical space; WI represents the infected individuals in physical space; WR represents the temporarily immune individuals in physical space; XS represents the susceptible individuals in cyberspace; XI represents the infected individuals in cyberspace; XR represents the temporarily immune individuals in cyberspace.

Emotional contagion rules We first define the parameters for our emotional contagion rules. The probability of susceptible individuals in cyberspace being contagious to infected individuals is β_V. The probability of susceptible individuals in physical space being contagious to infected individuals is β_P. The probability of infected individuals in cyberspace being cured to become temporarily immune individuals is μ_V. The probability of infected individuals in physical space being cured to become temporarily immune individuals is μ_P. The probability that a temporarily recovered individual in cyberspace will recur as a susceptible individual is δ_V. The probability that a temporarily recovered individual in physical space will recur as a susceptible individual is δ_P. The parameter descriptions of the Mean-Field Equations are shown in Table A5.

$$\frac{dW^I(t)}{dt} = \beta_P \langle k_0 \rangle W^S(t) W^I(t) + P_0 W^S(t) + \beta_V \langle k_1 \rangle W^S(t) X^I(t) - \mu_V W^I(t) - \mu_P W^I(t) \tag{A2}$$

$$\frac{dW^S(t)}{dt} = \delta_P \langle k_0 \rangle W^R(t) - \beta_P \langle k_0 \rangle W^S(t) W^I(t) - \beta_V \langle k_1 \rangle W^S(t) X^I(t) - P_0 W^S(t) \tag{A3}$$

$$\frac{dW^R(t)}{dt} = \mu_P W^I(t) + \mu_V W^I(t) - \delta_P W^R(t) \tag{A4}$$

$$\frac{dX^I(t)}{dt} = \beta_V \langle k_1 \rangle X^S(t) X^I(t) + P_0 X^S(t) + \beta_P \langle k_0 \rangle X^S(t) W^I(t) - \mu_V X^I(t) - \mu_P X^I(t) \tag{A5}$$

$$\frac{dX^S(t)}{dt} = \delta_V X^R(t) - \beta_V \langle k_1 \rangle X^S(t) X^I(t) - \beta_P \langle k_0 \rangle X^S(t) W^I(t) - P_0 X^S(t) \tag{A6}$$

$$\frac{dX^R(t)}{dt} = \mu_V X^I(t) + \mu_P X^I(t) - \delta_V X^R(t) \tag{A7}$$

Table A5. The parameter descriptions of the Mean-Field Equations.

Parameters	Description
N	the number of individuals in CPS
$W^S(t)$	the proportion of WS in the crowd at moment t
$W^I(t)$	the proportion of WI in the crowd at moment t
$W^R(t)$	the proportion of WR in the crowd at moment t
$X^S(t)$	the proportion of XS in the crowd at moment t
$X^I(t)$	the proportion of XI in the crowd at moment t
$X^R(t)$	the proportion of XR in the crowd at moment t
$\langle k_0 \rangle$	average degree in physical space
$\langle k_1 \rangle$	average degree in cyberspace
P_0	probability of spontaneous infection

Appendix C.2. SEEC

The construction of the SEEC model primarily aims to facilitate optimal intervention in the emotional state of the population, with the ultimate goal of controlling emotional transmission within the population. Consequently, the SEEC model has been simplified to serve this purpose. While intervention in crowd emotions is mentioned, it is not the central focus of the model proposed in this article. Instead, it represents one of the potential avenues for future research. As a result, the model presented in this article offers a more diverse framework for categorizing emotional states.

In Equation (A8), $\beta(t)$ represents the infection rate at time t (i.e., the number of effective infections per unit time of a single infected individual), and $I(t)\beta(t)$ represents the number of individuals who can be infected per unit time. $\frac{S(t)}{N}$ represents the proportion of susceptible individuals in the crowd at time t. The product of $I(t)\beta(t)$ and $\frac{S(t)}{N}$ represents the number of susceptible individuals among those who could be infected, that is, the number of times that the infection event occurs per unit time. In Equation (A9), $\gamma(t)$ is the recovery rate per unit time, and $\gamma(t)I(t)$ is the number of individuals who recover to normal per unit time, that is, the occurrence number of the recovery events per unit time.

$$\psi^I(R) = I(t)\beta(t)\frac{S(t)}{N} \tag{A8}$$

$$\psi^R(R) = \gamma(t)I(t) \tag{A9}$$

Let $T(t)$ be the history of emotional contagion process. Under the historical conditions, let $Pr\{d(I(t) = 1|T(t))\}, Pr\{d(I(t) = -1|T(t))\}, Pr\{d(I(t) = 0|T(t))\}$ as the occurrence probability of only one infection event, the occurrence probability of only one recovery event, and the occurrence probability of no event, respectively. According to Equations (A8) and (A9), the occurrence probability of event is shown as follows

$$\begin{cases} Pr\{d(I(t) = 1|T(t))\} \approx I(t)\beta(t)\frac{S(t)}{N}dt \\ Pr\{d(I(t) = -1|T(t))\} \approx \gamma(t)I(t)dt \\ Pr\{d(I(t) = 0|T(t))\} \approx 1 - I(t)\beta(t)\frac{S(t)}{N}dt - \gamma(t)I(t)dt \end{cases} \tag{A10}$$

We define i as the crowd state. i means that there are i infected individuals at time t, that is, $I(t) = i$ and $S(t) = N - i$. If an infection event occurs, the state of the crowd will change from i to $i + 1$. If a recovery event occurs, the state of the crowd will change from i to $i - 1$. Given two states i and j, Equation (A10) shows the state transition probability.

$$(t, \Delta t) = \begin{cases} i\beta(t)\frac{N-1}{N}\Delta t, & \text{if } j = i+1, \\ \gamma(t)i\Delta t, & \text{if } j = i-1, \\ 1 - i\beta(t)\frac{N-1}{N}\Delta t - \gamma(t)i\Delta t, & \text{if } j = i, \\ 0, & \text{otherwise,} \end{cases} \tag{A11}$$

where $P_{(j,i)}(t, \Delta t)$ is the transition probability from i to j in the time interval $(t, t + \Delta t)$. Let $g(t, i) = \beta(t)\frac{N-1}{N}\Delta t$ and $h(t, i) = \gamma(t)i\Delta t$. Thus, the state transition matrix in a time interval Δt is given by Equation (A12)

$$P(t, \Delta t) = \begin{bmatrix} 1 & h(t,1) & 0 & \cdots & 0 & 0 \\ 1-g(t,1)-h(t,1) & g(t,2) & & & & \\ g(t,1) & 1-g(t,2)-h(t,2) & & & & \\ & h(t,2) & \cdots & h(t,N-1) & & \\ & & & 1-g(t,N-1)-h(t,N-1) & h(t,N) \\ & & & g(t,N-1) & 1-h(t,N) \end{bmatrix} \tag{A12}$$

where $p_{(j,i)}(t, \Delta t)$ is the (j,i) element of the matrix. $P(t, \Delta t)$ is a $(N+1)(N+1)$ matrix, since the states of the crowd are ordered from 0 to N. The subscripts of the matrix indicate the form of state transition.

$$[I(t)] = \sum_{i=0}^{N} iP_i(t) \tag{A13}$$

$$\mathbf{P}(t_1) = \mathbf{P}(t_0 + \Delta t) = \mathbf{P}(t_0, \Delta t)p(t_0) \tag{A14}$$

$$\mathbf{P}(t_n + \Delta t) = \mathbf{P}(t_n, \Delta t)p(t_n) = \mathbf{P}(t_n, \Delta t)\ldots \mathbf{P}(t_2, \Delta t)\mathbf{P}(t_1, \Delta t)\mathbf{P}(t_0, \Delta t)p(t_0) \tag{A15}$$

Appendix C.3. ACSED

An advantageous aspect of the ACSED model lies in its incorporation of a computational formula for external emotional contagion within the emotion prediction module, seamlessly integrated with reinforcement learning theory. This aspect holds valuable lessons for the current article. Because of the intricacies involved in calculating emotional contagion, the ACSED model simplifies matters by categorizing the population into two distinct states. In contrast, the model proposed in this article adopts a more nuanced approach by dividing the population into multiple emotional states. This refined segmentation proves to be a more practical and robust strategy for studying emotional contagion within a population.

The calculation of emotional contagion is shown in Formula (A16):

$$E_i = E_i^{ex} + E_i^{se} \quad \text{(A16)}$$

The changing values of emotional contagion of *Agent i* is defined in Formula (A17):

$$\Delta E_{i,j}^{ex}(t) = \left[1 - \frac{1}{1 + \exp(-D)}\right] \times E_i(t) \times A_{j,i} \times B_{i,j} \quad \text{(A17)}$$

where D represents the distance between *Agent i* and other *Agent j*, E_i represents the emotion of *Agent i*, $A_{j,i}$ is the intensity of emotion received by the affected *Agent i* from the influencing *Agent j*, and $B_{i,j}$ refers to the emotional intensity sent from *Agent j* to *Agent i*.

Formula (A18) is to calculate the external emotional contagion of the righteous at time t. Formula (A19) is to calculate the external emotional contagion of the opposite at time t.

$$\Delta E_r^{ex} = \sum_{i=1}^{m} \Delta E_{r,r_i}^{ex}(t) + \sum_{j=1}^{n} \Delta E_{r,o_j}^{ex}(t) \quad \text{(A18)}$$

$$\Delta E_o^{ex} = \sum_{i=1}^{n} \Delta E_{o,o_i}^{ex}(t) + \sum_{j=1}^{m} \Delta E_{o,r_j}^{ex}(t) \quad \text{(A19)}$$

The mental emotion calculation method is as follows:

$$\Delta E_i^{se}(t) = 0.1 \times \left(\frac{1}{\delta + \exp(\gamma/r_i(t))}\right), \quad r_i(t) \geq \gamma \quad \text{(A20)}$$

$$\Delta E_i^{se}(t) = -0.1 \times \left(\frac{1}{\delta + \exp(r_i(t)/\gamma)}\right), \quad r_i(t) \leq -\gamma \quad \text{(A21)}$$

where $r_i(t)$ represents the difference between the reward values of two consequent time steps, δ is an empirical parameter. When $r_i(t) \in (-\gamma, \gamma)$, the action of *Agent i* has less effect on its emotions and can be ignored. When $r_i(t) \geq \gamma$, it means that *Agent i* performs the action to promote the battle result. If *Agent i* is righteous, its emotions will become positive, otherwise if it is opposite, it will become negative. When $r_i(t) \leq -\gamma$, it means that the action performed by *Agent i* is not conducive to the current combat situation.

The amount of emotional contagion of *Agent i* is shown as Equation (A22)

$$E(i, t) = E(i, t-1) + \Delta E_i^{ex}(t) + \Delta E_i^{se}(t) \quad \text{(A22)}$$

References

1. International Maritime Organization. *SOLAS: The International Convention for the Safety of Life at Sea*; International Maritime Organization: London, UK, 2020.
2. Tripathi, G.; Singh, K.; Vishwakarma, D.K. Crowd Emotion Analysis Using 2D ConvNets. In Proceedings of the 2020 Third International Conference on Smart Systems and Inventive Technology (ICSSIT), Tirunelveli, India, 20–22 August 2020.
3. Varghese, E.B.; Thampi, S.M. A Deep Learning Approach to Predict Crowd Behavior Based on Emotion. In *Smart Multimedia*; Indian Institute of Information Technology and Management-Kerala (IIITM-K): Thiruvananthapuram, India; Cochin University of Science and Technology: Kochi, India, 2018.
4. Sanchez, F.L. Revisiting crowd behaviour analysis through deep learning: Taxonomy, anomaly detection, crowd emotions, datasets, opportunities and prospects. *Inf. Fusion* **2020**, *64*, 318–335. [CrossRef] [PubMed]
5. Varghese, E.; Thampi, S.M.; Berretti, S. A Psychologically Inspired Fuzzy Cognitive Deep Learning Framework to Predict Crowd Behavior. *IEEE Trans. Affect. Comput.* **2020**, *13*, 1005–1022. [CrossRef]
6. Singh, N.; Roy, N.; Gangopadhyay, A. Analyzing the Emotions of Crowd for Improving the Emergency Response Services. *Pervasive Mob. Comput.* **2019**, *58*, 101018. [CrossRef]
7. Watanabe, K.; Inoue, K. Learning State Transition Rules from High-Dimensional Time Series Data with Recurrent Temporal Gaussian-Bernoulli Restricted Boltzmann Machines. *Hum.-Centric Intell. Syst.* **2023**, *3*, 296–311. [CrossRef]
8. Amato, F.; Guignard, F.; Robert, S.; Kanevski, M. A novel framework for spatio-temporal prediction of environmental data using deep learning. *Sci. Rep.* **2020**, *10*, 22243. [CrossRef]

9. Xu, Y.; Tian, Y.; Li, H. Unsupervised deep learning method for bridge condition assessment based on intra-and inter-class probabilistic correlations of quasi-static responses. *Struct. Health Monit.* **2023**, *22*, 600–620. [CrossRef]
10. Tian, Y.; Xu, Y.; Zhang, D.; Li, H. Relationship modeling between vehicle-induced girder vertical deflection and cable tension by BiLSTM using field monitoring data of a cable-stayed bridge. *Struct. Control Health Monit.* **2021**, *28*, e2667. [CrossRef]
11. Lv, P.; Xu, B.; Li, C.; Yu, Q.; Zhou, B.; Xu, M. Antagonistic Crowd Simulation Model Integrating Emotion Contagion and Deep Reinforcement Learning. *arXiv* **2021**, arXiv:2015.00854. [CrossRef]
12. Xue, J.; Yin, H.; Lv, P.; Xu, M.; Li, Y. Crowd queuing simulation with an improved emotional contagion model. *Sci. China* **2019**, *62*, 193–195. [CrossRef]
13. Rao, M.Y. Crowd evacuation simulation based on emotion contagion. *Int. J. Simul. Process Model.* **2018**, *13*, 43–56. [CrossRef]
14. Xu, T.; Shi, D.; Chen, J.; Li, T.; Lin, P.; Ma, J. Dynamics of emotional contagion in dense pedestrian crowds. *Phys. Lett. A* **2019**, *384*, 126080. [CrossRef]
15. Shi, Y.; Zhang, G.; Lu, D.; Lv, L.; Liu, H. Adaptive Intervention for Crowd Negative Emotional Contagion. In Proceedings of the 2021 IEEE 24th International Conference on Computer Supported Cooperative Work in Design (CSCWD), Dalian, China, 5–7 May 2021; IEEE: Washington, DC, USA, 2021.
16. Hethcote, H.W. The mathematics of infectious diseases. *SIAM Rev.* **2000**, *42*, 99–653. [CrossRef]
17. Bairagi, N.; Adak, D. Global analysis of hiv-1 dynamics with hill type infection rate and intracellular delay. *Appl. Math. Model.* **2014**, *38*, 5047–5066. [CrossRef]
18. Feng, L.; Liao, X.; Han, Q.; Li, H. Dynamical analysis and control strategies on malware propagation model. *Appl. Math. Model.* **2013**, *37*, 8225–8236. [CrossRef]
19. Ji, C.; Jiang, D. Threshold behavior of a stochastic sir model. *Appl. Math. Model.* **2014**, *38*, 5067–5079. [CrossRef]
20. Liu, H.; Lu, D.; Zhang, G.; Hong, X.; Liu, H. Recurrent emotional contagion for the crowd evacuation of a cyber-physical society. *Inf. Sci.* **2021**, *10*, 155–172. [CrossRef]
21. Qiu, L.; Liu, S. SVIR rumor spreading model considering individual vigilance awareness and emotion in social networks. *Int. J. Mod. Phys. C* **2021**, *32*, 2150120. [CrossRef]
22. Chen, Y.H.; Zhang, X.Q. Research on Netizen Group Emotion Contagion Model and the Simulation under Network Group Emergencies. *Inf. Sci.* **2018**, *36*, 151–156. [CrossRef]
23. Nizamani, S.; Memon, N.; Galam, S. From public outrage to the burst of public violence: An epidemic-like model. *Phys. A Stat. Mech. Appl.* **2014**, *416*, 620–630. [CrossRef]
24. Zhu, L.; Wang, B. Stability analysis of a SAIR rumor spreading model with control strategies in online social networks. *Inf. Sci.* **2020**, *526*, 1–19. [CrossRef]
25. Tian, S.H.; Sun, M.Q.; Zhang, J.Y. Research on the Emotion Evolution of Network Public Opinion Based on Improved SIR Model. *Inf. Sci.* **2019**, *37*, 52–57,64. [CrossRef]
26. Xu, M.; Li, C.; Lv, P.; Chen, W.; Deng, Z.; Zhou, B.; Manocha, D. Emotion-based crowd simulation model based on physical strength consumption for emergency scenarios. *IEEE Trans. Intell. Transp. Syst.* **2020**, *22*, 6977–6991. [CrossRef]
27. Song, J.; Zhang, M.G. Dynamic Simulation of the Group Behavior under Fire Accidents Based on System Dynamics. *Procedia Eng.* **2018**, *211*, 635–643. [CrossRef]
28. Song, B.W.; Li, J.; Li, J. Considering Trust Parameters the Evolution Model of Network Negative Emotion under Public Emergencies. In Proceedings of the 2020 4th International Conference on Electronic Information Technology and Computer Engineering, Xiamen, China, 6–8 November 2020.
29. Yao, J.J.; Liang, J.; Yao, H.X. Research on Emotional Information Communication Based on SIR Model. *Inf. Sci.* **2018**, *36*, 25–29. [CrossRef]
30. Cao, M.; Zhang, G.; Wang, M.; Lu, D.; Liu, H. A method of emotion contagion for crowd evacuation. *Phys. A Stat. Mech. Appl.* **2017**, *483*, 250–258. [CrossRef]
31. Fan, R.; Xu, K.; Zhao, J. An agent-based model for emotion contagion and competition in online social media. *Phys. A Stat. Mech. Appl.* **2017**, *495*, 245–259. [CrossRef]
32. Li, X.; Zhang, J. Research on SIRS Information Dissemination Model Based on System Dynamics. *Inf. Sci.* **2017**, *35*, 17–22. [CrossRef]
33. Hatfield, E.; Cacioppo, J.T.; Rapson, R.L. Emotional Contagion. *Curr. Dir. Psychol. Sci.* **1993**, *2*, 96–100. [CrossRef]
34. Shi, Y.; Zhang, G.; Lu, D.; Lv, L.; Liu, H. Intervention Optimization for Crowd Emotional Contagion. *Inf. Sci.* **2021**, *576*, 769–789. [CrossRef]
35. Guttman, L. A basis for analyzing test-retest reliability. *Psychometrika* **1945**, *10*, 255–282. [CrossRef]
36. Lopez-Odar, D.; Alvarez-Risco, A.; Vara-Horna, A.; Chafloque-Cespedes, R.; Sekar, M.C. Validity and reliability of the questionnaire that evaluates factors associated with perceived environmental behavior and perceived ecological purchasing behavior in Peruvian consumers. *Soc. Responsib. J.* **2020**, *16*, 403–417. [CrossRef]
37. Chung, R.H.; Kim, B.S.; Abreu, J.M. Asian American multidimensional acculturation scale: Development, factor analysis, reliability, and validity. *Cult. Divers. Ethn. Minor Psychol.* **2004**, *10*, 66–80. [CrossRef] [PubMed]
38. Elliott, A.C.; Hynan, L.S. A SAS(R) macro implementation of a multiple comparison post hoc test for a Kruskal-Wallis analysis. *Comput. Methods Programs Biomed.* **2011**, *102*, 75–80. [CrossRef] [PubMed]
39. Breiman, L. Random forests. *Mach. Learn.* **2001**, *45*, 5–32. [CrossRef]

40. Goel, S.; Watts, D.J.; Goldstein, D.G. The Structure of Online Diffusion Networks. In Proceedings of the 13th ACM Conference on Electronic Commerce, Valencia, Spain, 4–8 June 2012.
41. Hu, C.; Chen, Y.; Hu, L.; Peng, X. A novel random forests based class incremental learning method for activity recognition. *Pattern Recognit.* **2018**, *78*, 277–290. [CrossRef]
42. Abell'an, J.; Mantas, C.J.; Castellano, J.G.; Moral-García, S. Increasing diversity in random forest learning algorithm via imprecise probabilities. *Expert Syst. Appl.* **2018**, *97*, 228–243. [CrossRef]
43. Gomes, H.M.; Bifet, A.; Read, J.; Barddal, J.P.; Enembreck, F.; Pfharinger, B.; Holmes, G.; Abdessalem, T. Adaptive ra-ndom forests for evolving data stream classification. *Mach. Learn.* **2017**, *106*, 1469–1495. [CrossRef]
44. Genuer, R.; Poggi, J.; Tuleau-Malot, C.; Villa-Vialaneix, N. Random forests for big data. *Big Data Res.* **2017**, *9*, 28–46. [CrossRef]
45. Zhu, M.; Xia, J.; Jin, X.; Yan, M.; Cai, G.; Yan, J.; Ning, G. Class weights random forest algorithm for processing class imbalanced medical data. *IEEE Access* **2018**, *6*, 4641–4652. [CrossRef]
46. Anylogic. Available online: https://www.anylogic.com// (accessed on 15 February 2023).

Disclaimer/Publisher's Note: The statements, opinions and data contained in all publications are solely those of the individual author(s) and contributor(s) and not of MDPI and/or the editor(s). MDPI and/or the editor(s) disclaim responsibility for any injury to people or property resulting from any ideas, methods, instructions or products referred to in the content.

MDPI
St. Alban-Anlage 66
4052 Basel
Switzerland
www.mdpi.com

Mathematics Editorial Office
E-mail: mathematics@mdpi.com
www.mdpi.com/journal/mathematics

Disclaimer/Publisher's Note: The statements, opinions and data contained in all publications are solely those of the individual author(s) and contributor(s) and not of MDPI and/or the editor(s). MDPI and/or the editor(s) disclaim responsibility for any injury to people or property resulting from any ideas, methods, instructions or products referred to in the content.

www.ingramcontent.com/pod-product-compliance
Lightning Source LLC
LaVergne TN
LVHW070228100526
838202LV00015B/2107